Michael Waring

LABORATORY COPY

# MOLECULAR ASPECTS OF ANTICANCER DRUG–DNA INTERACTIONS

## Volume 2

# TOPICS IN MOLECULAR AND STRUCTURAL BIOLOGY

**Series Editors**

Stephen Neidle
*Institute of Cancer Research*
*Sutton, Surrey, UK*

Watson Fuller
*Department of Physics*
*University of Keele, UK*

Jack S. Cohen
*Georgetown University*
*USA*

**Recent titles**

Protein–Nucleic Acid Interaction
*Edited by* Wolfram Saenger and Udo Heinemann (1989)

Calcified Tissue
*Edited by* David W. L. Hukins (1989)

Oligodeoxynucleotides: Antisense Inhibitors of Gene Expression
*Edited by* Jack S. Cohen (1989)

Molecular Mechanisms in Muscular Contraction
*Edited by* John M. Squire (1990)

Connective Tissue Matrix, Part 2
*Edited by* David W. L. Hukins (1990)

New Techniques of Optical Microscopy and Microspectroscopy
*Edited by* Richard J. Cherry (1990)

Molecular Dynamics: Applications in Molecular Biology
*Edited by* Julia M. Goodfellow (1990)

Water and Biological Macromolecules
*Edited by* Eric Westhof (1993)

Topics in Molecular and Structural Biology

# MOLECULAR ASPECTS OF ANTICANCER DRUG–DNA INTERACTIONS
# Volume 2

*Edited by*

**Stephen Neidle**
*Institute of Cancer Research*
*Sutton, Surrey, UK*

*and*

**Michael Waring**
*Dept of Pharmacology*
*University of Cambridge*

MACMILLAN

First published 1994 by
THE MACMILLAN PRESS LTD
Houndmills, Basingstoke, Hampshire RG21 2XS
and London
Companies and representatives
throughout the world

ISBN 0–333–60238–2
ISSN 0265–4377

A catalogue record for this book is available
from the British Library

Printed in Great Britain by
Mackays of Chatham, PLC
Chatham, Kent

# Contents

# The Contributors

**Bruce C. Baguley**
Cancer Research Laboratory
University of Auckland School of Medicine
Private Bag 90192
Auckland
New Zealand

**Christian Bailly**
Department of Pharmacology
University of Cambridge
Tennis Court Road
Cambridge CB2 1QJ
UK

**William A. Denny**
Cancer Research Laboratory
Auckland Division Cancer Society of New Zealand Inc.
Auckland Medical School
University of Auckland
Private Bag 90192
Auckland
New Zealand

**Wei-dong Ding**
Infectious Disease Research Section
Medical Research Division
Lederle Laboratories
American Cyanamid Company
Pearl River
New York 10965
USA

**George A. Ellestad**
Infectious Disease Research Section
Medical Research Division
Lederle Laboratories
American Cyanamid Company
Pearl River
New York 10965
USA

**Sidney M. Hecht**
Department of Chemistry
University of Virginia
Charlottesville
Virginia 22901
USA

**Jean-Pierre Hénichart**
Centre de Recherche INSERM
Place de Verdun
59045 Lille Cedex
France

**Warren Judd**
School of Biological Sciences
University of Auckland
Private Bag 92019
Auckland
New Zealand

**Kurt W. Kohn**
Laboratory of Molecular Pharmacology
Development Therapeutics Program
National Cancer Institute
National Institutes of Health
Building 37 Room 5C25
Bethesda
Maryland 20892
USA

**Anand Natrajan**
Department of Biology
University of Virginia
Charlottesville
Virginia 22901
USA

**David R. Newell**
Division of Oncology
University of Newcastle upon Tyne
Cancer Research Unit
The Medical School
Framlington Place
The University
Newcastle-upon-Tyne NE2 4HH
UK

**Laurence H. Patterson**
Department of Pharmacy
School of Applied Sciences
De Montfort University
The Gateway
Leicester LE1 9BH
UK

**Yves Pommier**
Laboratory of Molecular Pharmacology
Development Therapeutics Program
National Cancer Institute
Bethesda
Maryland 20892
USA

**Raymond K. Ralph**
School of Biological Sciences
University of Auckland
Private Bag 92019
Auckland
New Zealand

**Farial A. Tanious**
Laboratory for Chemical and Biological Sciences
Georgia State University
Atlanta
Georgia 30303
USA

**Maria Tomasz**
Department of Chemistry
Hunter College
City University of New York
695 Park Avenue
New York
New York 10021
USA

**W. David Wilson**
Department of Chemistry
Georgia State University
University Plaza
Atlanta
Georgia 30303
USA

# Preface

DNA has long been a key target for cancer chemotherapy. Indeed, the first agents to be employed clinically in the treatment of human cancer (the nitrogen mustards) are DNA cross-linking agents. Spectacular advances have occurred during recent years in the treatment of childhood leukaemia and testicular cancer, largely as a result of the development of better DNA-interactive agents. Even though the majority of solid tumours remain resistant to chemotherapy, there is real promise that a new, third-generation of platinum compounds will prove successful in the treatment of ovarian cancer. Clinical advances in such key areas are the ultimate objective of much current research in cancer chemotherapy and biology. Future progress must surely result from wise application of the large body of fundamental knowledge being accumulated from studies in a whole range of disciplines. No one doubts that clinical success will increasingly depend upon the exploitation of such knowledge and on the interplay between it and more applied disciplines. This is especially important as the molecular and cellular bases of malignant cell growth become better understood. So the study of drug–DNA interactions has moved on from the position of a dozen years ago, when our understanding of the molecular basis of drug action was relatively poor, as were the prospects for rational design of new drugs, to a much more positive position with new horizons.

These two volumes survey our current knowledge about the mode of action of the major classes of DNA-interactive antitumour agents, and in so doing provide pointers for the discovery of new therapeutic substances. The reader will notice that certain related topics have been grouped together; indeed in one instance (that of topoisomerase inhibitors), what were originally planned as two separate chapters by different authors have been amalgamated into one (by mutual consent!) so as to produce a more balanced and co-ordinated treatment. Elsewhere the relationships between topics may be less obvious, but we hope that our choices will stimulate cross-fertilization of ideas.

An enterprise involving many authors such as this requires the cooperation of all the contributors if it is to succeed. We are grateful to everyone for their efforts in ensuring delivery of their manuscripts promptly and for making our task as editors such a pleasurable one. Both of us are indebted to the Cancer Research Campaign for supporting work on drug–DNA interactions in our own laboratories over a number of years. To the hard-working staff of the Campaign, as well as to those who devote their lives to the alleviation of cancer at the bedside and in the laboratory, we dedicate this pair of volumes.

*Sutton and Cambridge, 1993*

S. N.
M. W.

# 1
# DNA Topoisomerases

Raymond K. Ralph, Warren Judd, Yves Pommier and Kurt W. Kohn

## 1 Introduction

The simple elegance of the antiparallel, plectonemic DNA double helix revealed by Watson, Crick and Wilkins gave substance to the concept of DNA as a repository of genetic information and a rationale for its replication (Watson and Crick, 1953). Subsequently the complexity of chromosomes has been slowly revealed (Gasser *et al.*, 1989; Filipski *et al.*, 1990), bringing with it the realization that simple models for DNA replication must be modified to explain the replication and resolution of knotted DNA, concatemeric DNA, radial looped DNA, nucleosome-coiled DNA or other phenomena such as DNA replication from multiple origins in eukaryotic cells (Hamlin *et al.*, 1991). In addition, the progression of RNA or DNA polymerase produces topological effects on DNA templates which, if left unresolved, inhibit RNA and DNA synthesis (Brill *et al.*, 1987; Uemura *et al.*, 1987; Yamagishi and Nomura, 1988). The discovery of topoisomerases, enzymes that can resolve topological constraints in DNA, gave a clue to the mechanism(s) used by cells to overcome some of the problems resulting from DNA twisting (Vosberg, 1985; Wang, 1985). As their functions are revealed, these apparently magical enzymes can be seen as central to most of the events involving DNA, gene expression and growth of cells. Consequently, they are good potential targets for anticancer drugs.

Two main types of topoisomerases exist in eukaryotic cells and both types bind to DNA. They also recognize and bind to cross-overs in DNA, which are more abundant in supercoiled DNA (Zechiedrich and Osheroff, 1990). Type I DNA topoisomerases transiently nick and reseal one strand of double-stranded DNA. Passing the intact DNA strand through the nick in the other permits relaxation of a circularly constrained double-helical supercoiled DNA (Vosberg, 1985; Wang, 1985). Other activities of type I topoisomerases have also been suggested involving DNA strand rearrangements (Halligan *et al.*, 1982). Type II DNA topoisomerases nick and reseal both strands of double-stranded supercoiled DNA and they relax the DNA by passing one double-stranded section of the DNA through the transient break to reduce supercoiling. Eukaryotic type II topoisomerases require ATP to function, although some organisms (e.g. trypanosomes) may have type II topoisomerases that do not need ATP (Douc Rasy *et al.*, 1986). Bacterial DNA gyrase, a type II topoisomerase, has the additional feature that it will introduce negative supercoils into DNA when provided with ATP (Vosberg, 1985). No equivalent activity has been reported in eukaryotes to date. Preliminary evidence for additional topoisomerases or topoisomerase variants exists, but these are not yet well characterized (Fink, 1989).

DNA topoisomerases are targets for a number of anticancer drugs. The molecular mechanisms of anticancer drug action upon topoisomerases and DNA were previously reviewed by Marshall and Ralph (1985), Wang (1985), Maxwell and Gellert (1986), Kohn *et al.* (1987), Ralph and Schneider (1987), Lock and Ross (1987), D'Arpa and Liu (1989), Liu (1989) and Schneider *et al.* (1990).

It is our intention to focus upon more recent studies of topoisomerases that are pertinent to the action of anticancer drugs on cells. However, this inevitably requires some appreciation of the mechanism of topoisomerase cleavage of DNA which proceeds via transient protein–DNA complexes (PDCs) in which the 3′ ends (top I) or 5′ ends (top II) of nicked DNA strands are temporarily covalently linked via phosphate ester bonds to tyrosine hydroxyl residues in the respective topoisomerases. These open protein–DNA complexes are the targets for anticancer drugs that act upon topoisomerases and inhibit the resealing of DNA, sometimes contributing to successful chemotherapy.

Precisely why or how topoisomerase inhibitors cause cancer cell death has still to be defined, although various drugs that inhibit topoisomerases can cause DNA scission, affect RNA or DNA synthesis and produce sister-chromatid exchange, chromosomal rearrangements, recombination and other chromosome aberrations which undoubtedly disrupt many normal cellular processes. Some of the recent information related to topoisomerases has come from studies with yeasts, where genetic modification is possible, or from other organisms such as *Drosophila* or *Xenopus*, where the advantages of such systems can be exploited. In general, it is believed that the information obtained from these systems reflects events in other eukaryotic cells.

## 2 Topoisomerase I (top I)

### *General Properties and Overview*

The principal type I topoisomerase (top I) in human cells is a monomeric protein of $M_r$ 90 649 Da calculated from the sequence of the cloned gene (D'Arpa *et al.*, 1988) or $M_r$ 100 000 Da in SDS–polyacrylamide gels (Liu and Miller, 1981).

The major source of the enzyme in cells is the nucleus, where it is associated particularly with actively transcribed regions of DNA in different types of cells (Fleischmann *et al.*, 1984; Bonven *et al.*, 1985; Gilmour and Elgin, 1987; Stewart and Schütz, 1987). However, mitochondria also contain an ATP-independent type I topoisomerase with a molecular weight of 63–64 kDa estimated in polyacrylamide gels (Castora and Lazarus, 1984; Castora *et al.*, 1985; Lazarus *et al.*, 1987). The mitochondrial enzyme from platelets is inhibited by camptothecin (Kosovsky and Soslau, 1991). Recently a $M_r$ 165 kDa tissue-specific top I variant in *Xenopus* oocytes which disappeared during maturation was described, which, it was conjectured, plays a role in ribosomal DNA excision and amplification (Richard and Bogenhagen, 1991).

The gene coding for top I is a single-copy gene located on human chromosome 20q12-13.2 (Juan *et al.*, 1988; Kunze *et al.*, 1989). The gene is composed of 21 exons spread over 85 kilobase pairs (kb) of DNA, and with SP1, cAMP and octamer motifs in its promoter region (Kunze *et al.*, 1991). In addition, there are at least two truncated, processed and hypomethylated pseudogenes on chromosomes 1 and 22, respectively, in different human cell lines (Kunze *et al.*, 1989; Zhou *et al.*, 1989; Yang *et al.*, 1990; Hsieh, 1990).

Top I–DNA complexes are targets for the alkaloid camptothecin and certain analogues, which bind reversibly to top I–DNA complexes but not to isolated DNA or top I alone (Hsiang *et al.*, 1985; Hsiang and Liu, 1988; Hsiang *et al.*, 1989a,b; Bjornsti *et al.*, 1989). The bound drug reversibly traps and stabilizes the top I–DNA intermediate involved in DNA unwinding, ultimately causing DNA damage and cell death (Hertzberg *et al.*, 1989a). There is good evidence that camptothecin prevents the resealing of transient top I-induced single-strand breaks in DNA, since an increase in open or 'cleavable' complexes can be demonstrated with protein denaturants in cells or nuclei treated with the drug (Hsiang *et al.*, 1985; Hsiang and Liu, 1988; Covey *et al.*, 1989). Studies with yeast topoisomerase mutants have shown that the effects of camptothecin on cells are due to its action on top I (Nitiss and Wang, 1988; Eng *et al.*, 1988; Bjornsti *et al.*, 1989). Moreover, top I purified from camptothecin-resistant leukaemia and other cells is resistant to camptothecin (Andoh *et al.*, 1987; Kjeldsen *et al.*, 1988a; Gupta *et al.*, 1988).

Camptothecin and related drugs that inhibit the action of top I can have strong antitumour activity against a variety of experimental tumours (Gallo *et al.*, 1971; Neil and Homan, 1973; Tsuruo *et al.*, 1988; Giovanella *et al.*, 1989). However, problems with non-specific cytotoxicity seem to have precluded their general use as anticancer agents, although extracts of *Camptotheca accuminata* have been used to treat some solid tumours and leukaemias in China (Gottlieb and Luce, 1972; Moertel *et al.*, 1972; Muggia *et al.*, 1972). New derivatives of camptothecin that might be exploited as anticancer drugs have recently led to renewed interest in top I inhibitors, while a novel new drug, saintopin, inhibits both types I and II topoisomerases, causing DNA cleavage (Hsiang *et al.*, 1989b; Yamashita *et al.*, 1991). Acidic phospholipids have also been shown to inhibit top I by blocking binding of the enzyme to DNA (Tamura *et al.*, 1990). Whether this observation can be exploited to produce a new class of drugs active against the enzyme remains to be established, as does the significance of the observation *in vivo*.

*Reaction Mechanisms*

The enzymatic function of top I is to change the number of twists of one DNA strand about the other. This is accomplished in four steps: (1) binding of top I to DNA; (2) cleavage of one DNA strand and covalent linkage of top I to the 3′ terminus of the strand break; (3) rotation of the free strand segment around the intact strand; and (4) resealing of the strand break.

The binding of top I is only mildly constrained by DNA sequence and therefore can occur in most regions of the genome, although regions of bent or supercoiled DNA are preferred (Muller *et al.*, 1985; Camilloni *et al.*, 1989; Caserta *et al.*, 1989, 1990; Krogh *et al.*, 1991). Binding appears to occur over a span of at least 20 base pairs of duplex DNA and to be tighter for the DNA segment upstream from the cleavage site than for the downstream segment, as indicated by DNA footprinting (Stevnsner *et al.*, 1989), base-sequence specificity (Been *et al.*, 1984; Jaxel *et al.*, 1991a,b; Porter and Champoux, 1989b) and oligonucleotide cleavage experiments (Jaxel *et al.*, 1991a; Svejstrup *et al.*, 1990). Top I does not bind to single-stranded DNA (unless the strand can form a duplex region due to diadic symmetry of sequence) (Been and Champoux, 1984; Jaxel *et al.*, 1991a). Top I can bind to duplex DNA at the site of a single-strand break (McCoubrey and Champoux, 1986) and can cleave the intact strand to produce a double-strand cut which may promote DNA recombination.

Strand scission occurs through a transesterification in which a tyrosyl hydroxy group (Tyr 723 in human top I) links to the 3′ oxygen of a phosphodiester bond, displacing the 5′ phosphate to generate a DNA

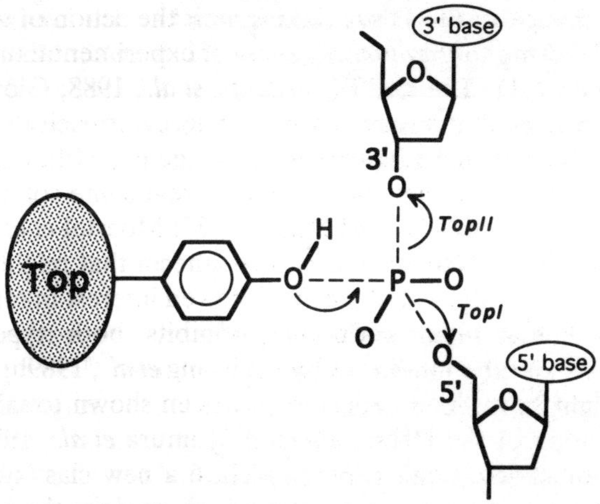


**Figure 1.1** Reaction between topoisomerase tyrosyl and DNA phosphodiester bond. The tyrosyl OH group attacks the phosphodiester bond; a curved arrow represents an unshared electron pair moving from the tyrosyl O atom towards the P atom during covalent bond formation. As this new O–P bond forms, a P–O bond of one or other phosphodiester bond must cleave so as to leave the P atom linked to 4 oxygens. In the case of top I, the P–O bond to the 5′ DNA strand terminus is cleaved; a curved arrow shows an electron pair being sucked into the 5′ oxygen. The tyrosyl residue then remains bound to the 3′ strand terminus. Conversely, in the case of top II, the P–O to the 3′ terminus is cleaved and the enzyme tyrosyl residue becomes bound to the 5′ terminus. The reactions are reversible

strand break (Figure 1.1) (Champoux, 1981; Lynn *et al.*, 1989; Tse *et al.*, 1980; Wang, 1987). The easy reversibility of this reaction evoked the term 'cleavable complex' to designate a topoisomerase–DNA complex which readily converts between a state in which the DNA is intact and a state in which a strand is cleaved (Wang, 1985, 1987). The sites of cleavable complex formation have a strong preference for T at the 3′ terminus (Been *et al.*, 1984; Porter and Champoux, 1989b; Jaxel *et al.*, 1991a,b); it is interesting to note that this preferred T could remain in close proximity to the incoming tyrosyl residue.

A change in the number of twists of one strand about the other ('linking number') could occur either by free rotation of the 3′ side of the cleaved strand around the intact strand or by passing the intact strand through a gap in the cleaved strand (Champoux, 1990). In the latter model, the two termini of the cleaved strand would remain fixed to the enzyme, while the intact strand passes through the gap. There is, however, evidence against strong binding of the 3′ side of the cleaved strand, in that this segment can be exchanged for other single-strand fragments bearing 5′-OH termini (Halligan *et al.*, 1982; McCoubrey and Champoux, 1986). This strand exchange process may reflect the ability of top I to promote genetic recombination.

Following change in the DNA linking number, the cleaved strand can readily reseal, owing to the intrinsic reversibility of the transesterification reaction (Figure 1.1). The rate of resealing is affected by the local DNA sequence, and sites that reseal slowly are more effectively inhibited by camptothecin (Porter and Champoux, 1989a,b).

*Mechanisms of Drug Actions*

Top I has become an important target for cancer chemotherapy since the discovery of its specific inhibition by camptothecin (Hsiang *et al.*, 1985). Inhibitors might have been expected to act by blocking top I functions, such as relaxation of the DNA supercoiling stress generated during transcription or replication. Some recently identified top I inhibitors may act in this way. However, the major action of camptothecin is to form stabilized cleavable complexes. Two actions of top I inhibitors that are likely sources of cytotoxicity are (1) inhibition of RNA synthesis due to the accumulation of supercoiling stress in transcribed regions of the genome; and (2) interaction with top I–DNA so as to cause stabilization of cleavable complexes which, when encountered by a polymerase, particularly a replication fork, may become an unrepairable DNA defect. These processes are discussed further in later sections.

Clues to the manner of interaction of camptothecin with the top I–DNA complex were obtained from structure–activity studies. The single chiral position in camptothecin (position 20, Figure 1.2) must be in the *S* configuration, both for action on top I and for antitumour activity. The alternative isomer, *R*-camptothecin, is totally inactive. Studies of derivatives bearing substituents on the A ring gave further indications of the steric specificity of the camptothecin binding site (Wani *et al.*, 1987; Jaxel *et al.*, 1989; Kingsbury *et al.*, 1991; Pommier *et al.*, 1991c). Substituents at position 12, and to a lesser degree bulky groups at position 11, abolish activity, suggesting that this region is in close proximity to the binding site on the enzyme–DNA complex. Similar substituents at the 9 or 10 position, on the other hand, generally increase activity (Hsiang *et al.*, 1989b; Jaxel *et al.*, 1989; Pommier *et al.*, 1991c).

The close parallel between the activities of camptothecin derivatives against top I and against murine tumours strongly supports top I as an effective antitumour target (Jaxel *et al.*, 1989; Pommier *et al.*, 1991c).

Although camptothecin binds neither to DNA (Fukada, 1985; Hsiang *et al.*, 1985; Kuwahara *et al.*, 1986) nor to top I alone, radiolabelled drug has been found to bind to complexes of DNA plus top I (Hertzberg *et al.*, 1989a). A camptothecin derivative bearing an alkylating group at the 9 position was found to bind covalently to the top I but not to the DNA component of the complex (Hertzberg *et al.*, 1990), indicating that a

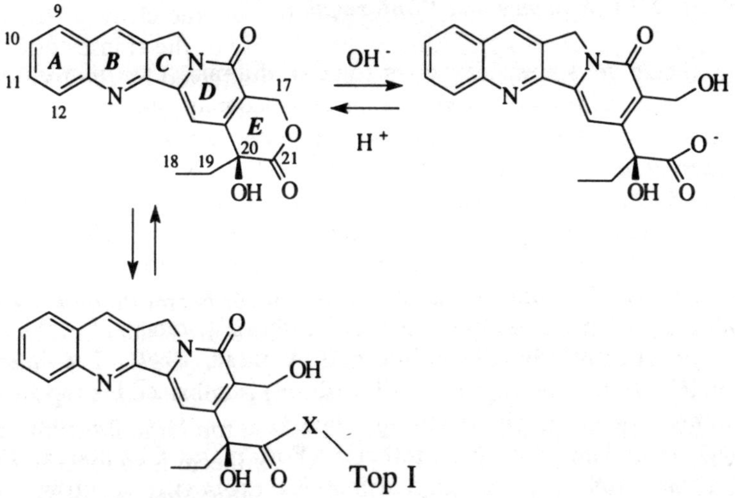


**Figure 1.2** Chemistry of camptothecin. The sole chiral position, C20, has the *S* configuration in the active form of the drug. The base-catalysed opening of the lactone ring (ring E) is reversible and facilitated by the OH group at position 20. The kinetics of ring opening and closing depend on pH (half-times are of the order of an hour under physiologic conditions). Lactone ring opening may occur by way of attack by hydroxide ion at position 21. The attack may also come from a suitably placed nucleophile, such as −SH or $-NH_2$ on the top I protein, leading to reversible covalent bonding between drug and enzyme

nucleophilic group of top I is within reach of position 9 of bound camptothecin.

The lactone ring of camptothecin is critical to activity. If the O in the lactone ring is replaced by an N to make a lactam, all activity against top I as well as against tumours is lost (Hertzberg *et al.*, 1989b; Jaxel *et al.*, 1989; Pommier *et al.*, 1991c). The lactone ring opens spontaneously (more rapidly at higher pH) to form a carboxylate salt which is itself inactive. The salt form can, however, convert spontaneously (more rapidly at lower pH) to the active lactone. The opening and closing of the lactone ring is facilitated by the OH group at position 20. This interconversion chemistry suggests the possibility that camptothecin may bind covalently and reversibly with a sulfhydryl, amino or other reactive group on the enzyme (Figure 1.2). Evidence for reversible covalent binding of camptothecin to top I has been reported by Hertzberg *et al.* (1990).

There is good evidence that bound camptothecin inhibits the resealing of the cleaved form of top I–DNA complexes (Hsiang *et al.*, 1985; Hsiang and Liu, 1988; Champoux and Aronoff, 1989; Covey *et al.*, 1989; Porter and Champoux, 1989a; Svejstrup *et al.*, 1990).

### *Effects of DNA Sequence and Conformation*

Camptothecin enhances cleavage at some of the same sites cleaved by the enzyme alone, although the degree of enhancement may vary. At other sites, however, camptothecin neither enhances nor suppresses cleavage (Thomsen *et al.*, 1987; Jaxel *et al.*, 1988a, 1991a,b; Kjeldsen *et al.*, 1988b; Porter and Champoux, 1989a,b). The sites of top I cleavage are influenced by the base pairs in the immediate vicinity of the cleavage site. The most prominent dependence is for the base immediately 5′ to the cleavage site (designated as −1 position), which is T at approximately 90% of the sites, in both the presence and absence of camptothecin (Been *et al.*, 1984; Jaxel *et al.*, 1988a, 1991a,b; Champoux and Aronoff, 1989). The base immediately 3′ to the cleavage site (+1 position) is unbiased for top I sites in the absence of drug, but is strongly biased towards G for sites where cleavage is enhanced by camptothecin (Porter and Champoux, 1989b; Jaxel *et al.*, 1991a,b). In oligonucleotides containing a strong camptothecin-stimulated site changes at the +1 position did not affect cleavage in the absence of drug, while in the presence of camptothecin cleavage was enhanced in the order $G \gg A, C > T$ (Jaxel *et al.*, 1991b). As will be discussed in more detail in the context of analogous findings with top II inhibitors, the predominant influence of the base pairs immediately adjacent to the cleavage site suggests a model for the top I complex in which camptothecin is stabilized by stacking against the flanking base pairs, as well as by covalent binding of the drug to the enzyme (Figure 1.3).

### *DNA Site Selectivity*

Westergaard and colleagues identified a particularly active sequence for top I in the amplified genes for ribosomal RNA in the ciliate macronucleus. The preferred top I sites were located in presumptive regulatory regions in the central spacer between divergently transcribed gene pairs, and had the sequence: 5′-ARACTT ↓ AGARAAAWWW-3′ (where R is purine, W is A or T and the arrow indicates the cleavage site) (Bonven *et al.*, 1985; Christiansen *et al.*, 1987; Svejstrup *et al.*, 1990). The sites were located in DNase I-hypersensitive regions which are probably nucleosome-free and accessible to enzymes. Accessibility alone, however, did not account for the preference of top I for these sequences, because the same selective cleavage sites were demonstrated in purified DNA. More remarkably still, incorporation of one copy of this sequence into the plasmid pBR322 significantly increased the rate of relaxation of the plasmid DNA by purified top I, while incorporation of a sequence modified in one base did not have this effect (Busk *et al.*, 1987). This sequence seems to be effective for phylogenetically diverse organisms, including humans (Kjeldsen *et al.*,

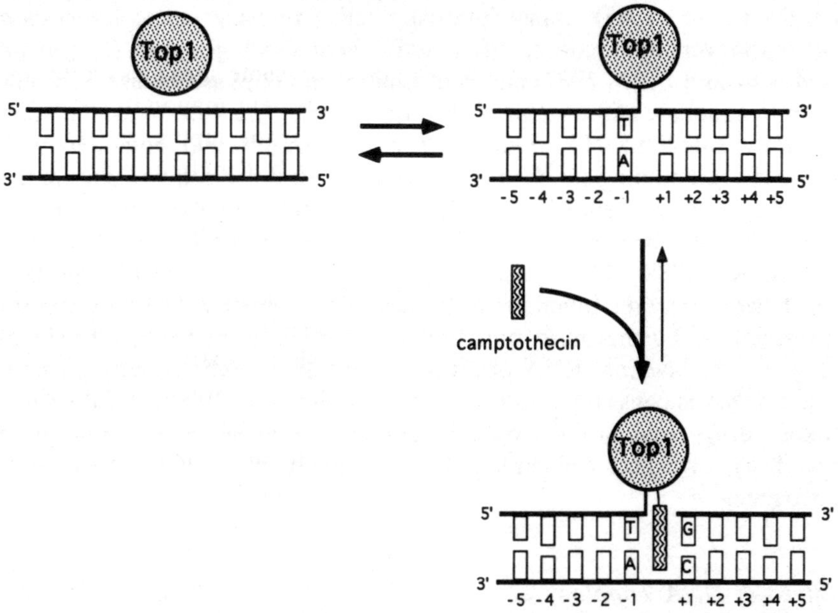


**Figure 1.3** Hypothetical model of top I cleavage complexes. In the absence of drug, top I forms transient cleavage complexes preferentially at sites having a T immediately upstream from the cleavage site. Camptothecin may bind to this complex by forming a reversible covalent bond to the enzyme and by stacking against the immediately adjacent base pairs. Base preferences for upstream T and downstream G suggest possible interactions simultaneously with the base pairs on both sides

1988b). It could, however, be one of a number of different types of hypereffective sequences. This preferred sequence is not significantly affected by camptothecin (Kjeldsen *et al.*, 1988b), possibly because camptothecin prefers G on the 5′ flank of the cleavage site (Jaxel *et al.*, 1991a,b).

The ribosomal RNA genes in *Dictyostelium* contain several clustered top I recognition sequences located upstream of the start point for transcription in nucleosome-free chromatin and close to putative promoters (Bettler *et al.*, 1988). Because RNA polymerase must pass through these sequences, the authors speculated that the passage of RNA polymerase I would dislodge the topoisomerases but that cooperativity of binding at the clustered recognition sequences assists rapid refilling of the vacant sites.

An association of top I with the regulatory region of transcriptionally active SV40 minichromosomes near the start of late transcription has also been described (Vassetzky *et al.*, 1990) as well as with puffs and transcriptionally active regions of *Drosophila* polytene chromosomes (Fleischmann *et al.*, 1984), DNAse I-hypersensitive sites in *Tetrahymena R*-chromatin

(Bonven *et al.*, 1985), transcriptionally active tyrosine amino transferase genes (Stewart and Schütz, 1987), active heat shock genes in *Drosophila* (Gilmour and Elgin, 1987) and centromeres in G2 phase mouse 3T6 cells prior to anaphase (Maul *et al.*, 1986). In the latter instance an examination of mouse satellite DNA sequences revealed considerable homology with the preferred binding site for top I. Evidence for top I cleavage sites in the linker regions of *Xenopus laevis* chromatin and a role in chromatin assembly has also been presented (Culotta and Sollner-Webb, 1988; Sekiguchi and Kmiec, 1988). These and numerous other observations indicate that top I has important functions in processes that involve DNA or control transcription. Furthermore, enrichment of top I in the nucleolus, the site of very active ribosomal RNA production, and its catalytic activity on ribosomal DNA is consistent with this view (Muller *et al.*, 1985). By the same token, drugs that inhibit top I would not be expected to show great specificity for cancer cells unless they selectively inhibit the expression of oncogenes.

### *Effects of DNA Topology*

There is increasing evidence that changes in the topology of DNA templates can affect or control the expression of genes, in some instances increasing and in others decreasing gene expression (Gellert, 1981; Wang, 1985; Tabuchi and Hirose, 1988; Mizutani *et al.*, 1991). Conversely, RNA transcription or DNA synthesis alters the topology of DNA, so that Liu and Wang (1987) have proposed a model to explain the effects of RNA polymerase on DNA in which advancing polymerases generate positive supercoils in DNA templates ahead of the polymerase with negative supercoils behind, and top I and II remove negative or positive supercoils respectively (Figure 1.4).

This model is supported by experiments demonstrating that pBR322 DNA has a higher degree of negative supercoiling when top I is inhibited, or positive supercoiling when DNA gyrase is inhibited. Consequently, drugs that inhibit top I could significantly alter cellular metabolism through effects on DNA topology and gene expression. For example, in bacteria high concentrations of DNA gyrase inhibitors decreased DNA supercoiling and increased the expression of the *gyr*A gene of DNA gyrase (Franco and Drlica, 1989). In contrast, inhibiting top I with camptothecin inhibited the expression of human top I in yeast (Bjornsti *et al.*, 1989).

From studies of the TetA protein in pBR322, which is thought to insert into the inner bacterial membrane, Lodge *et al.* (1989) suggested that supercoiling domains are created when DNA segments are anchored to a large cellular structure via coupled transcription–translation and membrane insertion of nascent proteins. By analogy, anchoring RNA or DNA

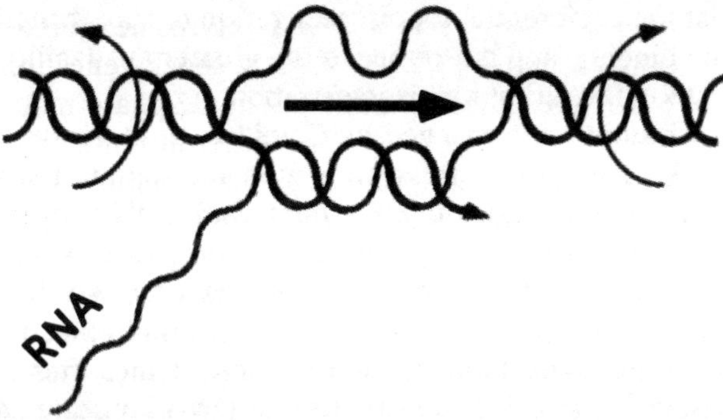


**Figure 1.4** DNA supercoiling effects of transcription. As transcription proceeds from left to right, the DNA helix must untwist ahead and retwist behind. If the transcription complex is fixed and cannot rotate, then the DNA helix must rotate. If its ability to rotate is limited, positive supercoils will accumulate in front and negative supercoils behind. This impediment to transcription is relieved by topoisomerase action

polymerase to the nuclear matrix in mammalian cells might also introduce supercoiling into transcribed or replicating loops of DNA, with consequent effects upon the expression of genes (Gasser *et al.*, 1989). In multigene loops of active chromatin the consequences of topological changes in DNA due to transcription of one gene could be important for the expression of other genes within the loop if they are not isolated by attachment of the intervening DNA to the structural matrix. In these circumstances drugs that affect topoisomerases could have extremely complex effects upon the expression of even distant genes.

Ljungman and Hanawalt (1992) recently reported evidence in mammalian cells for positive torsional tension downstream of an actively transcribed gene and negative torsional tension in the upstream region. The ability of negative torsional stress to stimulate transcription was demonstrated in yeast by Schultz *et al.* (1992). If the negative torsional tension that is created in the wake of an actively transcribed gene is not relaxed by the action of a topoisomerase, then the transcription of another gene located some distance upstream might conceivably be stimulated.

Caserta *et al.* (1989) analysed the steps of the topoisomerization reaction catalysed by calf thymus top I on a DNA domain containing an intrinsically bent segment of DNA, and found that the bent DNA segment was preferentially cleaved by the enzyme, demonstrating the importance of local DNA conformation for topoisomerase action. The interaction of top I with DNA is also stabilized by helix curvature, which is a feature of DNA supercoiling (Krogh *et al.*, 1991), while supercoiled DNA is a more efficient substrate for the enzyme than is relaxed DNA (Caserta *et al.*, 1990). Because in competition experiments supercoiled DNA bound top I more slowly than relaxed DNA, and relaxation showed first-order kinetics, Caserta *et al.*

suggested that the preferential topoisomerization of supercoiled DNA was not due to the binding step but rather to the greater availability of sites on supercoiled DNA favouring topoisomerization.

Similar conclusions were reached by Camilloni *et al.* (1988, 1989), who found that lack of torsional stress prevented the topoisomerization of a segment of mouse immunoglobulin κ-light chain DNA by a variety of eukaryotic type I topoisomerases. They concluded that topology was more critical than sequence effects for the cleavage of DNA by top I and proposed a model for the regulation of top I such that relaxation of DNA by top I causes topological inactivation of the enzyme. This would serve the major purpose of keeping the structure of DNA constant and preventing a continuous futile activity of the enzyme on relaxed DNA. Consequently, they suggested that top I is a biological sensor keeping constant the conformation of DNA domains placed under its control.

Direct support for the preferential relaxation of supercoiled DNA containing top I recognition sites comes from Busk *et al.* (1987), who compared the relaxation of supercoiled DNA containing a hexadecameric recognition sequence for top I with relaxation of a control DNA and found that the DNA containing the hexadecameric sequence was preferentially relaxed compared with DNA containing a single base pair mutation in the recognition sequence.

Kjeldsen *et al.* (1988b) also showed that human top I primarily stimulated DNA cleavage at a hexadecameric recognition sequence inserted into plasmid DNA, whereas camptothecin stimulated cleavage at additional sites by up to 200-fold with little (twofold) effect on cleavage at the hexadecameric site. Camptothecin also eliminated cleavage at some other sites. The authors concluded that at least three types of potential eukaryotic type I topoisomerase cleavage sites can be distinguished using camptothecin, with little sequence consensus evident, apart from a most frequently cleaved TG dinucleotide. These observations may explain the fact that the location and frequency of DNA sequences in cloned human Ha-*ras* and p53 genes cleaved by calf thymus top I changed when the enzyme was inhibited by camptothecin (Bronshtein *et al.*, 1989; Camilloni *et al.*, 1989).

Other evidence from studies on reconstituted minichromosomes suggests that cruciform structures exist in internucleosomal regions of DNA and that the persistence of such structures is influenced by effects of top I on the topology of DNA (Battistoni *et al.*, 1988). Whether cruciforms, like bent DNA, directly influence the function of top I is unknown.

Stimulatory and inhibitory effects of DNA minor groove binding drugs on top I activity were described by McHugh *et al.* (1989); these may also result from effects of the drugs on DNA topology.

Preliminary evidence has been presented that interference with *cis*-acting gene-to-gene interactions can alter cell transformation and reverse the cancerous nature of cells when the genes involved have short intergenic distances (Xu *et al.*, 1991). If this is substantiated, drugs that modify topoisomerase action on one gene might also affect adjacent gene expression and revert tumour cells towards normality. Inhibitors of top I or II do appear to induce some differentiation of human and mouse myeloid leukaemia cells (Chou *et al.*, 1990; Rappa *et al.*, 1990; Ling *et al.*, 1991).

### *Phosphorylation*

Recent research has emphasized the importance of protein kinase cascades and protein phosphorylation in the control of many cellular processes, including growth (Ralph *et al.*, 1990; Hall and Vulliet, 1991). Top I is no exception, since phosphorylation of calf thymus type I topoisomerase by virus- or cell-associated tyrosine protein kinases produced a tenfold reduction in activity (Tse-Dinh *et al.*, 1984). In contrast, Novikoff hepatoma and CHO cell top I were both phosphorylated on serine by nuclear casein kinase II, which stimulated their activity (Durban *et al.*, 1985; Caizergues-Ferrer *et al.*, 1987). Phosphorylation also regulates the activity of *Xenopus laevis* top I (Kaiserman *et al.*, 1988). In the latter case studies *in vitro* showed that dephosphorylation of the enzyme with calf intestinal alkaline phosphatase eliminated topoisomerase activity by blocking the formation of the initial protein–DNA complex, whereas preincubation of the enzyme with *Xenopus laevis* casein kinase II plus ATP restored topoisomerase activity coincident with serine phosphorylation. Gorsky *et al.* (1989) also found increased DNA top I activity in HL60 promyelocytic leukaemia cells treated with phorbol ester to activate protein kinase C, although they did not ascertain whether the topoisomerase was phosphorylated by protein kinase C or another kinase activated by protein kinase C. However, Pommier *et al.* (1990b) showed that dephosphorylating the top I from a Chinese hamster cell line (DC3F/9-0HE) resistant to 9-hydroxyellipticine with calf intestinal phosphatase abolished its DNA relaxing activity along with its sensitivity to camptothecin. Since relaxing activity and sensitivity to camptothecin were both restored after rephosphorylation by brain protein kinase C, Pommier *et al.* (1990b) concluded that top I was probably regulated by phosphorylation–dephosphorylation *in vivo*, and that this might alter gene expression through effects on DNA swivelling and consequent modulation of transactivator binding. Dephosphorylation of calf thymus top I also completely inhibited the activity of the enzyme and it blocked any effect of camptothecin, whereas rephosphorylation with calf thymus nuclear casein kinase II and ATP fully restored activity (Coderoni *et al.*, 1990).

Regulation of top I by phosphorylation or dephosphorylation could link the enzyme to control of growth signal transduction via activation of various protein kinases or protein phosphatases, since Ackerman and Osheroff (1989) have shown that epidermal growth factor increased cytosolic casein kinase activity fourfold in human A431 carcinoma cells and that protein kinase C mediated this effect, which was dependent upon phosphorylation of casein kinase II. Clearly, the nature and effects of phosphorylation of DNA topoisomerase on the activity of the enzyme are complex, and precisely how they alter the response to drugs such as camptothecin remains to be determined. Direct phosphorylation of the tyrosine residue that links to DNA during top I action by protein tyrosine kinases would presumably block all enzyme activity. However, casein kinase II and protein kinase C phosphorylate serine or threonine; therefore, more subtle effects of these kinases at other sites must be involved. Nuclear casein kinase II appears to have a general role in the regulation of events involved in cell growth in addition to effects on top I (Ralph *et al.*, 1990; Meisner and Czech, 1991) and its activity declines dramatically in confluent cells, consistent with a function in growth. It also regulates ribosomal DNA transcription in bovine aortic endothelial cells, and presumably other cells (Belenguer *et al.*, 1989). Therefore, casein kinase II may be a good potential target for future anticancer drugs, which would thereby inactivate top I.

Camptothecin has also been shown to inhibit histone H1 phosphorylation (Roberge *et al.*, 1990) and entry of BHK cells into mitosis, apparently through indirect effects, since the drug did not affect the H1 kinase activity in nuclear extracts (Lewin, 1990). There was also no effect on metaphase chromosomes. The effects of camptothecin were observed when the drug was applied in early G2 phase but not when the drug was applied late in S phase. However, these findings emphasize the fact that top I has multiple effects on cells, so that drugs inhibiting top I could also have widespread effects. For example, camptothecin blocks steroidogenesis and the differentiation of rat granulosa cells induced by follicle stimulating hormone (Barañao *et al.*, 1991). In general, there has been little investigation of the secondary effects of top I inhibitors, possibly because they are presumed to arise indirectly from effects on gene expression.

### *Poly(ADP) Ribosylation*

It has been proposed that human top I undergoes post-translational modification by poly(ADP) ribosylation, with consequent inhibition of its activity (Ferro and Olivera, 1984; Kasid *et al.*, 1989). Reagents that inhibit poly(ADP) ribosylation, such as 3-aminobenzamide, increase the incidence of DNA scission and cytotoxicity in response to drugs such as camptothecin (Mattern *et al.*, 1987). Apart from suggesting that top I may be regulated by poly(ADP) ribose polymerase, these observations raise the interesting possibility that the effectiveness of anticancer drugs directed against top I might be increased by cotreatment with agents that block ADP ribosylation. Hypersensitivity to camptothecin also occurs in poly(ADP) ribose polymerase-deficient cell lines, which develop increased sister-chromatid exchanges and single-strand DNA scissions compared with normal cells (Chatterjee *et al.*, 1989). These effects may reflect roles of top I and poly(ADP) ribosylation in the repair of DNA, since poly (ADP) ribose polymerase-deficient cells are also hypersensitive to clinically useful alkylating agents (Chatterjee *et al.*, 1990a, 1991).

Boothman *et al.* (1989) used beta-lapachone, an activator of top I, to sensitize human neoplastic cells to X-rays and neocarzinostatin (agents that predominantly cause single-strand DNA breaks), again suggesting that top I is involved directly or indirectly in DNA repair processes, including repair of X-ray damage. However, precisely how beta-lapachone functions is unknown, and Gedik and Collins (1990) found no evidence of a role for top I or II in UV repair of HeLa cell DNA.

An increase in poly(ADP) ribosylation occurs in mitogen-activated lymphoid cells (Boulikas *et al.*, 1990), yet there is little change in top I activity throughout the cell cycle (see below); therefore, the true significance of effects of poly(ADP) ribosylation on top I is unclear.

Top I from calf thymus is also inhibited *in vitro* by poly(ADP) ribosylation (Jongstra-Bilen *et al.*, 1983).

## *Role in RNA Synthesis*

The high cytotoxicity of drugs that inhibit top I and cleave DNA is not unexpected, in view of numerous reports of involvement of the enzyme in RNA synthesis and DNA replication, largely revealed through the action of camptothecin (Kann and Kohn, 1972; Fleischmann *et al.*, 1984; Muller *et al.*, 1985; Gilmour *et al.*, 1986; Brill *et al.*, 1987; Egyhazi and Durban, 1987; Garg *et al.*, 1987; Gilmour and Elgin, 1987; Stewart and Schütz, 1987; Champoux, 1988; Zhang *et al.*, 1988; Annunziato, 1989; Bendixen *et al.*, 1990; Charron and Hancock, 1990a).

It is widely believed that top I in eukaryotic cells plays a major role in maintaining transcription of RNA by relieving DNA supercoiling produced by advancing RNA polymerase molecules (Figure 1.4) (Giaever and Wang, 1988; Wu *et al.*, 1988; Tsao *et al.*, 1989). This conclusion is largely based on evidence from bacterial systems which suggests that top I removes negative supercoils in DNA generated during RNA transcription (Wu *et al.*, 1988; Wang, 1991). However, yeast strains with *ts* mutations in the gene for top I (*top* 1) are viable and exhibit no obvious growth defects; therefore top I does not appear to be essential for viability. In contrast, yeast *ts* mutants of top II (*top* 2) are conditional-lethal temperature-sensitive mutants, and *top* 1, *top* 2 double mutants grow poorly at permissive temperatures and rapidly cease to grow at non-permissive temperatures (Uemura and Yanagida, 1984; Goto and Wang, 1985; Brill *et al.*, 1987). These data suggest either that topoisomerase can relax torsionally stressed DNA or that top II has additional functions necessary for growth. Consequently, the cytotoxic action of drugs, such as camptothecin, that inhibit top I is unlikely to arise solely from preventing the relaxation of DNA, but is a further consequence of the drug's inhibiting top I.

Evidence supporting the view that camptothecin blocks elongation of nascent RNA by impeding the progression of RNA polymerase was presented by Zhang *et al.* (1988), who used camptothecin and nuclear run-off to show a decrease in RNA polymerase density towards the 3′ end of transcriptionally active ribosomal RNA genes, consistent with camptothecin inhibition of top I blocking the progress of RNA polymerase.

Early studies of drug effects on RNA synthesis indicated that camptothecin, like nitrogen mustard, can prematurely terminate the elongation of nascent ribosomal and nucleoplasmic RNA chains in mammalian cells (Kann and Kohn, 1972). Bendixen *et al.* (1990) also studied the effect of camptothecin on RNA synthesis, using an *in vitro* transcription system with SP6 RNA polymerase, a plasmid DNA containing a hexadecameric top I recognition sequence, and supplemented with top I. They located the site of premature termination of RNA synthesis 10 base pairs upstream of the top I-linked nucleotide residue on the coding strand which corresponded closely to the border of the area of DNA protected from microsomal nuclease digestion by top I. Transcription was not impeded in the absence of camptothecin or by top I–DNA complexes attached to the non-coding strand, while top I from a camptothecin-resistant cell line did not inhibit RNA synthesis in the presence of camptothecin.

Additional evidence for the importance of top I in RNA synthesis was provided by Kroeger and Rowe (1989) from studies on the transcribed heat shock protein gene (HSP70) in *Drosophila melanogaster*, using camptothecin. Mapping of top I cleavage sites *in vivo* showed that cleavage of the HSP70 gene only occurred when the gene was transcriptionally active. Single-strand DNA cleavages were produced on both transcribed and non-transcribed DNA strands, with the majority of cleavage on the transcribed strand, while some close single-strand breaks produced double-strand DNA scissions. Inhibiting HSP70 transcription with actinomycin D or 5,6-dichloro-1-β-D-ribofuranosylbenzimidazole inhibited most top I cleavage except at the ends of the gene, and camptothecin inhibited RNA synthesis by 95%, confirming that top I plays an integral part in transcription. Reduced cleavage in the presence of actinomycin D was attributed to the inability of RNA polymerase to alter DNA structure and produce sites recognized by the topoisomerases. Related studies with metallothionein genes also support a role for top I in active gene expression (Mattern *et al.*, 1989).

The sites at which top I interacts with transcriptionally active ribosomal chromatin of *Xenopus* oocytes were mapped, using camptothecin to trap ribosomal DNA–top I cleavable complexes in oocyte nuclei. Topoisomerase cleavage sites were concentrated in the region encoding rRNAs and spaced with a periodicity of approximately 200 nucleotides, reflecting the chromatin structure. A few sites were also present in the spacer regions separating ribosomal genes. Plasmid rDNA assembled into a nucleoprotein structure in oocyte extracts gave the same restricted pattern of cleavage sites, suggesting that active rRNA genes were organized into nucleosome-like arrays, with sequences cleaved by top I located in the linker regions (Culotta and Sollner-Webb, 1988).

Inhibition of top I by camptothecin markedly diminished the transcription of supercoiled ribosomal DNA (rDNA) by extracts from rat mammary adenocarcinoma cells but had little effect on transcription of linear rDNA, suggesting that relaxation of supercoiling is essential for ribosomal gene expression. Therefore, preferential effects of camptothecin on rDNA transcription *in vivo* could also be due to inhibition of top I affecting rDNA supercoiling (Garg *et al.*, 1987). In yeast cells reduced levels of topoisomerases lead to excision of ribosomal DNA as extrachromosomal rings which reintegrate upon expression of top I or II, indicating unusual effects of DNA topology on these genes (Kim and Wang, 1989a).

Altogether, the preceding data support the idea that at least part of the cytotoxicity of camptothecin results from stabilization of cleavable complexes formed on transcribed DNA which interfere with the elongation of nascent RNA. However, the topology of DNA templates may also contribute to inhibition of RNA synthesis, since effects of camptothecin on RNA synthesis from supercoiled or linear ribosomal DNA were quite different.

### *Role in Genetic Recombination*

Top I has been implicated in recombination (Halligan *et al.*, 1982; Bullock *et al.*, 1985; McCoubrey and Champoux, 1986; Wang *et al.*, 1990). Christman *et al.* (1988) found that null mutants of top I in *Saccharomyces cerevisiae*, or temperature-sensitive top II mutants when grown at semi-permissive temperatures, show a 50–200-fold higher frequency of mitotic recombination of ribosomal DNA relative to *top*$^+$ controls. This effect was not seen with other genes, and it has been suggested that where multiple copies of genes exist (yeast has roughly 200 tandem copies of a 10 kb repeat rDNA), topoisomerases may be needed to relax supercoils in rDNA resulting from rRNA transcription and thereby suppress homologous recombination (Fink, 1989; Wang *et al.*, 1990).

From an analysis of the unoccupied site of an integrated human papillomavirus 16 sequence in a cervical carcinoma, Choo *et al.* (1990) concluded that short patches of homologous sequences in the virus and host genomes were involved in integration and that preferred top I cleavage sites and alternating purine–pyrimidine bases (which favour Z-DNA) present in the integration region indicated a role for top I in illegitimate recombination during viral integration.

A preferred top I nicking region, 5′-CT(A/T)T(C/T)3′, has also been implicated in the non-homologous integration of hepatitis B virus (Hino *et al.*, 1989), and a similar region, 5′CT(A/T)TT(C/T)3′, in non-homologous recombination and integration of Parvovirus chromosomes (Hogan and Faust, 1986). In addition, sites of top I-mediated illegitimate recombination of woodchuck hepatitis virus DNA with cellular DNA *in vitro* were nearly identical with a subset of integrations cloned from hepatocellular carcinomas (Wang and Rogler, 1991).

Konopka (1988) compiled a list of DNA strand exchange sites for non-homologous recombination in somatic cells and found that 92% had trinucleotide homologies with top I cleavage sites. Other evidence implicating top I in non-homologous recombination comes from Legouy *et al.* (1989), who examined the structure of AT-rich non-homologous recombination joints in amplified DNA in a Syrian hamster cell line. Eight out of ten of the amplified DNAs contained a putative top I cleavage site. Finally, Shuman (1989) has shown that an essential 32 kDa top I from vaccinia virus (Schaffer and Traktman, 1987) promotes illegitimate recombination in *Escherichia coli*.

In total, these data strongly support a role for top I in non-homologous recombination of DNA. Studies with top I inhibitors that increase recombination suggest that increased aberrant recombination is an important component of the cytotoxic action of the drugs.

### *Sister-chromatid Exchange*

Drugs that affect topoisomerases have been implicated in the production of sister-chromatid exchanges (SCEs) or other chromosomal aberrations involving recombination (Lim *et al.*, 1986), and similar responses to camptothecin or amsacrine (*m*-AMSA) obtained in yeast cells suggest that formation of cleavable complexes is the common denominator for these effects (Nitiss and Wang, 1988).

In human lymphocytes stimulated to grow with phytohaemagglutinin, camptothecin produced dose-dependent sister-chromatid exchanges in S phase cells, whereas cells treated in G2 phase mainly developed chromatid breaks, suggesting that DNA replication was involved in chromatid exchange (Degrassi *et al.*, 1989). In simian virus 40-infected CV-1 cells camptothecin produced broken replication forks in growing Cairns-type structures as well as other abnormal viral DNAs which, it was suggested, are intermediate products of sister-chromatid exchange between replicating viral genomes (Snapka, 1986).

Dillehay *et al.* (1989) have discussed various models to explain the production of sister-chromatid exchanges involving exchange of topoisomerase subunits in protein–DNA complexes (Pommier *et al.*, 1985), or homologous displacement with strand switching during the removal of parental helical turns by topoisomerases at or near replication forks. However, strong evidence favouring a specific model is still lacking, although it does appear that DNA replication forks are important entities in the formation of SCEs (Snapka, 1986).

### *Cell Cycle Relationships*

Early measurements of top I activity in quiescent or cycling mouse embryo fibroblasts or human lymphocytes suggested that the enzyme was substantially increased in cycling cells immediately before or during S phase and as a result of events in G1 phase (Rosenberg *et al.*, 1976; Tricoli *et al.*, 1985). However, when Heck *et al.* (1988) used antibody probes and centrifugal elutriation to follow the expression of top I in various normal and transformed chicken cells as the cells progressed through the cell cycle, they found that the ratio between top I and total protein remained essentially constant, although the half-life of the enzyme in the transformed cells (15.9 ± 1.1 h) decreased compared with that in primary chicken embryo fibroblasts (23.1 ± 0.8 h). In contrast, DNA top II began to increase just prior to S phase and peaked at G2 + M, before declining rapidly in early G1 phase. The half-life of top II was also increased fourfold in the transformed cells.

Romig and Richter (1990) measured the levels of human top I messenger RNA in resting versus proliferating Hel-299 (human embryonic lung) fibroblasts, and showed that addition of serum to growth-arrested cells caused a continuous increase in top I mRNA which reached a sixfold maximum after 25 h (late S phase), while nuclear run-off studies detected a fourfold increase in *de novo* mRNA synthesis after 15 h. However, the amount of top I and top I activity only increased twofold in concert with total cellular protein as the cells doubled in size prior to cell division, so that the specific activity (enzymatic activity per mg protein) of the enzyme did not change during the cell cycle.

Hwong *et al.* (1989) also measured the top I mRNA in human skin fibroblasts after brief treatment with phorbol ester and found a transient increase after 3 h followed by a decrease to basal levels after 12 h. Actinomycin D abolished the rise in mRNA consistent with transcriptional activation. Unfortunately, there was no comparison with enzyme activity to assess whether top I also increased. Transient, 10 min maximum activation of top I in human U937 cells in response to leukotriene D4 has also been described, which was blocked by camptothecin, along with arachidonic acid release, possibly implicating top I in leukotriene D4 signal transduction (Mattern *et al.*, 1990).

No major changes in top I activity occurred in rat liver as a function of regeneration time after partial hepatectomy (Champoux *et al.*, 1978; Duguet *et al.*, 1983), although top I mRNA levels increased 5–10-fold between 3 h and 24 h after partial hepatectomy (Sobczak *et al.*, 1989). There was also little change in top I activity in mouse 3T3 cells at different stages of the cell cycle or after infection with polyoma virus (Champoux *et al.*, 1978), during the growth and differentiation of murine erythroleukaemia cells (Bodley *et al.*, 1987) or during spermatogenesis in chickens (Roca and Mezquita, 1989). Therefore, most of the evidence suggests that top I activity does not change dramatically during the cell cycle or after differentiation, which is consistent with its proposed role as a constant sensor of DNA topology.

Because top I mRNA increases after serum stimulation (Romig and Richter, 1990), adenovirus 5 infection (Romig and Richter, 1990) or phorbol ester treatment (Hwong *et al.*, 1989) yet the amount and specific activity of top I is constant in cycling cells (Romig and Richter, 1990), gross regulation of the enzyme during the cell cycle apparently occurs at the level of translation. However, finer control may be exerted by enzyme modification such as phosphorylation or poly(ADP) ribosylation.

Crespi *et al.* (1988) have shown that there are increases in DNA top I and II activity in a variety of chemically or virally transformed cells, and in spontaneously transformed A31 cells. However, it is not clear whether these increases resulted from increased transcription, decreased enzyme turnover, enzyme modification or other causes.

### *Effect of DNA Synthesis on Cytotoxicity*

Although camptothecin inhibits RNA synthesis, it is most cytotoxic to S phase cells, suggesting a vital role for top I during DNA synthesis (Kessel *et al.*, 1972; Li *et al.*, 1972; Bhuyan *et al.*, 1973; Hsiang *et al.*, 1985; Charron and Hancock, 1990a; Del Bino and Darzynkiewicz, 1991; Del Bino *et al.*, 1991). However, the DNA polymerase inhibitors aphidicolin or hydroxyurea partially protected Chinese hamster lung fibroblasts or L1210 cells from the cytotoxic effects of camptothecin without preventing cleavable complex stabilization; therefore, DNA breakage alone was not sufficient to kill the cells (Holm *et al.*, 1989a; Hsiang *et al.*, 1989a). Other evidence indicates that top I need not function during G2 phase or mitosis of CHO cells (Charron and Hancock, 1990a).

These results suggest that progression of DNA replication forks and their interaction with cleavable complexes is involved in camptothecin-induced effects on DNA synthesis which subsequently cause cell death. Indirect support for this proposal has been obtained by studying SV40 replication. Analysis of the SV40 DNA replication intermediates in virus-infected cells treated with camptothecin showed that cleavable complexes are preferentially associated with replicating intermediates (Champoux, 1988) and that strand breaks occur preferentially at replication forks (equivalent to a duplex break) (Avemann *et al.*, 1988; Shin and Snapka, 1990a,b). Furthermore, in cell-free SV40 replication systems, camptothecin and top I arrested DNA synthesis and produced linearized SV40 molecules with cleavable complexes that were not dissociated by high salt. These linear DNAs are thought to arise from camptothecin-induced strand breaks when top I forms cleavable complexes near single-stranded regions at arrested replication forks (Hsiang *et al.*, 1989a).

The manner in which a moving replication fork may interact with a camptothecin–top I complex is suggested in Figure 1.5, which shows how a double-strand termination may arise in daughter DNA. Evidence consistent with the production of such double-strand cuts in newly replicated DNA was presented by Ryan *et al.* (1991), who found that the cuts are persistent and that they are preventable by inhibiting DNA synthesis with aphidicolin. This model suggests the possibility of strand termination events, associated not only with replication, but also with transcription. In the latter case strand termination would only be expected when the camptothecin–top I complex is on the template strand.

Charron and Hancock (1990a) showed that 1 μg/ml camptothecin inhibited DNA synthesis by >80% in synchronously growing Chinese hamster ovary (CHO) cells. Inhibition increased if the drug was also present during G1 phase, suggesting that effects on events in G1 phase, presumably transcription, contributed to the inhibition of DNA replication. The progression of CHO cells from S phase through mitosis to cell division was not affected when camptothecin was present during this period, implying that top I function is not required for these steps. However, it is possible that top II substituted for top I to allow mitosis to proceed.

Evidence also exists that different cell types respond differently to top I inhibitors. For example, Del Bino *et al.* (1990) treated L1210 or MOLT-4 lymphocytic cells with low concentrations of camptothecin (0.02–0.5 μg/ml) which slowed growth of the cells through S and G2 phase, with a 'terminal' point of camptothecin action 1 h before mitosis. Some of the cells became hyperploid and progressed as such through the cell cycle. In contrast, human promyelocytic HL60 and KG1 cells exposed during S and

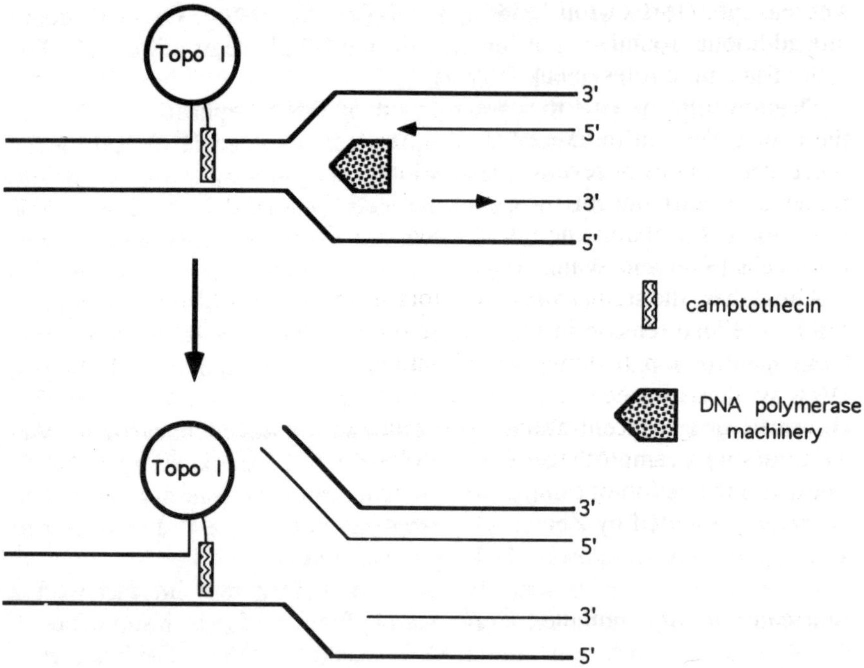


**Figure 1.5** Model of possible interaction between replication fork and camptothecin-trapped top I–DNA complex. Topoisomerase I induces a DNA single-strand break while bonding to the 3′ terminus of the interrupted strand. Camptothecin, which stabilizes the top I–DNA cleavage complex, is shown stacked against DNA bases at the cleavage site and covalently linked to top I. The leading daughter DNA strand, elongating in the 5′ to 3′ direction, may dislodge the template strand from the top I complex and produce a double-strand termination. The resulting configuration would probably be lethal

G2/M phase to the same drug concentrations were killed and exhibited considerable DNA degradation as early as 2 h after drug addition. In addition to indicating tissue specificity in response to camptothecin, these results suggested that tumour cells with higher proliferation rates might be more sensitive to the drug, especially if they can be synchronized in S phase.

Zucker and Elstein (1991) detected an increase in the ploidy of murine erythroleukaemia cells treated with camptothecin or top II inhibitors.

Annunziato (1989) found that a combination of camptothecin and the top II inhibitor VM-26 inhibited DNA synthesis in HeLa cells more effectively than either drug alone, and that the drugs did not affect maturation of non-nucleosomal replication intermediates as assessed by micrococcal nuclease protection. Schaak *et al.* (1990a) also concluded that both topoisomerase activities were required for the replication of adenovirus DNA and that inhibitors of top I caused immediate cessation of DNA synthesis, whereas inhibitors of top II blocked DNA synthesis after completion of one additional round.

Studies on the roles of eukaryote DNA topoisomerases I and II in DNA replication using yeast topoisomerase mutants have shown that the synthesis of DNA can initiate without topoisomerases I and II and produce short DNA chains of several thousand nucleotides, but further elongation requires at least one of the two topoisomerases. It was concluded that DNA top I is probably the major swivel for DNA replication in wild-type yeast cells (Kim and Wang, 1989b).

Altogether, the studies with inhibitors seem to support a requirement for top I to relieve tension in DNA near replication forks during DNA synthesis, and for top II during the resolution of replicated DNA. A role for DNA synthesis in the cytotoxic action of camptothecin in also indicated. However, drug concentrations may influence these effects, since at high concentrations camptothecin is not S-phase-specific (Liu, 1989). A model to explain the action of camptothecin on advancing DNA replication forks has been presented by Zhang *et al.* (1990).

### *Resistance to top I-inhibiting Drugs*

Camptothecin and its synthetic derivatives are the main group of drugs active against top I and several reports of resistance to camptothecin have appeared. Commonly, the level of top I is reduced in drug-resistant cells compared with parental controls (Andoh *et al.*, 1987; Gupta *et al.*, 1988; Eng *et al.*, 1990; Kanazawa *et al.*, 1990; Sugimoto *et al.*, 1990). In three of four resistant cell lines studied by Sugimoto *et al.* (1990), the reduction in enzyme level seemed commensurate with the degree of resistance, but a threefold reduced enzyme level did not seem adequate to explain the 45-fold resistance they reported in camptothecin-resistant P388 cells.

Altered, as well as reduced, top I was implicated in the reports of Andoh *et al.* (1987), Gupta *et al.* (1988) and Kanazawa *et al.* (1990). In the three cell lines analysed by these groups, top I showed resistance to camptothecin in DNA relaxation assays *in vitro*. Kjeldsen *et al.* (1988a) demonstrated by various criteria that the altered top I from the camptothecin-resistant version of human T lymphoblasts first described by Andoh *et al.* (1987) formed a more stable cleavable complex with DNA, even in the absence of camptothecin. Recently Tamura *et al.* (1991) have compared the top I DNA sequences from the drug-sensitive and drug-resistant versions of this cell line, and identified two mutations, each changing aspartic acid to glycine at residues 533 and 583. The altered enzyme expressed in bacteria showed the same 125-fold resistance to camptothecin as the cell line it was derived from, proving that these changes were sufficient to account for drug resistance in the cells.

The top I gene in camptothecin-resistant P388 cells studied by Tan *et al.* (1989) was hypermethylated, which, they postulated, would reduce RNA transcription. Subsequently this was confirmed by Eng *et al.* (1990), and a compensatory increase in top II activity was also noted.

Camptothecins are not cross-resistant with top II inhibitors, nor do they bind to P glycoprotein, so they are not subject to classical multidrug resistance (Naito *et al.*, 1988). Furthermore, top I (unlike top II) is expressed at a relatively constant level throughout the cell cycle. These advantages are leading to new interest in camptothecin as a possible chemotherapeutic agent. A new water-soluble analogue of camptothecin, CPT-II, with fewer side-effects is currently undergoing clinical trials (Ohno *et al.*, 1990), and other soluble derivatives, such as topotecan, also look promising.

## 3 Topoisomerase II (top II)

### *General Properties and Overview*

The major type II topoisomerase in mammalian cells is located in the cell nucleus, where it is closely associated with metaphase chromosome scaffolds or interphase nuclear matrices, suggesting that it has a functional role in events involving both metaphase and interphase chromosomes (Gasser *et al.*, 1989). *Drosophila* top II is also distributed along the chromosomes, confirming that localization of top II in association with chromatin is a general phenomenon (Heller *et al.*, 1986).

Human top II forms a homodimer with a monomer molecular weight of about 170 kDa (Miller *et al.*, 1981; Sander and Hsieh, 1983). A HeLa cell top II gene was cloned and sequenced by Tsai-Pflugfelder *et al.* (1988). Hybridization between the cloned sequences, mRNA and genomic DNA showed that the enzyme was encoded by a single-copy gene which was located in chromosome region 17q 21.22. Subsequent DNA sequence studies with top II partial cDNA clones obtained from a human Raji-Hn2 cell cDNA library detected two classes of related nucleotide sequences which had extensive sequence similarities to HeLa top II, and antibodies raised against synthetic peptides encoded by the individual cDNAs distinguished two species of top II in extracts of U937 cells (Chung *et al.*, 1989). The two isoenzymes with molecular weights of 170 kDa and 180 kDa were designated top II$\alpha$ and II$\beta$, respectively, and shown to be present in several other cell lines (Drake *et al.*, 1987, 1989a; Tan *et al.*, 1988). Hybridization of each class of cDNA to unique non-overlapping DNA restriction fragments in Southern blots suggested that top II$\alpha$ and II$\beta$ were products of two top II genes. The two enzymes differ in several properties such as thermal stability, salt optima, drug sensitivity and concentration during the cell cycle, leading to speculation that they may have different roles during cell proliferation and quiescence or in the development of drug resistance (Drake *et al.*, 1989a).

Possible evidence for another type II topoisomerase isoenzyme in human cells was recently obtained from a study of a HeLa cell cDNA library probed with a C-terminal region of the p170 form of human top II DNA instead of the *Drosophila* top II probe used by Chung *et al.* which had considerable homology with the N-terminal human top II sequence (Austin and Fisher, 1990). However, the relationship, if any, to the $M_r$ 180 kDa enzyme was not clearly established.

Type II topoisomerases will relax positively or negatively supercoiled DNA. The enzyme binds reversibly and covalently to DNA and catalyses a transient 4 base pair staggered cleavage of both strands of double-stranded DNA and the passage of another intact segment of double-stranded DNA through the gap to relieve torsional stress in the DNA. Religation of the DNA termini then restores intact DNA. Normally, the equilibrium between cleaved or ligated DNA strongly favours the ligated intact DNA species (Liu, 1983). However, various anticancer drugs interact with top II to stabilize the topoisomerase–DNA complex in the open or 'cleavable' state, thereby disrupting DNA-related cellular processes and culminating eventually in cell death (Nelson *et al.*, 1984). Removing the drugs usually allows the DNA to religate, although differences in the effects of amsacrine and mitoxantrone on cleavable complex longevity and cytotoxicity may indicate that different drugs stabilize different forms of the cleavable complex (Fox and Smith, 1990).

In contrast to type I topoisomerases, the type II topoisomerases require ATP to function fully. ATP and $Mg^{2+}$ facilitate DNA passage, while ATP hydrolysis assists the religation of DNA and enzyme release (Robinson and Osheroff, 1991). However, Halligan *et al.* (1985) obtained some DNA relaxation with calf thymus top II in the absence of ATP, although the enzyme was much more active in the presence of ATP. The top II inhibitors novobiocin or coumermycin bind to the ATP site on the enzyme and inhibit its function (Vosberg, 1985).

*Drosophila* top II protected a region of 25 nucleotides on both strands of a 21 base pair consensus cleavage sequence cloned into plasmid DNA from pancreatic deoxyribonuclease I digestion, with the cleavage site located near the centre of the protected region (Lee *et al.*, 1989). Calf thymus top II bound a 28 base segment of plasmid DNA, strongly protecting a 22 base pair core fragment during transcriptional footprinting using Ro15-0216 to stabilize topoisomerase–DNA complexes. Fragments less than 28 base pairs long were progressively less protected (Thomsen *et al.*, 1990) and the minimal DNA requirement for top II-mediated DNA cleavage *in vitro* was a 16 base pair double-stranded segment of DNA located symmetrically around the 4 base pair staggered cleavage site (Lund *et al.*, 1990).

*Reaction Mechanisms*

Top II is made up of two identical subunits, each able to react reversibly with a phosphodiester bond on one of the strands of a DNA duplex. The DNA strand is cleaved and one terminus of the strand break becomes covalently linked to a reactive tyrosyl residue of the enzyme. The local chemistry of the DNA strand breakage and enzyme linkage equilibria are essentially the same for top I and top II (Figure 1.1). In both cases a DNA phosphodiester bond undergoes an exchange reaction to form a phosphotyrosine bond between DNA and enzyme. At this chemical level, the only qualitative difference between eukaryotic top I and II is that the reactive tyrosine links to the 3′ DNA strand terminus in the case of top I and to the 5′ terminus in the case of top II.

The quintessential function of top II is to open one DNA duplex segment, pass another duplex segment through the gap and then reseal the first duplex. Although all the details of this remarkable process are not yet elucidated, the scheme shown in Figure 1.6 is consistent with available information. The overall process changes the topology of the DNA only so far as is dictated by the duplex passage event itself; the topological supercoiling of the opened duplex is otherwise preserved. In order to accomplish this feat, the enzyme must interact simultaneously with both DNA duplex segments. This raises two initial questions about the binding steps. Does the enzyme bind first to the cleaved or to the passed duplex, or can the enzyme recognize both simultaneously? Does binding (or cleavage) depend kinetically on one or both duplexes? There is evidence bearing on the second question, but the answer to the first remains conjectural.

Corbett *et al.* (1992) found that the kinetics of binding of top II to an appropriately designed oligonucleotide duplex depended linearly on the concentration of duplex, while cleavage exhibited a sigmoidal dependence. This indicates that binding, at least initially, involves an interaction with a single duplex. The rate of cleavage, on the other hand, depends both on the cleaved and on the passed duplex, which indicates that binding of the passed duplex helps in the cleavage of the other duplex.

Electron microscopic images have shown the enzyme bound at the intersection of two crossing DNA duplexes (Zechiedrich and Osheroff, 1990; Howard *et al.*, 1991). This can occur at duplex cross-overs of supercoiled circular DNA molecules, and could represent a specific recognition of supercoil cross-overs.

Another view, illustrated in Figure 1.6, envisions the enzyme binding first to the duplex that will be passed (step 1). The enzyme may be able to glide along this duplex until it encounters a suitable second duplex segment, perhaps at a supercoil cross-over (step 2). The second duplex is cleaved (step 3), and the first duplex passes through the gap (step 4). To complete this scheme, the enzyme then reseals the cleaved strand (step 5)

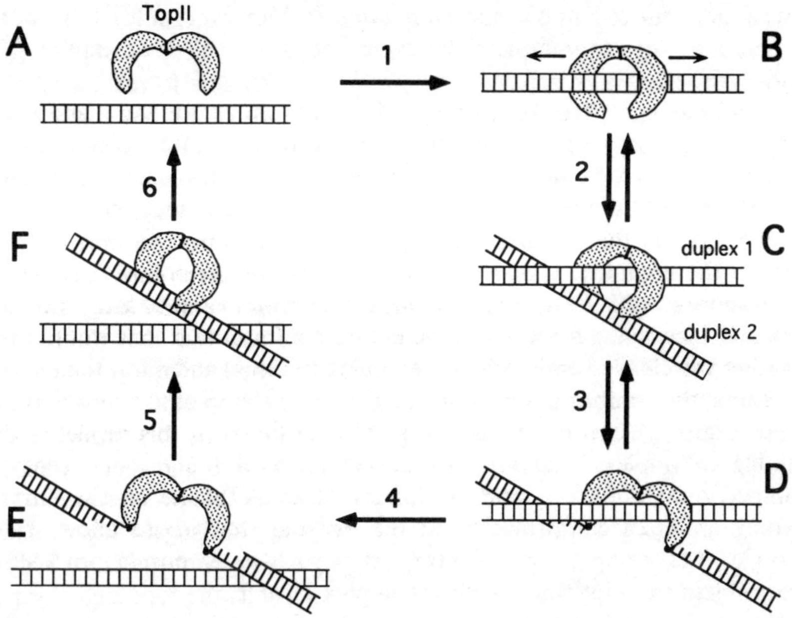


**Figure 1.6** DNA strand passage cycle of top II. The reaction changes the topology of DNA duplex 1 relative to duplex 2 (assuming that the ends of each duplex are either linked in a circle or fixed to a nuclear structure). Duplex 1 is initially in front of duplex 2 (complex C) and, after strand passage, is behind duplex 2 (complex F). Top II-targeted drugs often act by stabilizing complex D or E. ATP (or a non-hydrolysable analogue) is required for steps 3 and 4; ATP hydrolysis stimulates step 5 and is required for step 6

and reverts to its starting configuration, from which a new cycle can be initiated (step 6). This is an energy-utilizing process and requires ATP.

The cleavage reaction (step 3) is stimulated by ATP binding, but does not require ATP hydrolysis, since non-hydrolysable ATP analogues can support this step. This step is reversible without strand passage and requires $Mg^{2+}$ or $Ca^{2+}$ ions (or some other divalent metal ions, which, however, are inefficient) (Osheroff, 1987).

The cleavage/religation equilibrium prior to strand passage (step 3) is the likely point of action of a variety of top II blocking drugs, such as etoposide, *m*-AMSA and doxorubicin. The drugs stabilize the cleaved complex (complex D in Figure 1.6). As a result, reaction 3 is shifted towards complex D and, in addition, strand passage (reaction 4) is usually suppressed.

The binding of ATP (or analogue) to an allosteric site on the enzyme may bring about the major conformational change required to cleave duplex 2 and open a gap through which duplex 1 can pass (steps 3 and 4). In accord with this idea, it is found that a non-hydrolysable ATP analogue can support one strand passage event (step 4). However, at least one of the subsequent steps to complete the cycle (steps 5 or 6) does require ATP hydrolysis.

Strand passage (step 4) has a specific requirement for $Mg^{2+}$ and is not supported by $Ca^{2+}$ ions (Osheroff, 1987). Religation after strand passage (step 5) appears to be slower than religation prior to strand passage (step 3) (Robinson and Osheroff, 1991). Under either circumstance, however, the equilibrium (in the absence of drugs) strongly favours the intact duplex form of the complex. It may be that hydrolysis of the bound ATP causes the resulting ADP to dissociate from the enzyme, destabilizing the complex and facilitating a return to the closed conformation that would bring together the cleaved ends, allow the duplex to reseal and allow the enzyme to assume the conformation from which it could clamp onto a new duplex 1 (steps 6 and 1). An essential structural hypothesis in this model is that binding to duplex 1 favours the closed form of the enzyme (enzyme clamped onto duplex 1) and occupancy of an ATP site on the enzyme favours an open conformation of the enzyme that would allow strand passage. According to this picture, drugs such as etoposide, *m*-AMSA, doxorubicin or ellipticine stabilize complex D or E.

This simple and attractive picture is as yet unproved in regard to the order of initial binding of the passed and cleaved duplexes and the exact functions of ATP binding and hydrolysis.

Top II appears to interact preferentially with supercoiled compared with relaxed circular molecules. Osheroff *et al.* (1991) have noted that this property could allow the enzyme to distinguish between substrate and product of the duplex passing event. Indeed, without some way to make this distinction, it is hard to see how intertwined sister chromatids could be efficiently resolved.

### *Site Selectivity*

The localization of topoisomerase sites is pertinent to the functional role of topoisomerases in the genome. Since different classes of drugs trap topoisomerase cleavable complexes preferentially at different sites, the site selectivity could be an important aspect of drug action. Topoisomerase site selection has three parts: binding, cleavage and strand passage. Most studies have been on cleavage and this is not always congruent with sites of binding. Moreover, sites of binding and cleavage may not be indicative of where strand passage most often tends to occur.

Andersen *et al.* (1989) analysed DNA cleavage produced by type II topoisomerases from *Drosophila*, calf thymus and *Tetrahymena thermophilia*; the substrate was a recognition site isolated from the promoter region of the *Tetrahymena* extrachromosomal rDNA genes. The enzymes from these widely divergent species cleaved at an identical major site. All of the topoisomerases introduced single- and double-strand breaks in the DNA; changing the reaction conditions (e.g. pH or divalent cations) altered the ratio between single- and double-strand breaks without affecting the cleavage sites on either strand of the rDNA. A preference for cleaving the non-coding DNA strand was attributed to the forward cleavage reaction, since both strands religated at the same rate. Presumably differential cleavage of the two strands of DNA occurs in other segments of DNA, especially at low drug concentrations (Liu, 1989), but whether cleavage of one DNA strand is more detrimental than the other during anticancer drug treatment is unknown. A top II cleavable complex plausibly would interfere with the progress of RNA synthesis more severely if the complex were bound to the transcribed strand than if it were bound to the non-transcribed strand.

### *Dependence on DNA Sequence*

Although site selection is influenced also by chromatin structure, cleavage sites are most easily determined in purified DNA. The phenomena in pure DNA would basically still be operative in the more complex systems, but the site selection rules are modified by superimposed effects of DNA-binding proteins and altered DNA conformation. Studies with purified DNA can give clues to the structure of the topoisomerase–DNA complexes and of the site-specific drug-stabilized complexes.

Sander and Hsieh (1985) determined a consensus sequence for strong cleavage sites produced by *Drosophila* top II. More recently, Pommier *et al.* (1991a) determined cleavage sites of mammalian top II in SV40 DNA, using a statistical method to estimate site selectivity in the context of nucleotide triplet frequencies in the region of DNA studied. Cleavage complexes were trapped by means of sodium dodecyl sulfate. Results obtained in the absence of drugs in these two studies show essential coherence together with some differences in interpretation (Table 1.1). In position −1 (immediately 5′ to the break), the Sander and Hsieh consensus calls for C or T. The data of Pommier *et al.* are consistent with this, but show that a statistically more significant bias at this position is for the exclusion of A. Similarly, at positions −2 and −3, the Sander and Hsieh consensus is reinterpreted by Pommier *et al.* in terms of greater statistical bias against a particular base; thus, not-A at position −1, not-T at −2 and not-G at −3. The tendency to exclude particular bases suggests that, rather than being designed to recognize a base sequence, the enzyme shuns certain bases that happen to interfere with tight binding or cleavage.

The Pommier *et al.* findings also show a symmetry consistent with a 4 base pair 5′ overhang stagger in cleavage sites on the complementary strands. Thus, not-A at −1 corresponds with not-T at +5, not-T at −2

**Table 1.1** Top II cleavage preferences in absence of drugs

| | Drosophila *top II, plasmid DNA (Sander and Hsieh, 1985): 16 strong sites* | | *Mammalian top II, SV40 DNA (Pommier* et al., *1991a): 98 strong/medium sites* | |
|---|---|---|---|---|
| −6 | G | (50%) | — | |
| −5 | T | (50%) | C | (31%) |
| −4 | — | | G | (30%) |
| −3 | A/T | (100%) | not G | (8% G, 78% A/T) |
| −2 | A | (63%) | not T | (14% T, 39% A) |
| −1 | C/T | (88% C/T, 0% A) | not A | (5% A, 78% A/T) |
| +1 | A | (56%) | A | (48%) |
| +2 | T | (69%) | — | |
| +3 | T | (69%) | — | |
| +4 | — | | — | |
| +5 | A | (50%) | not T | (11% T, 36% A) |
| +6 | T | (69%) | T | (43%) |
| +7 | — | | A | (42%) |
| +8 | — | | C | (31%) |
| +9 | G | (56%) | — | |

The consensus sequence of Sander and Hsieh (1985) is compared with the statistical bias deduced by Pommier *et al.* (1991a). The latter was calculated from the observed frequency of a given nucleotide relative to its expected occurrence. The expected occurrence of a given base *x* flanked by bases *a* and *b* was determined from trinucleotide frequencies as

$$f(a, x, b) = n(a, x, b) / \sum_{i=1}^{4} n(a, i, b)$$

where $n(a, i, b)$ is the number of occurrences of the base sequence *a*, *i*, *b* in a region of 2000 nucleotides in which the observed sites are located. (The nucleotides A, C, G, T are numbered 1, 2, 3, 4.) A deviation greater or less than the expected frequency is listed if the probability of observing an equal or greater deviation was <0.003.

The observed cleavage site is indicated by the dashed line between positions −1 (upstream) and +1 (downstream). The assumed dyadic cleavage on the complementary strand is indicated by the dashed line between positions +4 and +5.

corresponds with T at +6, and G at −4 corresponds weakly with C at +8 (Table 1.1).

In the absence of drug, bias towards exclusion of particular bases at some positions is more prominent than bias towards preference for particular bases. For drug-stimulated sites, however, the bases immediately flanking the cleavage show strong preferential bias. The preferential bias is for a particular base immediately on one side or other of the cleavage site and is unique for each class of drugs so far analysed. This suggested the possibility that in the drug–DNA–topoisomerase complex the drug interacts directly by stacking against the base pair on one side or other of the cleavage site (Capranico *et al.*, 1990b; Pommier *et al.*, 1991a). For example, doxorubicin would stack against −1 A and *m*-AMSA would stack against +1 A (Figure 1.7). Such stacking interactions could explain why drugs stabilize cleavable complexes at drug-specific sites.

The doxorubicin and VM-26 site preferences flanking the cleavage sites

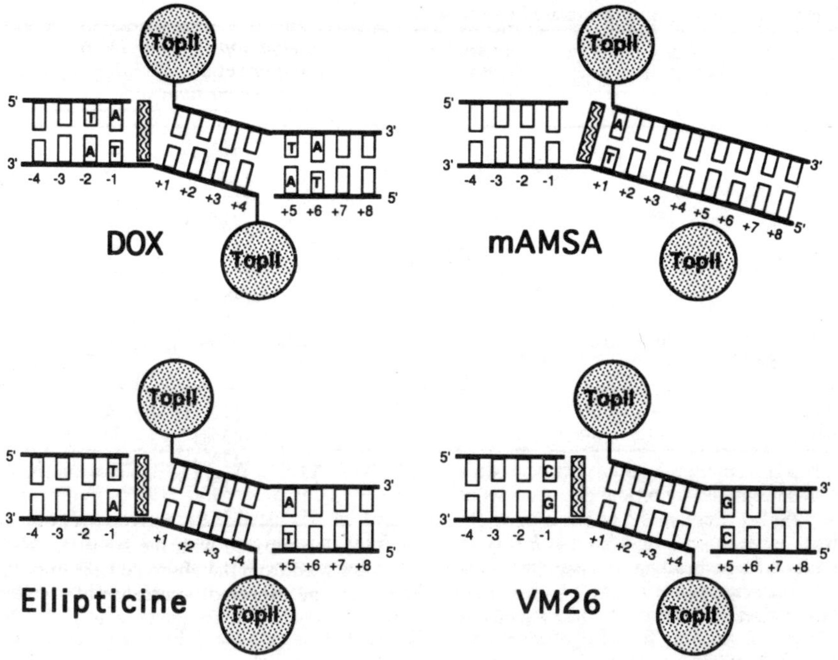


**Figure 1.7** Model of drug–DNA–top II complexes, proposed by two of us (K.W.K and Y.P.) (Capranico *et al.*, 1990b; Pommier *et al.*, 1991a), in which the flat polycyclic ring system of the drug, represented by a hatched rectangle, is stabilized by stacking against a base pair on the 5′ side of the strand break (DOX, VM-26, ellipticine) or on the 3′ side (*m*-AMSA). The absence of base pair preferences at both sides simultaneously for the same drug suggests that, in the stabilized open complex, drug is more closely associated with one side of the DNA break than with the other. Although all of the drugs probably can stabilize both single-strand and double-strand cleaved complexes, the single-strand type are most evident with *m*-AMSA. The base preference data for an ellipticine derivative are taken from Fossé *et al.* (1991)

exhibit the dyadic symmetry of 4 base pair 5′ overhangs typical of topoisomerase double-strand breaks (Table 1.2). The dyadic symmetry is shown by the complementary equivalence, in the case of doxorubicin, of the TA preference at positions −2 and −1 with the AT preference at +5 and +6. In the case of VM-26, preference for C at −1 corresponds with preference for G at +5. For *m*-AMSA, preference for A at +1 corresponds with preference for T at +4. These correspondences can be understood as reflecting frequent DNA double-strand cleavage in the drug complexes. If all sites were double-strand cleaved, however, the degree of bias in the complementary strand should have been similar to the degree of bias in the strand containing the observed break. Since the bias in the complementary strand was invariably weaker than in the observed strand (especially in the case of *m*-AMSA), many sites probably are of the single-strand cleaved type.

**Table 1.2** Bias at drug-stimulated top II sites

| | *DOX*[a] (97 sites) | | *m-AMSA*[a] (89 sites) | | *VM-26*[a] (112 sites) | | *VM-26*[b] (8 sites) |
|---|---|---|---|---|---|---|---|
| | + | − | + | − | + | − | |
| −4 | — | — | — | — | — | 13% T | C |
| −3 | — | 4% C | (48% A) | 7% G | — | 9% G | — |
| −2 | 57% T | — | 6% C | — | — | — | T |
| −1 | 71% A | (5% T) | 45% T | (13% A) | 55% C | (5% G) | A |
| +1 | — | — | 76% A | — | — | — | C |
| +2 | — | — | — | — | — | — | C/G |
| +3 | — | — | 38% C | 15% A | — | — | C/T |
| +4 | — | — | 7% T | — | — | — | C |
| +5 | 67% T | (10% A) | — | — | 41% G | (14% T) | C |
| +6 | 50% A | — | 42% T | — | — | — | G/T |
| +7 | | | | | | | C/T |
| +8 | | | | | | | C/T |
| +9 | | | | | | | T |

[a] Top II from mouse leukaemia cells acting on SV40 DNA (59% A + T). Significance of bias was determined by probability analysis as in Table 1.1 (Capranico *et al.*, 1990b; Pommier *et al.*, 1991a). Percentage occurrence of the indicated base is listed if the observed deviation from expectation is significant at $P < 0.003$. Base frequencies greater than expected are listed under column heading +; frequencies lower than expected are listed under column heading −. When entries are listed for both + and −, the less significant value is in parentheses (unless both values are nearly of equal significance).
[b] Partial consensus for sites observed in VM26-treated *Drosophila* cells *in vivo* (Käs and Laemmli, 1992) (see text).
Nucleotides biased towards preference or exclusion are listed under columns labelled + or −. Parentheses indicate that the bias is of lesser statistical significance.

The case of doxorubicin is particularly striking because the presence of an A at the 3′ terminus of at least one of the complementary break sites was observed without exception in 97 sites (Capranico *et al.*, 1990b). This was true for weak sites, as well as for strong sites, although strong sites more frequently had the 3′ terminal A at both dyadic sites on the complementary strands. It is interesting that, in the absence of drug, A at these positions is the strongest exclusionary factor.

Another approach to the analysis of site selection, taken by Spitzner and Muller (1988), assumes *a priori* that all sites in the set studied were of the double-strand break type. Using information theory to select which strand is the better match to a preliminary consensus sequence, these authors carried out repeated cycles of iteration with the view that each step would improve the strength of the consensus until a limit was achieved. Using the derived consensus, Spitzner and Muller (1989) achieved some degree of success in discriminating top II cleavage sites from non-sites. These studies, however, failed to discriminate between sites observed in the presence of drug from sites observed with the enzyme alone. It is not clear what features of the consensus sequence produced the observed partial discrimination between sites and non-sites.

In studies of the effect of an ellipticine derivative on calf thymus top II, Fossé *et al.* (1991) also investigated information content as a means to select which of two assumed dyadic strand cleavages were selected in the derivation of a consensus sequence. Rather than carrying out repeated iterations, however, they calculated the information content in all possible combinations of selected strands (observed or complementary). They then determined the sequence pattern in the combination that had the highest information content. Thus, each site was represented by one or other strand in such a way as to maximize the information signal. The resulting sequence had T at a 3′ terminus (−1 position in at least one of the strands) in all of 25 strong cleavage sites. The base at the −2 position was invariably a purine in at least one of the complementary strands, and the base at −3 was usually T.

This ellipticine derivative thus strongly preferred an alternating purine/pyrimidine sequence (TAT or TGT) on the immediate upstream flank of the cleavage site (Figure 1.7). This preference might be related to the avoidance of runs of purines or pyrimidines in the binding of an ellipticine derivative to DNA, reported by Bailly *et al.* (1990). The pyrimidine–purine sequence TA appears on both upstream flanks of DOX-stabilized top II complexes (Figure 1.7). A local pyrimidine–purine alternation thus may favour the complexes, possibly by enhancing the stacking energy of some drugs. It should be noted, however, that the strongest bias was always in the base pair immediately on the flank of the cleavage site, suggesting a direct drug interaction at this location.

A somewhat different sequence preference for VM-26-stimulated sites of cleavage by porcine spleen top II was recently reported by Huang *et al.* (1992). Instead of C as the preferred base at the 3′ terminus, they report a preference for A. However, there is agreement that G is excluded at this site. Huang *et al.* conclude that the 5′ stagger in their experiments was of 2 base pairs rather than the more usually observed 4 base pairs. Direct comparison of the different top II enzymes under the same experimental conditions has not yet been reported.

An unusually selective site enhancement was reported for an apparently new type of inhibitor, the trypanocide Ro15–0216 (Sørensen *et al.*, 1990). The only cleavage enhancement observed in pBR322 DNA was a twenty-fold enhancement of a double-strand cleavage site that was prominent even in the absence of drug. Another unusual feature of this inhibitor was that the enhanced cleavage was reversed by high salt much more rapidly than is usual for top II blockers such as amsacrine. High selectivity and rapid reversibility may be desirable features for clinical drugs.

*Dependence on DNA Conformation*

Electron microscopic observations have shown top II (in the absence of drugs) to bind preferentially to regions of DNA that have a bent conformation, such as exist at the SV40 terminus of replication and in the DNA minicircles of *Crithidia* kinetoplasts (Howard *et al.*, 1991). In the SV40 terminus preferential binding of top II was accompanied by a corresponding cluster of DNA cleavage sites (trapped by stopping the reaction with sodium dodecyl sulfate). However, the preferential binding to the bend in kinetoplast DNA did not correspond to strong cleavage.

Using SV40-infected cells, in which circular SV40 DNA is associated with nucleosomes, Yang *et al.* (1985) mapped the approximate positions of *m*-AMSA-induced breaks in the SV40 DNA. The breaks tended to be in or near regions of DNAase I hypersensitivity.

*Site Selectivity* in vivo

The binding and reactions of topoisomerases *in vivo* depend also on chromatin structure and possibly on other DNA binding proteins.

The effect of nucleosome positioning on top II cleavage sites was determined by Capranico *et al.* (1990a) in reconstituted SV40 chromatin. The reconstitution of nucleosomes in SV40 DNA restricted the location of top II sites to the internucleosomal linker regions, similar to the restriction of micrococcal nuclease sites to these regions.

A picture of how top II sites are organized in a multigene functional region was recently obtained by Käs and Laemmli (1992), who mapped top II cleavage *in vivo* in the histone gene cluster of *Drosophila* cells (Figure 1.8). Since teniposide (VM-26) was used to enhance the intensity of cleavage, it should be kept in mind that the sites detected may not be the same as the set of most intense sites in the absence of drug (Pommier *et al.*, 1991a). The number of sites detected *in vivo* was much smaller than the number in the same region in purified DNA, presumably because nuclear proteins block most sites. The *Drosophila* histone gene cluster contains one copy each of the H1, H2A, H2B, H3 and H4 genes, arranged as shown in Figure 1.8. The cluster covers about 5 kb and is repeated in tandem. Between the H1 and H3 genes there is a region of phased nucleosomes which also contains a region associated with the nuclear matrix or scaffold (MAR). Top II sites were located within each of the linker regions between nucleosomes, as were micrococcal nuclease sites, consistent with results in reconstituted chromatin (Capranico *et al.*, 1990a). In other places in the histone cluster, however, top II sites were not usually associated with micrococcal nuclease sites, but were instead located at DNase I-hypersensitive sites. Käs and Laemmli conclude that there are two classes of top II sites *in vivo*: those located in scaffold-associated regions and those

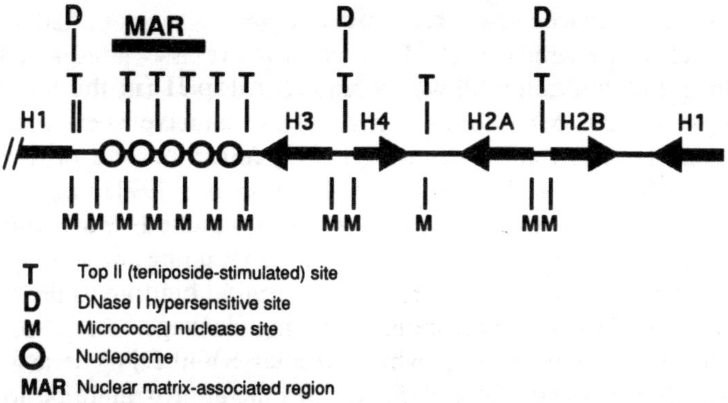


**Figure 1.8** Two classes of top II cleavage sites observed by Käs and Laemmli (1992) in the histone gene cluster in *Drosophila* cells. Top II sites were cleavage sites that were detected in cells treated with teniposide (VM-26). The *Drosophila* histone genes are arranged as shown in a unit that spans approximately 5 kb and is repeated in tandem. One class of top II sites was located within the linker regions between nucleosomes; these sites were suppressed when cells were subjected to heat shock. Another class of top II sites coincided with DNase I-hypersensitive regions; these were strongly intensified by heat shock. (Redrawn and modified from Käs and Laemmli, 1992)

located in open chromatin regions defined by DNase I-hypersensitivity.

The histone MAR region in isolated nuclei, however, seems to be much less susceptible to top II cleavage than that in intact cells (Udvardy *et al.*, 1986; Villeponteau, 1989), perhaps because gene activation in intact cells increases accessibility. Another observation pointing to the role of gene function is the effect of heat shock in *Drosophila* cells (Käs and Laemmli, 1992). Heat shock reduced top II cleavage intensity in the nucleosomal linker sites in the MAR region but strongly intensified cleavage in some DNase I-hypersensitive regions. However, the transcribed regions themselves remained free of major cleavage sites.

Käs and Laemmli determined the DNA sequence location of the major top II cuts between the H1 and H3 genes, and found a quite different configuration from what is usually observed in purified systems. To begin with, the 5′ overhang stagger of cuts on complementary strands was now 6 base pairs rather than the usual 4 base pairs. This difference might be due to an altered DNA twist in the nucleosomal spacer regions or to an alternative form of the active top II enzyme *in vivo*. Moreover, the DNA sequences about the cleavage sites (determined in cells treated with teniposide) differed from what has been observed in purified systems. Rather than the preference for C at $-1$ observed *in vitro* in the presence of VM-26, the major *in vivo* preference was for TA at $-2$, $-1$, which happens to be the same preference as that found for doxorubicin sites *in vitro* (Table 1.2). The *in vivo* sites exhibited no dyadic symmetry and the overhang region tended to consist of pyrimidines, usually C. Consistent with this pattern in the nucleosomal spacer sites, Käs and Laemmli also identified a major cleavage site in the repeat unit of the centromere-associated satellite III DNA, dimers or multimers of which bind to nuclear scaffold.

Investigating the beta-globin locus during development of chicken erythrocytes, Reitman and Felsenfeld (1990) found a remarkable dependence of top II sites on the stage of development, as well as on position within the locus. The top II sites observed were those that were made detectable by teniposide. A region of about 25 kb was surveyed, including four genes in the locus, and 12 major teniposide-induced cleavage sites were observed. This is a small fraction of the number of sites observed in purified DNA. The sites were localized in low-resolution agarose gels and might actually represent clusters of sites at the DNA nucleotide level. In the regions of the four beta-globin-type genes, the teniposide-induced sites were seen (with few exceptions) only when the respective genes were expressed during development. There was also a close, although not invariant, agreement between teniposide-induced cleavage sites and DNase I-hypersensitive sites. The latter are determined by digestion of isolated cell nuclei with DNase I, whereas the teniposide-induced sites are measured in living cells, possibly accounting for the exceptional differences between the two assays. Since DNase I-hypersensitive sites are attributable to nucleosome-free regions in chromatin, it may be that the accessibility of these regions also allows them to be acted upon by top II. Treatment of cells with top II-blocking drugs may therefore selectively block DNA (through stabilization of cleavable complexes) in regions that are actively transcribed in particular cell types. Comprehensive studies remain to be carried out on different types of top II-blocking drugs in living cells.

### *Structure of the Cleavable Complex*

Electron microscopic observations by Howard *et al.* (1991) indicated that binding of top II foreshortened the DNA by about 15 base pairs. This was consistent with a footprint of 20–30 base pairs protected by the binding of *Drosophila* top II (Lee *et al.*, 1989). The small extents of foreshortening and protection argue against models in which the DNA is wrapped around the top II molecule.

Bacterial DNA gyrase, however, does appear to wrap DNA around itself. Such a fundamental difference from eukaryotic top II is surprising in view of the sequence homology, as well as the analogy of mechanism, between the two enzymes.

The preferential binding of top II at DNA bends, however, suggests a curved or angled region of contact between DNA and enzyme. General DNA intercalation tends to straighten DNA bends and could destabilize complexes of curved DNA, as occurs, for example, in the dissociation of nucleosomes from DNA by ethidium. A similar mechanism might explain the reversal of cleavable complexes by ethidium and other strong intercalators. This could also explain the biphasic behaviour of many strongly intercalating top II blockers, such as doxorubicin and ellipticines, which at high drug concentrations show anomalous reduction of cleavable complex formation with increasing drug concentration (Monnot *et al.*, 1991).

### *Relationship between DNA Cleavage and Cytotoxicity*

Top II-mediated DNA cleavage can be measured in drug-treated cells or isolated nuclei by DNA filter elution methods (reviewed by Kohn *et al.*, 1987). The relationship between top II-mediated DNA cleavage and cytotoxicity, however, can only be evaluated if drug exposure can be terminated promptly, so that subsequent cell growth would occur in the absence of drug. Doxorubicin, for example, presents a problem because it does not diffuse readily through cell membranes and therefore cannot be rapidly washed out of cells. For this type of study, 5-iminodaunorubicin was found to be more suitable, because it induces top II-mediated DNA cleavage which is readily reversed upon washing the cells. Different types of top II blockers were compared: amsacrine, 2-methyl-9-hydroxyellipticinium and 5-iminodaunorubicin, all of which produce top II-mediated DNA breaks that are readily reversed by washing the cells (reviewed by Kohn *et al.*, 1987). Cytotoxicity was determined by colony assays in mouse leukaemia L1210 cells. Neither DNA single- nor double-strand break assays correlated with the cytotoxicity differences between these types of drugs. Relative to the amount of DNA breakage, cytotoxic potency was in the order: 5-iminodaunorubicin > 2-methyl-9-hydroxyellipticinium > amsacrine.

Assuming that the action of these compounds on top II is the main cause of cytotoxicity, the cytotoxicity differences could be due in part to differences in the genomic sites where the lesions tend to be concentrated, depending upon DNA sequence and conformation.

Within a series of anthracyclines, the DNA sequence dependence of top II cleavable complex production remained essentially unchanged (Capranico *et al.*, 1990c) and therefore is not likely to contribute to cytotoxicity differences. Cytotoxic potency correlated well with the ability of the compounds to induce top II cleavable complexes *in vitro* (Figure 1.9; compare circles and triangles). Both correlated with DNA binding affinity, but not as well as with each other. For example, switching the methoxy and hydroxy groups on positions 4 and 6 of doxorubicin (compounds 1 and 3 in Figure 1.9) reduced the DNA affinity constant approximately twentyfold, but reduced top II-mediated DNA cleavage only about twofold. The cytotoxic potency was nearly unchanged, thus seeming more closely related to the top II effect than to general DNA binding.

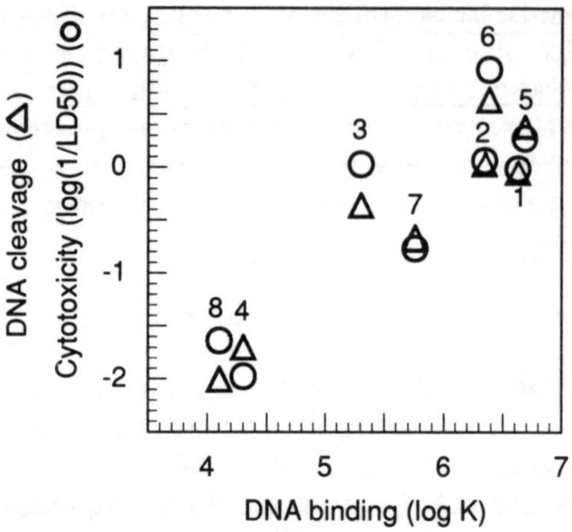


**Figure 1.9** For a series of anthracyclines, cytotoxic potency (circles) and top II-induced DNA cleavage *in vitro* (triangles) are plotted against DNA affinity constant (at or estimated for ionic strength 0.1 M). Data are from Table 1 of Capranico *et al.* (1990c). DNA cleavage was determined as the effect of 1 μM drug at a major top II site in end-labelled SV40 DNA and expressed relative to the effect of doxorubicin (DOX). Numbering of compounds: (1) DOX, (2) 4′-epi-DOX, (3) 4-demethyl-6-*O*-methyl-DOX, (4) DOX-beta-anomer, (5) daunorubicin (DNR), (6) 4-demethoxy-DNR, (7) 11-deoxy-DNR, (8) 6-*O*-methyl-DNR

The beta-anomer of doxorubicin, which is isomeric about the glycosidic bond, binds very weakly to DNA and has little effect on either top II or cytotoxicity. Similarly, 6-*O*-methyldaunorubicin lacks DNA binding, top II effect and cytotoxicity. (These are compounds 4 and 8 in Figure 1.9.)

### *Mechanisms of Drug Action*

Most top II inhibitors can be classified as shown in Table 1.3. The structures of key compounds are shown in Figure 1.10. Many are avid DNA binders, but a few, most notably etoposide (VP-16) and teniposide (VM-26), exhibit little or no binding to pure DNA. Of the DNA-binding top II inhibitors, all so far have been DNA intercalators.

Not all DNA intercalators block top II by stabilizing cleavable complexes. For example, *o*-AMSA intercalates as strongly as *m*-AMSA, but does not stabilize cleavable complexes. Moreover, *o*-AMSA is much less toxic than *m*-AMSA and lacks activity against experimental tumours in mice. Other strong DNA intercalators that affect top II similarly to *o*-AMSA include-9-aminoacridine and ethidium. These compounds do inhibit top II function, presumably through their DNA binding. Rather than stabilizing cleavable complexes, however, these compounds actually reverse cleavable complexes and in some cases reduce the toxicity of drugs such as *m*-AMSA and VM-26. This curious behaviour is also reflected in

**Table 1.3** Topoisomerase inhibitor types

| Drug | *DNA binding* | *Top* | *Cleavable complex* | *Reverses complexes* |
|---|---|---|---|---|
| *m*-AMSA | Intercal. | II | Reversible | No |
| *o*-AMSA | Intercal. | I and II | None | Yes |
| DOX | Intercal. | II | Reversible | Biphasic |
| VM-26 | Slight | II | Reversible | No |
| CAMPTO | Slight | I | Reversible | No |

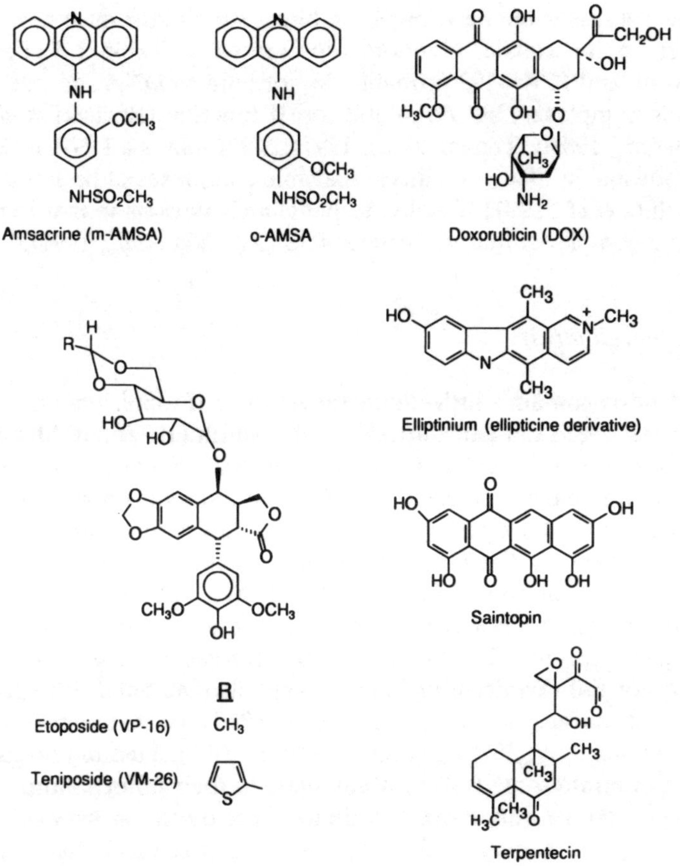


**Figure 1.10** Structures of some top II inhibitors discussed in the text

the biphasic concentration dependence of some cleavable complex formers, such as doxorubicin and ellipticine derivatives. As drug concentration is increased, cleavage at individual sites increases and then decreases. This biphasic effect is seen over different concentration ranges at individual cleavage sites.

**Table 1.4** New topoisomerase inhibitors

| *Drug* | *DNA binding* | *Top* | *Cleavable complex* | *Reverses complexes* | *Reference* |
|---|---|---|---|---|---|
| Fostriecin | None (?) | II | None | Yes | Boritzki *et al.* (1988) |
| Merbarone | None (?) | II | None | Yes | Drake *et al.* (1989b) |
| ICRF-193 | None | II | None | Yes | Tanabe *et al.* (1991) |
| Saintopin | Intercal. | I and II | Reversible | No | Yamashita *et al.* (1991) |
| Terpentecin | None | II | Irreversible | No | Kawada *et al.* (1991) |

Table 1.4 lists some new top II inhibitors which introduce new features not seen in the classic inhibitors represented in Table 1.3. Fostriecin, merbarone and ICRF-193 probably do not bind to DNA, do not stabilize cleavable complexes, but do inhibit top II function (Boritzki *et al.*, 1988; Drake *et al.*, 1989b; Tanabe *et al.*, 1991). Saintopin is a DNA intercalator that is unusual in that it stabilizes cleavable complexes of both top I and II (Yamashita *et al.*, 1991). Finally, terpentecin is unusual in that it produces cleavable complexes that are irreversible (Kawada *et al.*, 1991).

### *Mitochondrial top II*

Mitochondria contain a little-characterized type II topoisomerase (Castora *et al.*, 1982, 1985; Lin and Castora, 1991). Smith (1977) and Marshall and Ralph (1985) treated HeLa and mouse cells with 20 μM ethidium bromide or 4 μM amsacrine but detected no effects on the sedimentation properties of covalently closed circular mitochondrial DNA in neutral or alkaline sucrose gradients or CsCl buoyant density gradients. However, Shapiro *et al.* (1989) and Shapiro and Englund (1990) have demonstrated cleavage of trypanosome kinetoplast DNA minicircles by anticancer and antitrypanosomal agents *in vivo*. Type II topoisomerases isolated from *Crithidia* ($M_r$ 132 kDa) or calf thymus mitochondria were also inhibited by high concentrations of amsacrine, etoposide (VP-16-213) or teniposide (VM-26) (Melendy and Ray, 1989; Lin and Castora, 1991). Possibly drugs do not readily penetrate some cells or accumulate in their mitochondria, since in our hands 100 μM amsacrine is without effect on the growth of *Crithidia fasciculata*. Alternatively, inhibiting mitochondrial top II may have little effect on growth.

### *Nuclear Matrix*

The DNA in mammalian cell nuclei is tightly constrained by winding around nucleosomes and further condensation into higher-order structures dictated by the different phases of the cell cycle. The most tightly condensed structures, metaphase chromosomes, are composed of numerous loops of nucleosome-bound DNA folded into a supercoiled 'radial' loop structure (Gasser *et al.*, 1989; Filipski *et al.*, 1990). Individual loops contain of the order of 50–200 kb of DNA anchored at the base by a chromosomal 'scaffold' or 'nuclear matrix' composed of a number of different proteins (van Kries *et al.*, 1991). Two proteins designated SC1 ($M_r$ 170 kDa) and SC2 ($M_r$ 135 kDa) predominate among metaphase chromosome scaffold proteins. The protein SC1 was identified as top II and its location at the base of the loops in the axial core of chromosome scaffolds was identified by use of immunofluorescence and antibodies to top II (Gasser and Laemmli, 1986a,b; Gasser *et al.*, 1986, 1989). From the limited data available, it appears that top II comprises 1–2% of total chromosomal proteins, on average representing about three top II molecules per loop of DNA (Gasser *et al.*, 1986).

In interphase nuclei individual chromosomes are dispersed throughout the nucleus as interconnected loops or domains attached to the nuclear matrix at multiple points and to the nuclear lamina (Berrios *et al.*, 1985; Razin, 1987; Berrios and Fisher, 1988; Verheijen *et al.*, 1988; Mirkovitch *et al.*, 1988). Individual looped domains can contain multiple unrelated genes including more than one transcribed gene (Mirkovitch *et al.*, 1986). The nuclear matrix also contains more or less firmly associated top II according to its method of preparation, possibly indicating different subpopulations of the enzyme (Gasser *et al.*, 1989) or effects of intermolecular disulphide bonds (Kaufmann and Shaper, 1991; Kaufmann *et al.*, 1991a). Regions of DNA associated with nuclear matrices or scaffolds have the following characteristics: they are confined to non-coding sequences in DNA, usually non-transcribed regions or introns; they are separated by variable lengths of DNA ranging from 3 kb to 140 kb, and they often flank one or more coordinately regulated genes and coincide with the boundaries of nuclease-sensitive domains produced after gene activation (Gasser *et al.*, 1989).

Analysis of the scaffold-associated regions (SARs) in *Drosophila*, mouse, yeast, CHO and human cells showed that putative sequences for cleavage by top II are clustered in SARs relative to non-SARs (Cockerill and Garrard, 1986a; Gasser and Laemmli, 1986a,b; Käs and Chasin, 1987; Amati and Gasser, 1988; Jarman and Higgs, 1988; Blasquez *et al.*, 1989; Vassetzky *et al.*, 1989). Furthermore, the accessibility of genes to top II may change when they become activated (Udvardy *et al.*, 1985, 1986), and some SARs have been mapped to regulatory regions of genes that contain enhancer promoter sequences (Cockerill and Garrard, 1986a; Gasser and Laemmli, 1986a; Cockerill *et al.*, 1987; Jarman and Higgs, 1988; Lake *et al.*, 1990). The chicken alpha globin gene has a 1 kb A/T-rich DNA fragment at the 3′ boundary containing sites for transcription termination and matrix attachment and four variable top II recognition sites (Farache *et al.*, 1990), while the limits of the human apolipoprotein B gene coincide with chromosomal anchorage loops that define the 5′ and 3′ boundaries of the gene (Levy and Fortier, 1989).

### *Chromosomes and top II*

Using HeLa or chicken erythrocyte cell nuclei and a mitotic extract from *Xenopus laevis* eggs, Adachi *et al.* (1991) showed that top II is required to condense and assemble chromosomes. Without top II, chicken erythrocyte nuclei only progressed to precondensation clusters of chromatids, whereas addition of top II stimulated normal chromosome condensation. It is also possible that top II is involved in relaxing metaphase chromosomes as cells enter G1 phase of the cell cycle (Roberge *et al.*, 1990). However, Downes *et al.* (1987) did not detect much effect of etoposide on the chromosome decondensation that accompanies incomplete UV excision repair in mitotic ataxia telangiectasia cells.

A number of anticancer drugs that act on top II inhibit the mammalian cell cycle, causing cells to accumulate in G2 phase, consistent with inhibition of daughter chromosome untangling by top II (Grieder *et al.*, 1974; Misra and Roberts, 1975; Tobey *et al.*, 1978; Charron and Hancock, 1990a). Studies with yeast have also indicated essential functions for top II in the segregation of intertwined daughter chromosomes at the end of DNA replication and at mitosis (Uemura and Yanagida, 1984; Di Nardo *et al.*, 1984; Holm *et al.*, 1985, 1989b). Furthermore, inhibition of decatenation of SV40 virus daughter DNA molecules by drugs that interfere with top II has been demonstrated both *in vitro* and *in vivo*, and a role proposed for top II in replicating the terminal region of the viral DNA and resolving the daughter molecules (Snapka, 1986; Richter *et al.*, 1987; Yang *et al.*, 1987; Snapka *et al.*, 1988). Effects of top II inhibitors on chromosome segregation are supported by Charron and Hancock (1990b), who studied the action of VM-26 on the formation of mitotic chromosomes in Chinese hamster ovary cells and concluded that VM-26 did not inhibit DNA replication in S phase, whereas it delayed completion of DNA replication and chromosome condensation.

Fields-Berry and De Pamphlis (1989) investigated the formation of catenated intertwines between plasmids containing yeast centromeric sequences replicating in mammalian cells. Catenated dimers and late-replicating intermediates accumulated when the yeast sequences were located in the termination region (180° from *ori*) and the cells were subjected to hypertonic shock to reduce top II activity. This led to the suggestion that top II acts behind replication forks in the termination region to remove intertwines generated by unwinding DNA rather than after replication is completed and catenanes are formed.

The preceding observations from different systems implicate top II in the processes that resolve daughter DNA molecules after DNA replication in cell nuclei. In mammalian cells, which replicate their DNA from multiple origins (Alberts *et al.*, 1989), presumably similar topological problems arise which are normally resolved by top II and drugs that inhibit top II cause chromosomal abnormalities which eventually kill the cells (Hancock *et al.*, 1989; Charron and Hancock, 1990a,b; Charron and Hancock, 1991).

The human c-*fos* protooncogene also contains clusters of sites cleaved by calf thymus top II in the 5′-enhancer-promoter region, and c-*fos* and c-*jun* gene expression is induced by etoposide, confirming that growth-related protooncogenes are sensitive to top II (Darby *et al.*, 1986; Rubin *et al.*, 1991). When *m*-AMSA or VM-26 was allowed to stimulate DNA cleavage in various human tumour cell lines and peripheral lymphocytes, overall cleavage was twentyfold lower than *myc* gene cleavage, while topoisomerase activity was threefold lower in quiescent compared with growing cells, although *m*-AMSA-induced cleavages were the same. Consequently, Riou *et al.* (1989) concluded that increased *myc* gene cleavage was related to *myc* gene transcription or accessibility rather than top II content, and they suggested that top II was a good target for anticancer drugs, because oncogenes are usually overexpressed in tumour cells. However, top II inhibitors may not necessarily reduce oncogene expression (see above).

### *Nucleosomes and top II*

There appears to be no requirement for top II in nucleosome formation, since neither ATP nor $Mg^{2+}$ was needed for assembly of chromatin from nucleosomes and DNA in extracts from *Xenopus laevis* eggs or oocytes (Almouzni and Mechali, 1988). Neither does camptothecin or VM-26 inhibit nucleosome assembly *in vivo* (Annunziato, 1989). However, the assembly of physiologically spaced chromatin required ATP, and changes in linking number proceeding by steps of unity suggested that top I might be involved (Almouzni and Mechali, 1988).

To assess whether nucleosomes influence the action of top II, Capranico *et al.* (1990a) mapped the sites of cleavage of naked versus nucleosome-reconstituted SV40 DNA following the action of four unrelated top II inhibitors. Cleavage by top II was suppressed in nucleosomes, and persisted or was enhanced in linker DNA or a nucleosome-free region around the DNA replication origin. Thus, the presence of nucleosomes at preferred sites influences the DNA breaks induced by top II inhibitors.

In summary, the 5′- and 3′-flanking regions of many genes or coordinately controlled gene clusters contain SARs or MARs which constrain the intervening DNA or chromatin into topologically independent loops when they are attached to the nuclear matrix. Matrix-associated DNA contains recognition sequences for top II often adjoining enhancer or promoter binding elements. Initiation and maintenance of transcription presumably requires the action of top II (and top I) to overcome torsional effects on DNA and/or expose active genes in an appropriate state for RNA synthesis to begin and proceed. In these circumstances, drugs that inhibit top II and stabilize cleavable complexes should effectively separate adjacent loops of DNA from the continuity of chromatin, and studies have shown that DNA prepared from cells treated with drugs that inhibit top II consists of large segments of DNA of the order of size expected for the individual loops of chromatin (Marshall and Ralph, 1985) which contain clusters of genes (Razin *et al.*, 1991).

It is worth noting that the *E. coli* genome seems to be organized in a similar way. DNA gyrase is thought to be located at the base of looped domains in *E. coli* DNA (Yang and Ames, 1988) and the inhibitor oxolinic acid produces cleavage sites on the *E. coli* chromosome which are usually grouped in 5–19 kb clusters 50–100 kb apart, with a total number of 50–100 clusters per genome. Evidence that chromosomal loop anchorage sites are evolutionarily conserved in higher cells has also been presented (Cockerill and Garrard, 1986b). In *E. coli*, cleavage of DNA was strongly modulated by transcription occurring at distances up to 35 kb away, presumably as a result of topological changes caused by advancing polymerases (Condemine and Smith, 1990).

*Regulation*

There is now general recognition that mammalian cell growth occurs in response to external growth factors that react with cell surface receptors and initiate a protein kinase cascade involved in triggering the events required for cells to progress from quiescence into S phase (Lewin, 1990; Nurse, 1990; Ralph *et al.*, 1990). Therefore it was not surprising to learn that top II is a phosphoprotein. Immunoprecipitation of top II in nuclear extracts from $^{32}$P-labelled *Drosophila* cells showed that the enzyme was a phosphoprotein (Sander *et al.*, 1984) and $^{32}$P-labelling of chicken lymphoblastoid, human leukaemia or human HeLa cells has confirmed that top II is phosphorylated in cells (Constantinou *et al.*, 1989; Heck *et al.*, 1989; Kroll and Rowe, 1991).

Ackerman *et al.* (1985) found that *Drosophila melanogaster* top II DNA relaxing activity was stimulated by phosphorylation *in vitro* with casein kinase II and that alkaline phosphatase reversed the stimulation. One serine residue was phosphorylated per homodimer subunit and phosphorylation produced a threefold increase in DNA relaxation. Subsequent studies with *Drosophila* cells confirmed that casein kinase II was the enzyme primarily responsible for phosphorylating top II *in vivo* (Ackerman *et al.*, 1988). A type II casein kinase has also been demonstrated in the nucleolus of mammalian cells, and the activity present in the nucleus correlated with growth (Caizergues-Ferrer *et al.*, 1987; Belenguer *et al.*, 1989).

Saijo *et al.* (1990) purified a mouse type II topoisomerase and obtained two structurally related proteins ($M_r$ 167 and 151 kDa). A serine protein kinase isolated from the partially purified enzyme phosphorylated the top II and increased its activity 8.6-fold, while alkaline phosphatase almost completely abolished the activity. The properties of this protein kinase were similar to those of the protein kinase tightly associated with *Drosophila* top II (Sander *et al.*, 1984).

Phosphorylation and activation of *Drosophila* top II by protein kinase C has also been reported with about 0.85 mol of phosphate incorporated on serine residues per mole of topoisomerase and substantial increases in the unknotting and relaxation activities of the phosphorylated enzyme. In addition, calmodulin-dependent protein kinase phosphorylated the enzyme on serine, whereas it was not altered by cyclic AMP-dependent protein kinase (Sahyoun *et al.*, 1986). However, Constantinou *et al.* (1989) found that inducing differentiation of human promyelocytic HL60 leukaemia cell variants with the protein kinase C activator phorbol diester caused a rapid reduction in top II activity which was associated with increased phosphorylation of top II, while a delayed decrease in top II activity concomitant with loss of proliferative potential of phorbol diester-treated HL60 cells was observed by Gorsky *et al.* (1989). These results suggest that induction of differentiation in HL60 cells is associated with a reduction in top II activity mediated directly or indirectly by protein kinase(s). It has been proposed that phorbol ester treatment uncouples the mechanisms linking drug-induced top II–DNA cleavable complex stabilization and cytotoxicity in etoposide-treated HL60 cells (Zwelling *et al.*, 1990a).

Phosphorylation of top II has also been proposed during the response of *Geordia* sponge cells to aggregation factor which stimulates DNA synthesis by the cells, suggesting that regulation of top II by phosphorylation is a widespread phenomenon associated with growth (Rottman *et al.*, 1987).

In studies with Balb/c 3T3, CCRF-CEM, L1210 or K21 mastocytoma cells inhibitors of protein synthesis reduced the cytotoxic action of etoposide or amsacrine. Moreover, cytoprotection was not a result of decreased top II (Chow *et al.*, 1988; Schneider *et al.*, 1988a,b). This and other data led Schneider *et al.* (1988a) to suggest that factors in addition to top II were involved in the action of amsacrine, and a heat-labile, protease-sensitive factor was detected in extracts of K21 cells which enhanced the formation of top II–DNA complexes by amsacrine in isolated nuclei (Darkin and Ralph, 1989). The enhancing activity was shown subsequently to reside in a protein kinase ($M_r$ 70 kDa approx.) with properties reminiscent of casein kinase II (Darkin-Rattray and Ralph, 1991).

Evidence that phosphorylation of *Xenopus laevis* top I with a *Xenopus* casein kinase activates the enzyme and that dephosphorylation prevents the formation of camptothecin-induced top I–DNA complexes (Kaiserman *et al.*, 1988) raises the interesting question whether the resistance of some tumours to topoisomerase inhibitors could result from reduced phosphorylation, or increased dephosphorylation of topoisomerases in non-dividing cells and consequently decreased cleavable complex formation in response to drugs. Other mechanisms of resistance could involve phosphorylation effects on the association of top II with the nuclear matrix or matrix-associated regions of DNA. Fernandes *et al.* (1990) have presented evidence that the association of top II with the nuclear matrix is reduced in CEM leukaemia cells resistant to VM-26 or amsacrine. However, they did not investigate the state of phosphorylation of the enzyme.

Schroder *et al.* (1989) detected an age-dependent increase in DNA top II activity in quail oviduct tissue. Either nuclear protein kinase NII or protein kinase C increased the specific activity of the topoisomerase, while poly(ADP) ribosylation inhibited the enzyme. Changes in matrix-associated protein kinase and poly(ADP) ribose synthetase activity were conjectured to affect DNA or RNA production during ageing. Poly(ADP) ribosylation has also been shown to inhibit calf thymus top II (Darby *et al.*, 1985). However, the significance of poly(ADP) ribosylation *in vivo* is not known.

Recent recognition of a role for casein kinase II in the transmission of growth signals in mammalian cells, and the fact that top II is activated by casein-type kinases, suggests that top II phosphorylation plays a role in initiation of cell growth (Ralph *et al.*, 1990; Meisner and Czech, 1991). In support of such a role, Ackerman and Osheroff (1989) showed that epidermal growth factor (EGF) treatment of A431 human carcinoma cells produced a transient fourfold increase in cytosolic casein kinase activity which was modulated by protein kinase C, suggesting that the protein kinase cascade activated by EGF could potentially activate top II. Furthermore, histone H1 (or M phase) kinase is involved in the control of the cell cycle in yeast and other cells, and top II has potential phosphorylation sites for M phase kinase, suggesting that it may be involved in initiating growth (Lewin, 1990; Moreno and Nurse, 1990; Nurse, 1990). An indirect inhibitory effect of the top II inhibitor VM-26 on the phosphorylation of histone H1 and the entry of BHK cells into mitosis was reported by Roberge *et al.* (1990). In cells arrested in mitosis, VM-26 caused dephosphorylation of histones H1 and H3, DNA breaks and partial chromosome decondensation.

Inhibition of replicon initiation in human cells following stabilization of topoisomerase–DNA complexes with amsacrine has been described by Kaufmann *et al.* (1991b). In addition, etoposide has inhibitory effects on the growth-related protein kinase $p34^{cdc2}$ (Lock and Ross, 1990). Clearly, there is sufficient circumstantial evidence linking growth signal transduction, protein kinase activation and the control of top II to the cell cycle to warrant a more detailed investigation of the action of top II inhibitors on the initiation of cell growth.

### *Role in Replication*

Top II has been identified as a component of the 'replitase', a multienzyme complex associated with DNA replication (Noguchi *et al.*, 1983), and a number of investigators have attempted to determine whether inhibition of top II inhibits DNA synthesis. In general, there appears to be no direct effect of inhibitors of top II on DNA polymerase in human cells. However, Lönn *et al.* (1989) showed that treating human melanoma CRL1424 cells with high (100 μg/ml) concentrations of etoposide (VP-16) prevented the formation of large (10 kb) DNA intermediates from Okazaki fragments but allowed the formation of small intermediates (Okazaki fragments), which suggests that during DNA replication resolution of topological problems by top II is necessary before larger DNA intermediates can form.

In HeLa-S3 cells 100 μM VM-26 inhibited DNA synthesis by more than 80%, and camptothecin and VM-26 together stopped replication (Annunziato, 1989). However, in CHO cells DNA replication was not affected by low concentrations of VM-26 (0.05 μg/ml) and it continued at approximately 50% of the normal rate with 50 μg/ml VM-26, although the higher concentration inhibited the uptake of DNA precursors (Hancock *et al.*, 1989; Charron and Hancock, 1990a). These data suggest that secondary effects of high drug concentrations may explain some of the apparent effects of top II inhibitors on DNA synthesis.

Nelson *et al.* (1986) analysed the covalent complexes formed between the DNA of 90 s [$^3$H]-thymidine pulse labelled rat prostatic adenocarcinoma cells and top II following 20 μM VM-26 treatment, and found a direct association of the enzyme with nascent DNA. Additional support for an association between top II and DNA synthesis was provided by Fernandes *et al.* (1988), who used [$^3$H]-thymidine to study the effect of 0.1 or 0.2 μM teniposide (VM-26) and 0.5 μM amsacrine on [$^3$H]-DNA associated with nuclear matrix compared with non-nuclear matrix DNA in CCRF-CEM leukaemia cells. In 24 h [$^{14}$C]-thymidine prelabelled cells pulse-labelled for 45 s with [$^3$H]-thymidine the specific activity of nascent DNA associated with the nuclear matrix was fourfold greater than that of non-matrix DNA, consistent with DNA synthesis occurring in association with the nuclear matrix. Furthermore, preincubating the cells with topoisomerase inhibitors reduced the relative specific activity of the nascent DNA on the nuclear matrix by 50–60% compared with control cells, showing that top II assists DNA replication.

Other data possibly pertinent to the question of the role of top II in DNA synthesis are available from studies on the protective effects of various drugs. The DNA polymerase inhibitor aphidicolin partially protected DC3F cells against the cytotoxic effects of VP-16 (Holm *et al.*, 1989a) and S phase V79 cells from the effect of low doses of amsacrine (Schneider *et al.*, 1990). These and other similar observations indicate that top II cooperates in DNA synthesis, although they do not implicate top II directly and aphidicolin does not protect cells from high doses of top II inhibitors (Schneider *et al.*, 1990).

A comparison of the effects of 0.5 μM amsacrine or mitoxantrone on cleavable complex formation and DNA synthesis by human fibroblasts indicated little difference in cleavable complex formation, whereas mitoxantrone was twice as effective in inhibiting DNA synthesis, and cleavable complexes formed with mitoxantrone persisted longer and were more toxic than those produced by amsacrine (Fox and Smith, 1990). There was also no preferential localization of mitoxantrone on newly replicated DNA, suggesting that the forms of the cleavable complexes induced by mitoxantrone may be different and possibly they do not involve top II (Stetina and Vesela, 1991). As additional evidence for differences in the action of mitoxantrone (or doxorubicin) and amsacrine, teniposide or camptothecin, only the latter three drugs induced DNA degradation in S phase HL60 cells and there were differences in the response of different cells to the drugs (Del Bino and Darzynkiewicz, 1991). Possibly some of the differences in the effects of individual drugs are due to the use of different drug concentrations, since low drug concentrations produce more single-strand breaks in DNA (Liu, 1989). Topological effects on DNA as a result of top II inhibition or drug intercalation into DNA at high concentrations may also affect DNA synthesis.

### *Role in Transcription*

Although several inhibitors of top II have been reported to inhibit RNA synthesis, there are no convincing data that such reagents directly affect the polymerases involved. Most of the effects are probably explicable in terms of the drugs interfering with DNA supercoiling, since genetic studies with yeast have shown that either type I or type II topoisomerase will sustain ribosomal RNA synthesis, whereas rRNA synthesis is inhibited in double mutants, consistent with the ability of either topoisomerase to relax DNA. Furthermore, considerably less effect on poly $A^+$ RNA or transfer RNA synthesis was observed, indicating little effect on RNA polymerase II or III (Brill *et al.*, 1987).

Schaak *et al.* (1990b) concluded that VM-26 did not cause a direct reduction in transcription of adenoviral or HeLa cell heat shock genes, although it did affect other cellular processes. Also, injection of VM-26 into *Xenopus* oocytes failed to inhibit the transcription of rRNA or thymidine kinase genes or the action of the rRNA enhancer, although the drug inhibited decatenation of DNA rings. Therefore, neither RNA polymerase I nor II appeared to be sensitive to top II inhibitors (Dunaway, 1990). However, Thomsen *et al.* (1990) concluded that drug-stabilized calf thymus top II cleavable complexes impeded the progress of RNA polymerase during an investigation of the site of topoisomerase action *in vitro*.

Several of the studies suggesting that top II is involved in transcription used the inhibitors novobiocin or coumermycin to identify a role for top II (Glikin and Blangy, 1986; Mattern *et al.*, 1989). However, it has since been shown that novobiocin also inhibits the formation of preinitiation complexes during the initiation of RNA synthesis, which probably adequately explains the effects of these drugs on transcription (Gottesfield, 1986; Webb *et al.*, 1987; Van Dyke and Roeder, 1987; Webb and Jacob, 1988). Neither Webb *et al.* (1987) nor Mattern *et al.* (1989) detected effects of VM-26, amsacrine or VP-16 on the accumulation of metallothionein message *in vitro* or *in vivo*.

Top II may alter RNA synthesis by its effects on DNA supercoiling. Tarr and van Holden (1990) concluded that addition of plasmid DNA to *Xenopus* oocyte extracts relaxed the DNA in a few minutes and that loss of negative helicity caused a reduction in transcription *in vitro*. Addition of adriamycin restored the negative superhelicity of the DNA and its function in *in vitro* transcription. Others have also proposed that the state of supercoiling of the DNA template, or effects of topoisomerase inhibitors on the topology of DNA, can enhance or inhibit transcription of different genes (Kubo *et al.*, 1979; Borowiec and Gralla, 1985; Rowe *et al.*, 1986b; Brill *et al.*, 1987; Brill and Sternglanz, 1988; Hirose and Suzuki, 1988; Tabuchi and Hirose, 1988; Wong and Hsu, 1990). However, transcription initiation from the rat prolactin and adenovirus major late promoters was equally inhibited by VM-26, using supercoiled or linear templates in an assay with rat pituitary tumour GH3 cell nuclear extracts, possibly as a consequence of VM-26-induced cleavage in the promoter regions (Preston and White, 1990). Therefore, additional factors such as chromatin folding may be important *in vivo*.

Recently Franco and Drlica (1989) showed that under certain conditions DNA gyrase inhibitors can increase *gyr*A gene expression, and DNA supercoiling in *Escherichia coli*, while Carty and Menzel (1990) found that inhibiting *E. coli* DNA gyrase in an *in vitro* transcription–translation system stimulated expression of the *gyr*A gene in a DNA concentration-dependent manner. These results suggest the *gyr*A gene may control its own expression by regulating the degree of DNA supercoiling. Whether expression of human top II genes is also controlled by the degree of DNA supercoiling or increased by drugs that target top II remains to be established.

*Role in Recombination*

Topoisomerases I and II have both been implicated in the processes of non-homologous or illegitimate recombination. Bae *et al.* (1988) incubated calf thymus top II with two different phage lambda DNAs in an *in vitro* system and showed that the enzyme produced linear monomer recombinant DNA that could be packaged into phage particles *in vitro*. Production of the recombinant molecules required ATP and was inhibited by anti-calf thymus top II antibody. The resulting recombinant DNA molecules contained duplications or deletions, and sequencing the DNA junctions showed that recombination had occurred between non-homologous sequences of lambda DNA. Since non-homologous recombination also occurs with *E. coli* DNA gyrase or phage T4 DNA topoisomerase *in vitro* (Ikeda, 1986), the ability to catalyse illegitimate recombination appears to be a general property of type II topoisomerases.

Using a shuttle vector containing three bacterial markers to study illegitimate recombination in COS-1 cells, Bae *et al.* (1991) observed an increase in deletion formation in the presence of VM-26, suggesting that stabilization of cleavable complexes was involved in the deletion process.

Ikeda (1986) also found that non-homologous recombination in bacteria was facilitated by the DNA gyrase inhibitor oxolinic acid, and a model for top II-assisted recombination was proposed involving exchange of DNA-linked top II subunits. A similar but more complex subunit exchange model was suggested by Filipski (1983) to explain the action of top II inhibitors on recombination or sister-chromatid exchange. Further support for a role for topoisomerases in recombination comes from studies on mitotic recombination of the rDNA in yeast which showed that recombination in *S. cerevisiae* is suppressed by the combined action of DNA topoisomerases I and II, suggesting that relaxation of rDNA supercoiling hinders recombination (Christman *et al.*, 1988). Consequently, drugs that inhibit topoisomerase cleavable complexes should increase recombination, as is observed (but see Wang *et al.*, 1990).

Blasquez *et al.* (1989) identified an evolutionarily conserved class of nuclear matrix-associated DNA sequences about 200 base pairs long, rich in AT and top II consensus sequences, usually near *cis*-acting gene regulatory sequences. Subsequently, Sperry *et al.* (1989) localized the matrix-associated regions as sites of chromosomal insertion, deletion and translocation, and showed that a matrix-associated region was deleted from one of the two rabbit immunoglobulin κ genes. Because matrix-associated regions constitute one class of sites that are targets for illegitimate recombination, it was suggested that chromosomal loop attachment sites are associated with non-homologous recombination. Furthermore, top II is also localized to matrix attachment sites and top II-targeted drugs increase recombination and sister-chromatid exchange, which strongly implicates

DNA supercoiling, cleavable complex stabilization and recombination as important factors contributing to the action of inhibitors of top II.

Recombination involving flanking direct repeats and top II-related DNA recognition sequences has also been implicated as a major cause of large-scale deletions in the mitochondrial DNA of patients with progressive external ophthalmoplegia and ragged red fibres (Mita *et al.*, 1990).

### *Sister-chromatid Exchange*

Topoisomerases have been implicated in the production of sister-chromatid exchanges and other chromosomal aberrations involving recombination in response to drugs, such as amsacrine, that interact with top II (Crossen, 1979; Pommier *et al.*, 1985, 1988). Various models to explain the formation of sister-chromatid exchanges by exchange of top II subunits in protein–DNA cleavable complexes, or homologous displacement with strand switching during removal of parental helical turns by the topoisomerase at or near replication forks, have been considered by Dillehay *et al.* (1989). The ability of drugs that stabilize top II–DNA cleavable complexes to increase sister-chromatid exchange, and the inhibition of sister-chromatid exchange by novobiocin, which interferes with ATP binding and topoisomerase action, strongly support involvement of cleavable complexes in the exchange reaction. Also, in Chinese hamster V79 cells cleavable complexes were highest in S phase, the period when cells are most sensitive to induction of sister-chromatid exchanges (Dillehay *et al.*, 1987). These observations suggest that DNA replication forks participate in sister-chromatid exchange together with top II–DNA cleavable complexes formed in response to drugs. However, when Charron and Hancock (1991) treated synchronized Chinese hamster cells with very low concentrations of VM-26 (0.008 μM), they observed the formation of aberrant quadriradial chromosomes, whether the drug was present throughout the entire cell cycle or just during G2 phase, suggesting that DNA replication was not essential to disrupt normal mitosis in this instance, although it is possible that some residual DNA synthesis essential for chromosome segregation in G2 phase occurred in these experiments. Hancock *et al.* (1989) proposed that the cytostatic effect of top II inhibitors is due to the reversible arrest of cells in G2 phase, whereas cells which reach mitosis cannot segregate aberrant chromosomes successfully and this irreversible phenomenon is responsible for the toxicity of the inhibitors.

Hancock *et al.* (1989) also proposed that sister-chromatid exchanges may be a manifestation of a potential defect in DNA replication which only materializes as a strand exchange when the template DNA contains bromodeoxyuridine. A bromodeoxyuridine-dependent increase in sister-chromatid exchange formation described in Bloom's syndrome was also attributed to effects of bromodeoxyuridine substitution of template DNA on top II activity (Heartlein *et al.*, 1987). Moreover, Pommier *et al.* (1991d) have shown that bromodeoxyuridine produces alkali-labile sites in the DNA of normal and Bloom's syndrome fibroblasts which are independent of top II; presumably these reflect alterations in the DNA that favour sister-chromatid formation. Nevertheless, sister-chromatid exchanges do appear to involve top II, since they are produced by epipodophyllotoxins that target top II specifically.

Pommier *et al.* (1988) reported a correlation between top II inhibitor-induced sister-chromatid exchange and cytotoxicity in sensitive cells and reduced sister-chromatid exchange in drug-resistant cells, and similar conclusions were reached by Dillehay *et al.* (1987), studying the relationships between amsacrine-induced cleavable complexes and sister-chromatid exchange.

Chatterjee *et al.* (1990b) also investigated the mechanism of cell death in poly(adenosinediphosphate-ribose) synthesis-deficient V79 Chinese hamster cells that were resistant to VP-16-induced cytotoxicity. Although there was no correlation between DNA strand breaks and cytotoxicity, or evidence of altered DNA repair, a good correlation between drug-induced sister-chromatid exchange and cytotoxicity emerged. The time course of VP-16-induced cytotoxicity correlated better with formation of sister-chromatid exchanges than with cleavable complex formation, leading to the suggestion that drug-induced stabilization of top II–DNA complexes stimulates sister-chromatid exchanges which eventually cause cell death. In a further study evidence was presented that VP-16-induced non-homologous recombination also caused deletions and rearrangements of the HPRT gene and presumably of other genes in CH V79 cells which may also contribute to cell death in response to top II inhibitors (Berger *et al.*, 1991).

Clearly the roles of topoisomerase inhibitors in eliciting recombination, sister-chromatid exchanges or other chromosomal aberrations and the precise contributions of these various processes to cytotoxicity have still to be resolved.

### *Role in Differentiation*

There is increasing evidence that inhibitors of topoisomerases induce some cells to differentiate (Sahyoun *et al.*, 1986; Francis *et al.*, 1987; Gieseler *et al.*, 1990; Kiguchi *et al.*, 1990; Nakaya *et al.*, 1991). Amsacrine or VP-16 induced the differentiation of human promonocytic leukaemia U-937 cells, at the same time reducing c-*myc* and β-*actin* mRNA, increasing vimentin mRNA and inhibiting the cells in G2 phase (Rius *et al.*, 1991). Others have also described preferential effects of top II inhibitors on oncogene expression which could affect growth and permit differentiation (Darby *et al.*, 1986; Riou *et al.*, 1989). The non-intercalating top II inhibitor genistein (Markovits *et al.*, 1989) induced protein-linked DNA strand breaks and differentiation of human melanoma cells, suggesting that torsional effects on DNA or inactivation of growth-related genes might be involved (Constantinou *et al.*, 1990). However, genistein is also an inhibitor of protein tyrosine kinases, and it might act by inhibiting the transmission of growth signals (Watanabe *et al.*, 1991).

Effects of camptothecin or VP-16 on the differentiation of several types of leukaemia cells showed a parallelism between induction of differentiation and DNA-strand breaks, and these effects were synergistically or additively increased by tumour necrosis factor (Nakaya *et al.*, 1991). Presumably, effects on differentiation reflect effects of topoisomerase inhibitors on gene expression, but this remains to be demonstrated.

### *Cell Cycle Relationships*

Numerous studies *in vivo* and using cell-free extracts have shown that top II activity is greater in proliferating than in quiescent cells (Duguet *et al.*, 1983; Miskimins *et al.*, 1983; Taudou *et al.*, 1984; Sullivan *et al.*, 1986; Bodley *et al.*, 1987; Chow and Ross, 1987; Markovits *et al.*, 1987; Nelson *et al.*, 1987; Zwelling *et al.*, 1987). This has led to the suggestion that top II is a specific marker for cell proliferation which is lost (degraded) at mitosis (Heck and Earnshaw, 1986). However, studies with *Drosophila* have questioned the universality of this suggestion (Whalen *et al.*, 1991).

Using antibody probes, Heck *et al.* (1988) examined the expression of topoisomerases I and II and scaffold protein SC2 in normal and transformed chicken cells. Only top II showed significant fluctuations during the cell cycle, in accord with a major requirement for the enzyme during late S and G2/M phases when resolution of DNAs and condensation of chromosomes must occur. The half-life of top II decreased sevenfold after mitosis compared with that in asynchronous populations, and rapid loss of smaller enzyme fragments occurred, consistent with enzyme degradation. The half-life of top II in normal cells was also fourfold lower than in transformed cells, suggesting that transformation alters the control of top II stability.

Heck *et al.* (1989) also examined top II phosphorylation in chicken lymphoblastoid cells in different phases in the cell cycle. The level of phosphorylation in G2-M phase cells was 3.5-fold higher than in cells early in the cell cycle and the $M_r$ 170 kDa topoisomerase form was phosphorylated more than various antigenic fragments. It was suggested that phosphorylation activates the enzyme for its role in dysjunction of sister chromatids at anaphase, a proposal consistent with the action of anticancer drugs that stabilize cleavable complexes and cause increased sister-chromatid exchange or other chromosomal aberrations.

In proliferating HeLa cells Kroll and Rowe (1991) found that top II-associated phosphate was remarkably stable, having a half-life of 17 h, while top II had a half-life of 27 h, which was longer than the HeLa cell cycle time, and considerably longer than the half-lives of 3.3 or 12 h for top II in normal or transformed chicken cells reported by Heck *et al.* (1988). The possibility that two isoenzymes of top II might explain the differences between normal and transformed cells was considered; however, since top II is phosphorylated at multiple sites in HeLa cells (Kroll and Rowe, 1991), it seems equally possible that different sites or degrees of phosphorylation could affect the stability of the enzyme in normal versus transformed cells and its response to drugs.

Hsiang *et al.* (1988) measured the top II in primary human skin fibroblasts before and after stimulation by serum. After 27 h in the absence of serum, top II was undetectable, although it returned to normal levels (approx. $10^6$ molecules/cell) 24 h after adding 10% serum. Levels of top II were also sevenfold reduced in density-inhibited cells and similar effects were seen with mouse NIH 3T3 and 3T6 cells. In contrast, top II remained at high levels in transformed cells such as HeLa, L1210 and SV40 T-antigen transformed Cos-1 cell under the different growth conditions. Moreover, in synchronized HeLa cells there was little change in top II throughout the cell cycle. These observations suggest that top II inhibitors should have preferential effects on tumour cells which are not density inhibited.

Using antibodies to measure the levels of top II in developmentally regulated normal chicken cells, Heck and Earnshaw (1986) found that the level of top II fell from $7.8 \times 10^4$ copies per chicken erythroblast to $\leq 300$ copies per erythrocyte as mitosis ceased, while activation of $G_0$ chicken lymphocytes increased the percentage of immunopositive cells from undetectable levels to 5–30% positive cells, concomitant with the onset of DNA synthesis. Others have confirmed that the increase in top II in lectin-stimulated lymphocytes is accompanied by increased production of top II messenger RNA (Hwong *et al.*, 1990).

In nuclear extracts of normal human tissues the highest levels of top II activity (units per ng nuclear protein) were present in the spleen and thymus, sites of rapidly proliferating lymphocytes, while in neoplastic cells the highest enzyme activity was associated with aggressively proliferating tumours (breast carcinoma, lymphoma) and in the same range as that in normal spleen and thymus tissues (Holden *et al.*, 1990). Following mitogen stimulation, the sensitivity of human lymphocytes to etoposide rose in concert with top II levels; moreover, cells from 13 leukaemic patients showed much less sensitivity to etoposide than did human lymphoblastoid cell lines and they contained markedly less top II, possibly indicating a reduced proliferative status (Edwards *et al.*, 1987). These data strengthen the conclusion that drug sensitivity is dependent upon top II content in cells, which is in turn related to their proliferative status.

In synchronized HeLa cells the frequency of DNA cleavage by amsacrine was 4–15-fold greater in mitosis than in S phase, while G1 and G2 cells exhibited intermediate sensitivity to cleavage (Estey *et al.*, 1987). The hypersensitivity of mitotic DNA to cleavage increased suddenly in late G2, was lost abruptly in early G1 and was not paralleled by an increase in top II activity in mitotic cells compared with G1, S or G2 phase cell extracts. Possible explanations for these effects include changes in chromatin structure during the cell cycle, or in location or phosphorylation of top II. In Balb/c 3T3 (A31) cells the sensitivity to etoposide-induced DNA cleavage increased during S phase and peaked in G2 phase just before mitosis, in parallel with increasing top II. However, maximal cytotoxicity occurred during S phase and it was almost completely eliminated by treating the cells with cycloheximide, coincident with a reduction in $M_r$ 168 kDa top II (Chow and Ross, 1987). The production by amsacrine of sister-chromatid exchanges in Chinese hamster V79 cells was also strongly induced only in early to mid-S phase, consistent with the higher levels of top II in cycling cells and a possible role for sister-chromatid exchange in cytotoxicity (Dillehay *et al.*, 1987).

After studying the effects of novobiocin, nalidixic acid, etoposide or amsacrine on cell progression through G2 phase, Rowley and Kort (1989) concluded that progression of Chinese hamster ovary cells into, but not through, mitosis requires top II. The manner of recovery from the drugs was consistent with blockade at, or before, a specific point in G2 about 30 min before metaphase.

### *Cytotoxic Mechanisms*

#### Drug Antagonism

Good evidence exists that drugs that interact with DNA can alter the action of top II. For example, the DNA minor groove binder distamycin potentiated the cleavage of SV40 DNA by VM-26 (teniposide) *near* distamycin binding sites on the DNA but suppressed cleavage *at* the distamycin binding sites, presumably owing to effects on DNA secondary structure altering top II–DNA interactions (Fesen and Pommier, 1989). Chemoprotection against the cytotoxic effects of top II-targeted drugs in cell cultures and *in vivo* has also been described (Finlay *et al.*, 1989; Jensen *et al.*, 1990; Kaufmann, 1991). Thus, 9-aminoacridine protected a Lewis lung tissue culture cell line or human MM96 melanoma cells against the cytotoxic action of the acridine carboxamide derivative CI 921, amsacrine or etoposide but not against doxorubicin, and it also protected P388 leukaemia *in vivo* against acridine carboxamide, CI 921 and amsacrine, partially against etoposide, but not against mitoxantrone or doxorubicin. To explain the protection by 9-aminoacridine, Finlay *et al.* (1989) proposed that DNA intercalating chemoprotectors restrict the conformational flexibility of DNA and, hence, the ability of top II to form cleavable complexes. Since individual drugs that target top II cause the enzyme to cleave DNA at different sites (Liu, 1989), intercalation of 9-aminoacridine could presumably have a range of effects.

Antagonistic effects of aclarubicin, ethidium bromide or camptothecin on the cytotoxicity of etoposide and amsacrine to human small cell lung cancer cell lines and on top II-mediated DNA cleavage suggest that simultaneous administration of more than one drug can sometimes reduce effectiveness (Jensen *et al.*, 1990; Kaufmann, 1991). High concentrations of individual drugs (e.g. anthracyclines) can also suppress DNA cleavage by top II (Zunino and Capranico, 1990). Moreover, pibenzimol analogues that bind tightly to the minor groove in DNA stimulate (low concentrations) or inhibit (high concentrations) the cytotoxic effects of some top II-active drugs, presumably as a result of distortion of the DNA helix, so that one cannot easily predict which combinations of drugs will be efficacious (Finlay and Baguley, 1990).

### Protection against Inhibitors

Holm *et al.* (1989a) and D'Arpa *et al.* (1990) have conjectured that production of topoisomerase-mediated DNA strand breaks by anticancer drugs is not sufficient to kill Chinese hamster lung cells, and they showed that drugs such as aphidicolin, cordycepin or 5,6-dichloro-1-β-ribofuranosylbenzimidazole gave protection against camptothecin or *m*-AMSA, suggesting that DNA or RNA synthesis was necessary for cell death. Additional support for a role of DNA synthesis in cytotoxicity comes from a study of the combined effects of methotrexate pretreatment and VP-16 on U937 cells, which substantially increased the cytotoxicity of VP-16 by causing the cells to accumulate at the G1–S phase boundary of the cell cycle with increased top II. When methotrexate was removed, a wave of synchronization followed, producing a larger fraction of S phase cells containing a higher concentration of top II, and greater drug cytotoxicity (Lorico *et al.*, 1990).

Other groups have shown that the cytotoxic effects of various top II inhibitors can be blocked by reagents such as 2,4-dinitrophenol, ouabain, novobiocin or caffeine, or by glucose-regulated stress (Kupfer *et al.*, 1987; Kaufmann, 1989; Lawrence *et al.*, 1989; Shen *et al.*, 1989; Utsumi *et al.*, 1990; Shibuya *et al.*, 1991). How these different reagents act is unclear. An explanation for some effects may be that the terminal steps of DNA replication are required for chromosome recognition by top II, so that inhibiting DNA synthesis (by a variety of means) prevents top II from functioning. Under these circumstances, top II inhibitors cannot stabilize cleavable complexes and cause non-dysjunction, aneuploidy and loss of viability (Holm *et al.*, 1989b).

It has been suggested that prolonged treatment of proliferating splenocytes with etoposide or other top II inhibitors (e.g. novobiocin) may induce DNA fragmentation by a mechanism that does not involve top II (Jaxel *et al.*, 1988b), or that endonucleolytic DNA damage by cellular enzymes (apoptosis) occurs when cells are treated with a variety of cytotoxic agents, including inhibitors of top II (Zwelling, 1989; Schneider *et al.*, 1990; Walker *et al.*, 1991).

In evaluating all of the data related to top II, its regulation during the cell cycle and effects of drugs that stabilize cleavable complexes, it is clear that top II levels fluctuate in the normal cell cycle, whereas in at least some transformed cells the enzyme is more stable. Furthermore, the extent of cleavable complex formation in response to top II inhibitors reflects the availability of the enzyme and its activation, being greatest in late G2-M phases of the cell cycle, when topoisomerase levels and activity are maximal. However, the cytotoxicity of top II inhibitors apparently stems from effects initiated in S rather than G2-M phase cells, and evidence that S phase cells are most sensitive to drugs (also that aphidicolin can protect these cells) suggests that ongoing DNA synthesis is essential for cytotoxicity. The association of top II with DNA replication forks (and the nuclear matrix) is also consistent with drugs affecting events related to DNA replication. Therefore, indications that increased sister-chromatid exchange and other chromosome aberrations correlate best with drug-induced cytotoxicity suggest that when top II inhibitors stabilize cleavable complexes near advancing DNA replication forks associated with the nuclear matrix in S phase cells, they produce DNA replication-dependent chromosomal aberrations as a result of non-homologous recombination. Recombination presumably results from exchange of DNA-associated top II subunits or the free 3′-OH ends of DNA held in cleavable complexes in response to unrelieved tension and supercoiling of DNA ahead of DNA replication forks in the presence of drugs. Evidence that intermolecular religation can occur at sites of top II action has recently been obtained, using synthetic substrates *in vitro* (Andersen *et al.*, 1991) and a role for supercoiling in recombination is established (Christman *et al.*, 1988).

In this scenario transformation would play no particular role in top II-targeted drug action on tumour cells other than ensuring that more cells are cycling through S phase and more top II is available. Non-cycling cells are known to be more resistant to top II inhibitors than are cycling cells, and cycling normal cells (e.g. lymphocytes) are as sensitive to the drugs as are cycling transformed cells. In these circumstances the ability of topoisomerase inhibitors to eradicate tumours will rely on the active growth and increased top II content of tumour cells to create a differential sensitivity to the drugs, while solid or other tumours with slow or non-cycling cells will tend to be resistant to the drugs.

Epstein and Smith (1989) have shown that oestrogen induces top II synthesis preceding DNA synthesis in T47D human breast cancer cells, suggesting that, if cancer cells could be induced to cycle, they might become sensitive to eradication by top II inhibitors which are currently ineffective against many tumours. Consistent with this view, oestrogen potentiates top II-mediated cytotoxicity in activated human breast cancer cells (Epstein *et al.*, 1989) and granulocyte-macrophage colony stimulating factor stimulates the clonal growth of human ovarian cancer cells, potentiating the effect of top II-targeted drugs (Cimoli *et al.*, 1991). Learning how to express top II and ensure its activation in cancer cells may prove to be a significant advance in future cancer research.

### Tumour Necrosis Factor

As mentioned above, synergistic cytotoxic interactions between drugs that inhibit top I or II and tumour necrosis factor (TNF) have been described for a variety of drugs and cell lines, although there was no effect of TNF on human lung cancer cell lines (Alexander *et al.*, 1987a,b; Coffman *et al.*, 1989; Baloch *et al.*, 1990; Branellec *et al.*, 1990; Giaccone *et al.*, 1990; Vigani *et al.*, 1991). Topoisomerase inhibitors and TNF are required simultaneously to obtain any effect of TNF, and, since TNF reacts with receptors on the cell surface, Utsugi *et al.* (1990) have proposed that receptor activation causes a transient increase in protein kinase C activity (cf. Vilcek and Lee, 1991) which phosphorylates top II and increases its activity, thereby enhancing the action of drugs that stabilize cleavable complexes in DNA and ultimately toxicity. The transient nature of protein kinase C activation might then explain the concurrent requirement for TNF and drugs.

Despite the very substantial increases in cytotoxicity of combinations of TNF and topoisomerase inhibitors towards some cell lines, limited or negligible effects have usually been obtained *in vivo* (Alexander *et al.*, 1987a; Bahnson and Ratliff, 1990). Furthermore, novobiocin or coumermycin protected L929 cells from the action of TNF and top II inhibitors, but they enhanced their action on ME 180 cells, indicating a much more complex relationship (Baloch *et al.*, 1990).

Schneider *et al.* (1990) have invoked apoptosis as a possible cause of cell killing by topoisomerase inhibitors, while Kyprianou *et al.* (1991) observed that TNF alone causes fragmentation of L929 cell DNA and induction of expression of the programmed cell-death-associated gene TRP-$M_2$ leading to apoptosis, which is accelerated by top II inhibitors. Lassota *et al.* (1991) found that some anti-tumour drugs with affinity for nucleic acids release specific proteins from the nuclei of HL60 or MoLt-4 cells. This might explain accelerated apoptosis if endogenous nucleases then degrade the exposed DNA at higher drug concentrations. Further research on this topic is needed to clarify the exact nature of the events involving TNF, and their likely relevance to topoisomerases and improved treatment of cancer.

### Viruses

Indications that viruses affect expression of top II and probably the control of enzyme activity reflect an important role of top II in cell development and virus infection (Crespi *et al.*, 1988; Benson and Huang, 1990; Matthes *et al.*, 1990). Using phosphonoacetic acid to inhibit DNA synthesis, Ebert *et al.* (1990) uncovered a requirement for viral DNA replication before VM-26 and top II could begin to cleave herpes simplex virus type I DNA *in vivo*, again illustrating the link between DNA replication and top II action. However, the precise nature of these effects and their consequences in cells are still unresolved.

### Resistance

Drugs that poison top II constitute an important part of the oncologist's armoury against cancer, so resistance to these compounds may be a significant clinical problem. A considerable range of mechanisms of resistance to top II-inhibiting drugs have been described, although some also apply to drugs that act on different cellular targets. More than one mechanism may operate in a single cell. For convenience, we can distinguish resistance mechanisms that directly involve top II from those that operate by some other process.

#### Mechanisms Not Primarily Involving top II

The best-known (and best-characterized) mechanism of resistance is over-expression of the 170 kDa transport glycoprotein, known as P glycoprotein (Pgp). This phenomenon has been extensively reviewed elsewhere (Endicott and Ling, 1989), so will receive only brief mention here. Cells utilizing P glycoprotein are able to pump out a wide range of drugs, including vinca alkaloids, epipodophyllotoxins and anthracyclines, to reduce intracellular concentrations to tolerable levels. Calcium antagonists and calmodulin inhibitors (e.g. verapamil and trifluoperazine) generally disrupt resistance mediated by P glycoprotein (Kamath *et al.*, 1991). This type of resistance — often known as classical multidrug resistance (MDR) — has frequently been described for cells resistant to topoisomerase-inhibiting drugs, often in conjunction with other mechanisms (Ganapathi *et al.*, 1989; Minato *et al.*, 1990). Not all topoisomerase inhibitors are susceptible to MDR. Amsacrine, for instance, enters cells so rapidly that efflux pumping mechanisms of this type are generally ineffective (Baguley *et al.*, 1990).

It is worth noting that intracellular drug concentrations are frequently much higher than those in surrounding growth medium. For instance, cells incubated with 1 μM amsacrine may attain intracellular concentrations of 70–180 μM (Zwelling *et al.*, 1982; Kessel and Wheeler, 1984; Snow and Judd, 1991). Similarly, in cells incubated with adriamycin, the intracellular to extracellular ratio varies from 20 to 120 (Chang and Gregory, 1985; Snow and Judd, 1991). Zwelling *et al.* (1982) postulated that at higher amsacrine concentrations a process of 'cooperative sequestration' takes place, manifested by the intracellular to extracellular drug ratio increasing as drug concentration in the medium is raised. P glycoprotein overexpression typically reduces intracellular drug concentration 2–12-fold. For instance, Bhalla *et al.* (1985) reported that P glycoprotein-expressing HL-60 cells, 111-fold resistant to adriamycin and 50-fold resistant to daunorubicin, accumulated 2.6 times less [$^{14}$C]-daunorubicin than did sensitive HL-60 cells. Classical MDR cells has been recovered from patients (Fojo *et al.*, 1987) as well as generated *in vitro*.

In some resistant cells P glycoprotein expression was not increased, yet drug accumulation was modestly depressed, probably contributing to resistance (Ferguson *et al.*, 1988; de Jong *et al.*, 1990; Matsuo *et al.*, 1990). Other membrane proteins have been suggested to contribute to reduced drug accumulation. For instance, phosphorylation of a 150 kDa protein present on sensitive cells has been associated with decreased accumulation of adriamycin in resistant HL60 cells (Marsh and Center, 1987), as has a 190 kDa protein largely unrelated to P glycoprotein (Marquardt *et al.*, 1990) from the same cells.

The glutathione detoxification system reduces the sensitivity of cells to doxorubicin, possibly by inactivating transient free radicals (McLane *et al.*, 1983) or perhaps for yet unknown reasons. An MDR MCF7 breast cancer line was found to contain elevated levels of glutathione peroxidase activity and overexpressed a glutathione transferase compared with control cells (Batist *et al.*, 1986). Increased peroxidase activity was found to reduce levels of hydroxyl radicals in these drug-resistant cells after exposure to adriamycin (Sinha *et al.*, 1987).

Recently Zwelling *et al.* (1990b) detected elevated activity of several antioxidant enzymes, especially glucose-6-phosphate dehydrogenase in the drug-resistant variant DR4 or the HT1080 fibrosarcoma line, and Lefevre *et al.* (1991) have reported tenfold higher levels of glutathione-*S*-transferase mRNA in a breast adenocarcinoma cell line. In contrast, Snow and Judd (1991) failed to detect enhanced expression of any glutathione-related or other detoxifying enzymes in an adriamycin-resistant human T cell line. Recently MCF7 cells transfected with an expression vector giving 8–18-fold increased cellular expression of glutathione-*S*-transferase were found to be no more resistant to drugs than were sensitive parental cells (Fairchild *et al.*, 1990).

Evidence for altered metabolism of drugs was found by Vasanthakumar and Ahmed (1986) in three myelocytic cell lines showing differing degrees of resistance to daunomycin, but cross-resistant to the top II inhibitors adriamycin, VP-16, VM-26 and mitoxantrone. Daunorubicin reductase activities were reduced by 50–70% in the drug-resistant sublines compared with controls. Snow and Judd (1991) also reported diminished amounts of [$^3$H]-daunorubicinol in extracts of cells selected for resistance to either amsacrine or adriamycin. Some drug metabolites such as daunorubicinol are potentially more toxic to cells than the parental drugs (Ahmed, 1985), so metabolic changes to reduce their production confer resistance.

Blocking the $Na^+$, $K^+$ pump with ouabain has been shown to decrease doxorubicin-associated DNA strand breaks and also drug cytotoxicity in a variety of tumour cell lines. The phenomenon was not associated with altered drug influx or efflux, but it was thought that the altered intracellular ionic environment reduced the activity of top II (Lawrence *et al.*, 1989; Lawrence and Davis, 1990). Interestingly, the unavailability of calcium has recently been related to resistance to cytotoxicity caused by VP-16, camptothecin, hyperthermia, gamma radiation and nitrogen mustard (Bertrand *et al.*, 1991). Pretreatment of cells with EGTA conferred protection, without affecting DNA synthesis or cell cycle distribution.

DNA hypermethylation occurs in a variety of cell types in response to a wide range of chemotherapeutic drugs (Nyce, 1989). The top II inhibitors etoposide, nalidixic acid and doxorubicin reduced methylation at low to moderate concentrations but stimulated it at higher dosages. Nyce suggested that the methylase enzyme involved may require top II to manipulate DNA into an appropriate substrate conformation. Methylation is thought to turn genes on or off, including those whose products may be cytotoxic or protective to the cell, so it could well constitute a mechanism of resistance.

Elevated repair of damage to DNA is a possible mechanism of drug resistance which has been infrequently documented. Meijer *et al.* (1987) reported that a human small cell lung carcinoma line which had been made resistant to adriamycin repaired DNA damage induced by adriamycin, $H_2O_2$ or X-rays more rapidly than did the parental cell line. Deffie *et al.* (1988) noted a more rapid onset of repair in adriamycin-resistant P388 cells. Others found no evidence for enhanced repair in drug-resistant cells (Zwelling *et al.*, 1981; Robson *et al.*, 1987; Snow and Judd, 1991).

Certain extraneous stimuli, e.g. glucose deprivation (Shen *et al.*, 1989) or heat shock (Li, 1987; Kampinga *et al.*, 1989), have been shown to engender resistance to the cytotoxic effects of topoisomerase II-poisoning drugs. The stress factor is thought to induce a variety of cellular changes, including reduced top II expression, which in turn confers resistance to top

II-associated cytotoxicity. Solid tumours, most of which are refractory to chemotherapy, frequently contain hypoxic and glucose-starved cells.

More surprising were the results of Chatterjee *et al.* (1990b), who discovered that a series of Chinese hamster ovary (CHO) cell lines deficient in poly(ADP ribose) polymerase were significantly resistant to VP-16 cytotoxicity. These cell lines remained susceptible to cleavable complex formation, but showed a close relationship between levels of resistance and sister-chromatid exchange. The authors postulated that the mutant cells had developed some unknown compensatory activity that protected them against VP-16-induced SCE occurring subsequent to cleavable complex formation.

A number of authors have associated resistance to top II inhibitors with the presence of particular proteins, mostly on the cell surface. For instance, Hamada *et al.* (1988) noted increased abundance of an 85 kDa membrane protein on two adriamycin-resistant cell lines and Chen *et al.* (1990) developed a human breast cancer cell line 900-fold resistant to adriamycin which expressed a novel 95 kDa cell surface protein. The level of expression of the latter protein correlated with drug resistance and it was also present in clinical samples from patients resistant to adriamycin. Drug levels were not reduced inside these resistant cells, so the 95 kDa protein did not interfere with drug transport.

### Specific Mechanisms Involving top II

When a cell line grown *in vitro* is selected for resistance to a top II-blocking drug, patterns of cross-resistance to other chemotherapeutic drugs may provide clues to the mechanism of resistance. Classical MDR cells are typically cross-resistant to other top II inhibitors, but also to vinca alkaloids, and, perhaps, colchicine and mitomycin C. Another major group of resistant cells are only cross-resistant to other top II-inhibiting drugs, and Danks *et al.* (1987) coined the term 'atypical multidrug resistance' (AT-MDR) to describe cells which show no alterations in drug transport or accumulation.

Various assays — protein–DNA complex formation (PDC) (Rowe *et al.*, 1986a), the fluorescence assay for DNA unwinding (FADU) (Kanter and Schwartz, 1982), alkaline elution (Kohn *et al.*, 1981) and cytogenetic damage — have been used to show that AT-MDR cells are able to avoid DNA damage associated with drug cytotoxicity. For instance, Beck *et al.* (1987) used alkaline elution to show that VM-26 produced fewer DNA single-strand breaks in resistant CEM cells; Patet *et al.* (1986) discovered that adriamycin pulverized chromosomes of Friend leukaemia cells, but had little effect on the chromosomes of drug-resistant cells; and Glisson *et al.* (1986) documented reduced protein–DNA complex formation in CHO cells resistant to top II inhibitors, when exposed to etoposide or *m*-AMSA.

Snow and Judd (1991) used all the above methods except alkaline elution to demonstrate resistance of human T lymphoblastoid cell lines to DNA damage caused by *m*-AMSA and adriamycin, and noted that resistance factors derived from the three assay systems differed considerably.

Specific assays, complemented by Northern and Western blotting, may be used to ascertain top II enzyme levels. Activity assays enjoy the advantage that distinct actions of the enzyme can be studied by employing the appropriate assay (e.g. unknotting and catenation using P4 DNA, decatenation with K DNA, relaxation with PM2 or pBR322 DNAs), and drugs can be included in the reaction. Using these approaches, researchers have frequently correlated AT-MDR with a 2–8-fold reduction in top II levels in many drug-resistant cell lines (Ferguson *et al.*, 1988; Deffie *et al.*, 1989; Hong, 1989; De Isabella *et al.*, 1990; de Jong *et al.*, 1990; Matsuo *et al.*, 1990). Non-cycling or slow-growing cells such as may be found in the hypoxic interiors of solid tumours also have a naturally low content of top II, and are not particularly susceptible to top II poisons (Markovits *et al.*, 1987; Sullivan *et al.*, 1987; Robbie *et al.*, 1988). How can the cell content of an essential enzyme be considerably lowered like this without producing apparent adverse effects on the cell? One possibility is that drug-resistant cells may reduce their chromosome number and, hence, their DNA, and so their requirement for top II. This is possible because many cell lines are aneuploid and have excess chromosomes. In control JL cells, and cells undergoing active selection for resistance to amsacrine or adriamycin, we have noted chromosome numbers ranging from 42 to 205 with means of 51–86; however, in stable drug-resistant variants of the same cell line, the mean chromosome number was 45, with a range of 42–46 (Snow and Judd, 1991). Similarly, Harker *et al.* (1989) noted a mean chromosome number of 45 in mitoxantrone-resistant HL-60 cells compared with over 70 in control cells.

Since the actions of top I and II overlap to some extent, the question arises: is reduced top II accompanied by increases in top I? A slight increase in top I noted by Ferguson *et al.* (1988) and a threefold increase by Lefevre *et al.* (1991) represent the only reported increases of which we are aware, so this sort of compensation seems to be rare.

Using a decatenation assay, Takano *et al.* (1991) found that VP-16-resistant and -sensitive human KB cells contained similar levels of top II enzymatic activity. However, Northern and Western blotting revealed that the resistant cells contained only a tenth as much enzyme as did the sensitive cells. Phosphorylation of top II (on serine) proved to be 14–18-fold higher in the drug-resistant cells.

Levels of top II mRNA were found to correlate well with responsiveness of patients' tumour cells to adriamycin (Kim *et al.*, 1991). Drug-resistant cells showed low levels of top II mRNA, whereas greater amounts were present in sensitive cells. Levels of P glycoprotein and glutathionine-*S*-transferase π mRNA proved much less satisfactory as indicators of drug resistance and sensitivity.

The results of top II assays have been used to argue that drug-resistant cells contain an altered enzyme molecule. For instance, Patel *et al.* (1990) found that the top II from a CEM cell line made resistant to VP-16 was resistant to the drug in a DNA decatenation assay, suggesting that the enzyme was altered. Sullivan *et al.* (1989), using CHO cells resistant to VP-16, reported that DNA cleavage activity was not stimulated by drugs, and that the drug-resistant enzyme was unstable at 37 °C, leading them to postulate an altered enzyme. It is noteworthy that although most reports describe resistance to cleavable complex formation in drug-resistant cells, most fail to find resistance to drugs in enzyme activity assays using cell extracts, frequently leaving lower top II content as the sole explanation for the resistance.

Indirect evidence for an altered top II was provided by Deffie *et al.* (1989) from studies of an adriamycin-resistant P388 murine leukaemia cell line. On Northern blots they discovered a smaller top II mRNA transcript (5.5 kb rather than the full-size 6.6 kb message) present only in resistant cells, and this they correlated with new bands present on Southern blots. However, no unusual proteins were detected in the cells.

Unequivocal evidence for the existence of mutated type II topoisomerases in drug-resistant cells has recently come to hand. Danks *et al.* (1988) found that the top II from VM-26-resistant CEM cells showed altered unknotting and DNA cleavage, and twentyfold slower catenation activities than did enzyme from sensitive cells. Subsequently, they discovered that the drug-resistant top II required additional ATP for optimal activity (Danks *et al.*, 1989). Very recently they have reported (Bugg *et al.*, 1991) that the drug-resistant top II bears a single point mutation that results in an Arg→Glu change at residue 449. This change is in the ATP binding fold of the molecule. Since the drug-resistant cells were near triploid, it is unlikely that the resistant top II gene was present on all copies of chromosome 17, but the degree of resistance of the cells may well correlate with the level of expression of the mutated gene relative to the normal gene, which is still expressed. There was also evidence that the association of the enzyme with the nuclear matrix was altered (Fernandes *et al.*, 1990), as well as reduced amounts of enzyme in the nucleus of the resistant CEM cells.

HL-60 cells 50–100-fold resistant to *m*-AMSA have also been found to bear a point mutation in their top II gene compared with parental HL-60 cells (Hinds *et al.*, 1991). The mutation resulted in substitution of lysine for arginine at residue 486 of the molecule to produce a MseI restriction site that was detected on blots of genomic DNA. Analysis of DNA from eleven normal and eleven leukaemic volunteers revealed that none bore the MseI site, so the authors concluded that the site represents a mutation rather than an allelic variant.

An alternative explanation for drug-resistant type II topoisomerase that involves neither mutations nor altered enzyme levels has emerged over the last few years. Drake *et al.* (1987) purified top II from amsacrine-resistant P388 cells to discover two forms of the enzyme — p170 and p180. The top II poisons teniposide and merbarone both inhibited p170 (top IIα) much more effectively than p180 (top IIβ), which predominated in the amsacrine-resistant P388 cells. The same two forms could be purified from drug-sensitive P388 cells, but the relative amounts of the two forms differed between sensitive and resistant cells. In a subsequent paper (Drake *et al.*,, 1989a), real biochemical and pharmacological differences were found between the two forms of the enzyme, and both genes have since been cloned and sequenced (Chung *et al.*, 1989). Despite the clear evidence for the existence of top IIβ, its functional role in the cell is as yet unclear.

Recently amsacrine derivatives have been produced that show enhanced activity against drug-resistant cells, and it is thought that these drugs have enhanced effectiveness against top IIβ, although this remains to be proved (Baguley *et al.*, 1990; Finlay *et al.*, 1990).

Some authors have suggested that drug resistance in cells is mediated by a factor that is associated with or acts on top II. De Isabella *et al.* (1990) suggested this as a possibility when tenfold lower drug-stimulated cleavage activity observed in nuclear extracts of resistant cells could not be demonstrated in experiments with purified top II from the same resistant cells. Darkin-Rattray and Ralph (1991) suggested that modulation of top II could involve a kinase, such as casein kinase II, that phosphorylates top II. Such a factor could account for the results of Hill *et al.* (1991), where HN-1 cells, selected for low-level resistance to etoposide, displayed typical AT-MDR cross-resistance patterns, but no alterations to top II levels or activity.

Finally, it is worth noting that multiple mechanisms of resistance may operate in one cell line (and, presumably, in an individual cell). Ferguson *et al.* (1988) discovered both reduced accumulation of drug (although no P glycoprotein) and lowered top II levels in etoposide-resistant KB cells, while Ganapathi *et al.* (1989) noted reduced accumulation in cells with P glycoprotein as well as resistance to drug-induced DNA cleavage. Both Lawrence *et al.* (1989) and Matsuo *et al.* (1990) found a similar combination of reduced drug uptake and reduced top II activity in resistant cell lines. In the absence of selection pressure, the top II-based resistance was stable but that involving P glycoprotein was not (Lawrence *et al.*, 1989). Lefevre *et al.* (1991) have described a somewhat different group of mechanisms in amsacrine-resistant human breast cancer cells — viz. reduced top II, increased top I and increased glutathione-*S*-transferase. A small cell lung cancer line made resistant to adriamycin was cross-resistant to the drugs vinblastine and gramicidin D (which do not act on top II), but the cells failed to express any P glycoprotein, displayed no alterations in drug uptake or efflux, and had reduced top II levels (Cole *et al.*, 1991). An additional unrecognized mode of resistance must operate in these cells to account for the resistance to vinca alkaloid and gramicidin.

In contrast to what was thought a decade ago, there now appear to be a whole constellation of modes by which cells can acquire resistance, even to those drugs which target merely the topoisomerases. Although most of these mechanisms have been characterized from experiments, *in vitro* it seems likely that many (plus other as yet undiscovered mechanisms) will also occur *in vivo*.

Since a number of distinct modes of resistance may operate in the one cell, clinical circumvention of resistance will continue to prove challenging. However, the continuing emergence of new drugs (Finlay *et al.*, 1990; Yamashita *et al.*, 1991) and phenomena such as collateral sensitivity (where resistance to one drug is accompanied by hypersensitivity to others) (e.g. Waud *et al.*, 1991) offer some hope for improvements in chemotherapy.

## 4 Conclusions

Anticancer drugs that target the enzymes top I or top II in cells derive their cytotoxic activity from the fact that cells in many cancers continuously cycle and replicate their DNA as a consequence of mutations that remove regulatory controls on processes that coordinate growth or differentiation. Cycling cells are particularly sensitive to drugs such as camptothecin, which prevent top I action and RNA synthesis, or to top II inhibitors that interfere with DNA replication forks. Production of cleavable complexes during S phase in response to the drugs and the resulting inability to relax DNA topology lead to aberrant DNA rearrangements, presumably as cells try to subvert or overcome the impediments to further growth and survival. However, inability to segregate aberrant chromosomes correctly, combined with effects upon gene expression necessary for recovery, prove to be insurmountable and the cells ultimately succumb to the general disruption of their metabolism. Whether cell death is an active process (apoptosis) or merely due to the failure of cellular processes (necrosis) has yet to be resolved.

In considering the potential or limitations of topoisomerase inhibitors, it is clear that rapidly cycling normal cells are equally susceptible to their action (e.g. thymocytes), while slowly or non-cycling cancer cells (e.g. in the interior of solid tumours) often evade topoisomerase inhibitors. However, some selective action may arise from the fact that the turnover of top II appears to be reduced in tumour cells, which may render them more susceptible to topoisomerase inhibitors than are normal cells.

A drug that inhibited the normal cell cycle in G1 phase (or Go) without affecting tumour cell cycling or reducing top II content would provide an ideal adjunct when used in combination with existing top II inhibitors to treat cancers. However, knowledge of the control of the cell cycle is still insufficient to discern how to hold all normal cells in G1 phase (or Go), other than by removing stimuli to their growth such as growth factors. It is hoped that as knowledge of the events controlling the initiation and control of the cell cycle improves, it will become possible to retard normal cells in G1 (Go) phase and exploit the continuous cycling of tumour cells and the longer lifetime of their topoisomerase so as to destroy these cells selectively. Learning how to stimulate specifically the growth and, hence, topoisomerase activity in slowly cycling or dormant tumour cells will also contribute to improved chemotherapy. Then topoisomerase inhibitors will really display their inherent potential to combat cancer.

**References**

Ackerman, P., Glover, C. V. C. and Osheroff, N. (1985). Phosphorylation of DNA topoisomerase II by casein kinase II: modulation of eukaryotic topoisomerase II activity *in vitro*. *Proc. Natl Acad. Sci. USA*, **82**, 3164–3168

Ackerman, P., Glover, C. V. C. and Osheroff, N. (1988). Phosphorylation of DNA topoisomerase II *in vivo* and in total homogenates of *Drosophila* Kc cells. *J. Biol. Chem.*, **263**, 12653–12660

Ackerman, P. and Osheroff, N. (1989). Regulation of casein kinase II activity by epidermal growth factor in human A-431 carcinoma cells. *J. Biol. Chem.*, **264**, 11958–11965

Adachi, Y., Luke, M. and Laemmli, U. (1991). Chromosome assembly *in vitro*: topoisomerase II is required for condensation. *Cell*, **64**, 137–148

Ahmed, N. K. (1985). Daunorubicin reductase activity in human normal lymphocytes, myeloblasts and leukemic cell lines. *Eur. J. Cancer Clin. Oncol.*, **21**, 1209–1214

Alberts, B., Bray, D., Lewis, J., Raff, M., Roberts, K. and Watson, J. D. (1989) *Molecular Biology of the Cell*, Garland Publishing Inc., New York

Alexander, R. B., Isaacs, J. T. and Coffey, D. S. (1987a). Tumour necrosis factor enhances the *in vitro* and *in vivo* efficacy of chemotherapeutic drugs targeted at DNA topoisomerase II in the treatment of murine bladder cancer. *J. Urol.*, **138**, 427–429

Alexander, R. B., Nelson, W. G. and Coffey, D. S. (1987b). Synergistic enhancement by tumour necrosis factor of *in vitro* cytotoxicity from chemotherapeutic drugs targeted at DNA topoisomerase II. *Cancer Res.*, **47**, 2403–2406

Almouzni, G. and Mechali, M. (1988). Assembly of spaced chromatin involvement of ATP and DNA topoisomerase activity. *EMBO Jl*, **7**, 4355–4365

Amati, B. B. and Gasser, S. M. (1988). Chromosomal ARS and CEN elements bind specifically to the yeast nuclear scaffold. *Cell*, **54**, 967–978

Andersen, A. H., Christiansen, K., Zechiedrich, E. L., Jensen, P. S., Osheroff, N. and Westergaard, O. (1989). Strand specificity of the topoisomerase II mediated double-stranded DNA cleavage reaction. *Biochemistry*, **28**, 6237–6244

Andersen, A. H., Sørensen, B. S., Christiansen, K., Svejstrup, J. Q., Lund, K. and Westergaard, O. (1991). Studies of the topoisomerase II-mediated cleavage and religation reactions by use of a suicidal double-stranded DNA substitute. *J. Biol. Chem.*, **266**, 9203–9210

Andoh, T., Ishi, K., Suzuki, Y., Ikegami, Y., Kusunoki, Y., Takemoto, Y. and Okada, K. (1987). Characterisation of a mammalian mutant with a camptothecin resistant DNA topoisomerase I. *Proc. Natl Acad. Sci. USA*, **84**, 5565–5569

Annunziato, A. T. (1989). Inhibitors of topoisomerase I and II arrest DNA replication, but do not prevent nucleosome assembly *in vivo*. *J. Cell Sci.*, **93**, 593–603

Austin, C. A. and Fisher, L. M. (1990). Isolation and characterisation of a human cDNA clone encoding a novel DNA topoisomerase II homologue from HeLa cells. *FEBS Lett.*, **266**, 115–117

Avemann, K., Knippers, R., Koller, T. and Sogo, J. M. (1988). Camptothecin, a specific inhibitor of type I DNA topoisomerase, induces DNA breakage at replication forks. *Mol. Cell. Biol.*, **8**, 3026–3034

Bae, Y. S., Chiba, M., Ohira, M. and Ikeda, H. (1991). A shuttle vector for analysis of illegitimate recombination in mammalian cells: effects of DNA topoisomerase inhibitors on deletion frequency. *Gene*, **101**, 285–289

Bae, Y. S., Kawasaki, I., Ikeda, H. and Liu, L. F. (1988). Illegitimate recombination mediated by calf thymus DNA topoisomerase II *in vitro*. *Proc. Natl Acad. Sci. USA*, **85**, 2076–2080

Bailly, C., OhUigin, C., Rivalle, C., Bisagni, E., Hénichart, J. P. and Waring, M. J. (1990). Sequence-selective binding of an ellipticine derivative to DNA. *Nucleic Acids Res.*, **18**, 6283–6291

Baguley, B. C., Holdaway, K. M. and Fray, L. M. (1990). Design of DNA intercalators to overcome topoisomerase II-mediated multidrug resistance. *J. Natl Cancer Inst.*, **82**, 398–402

Bahnson, R. R. and Ratliff, T. L. (1990). *In vitro* and *in vivo* anti-tumour activity of recombinant mouse tumour necrosis factor (TNF) in a mouse bladder tumour (MBT-2). *J. Urol.*, **144**, 172–175

Baloch, Z., Cohen, S. and Coffman, F. D. (1990). Synergistic interactions between tumour necrosis factor and inhibitors of DNA topoisomerase I and II. *J. Immunol.*, **145**, 2908–2913

Barañao, J. L., Bley, M. A., Batista, F. D. and Gliken, G. C. (1991). A DNA topoisomerase I inhibitor blocks the differentiation of rat granulosa cells induced by follicle stimulating hormone. *Biochem. J.*, **277**, 557–560

Batist, G., Tulpule, A., Sinha, B. K., Katki, A. G., Myers, C. E. and Cowan, K. H. (1986). Overexpression of a novel anionic glutathione transferase in multidrug resistant human breast cancer cells. *J. Biol. Chem.*, **261**, 15544–15549

Battistoni, A., Leoni, L., Sampaolese, B. and Savino, M. (1988). Kinetic persistence of cruciform structures in reconstituted minichromosomes. *Biochim. Biophys. Acta*, **950**, 161–171

Beck, W. T., Cirtain, M. C., Danks, M. K., Felsted, R. L., Safa, A. R., Wolverton, J. S. and Suttle, D. P. (1987). Pharmacological, molecular and cytogenetic analysis of 'atypical' multidrug resistant human leukemic cells. *Cancer Res.*, **47**, 5455–5460

Been, M. D., Burgess, R. R. and Champoux, J. J. (1984). Nucleotide sequence preference at rat liver and wheat germ type 1 DNA topoisomerase breakage sites in duplex SV40 DNA. *Nucleic Acids Res.*, **12**, 3097–3114

Been, M. D. and Champoux, J. J. (1984). Breakage of single-stranded DNA by eukaryotic type 1 topoisomerase occurs only at regions with the potential for base-pairing. *J. Mol. Biol.*, **180**, 515–531

Belenguer, P., Baldin, V., Matheu, C., Prats, H., Bensaid, M., Bouche, G. and Amalric, F. (1989). Protein kinase NII and the regulation of rDNA transcription in mammalian cells. *Nucleic Acids Res.*, **17**, 6625–6636

Bendixen, C., Thomsen, B., Alsner, J. and Westergaard, O. (1990). Camptothecin-stabilized topoisomerase I-DNA adducts cause premature termination of transcription. *Biochemistry*, **29**, 5613–5619

Benson, J. D. and Huang, E. (1990). Human cytomegalovirus induces expression of cellular topoisomerase II. *J. Virol.*, **64**, 9–15

Berger, N. A., Chatterjee, S., Schmotzer, J. A. and Helms, S. R. (1991). Etoposide (VP-16-213)-induced gene alterations: potential contribution to cell death. *Proc. Natl Acad. Sci. USA*, **88**, 8740–8743

Berrios, M. and Fisher, P. A. (1988). Thermal stabilization of putative karyoskeletal protein enriched fractions from *Saccharomyces cerevisiae*. *Mol. Cell. Biol.*, **8** 4573–4575

Berrios, M., Osheroff, N. and Fisher, P. A. (1985). *In situ* localization of DNA topoisomerase II, a major polypeptide component of the *Drosophila* nuclear matrix fraction. *Proc. Natl Acad. Sci. USA*, **82**, 4142–4146

Bertrand, R., Kerrigan, D., Sarang, M. and Pommier, Y. (1991). Cell death induced by topoisomerase inhibitors: role of calcium in mammalian cells. *Biochem. Pharmacol.*, **42**, 77–85

Bettler, B., Ness, P. J., Schmidlin, S. and Parish, R. W. (1988). The upstream limit of nuclease-sensitive chromatin in *Dictyostelium* rRNA genes neighbours a topoisomerase I-like cluster. *J. Mol. Biol.*, **204**, 549–558

Bhalla, K., Hindenburg, A., Taub, R. N. and Grant, S. (1985). Isolation and characterisation of an anthracycline-resistant human leukemic cell line. *Cancer Res.*, **45**, 3657–3662

Bhuyan, B. K., Fraser, T. J., Gray, L. G., Kuentzel, S. L. and Neil, G. L. (1973). Cell-kill kinetics of several S-phase specific drugs. *Cancer Res.*, **33**, 888–894

Bjornsti, M.-A., Benedetti, P., Viglianti, G. A. and Wang, J. C. (1989). Expression of human DNA topoisomerase I in yeast cells lacking yeast DNA topoisomerase I: restoration of sensitivity of the cells to the antitumour drug camptothecin. *Cancer Res.*, **49**, 6318–6323

Blasquez, V. C., Sperry, A. O., Cockerill, P. N. and Garrard, W. T. (1989). Protein:DNA interactions at chromosomal loop attachment sites. *Genome*, **31**, 503–509

Bodley, A. L., Wu, H. Y. and Liu, L. F. (1987). Regulation of DNA topoisomerases during cellular differentiation. *N.C.I. Monograph*, **4**, 31–35

Bonven, B. J., Gocke, E. and Westergaard, O. (1985). A high affinity topoisomerase I binding sequence is clustered at DNAase I hypersensitive sites in *Tetrahymena* R-chromatin. *Cell*, **41**, 541–551

Boothman, D. A., Trask, D. K. and Pardee, A. B. (1989). Inhibition of potentially lethal DNA damage repair in human tumour cells by beta-lapachone, an activator of topoisomerase I. *Cancer Res.*, **49**, 605–612

Boritzki, T. J., Wolfard, T. S., Besserer, J. A., Jackson, R. C. and Fry, D. W. (1988). Inhibition of type II topoisomerase by fostriecin. *Biochem. Pharmacol.*, **37**, 4063–4068

Borowiec, J. A. and Gralla, J. D. (1985). Supercoiling response of the lac $p^S$ promoter *in vitro*. *J. Mol. Biol.*, **184**, 587–598

Boulikas, T., Bastin, B., Boulikas, P. and Dupuis, G. (1990). Increase in poly(ADP-ribosylation) in mitogen activated lymphoid cells. *Exp. Cell Res.*, **187**, 77–84

Branellec, D., Markovits, J. and Chouaib, S. (1990). Potentiation of TNF-mediated cell killing by VP-16: relationship to DNA single-strand break formation. *Int. J. Cancer*, **46**, 1048–1053

Brill, S. J., DiNardo, S., Voelkel-Meiman, K. and Sternglanz, R. (1987). Need for DNA topoisomerase activity as a swivel for DNA replication for transcription of ribosomal RNA. *Nature*, **326**, 414–416

Brill, S. J. and Sternglanz, R. (1988). Transcription-dependent DNA supercoiling in yeast DNA topoisomerase mutants. *Cell*, **54**, 403–411

Bronshtein, I. B., Gromova, I. I., Bukhman, V. L. and Kafiani, K. A. (1989). Effect of camptothecin on the DNA-relaxing and DNA-cleavage activity of calf thymus topoisomerase I. *Mol. Biol. (Mosk.)*, **23**, 491–501

Bugg, B. Y., Danks, M. K., Beck, W. T. and Suttle, D. P. (1991). Expression of a mutant topoisomerase II in CCRF-CEM human leukemic cells selected for resistance to teniposide. *Proc. Natl Acad. Sci. USA*, **88**, 7654–7658

Bullock, P., Champoux, J. T. and Botchan, M. (1985). Association of crossover points with topoisomerase I cleavage sites: a model for non-homologous recombination. *Science*, **230**, 954–958

Busk, H., Thomsen, B., Bonven, B. J., Kjeldsen, E., Nielsen, O. F. and Westergaard, O. (1987). Preferential relaxation of supercoiled DNA containing a hexadecameric recognition sequence for topoisomerase I. *Nature*, **327**, 638–640

Caizergues-Ferrer, M., Belenguer, P., Lapeyre, B., Amalric, F., Wallace, M. O. and Olson, M. O. J. (1987). Phosphorylation of nucleolin by a nucleolar type NII protein kinase. *Biochemistry*, **26**, 7876–7883

Camilloni, G., Martino, E. D., Caserta, M. and Di Mauro, E. (1988). Eukaryotic DNA topoisomerase I reaction is topology dependent. *Nucleic Acids Res.*, **16**, 7071–7085

Camilloni, G., Martino, E. D., Di Mauro, E. and Caserta, M. (1989). Regulation of the function of eukaryotic DNA topoisomerase I: topological conditions for inactivity. *Proc. Natl Acad. Sci. USA*, **86**, 3080–3084

Capranico, G., Jaxel, C., Roberge, M., Kohn, K. and Pommier, Y. (1990a). Nucleosome positioning as a critical determinant for the DNA cleavage sites of mammalian DNA topoisomerase II in reconstituted simian virus 40 chromatin. *Nucleic Acids Res.*, **18**, 4553–4559

Capranico, G., Kohn, K. W. and Pommier, Y. (1990b). Local sequence requirements for DNA cleavage by mammalian topoisomerase II in the presence of doxorubicin. *Nucleic Acids Res.*, **18**, 6611–6619

Capranico, G., Zunino, F., Kohn, K. W. and Pommier, Y. (1990c). Sequence-selective topoisomerase II inhibition by anthracycline derivatives in SV40 DNA: relationship with DNA binding affinity and cytotoxicity. *Biochemistry*, **29**, 562–569

Carty, M. and Menzel, R. (1990). Inhibition of DNA gyrase activity in an *in vitro* transcription–translation system stimulates gyrA expression in a DNA concentration dependent manner. Evidence for the involvement of factors which may be titrated. *J. Mol. Biol.*, **214**, 397–406

Caserta, M., Amadei, A., Camilloni, G. and Di Mauro, E. (1990). Regulation of the function of eukaryotic DNA topoisomerase I: analysis of the binding step and of the catalytic constants of topoisomerisation as a function of DNA topology. *Biochemistry*, **29**, 8152–8157

Caserta, M., Amadei, A., Di Mauro, E. and Camilloni, G. (1989). *In vitro* preferential topoisomerization of bent DNA. *Nucleic Acids Res.*, **17**, 8463–8474

Castora, F. J. and Lazarus, G. M. (1984). Isolation of mitochondrial DNA topoisomerase from human leukemia cells. *Biochem. Biophys. Res. Commun.*, **121**, 77–86

Castora, F. J., Lazarus, G. M. and Kunes, D. (1985). The presence of two mitochondrial DNA topoisomerases in human acute leukemia cells. *Biochem. Biophys. Res. Commun.*, **130**, 854–866

Castora, F. J., Sternglanz, R. and Simpson, M. V. (1982). A new mitochondrial topoisomerase from rat liver that catenates DNA. In Slonimski, P., Borst, P. and Attardi, G. (Eds), *Mitochondrial Genes*, Cold Spring Harbor Laboratory Press, Cold Spring Harbor, N.Y., pp. 143–154

Champoux, J. J. (1981). DNA is linked to the rat liver DNA nicking-closing enzyme by a phosphodiester bond to tyrosine. *J. Biol. Chem.*, **256**, 4805–4809

Champoux, J. J. (1988). Topoisomerase I is preferentially associated with isolated replicating simian virus 40 molecules after treatment of infected cells with camptothecin. *J. Virol.*, **62**, 3675–3683

Champoux, J. J. (1990). Mechanistic aspects of type-I topoisomerases. In Cozzarelli, N. R. and Wang, J. C. (Eds), *DNA Topology and Its Biological Effects*, Cold Spring Harbor Laboratory Press, Cold Spring Harbor, N.Y., pp. 217–242

Champoux, J. J. and Aronoff, R. (1989). The effects of camptothecin on the reaction and the specificity of the wheat germ type I topoisomerase. *J. Biol. Chem.*, **264**, 1010–1015

Champoux, J. J., Young, L. S. and Been, B. M. (1978). Studies on the regulation and specificity of the DNA untwisting enzymes. *Cold Spring Harbor Symp. Quant. Biol.*, **43**, 53–58

Chang, B. K. and Gregory, J. A. (1985). Comparison of the cellular pharmacology of doxorubicin in resistant and sensitive models of pancreatic cancer. *Cancer Chemother. Pharmacol.*, **14**, 132–137

Charron, M. and Hancock, R. (1990a). Roles of DNA topoisomerases I and II in DNA replication, mitotic chromosome formation and recombination in mammalian cells. In Harris, I. R. and Zbarsky, I. B. (Eds), *Nuclear Structure and Function*, Plenum Press, New York, pp. 405–411

Charron, M. and Hancock, R. (1990b). DNA topoisomerase II is required for formation of mitotic chromosomes in Chinese hamster ovary cells: studies using the inhibitor 4-demethylepipodophyllotoxin 9-(4,6-*o*-thenylidene-beta-D-glucopyranoside). *Biochemistry*, **29**, 9531–9537

Charron, M. and Hancock, R. (1991). Chromosome recombination and defective genome segregation induced in Chinese hamster cells by the topoisomerase II inhibitor VM-26. *Chromosoma*, **100**, 97–102

Chatterjee, S., Cheng, M. F. and Berger, N. A. (1990a). Hypersensitivity to clinically useful alkylating agents and radiation in poly(ADP-ribose) polymerase-deficient cell lines. *Cancer Commun.*, **2**, 401–407

Chatterjee, S., Cheng, M. F., Berger, S. J. and Berger, N. A. (1991). Alkylating agent hypersensitivity in poly(adenosine diphosphate-ribose) polymerase deficient cell lines. *Cancer Commun.*, **3**, 71–75

Chatterjee, S., Cheng, M. F., Trivedi, D., Petzold, S. J. and Berger, N. A. (1989). Camptothecin hypersensitivity in poly(adenosinediphosphate-ribose) polymerase-deficient cell lines. *Cancer Commun.*, **1**, 389–394

Chatterjee, S., Trivedi, D., Petzold, S. J. and Berger, N. A. (1990b). Mechanism of epipodophyllotoxin-induced cell death in poly(adenosinediphosphate-ribose) synthesis-deficient V-79 Chinese hamster cell lines. *Cancer Res.*, **50**, 2713–2718

Chen, Y. N., Mickley, L. A., Schwartz, A. M., Acton, E. M., Hwang, J. L. and Fojo, A. T. (1990). Characterization of adriamycin-resistant human breast cancer cells which display overexpression of a novel resistance-related membrane protein. *J. Biol. Chem.*, **265**, 10073–10080

Choo, K.-B., Lee, H.-H., Liew, L.-N., Chong, K.-Y. and Chou, H.-F. (1990). Analysis of the unoccupied site of an integrated human papillomavirus 16 sequences in a cervical carcinoma. *Virology*, **178**, 621–625

Chou, S., Kaneko, M., Nakaya, K. and Nakamura, Y. (1990). Induction of differentiation of human and mouse myeloid leukemia cells by camptothecin. *Biochem. Biophys. Res. Commun.*, **166**, 160–167

Chow, K. C., King, C. K. and Ross, W. E. (1988). Abrogation of etoposide-mediated cytotoxicity by cycloheximide. *Biochem. Pharmacol.*, **37**, 1117–1122

Chow, K. C. and Ross, W. E. (1987). Topoisomerase-specific drug sensitivity in relation to cell cycle progression. *Mol. Cell Biol.*, **7**, 3119–3123

Christiansen, K., Bonven, B. J. and Westergaard, O. (1987). Mapping of sequence-specific chromatin proteins by a novel method: topoisomerase I on tetrahymena ribosomal chromatin. *J. Mol. Biol.*, **193**, 517–525

Christman, M. F., Dietrich, F. S. and Fink, G. R. (1988). Mitotic recombination in the rDNA of *S. cerevisiae* is suppressed by the combined action of DNA topoisomerases I and II. *Cell*, **55**, 413–425

Chung, T. D. Y., Drake, F. H., Tan, K. B., Per, S. R., Crooke, S. T. and Mirabelli, C. K. (1989). Characterisation and immunological identification of cDNA clones encoding two human DNA topoisomerase II isozymes. *Proc. Natl Acad. Sci. USA*, **86**, 9431–9435

Cimoli, G., Venturini, M., Billi, G., Valentini, M., Rosso, R. and Russo, P. (1991). Human granulocyte-macrophage colony stimulating factor (GM-CSF) stimulates clonal growth of human ovarian cancer cell lines (IGROV-1) and potentiates the effect of topoisomerase II-targeted drugs. *International Symposium on DNA Topoisomerases in Chemotherapy*, Nagoya, Japan. Abstracts, p. 137

Cockerill, P. N. and Garrard, W. T. (1986a). Chromosomal loop anchorage of the kappa immunoglobulin gene occurs next to the enhancer in a region containing topoisomerase II sites. *Cell*, **44**, 273–282

Cockerill, P. N. and Garrard, W. T. (1986b). Chromosomal loop anchorage sites appear to be evolutionarily conserved. *FEBS Lett.*, **204**, 5–7

Cockerill, P. N., Yuen, M.-H. and Garrard, W. T. (1987). The enhancer of the immunoglobulin heavy chain locus is flanked by presumptive chromosomal loop anchorage elements. *J. Biol. Chem.*, **262**, 5394–5397

Coderoni, S., Paparelli, M. and Giafranceschi, G. L. (1990). Role of calf thymus DNA topoisomerase I phosphorylation on relaxation, activity, expression and on DNA–protein interaction. Role of DNA-topoisomerase I phosphorylation. *Mol. Biol. Reports*, **14**, 35–39

Coffman, F. D., Green, L. M., Godwin, A. and Ware, C. F. (1989). Cytoxicity mediated by tumour necrosis factor in variant subclones of the ME-180 cervical carcinoma line: modulation by specific inhibitors of DNA topoisomerase II. *J. Cell. Biochem.*, **39**, 95–105

Cole, S. P., Chanda, E. R., Dicke, F. P., Gerlach, J. H. and Mirski, S. E. (1991). Non-P-glycoprotein-mediated multidrug resistance in a small cell lung cancer cell line: evidence for decreased susceptibility to drug-induced DNA damage and reduced levels of topoisomerase II. *Cancer Res.*, **51**, 3345–3352

Condemine, G. and Smith, C. (1990). Transcription regulates oxolinic acid-induced DNA gyrase cleavage at specific sites on the *E. coli* chromosome. *Nucleic Acids Res.*, **18**, 7389–7396

Constantinou, A., Henning-Chub, C. and Huberman, E. (1989). Novobiocin and phorbol-12-myristate-13-acetate-induced differentiation of human leukemia cells associated with a reduction in topoisomerase II activity. *Cancer Res.*, **49**, 1110–1117

Constantinou, A., Kiguchi, K. and Huberman, E. (1990). Induction of differentiation and DNA strand breakage in human HL-60 and K-562 leukemia cells by genistein. *Cancer Res.*, **50**, 2618–2624

Corbett, A. H., Zechiedrich, E. L. and Osheroff, N. (1992). A role for the passage helix in the DNA cleavage reaction of eukaryotic topoisomerase II. A two-site model for enzyme-mediated DNA cleavage. *J. Biol Chem.*, **267**, 683–686

Covey, J. M., Jaxel, C., Kohn, K. W. and Pommier, Y. (1989). Protein-linked DNA strand breaks induced in mammalian cells by camptothecin, an inhibitor of topoisomerase I. *Cancer Res.*, **49**, 5016–5022

Crespi, M. D., Mladovan, A. G. and Baldi, A. (1988). Increment of DNA topoisomerases in chemically and virally transformed cells. *Exp. Cell Res.*, **175**, 206–215

Crossen, P. G. (1979). The effect of acridine compounds on sister-chromatid exchange formation in cultured human lymphocytes. *Mutation Res.*, **68**, 295–299

Culotta, V. and Sollner-Webb, B. (1988). Sites of topoisomerase I action on *X. laevis* ribosomal chromatin: transcriptionally active rDNA has an approximately 200 bp repeating structure. *Cell*, **52**, 585–597

Danks, M. K., Schmidt, C. A., Cirtain, M. C., Suttle, D. P. and Beck, W. T. (1988). Altered catalytic activity of and DNA cleavage by DNA topoisomerase II from human leukemic cells selected for resistance to VM-26. *Biochemistry*, **27**, 8861–8869

Danks, M. K., Schmidt, C. A., Deneka, D. A. and Beck, W. T. (1989). Increased ATP requirement for activity of and complex formation by DNA topoisomerase II from human leukemic CCRF-CEM cells selected for resistance to teniposide. *Cancer Commun.*, **1**, 101–109

Danks, M. K., Yalowich, J. C. and Beck, W. T. (1987). Atypical multiple drug resistance in a human leukemic cell line selected for resistance to teniposide (VM-26). *Cancer Res.*, **47**, 1297–1301

Darby, M. K., Herrera, R. E., Vosberg, H.-P. and Nordheim, A. (1986). DNA topoisomerase II cleaves at specific sites in the 5′ flanking region of c-*fos* protooncogenes *in vitro*. *EMBO Jl*, **5**, 2257–2265

Darby, M. K., Schmitt, J., Jongstra-Bilen, J. and Vosberg, H. P. (1985). Inhibition of calf thymus type II DNA topoisomerase by poly(ADP-ribosylation). *EMBO Jl*, **4**, 2129–2134

Darkin, S. J. and Ralph, R. K. (1989). A protein factor that enhances amsacrine-mediated formation of topoisomerase II–DNA complexes in murine mastocytoma cell nuclei. *Biochim. Biophys. Acta*, **1007**, 295–300

Darkin-Rattray, S. J. and Ralph, R. K. (1991). Evidence that a protein kinase enhances amsacrine mediated formation of topoisomerase II–DNA complexes in murine mastocytoma cell nuclei. *Biochim. Biophys. Acta*, **1088**, 285–291

D'Arpa, P., Beardmore, C. and Liu, L. F. (1990). Involvement of nucleic acid synthesis in cell killing mechanisms of topoisomerase poisons. *Cancer Res.*, **50**, 6919–6924

D'Arpa, P. and Liu, L. F. (1989). Topoisomerase-targeting antitumour drugs. *Biochim. Biophys. Acta*, **989**, 163–177

D'Arpa, P., Machlin, P. S., Ratrie, H., Rothfield, N. F., Cleveland, D. W. and Earnshaw, W. C. (1988). cDNA cloning of human DNA topoisomerase I: catalytic activity of a 67.7 kDa carboxylterminal fragment. *Proc. Natl Acad. Sci. USA*, **85**, 2543–2547

Deffie, A. M., Alam, T., Seneviratae, C., Beenken, S. W., Batra, J. K., Shea, T. C., Henner, W. D. and Goldenberg, G. J. (1988). Multifactorial resistance to adriamycin: relationships of DNA repair, glutathione transferase activity, drug efflux and P glycoprotein in cloned cell lines of adriamycin sensitive and resistant P388 leukemia. *Cancer Res.*, **48**, 3595–3602

Deffie, A. M., Bosman, D. J. and Goldenberg, G. J. (1989). Evidence for a mutant allele of the gene for DNA topoisomerase II in adriamycin-resistant P388 murine leukemia cells. *Cancer Res.*, **49**, 6879–6882

Degrassi, F., De Salvia, R., Tanzarella, C. and Palitti, F. (1989). Induction of chromosomal aberrations and SCE by camptothecin, an inhibitor of mammalian topoisomerase I. *Mutation Res.*, **211**, 125–130

De Isabella, P., Capranico, G., Binaschi, M., Tinelli, S. and Zunino, F. (1990). Evidence of DNA topoisomerase II-dependent mechanisms of multidrug resistance in P388 leukemia cells. *Mol. Pharmacol.*, **37**, 11–16

de Jong, S., Zijlstra J. G., de Vries, E. and Mulder, N. H. (1990). Reduced DNA topoisomerase II activity and drug-induced DNA cleavage activity in an adriamycin-resistant human small cell lung carcinoma cell line. *Cancer Res.*, **50**, 304–309

Del Bino, G. and Darzynkiewicz, Z. (1991). Camptothecin, teniposide or 4′-(9-acridinylamino)-3-methanesulfon-*m*-anisidide but not mitoxantrone or doxorubicin induces degradation of nuclear DNA in the S phase of HL60 cells. *Cancer Res.*, **51**, 1165–1169

Del Bino, G., Lassota, P. and Darzynkiewicz, Z. (1991). The S phase cytotoxicity of camptothecin. *Exp. Cell Res.*, **193**, 27–35

Del Bino, G., Skierski, J. S. and Darzynkiewicz, S. (1990). Diverse effects of camptothecin, an inhibitor of topoisomerase I, on the cell cycle of lymphocytic (L1210, MOLT-4) and myelogeneous (HL60, KG1) leukemic cells. *Cancer Res.*, **50**, 5746–5750

Dillehay, L. E., Denstman, S. C. and Williams, J. R. (1987). Cell cycle dependence of sister chromatid exchange induction by DNA topoisomerase II inhibitors. *Cancer Res.*, **47**, 206–210

Dillehay, L. E., Jacobson-Kram, D. and Williams, J. R. (1989). DNA topoisomerases and models of sister chromatid exchange. *Mutation Res.*, **215**, 15–23

Di Nardo, S., Voelkel, K. and Sternglanz, R. (1984). DNA topoisomerase II mutant of *Saccharomyces cerevisiae*: topoisomerase II is required for segregation of daughter molecules at the termination of DNA replication. *Proc. Natl Acad. Sci. USA*, **81**, 2616–2620

Douc-Rasy, S., Kayser, A., Riou, J. F. and Riou, G. (1986). ATP independent type II topoisomerase from trypanosomes. *Proc. Natl Acad. Sci. USA*, **83**, 7152–7156

Downes, C. S., Mullinger, A. M. and Johnson, R. T. (1987). Action of etoposide (VP-16-123) on human cells: no evidence for topoisomerase II involvement in excision repair of u.v. induced DNA damage, nor for mitochondrial hypersensitivity in ataxia telangiectasia. *Carcinogenesis*, **8**, 1613–1618

Drake, F. H., Hofmann, G. A., Bartus, H. F., Mattern, M. R., Crooke, S. T. and Mirabelli, C. K. (1989a). Biochemical and pharmacological properties of p170 and p180 forms of topoisomerase II. *Biochemistry*, **28**, 8154–8160

Drake, F. H., Hofmann, G. A., Mong, S. M., Bartus, J. O., Hertzberg, R. P., Johnson, R. K., Mattern, M. R. and Mirabelli, C. K. (1989b). *In vitro* and intracellular inhibition of topoisomerase II by the antitumor agent merbarone. *Cancer Res.*, **49**, 2578–2583

Drake, F. H., Zimmerman, J. P., McCabe, F. L., Bartus, H. F., Per, S. R., Sullivan, D. M., Ross, W. E., Mattern, M. R., Johnson, R. K., Crooke, S. T. and Mirabelli, C. K. (1987). Purification of topoisomerase II from amsacrine-resistant P388 leukemia cells. Evidence for two forms of the enzyme. *J. Biol. Chem.*, **262**, 16739–16747

Duguet, M., Lavenot, C., Harper, F., Mirambeau, G. and De Recondo, A.-M. (1983). DNA topoisomerases from rat liver: physiological variations. *Nucleic Acids Res.*, **11**, 1059–1075

Dunaway, M. (1990). Inhibition of topoisomerase II does not inhibit transcription of RNA polymerase I and II genes. *Mol. Cell. Biol.*, **10**, 2893–2900

Durban, E., Goodenough, M., Mills, J. and Busch, H. (1985). Topoisomerase I phosphorylation *in vitro* and in rapidly growing Novikoff hepatoma cells. *EMBO Jl*, **4**, 2921–2926

Ebert, S. N., Shtrom, S. S. and Muller, M. T. (1990). Topoisomerase II cleavage of *Herpes simplex* virus type I DNA *in vivo* is replication dependent. *J. Virol.*, **64**, 4059–4066

Edwards, C. M., Glisson, B. S., King, C. K., Smallwood, K. S. and Ross, W. E. (1987). Etoposide-induced DNA cleavage in human leukemia cells. *Cancer Chemother. Pharmacol.*, **20**, 162–168

Egyhazi, E. and Durban, E. (1987). Microinjection of anti topoisomerase I immunoglobulin G into nuclei of *Chironomus tentans* salivary gland cells leads to blockage of transcription elongation. *Mol. Cell. Biol.*, **7**, 4308–4316

Endicott, J. A. and Ling, E. (1989). The biochemistry of P glycoprotein-mediated multidrug resistance. *Ann. Rev. Biochem.*, **58**, 137–171

Eng, W. K., Faucette, L., Johnson, R. K. and Sternglanz, R. (1988). Evidence that DNA topoisomerase I is necessary for the cytotoxic effects of camptothecin. *Mol. Pharmacol.*, **34**, 755–760

Eng, W. K., McCabe, F. L., Tan, K. B., Mattern, M. R., Hofmann, G. A., Woessner, R. D., Hertzberg, R. P. and Johnson, R. (1990). Development of a stable camptothecin-resistant subline of P388 leukemia with reduced topoisomerase I content. *Mol. Pharmacol.*, **38**, 471–480

Epstein, R. J. and Smith, P. J. (1989). Mitogen-induced topoisomerase II synthesis precedes DNA synthesis in human breast cancer cells. *Biochem. Biophys. Res. Commun.*, **160**, 12–17

Epstein, R. J., Smith, P. J., Watson, J. V., Waters, C. and Bleehen, N. M. (1989). Oestrogen potentiates topoisomerase II-mediated cytotoxicity in an activated subpopulation of human breast cancer cells: implications for cytotoxic drug resistance in solid tumours. *Int. J. Cancer*, **44**, 501–505

Estey, E., Adlakha, R. C., Hittelman, W. N. and Zwelling, L. A. (1987). Cell cycle stage dependent variation in drug-induced topoisomerase II mediated DNA cleavage and cytotoxicity. *Biochemistry*, **26**, 4338–4344

Fairchild, C. R., Moscow, J. A., O'Brien, E. E. and Cowan, K. H. (1990). Multidrug resistance in cells transfected with human genes encoding a variant P glycoprotein and glutathione *S*-transferase-p. *Mol. Pharmacol.*, **37**, 801–809

Farache, G., Razin, S. V., Targa, F. R. and Scherrer, K. (1990). Organisation of the 3′-boundary of the chicken α globin gene domain and characterisation of a CR1-specific protein binding site. *Nucleic Acids Res.*, **18**, 401–409

Ferguson, P. J., Fisher, M. H., Stephenson, J., Li, D. H., Zhou, B. S. and Cheng, Y. C. (1988). Combined modalities of resistance in etoposide-resistant human KB cell lines. *Cancer Res.*, **48**, 5956–5964

Fernandes, D. J., Danks, M. K. and Beck, W. T. (1990). Decreased nuclear matrix DNA topoisomerase II in human leukemia cells resistant to VM-26 and mAMSA. *Biochemistry*, **29**, 4235–4241

Fernandes, D. J., Smith-Nanni, C., Paff, M. T. and Neff, T.-A. M. (1988). Effects of antileukemia agents on nuclear matrix bound DNA replication in CCRF-CEM leukemia cells. *Cancer Res.*, **48**, 1850–1855

Ferro, A. M. and Olivera, B. M. (1984). Poly(ADP) ribosylation of DNA topoisomerase I from calf thymus. *J. Biol. Chem.*, **259**, 547–554

Fesen, M. and Pommier, Y. (1989). Mammalian topoisomerase II activity is modulated by the DNA minor groove binder distamycin in simian virus 40 DNA. *J. Biol. Chem.*, **264**, 11354–11359

Fields-Berry, S. C. and De Pamphlis, M. L. (1989). Sequences that promote formation of catenated intertwines during formation of DNA replication. *Nucleic Acids Res.*, **17**, 3261–3273

Filipski, J. (1983). Competitive inhibition of nicking-closing enzymes may explain some biological effects of DNA intercalators. *FEBS Lett.*, **159**, 6–12

Filipski, J., Leblanc, J., Youdale, T., Sikorska, M. and Walker, P. R. (1990). Periodicity of DNA folding in higher order chromatin structures. *EMBO Jl*, **9**, 1319–1327

Fink, G. R. (1989). A new twist to the topoisomerase I problem. *Cell*, **58**, 225–226

Finlay, G. J. and Baguley, B. C. (1990). Potentiation by phenylbis-benzimidazoles of cytotoxicity of anticancer drugs directed against topoisomerase II. *Eur. J. Cancer Clin. Oncol.*, **26**, 586–589

Finlay, G. J., Baguley, B. C., Snow, K. and Judd, W. (1990). Multiple patterns of resistance of human leukemia cell sublines to amsacrine analogues. *J. Natl Cancer Inst.*, **82**, 662–667

Finlay, G. J., Wilson, W. R. and Baguley, B. C. (1989). Chemoprotection by 9-aminoacridine derivatives against the cytotoxicity of topoisomerase II-directed drugs. *Eur. J. Cancer Clin. Oncol.*, **25**, 1695–1701

Fleischmann, G., Pflugfelder, G., Steiner, E. K., Javahurian, K., Howard, G. C., Wang, J. C. and Elgin, S. C. R. (1984). *Drosophila* DNA topoisomerase I is associated with transcriptionally active regions of the genome. *Proc. Natl Acad. Sci. USA*, **81**, 6958–6962

Fojo, A. T., Ueda, K., Salmon, D. J., Poplack, D. G., Gottesman, M. M. and Pastan, I. (1987). Expression of a multidrug resistance gene in human tumours and tissues. *Proc. Natl Acad. Sci. USA*, **84**, 265–269

Fossé, P., Rene, B., Le, B. M., Paoletti, C. and Saucier, J. M. (1991). Sequence requirements for mammalian topoisomerase II mediated DNA cleavage stimulated by an ellipticine derivative. *Nucleic Acids Res.*, **19**, 2861–2868

Fox, M. E. and Smith, P. J. (1990). Long term inhibition of DNA synthesis and the persistence of trapped topoisomerase II complexes in determining the toxicity of the antitumour DNA intercalators mAMSA and mitoxantrone. *Cancer Res.*, **50**, 5813–5818

Francis, G. E., Berney, J. J., North, P. S., Khan, Z., Wilson, E. L., Jacobs, P. and Ali, M. (1987). Evidence for the involvement of DNA topoisomerase II in neutrophil–granulocyte differentiation. *Leukemia*, **1**, 653–659

Franco, R. J. and Drlica, K. (1989). Gyrase inhibitors can increase gyrA expression and DNA supercoiling. *J. Bacteriol.*, **171**, 6573–6579

Fukada, M. (1985). Action of camptothecin and its derivatives on deoxyribonucleic acid. *Biochem. Pharmacol.*, **34**, 1225–1230

Gallo, R. C., Whang-Peng, J. and Adamson, R. H. (1971). Studies on the antitumour activity, mechanism of action, and cell cycle effects of camptothecin. *J. Natl Cancer Inst.*, **46**, 789–795

Ganapathi, R., Grabowski, D., Ford, J., Heiss, C., Kerrigan, D. and Pommier, Y. (1989). Progressive resistance to doxorubicin in mouse leukemia L1210 cells with multidrug resistance phenotype: reductions in drug-induced topoisomerase II-mediated DNA cleavage. *Cancer Commun.*, **1**, 217–224

Garg, L. C., Di Angelo, S. and Jacob, S. T. (1987). Role of DNA topoisomerase I in the transcription of supercoiled rRNA gene. *Proc. Natl Acad. Sci. USA*, **84**, 3185–3188

Gasser, S. M., Amati, B. B., Cardenas, M. E. and Hofmann, J. F.-X. (1989). Studies on scaffold attachment sites and their relation to genome function. *Int. Rev. Cytol.*, **119**, 57–96

Gasser, S. M. and Laemmli, U. K. (1986a). Cohabitation of scaffold binding regions with upstream/enhancer elements of three developmentally regulated genes of *D. melanogaster*. *Cell*, **46**, 521–530

Gasser, S. M. and Laemmli, U. K. (1986b). The organisation of chromatin loops: characterization of a scaffold attachment site. *EMBO Jl*, **5**, 511–518

Gasser, S. M., Laroche, T., Falquet, J., Boy de la Tour, E. and Laemmli, U.K. (1986). Metaphase chromosome structure. Involvement of topoisomerase II. *J. Mol. Biol.*, **188**, 613–629

Gedik, C. M. and Collins, A. R. (1990). Comparison of effects of fostriecin, novobiocin and camptothecin inhibitors of DNA topoisomerases on DNA replication and repair in human cells. *Nucleic Acids Res.*, **18**, 1007–1014

Gellert, M. (1981). DNA topoisomerases. *Ann. Rev. Biochem.*, **50**, 879–910

Giaccone, G., Kadoyama, C., Maneckjee, R., Venzon, D., Alexander, R. B. and Gazdar, A. (1990). Effects of tumour necrosis factor, alone or in combination with topoisomerase II-targeted drugs, on human lung cancer cell lines. *Int. J. Cancer*, **46**, 326–329

Giaever, G. N. and Wang, J. C. (1988). Supercoiling of intracellular DNA can occur in eukaryotic cells. *Cell*, **55**, 849–856

Gieseler, F., Boege, F. and Clark, M. (1990). Alteration of topoisomerase II action is a possible molecular mechanism of HL60 cell differentiation. *Environ. Hlth Perspect.*, **88**, 183–186

Gilmour, D. S. and Elgin, S. C. R. (1987). Localisation of specific topoisomerase I interactions within the transcribed regions of active heat shock genes by using the inhibitor camptothecin. *Mol. Cell. Biol.*, **7**, 141–148

Gilmour, D. S., Pflugfelder, G., Wang, J. C. and Lis, J. I. (1986). Topoisomerase I interacts with transcribed regions in *Drosophila* cells. *Cell*, **44**, 401–407

Giovanella, B. C., Stehlin, J. S., Wall, M. E., Wani, M. C., Nicholas, A. W., Liu, L. F., Silber, R. and Potmesil, M. (1989). DNA topoisomerase I-targeted chemotherapy of human colon cancer in xenografts. *Science*, **246**, 1046–1048

Glikin, G. C. and Blangy, D. (1986). *In vitro* transcription by *Xenopus* oocytes. RNA polymerase III requires a DNA topoisomerase II activity. *EMBO Jl*, **5**, 151–155

Glisson, B., Gupta, R., Smallwood-Kento, S. and Ross, W. E. (1986). Characterisation of acquired epipodophyllotoxin resistance in a Chinese hamster ovary cell line: loss of drug stimulated cleavage activity. *Cancer Res.*, **46**, 1934–1938

Gorsky, L. D., Cross, S. M. and Morin, M. J. (1989). Rapid increase in the activity of DNA topoisomerase I but not topoisomerase II, in HL60 promyelocytic leukemia cells treated with a phorbol ester. *Cancer Commun.*, **1**, 83–92

Goto, T. and Wang, J. C. (1985). Cloning of yeast topo I, the gene encoding DNA topoisomerase I, and construction of mutants defective in both DNA topoisomerase I and DNA topoisomerase II. *Proc. Natl Acad. Sci. USA*, **82**, 7178–7182

Gottesfield, J. M. (1986). Novobiocin inhibits RNA polymerase III transcription *in vitro* by a mechanism distinct from DNA topoisomerase II. *Nucleic Acids Res.*, **14**, 2075–2087

Gottlieb, J. A. and Luce, J. K. (1972). Treatment of malignant melanoma with camptothecin (NSC-100880). *Cancer Chemother. Rep.*, **56**, 103–105

Grieder, A., Maurer, R. and Stahlin, H. (1974). Effect of an epipodophyllotoxin derivative (VP-16-213) on macromolecular synthesis and mitosis in mastocytoma cells *in vitro*. *Cancer Res.*, **34**, 1788–1793

Gupta, R. S., Gupta, R., Eng, B., Lock, R. B., Ross, W. E., Hertzberg, M. J. and Johnson, R. K. (1988). Camptothecin-resistant mutants of Chinese hamster ovary cells containing a resistant form of topoisomerase I. *Cancer Res.*, **48**, 6404–6410

Hall, F. L. and Vulliet, P. R. (1991). Proline-directed protein phosphorylation and cell cycle regulation. *Curr. Opinion Cell Biol.*, **3**, 176–184

Halligan, B. D., Davis, J. L., Edwards, K. A. and Liu, L. F. (1982). Intra- and intermolecular strand transfer by HeLa DNA topoisomerase I. *J. Biol. Chem.*, **257**, 3995–4000

Halligan, B. D., Edwards, K. A. and Liu, L. F. (1985). Purification and characterization of a type II DNA topoisomerase from bovine calf thymus. *J. Biol. Chem.*, **260**, 2475–2482

Hamada, H., Okochi, E., Watanabe, M., Oh-hara, T., Sugimoto, Y., Kawabata, H. and Tsurno, T. (1988). Mr 85,000 membrane protein specifically expressed in adriamycin-resistant human tumour cells. *Cancer Res.*, **48**, 7082–7087

Hamlin, J. L., Vaughn, J. P., Dijkwel, P. A., Leu, T.-H. and Ma, C. (1991). Origins of replication timing and chromosomal position. *Curr. Opinion Cell Biol.*, **3**, 414–421

Hancock, R., Charron, M., Lambert, H., Lemieux, M., Pankov, R. and Pépin, N. (1989). Topoisomerase II as a target of antitumour agents. In Borowski, E. and Shugar, D. (Eds), *Molecular Aspects of Chemotherapy*, Pergamon Press, Oxford, pp. 119–137

Harker, W. G., Slade, D. L., Dalton, W. S., Meltzer, P. S. and Trent, J. M. (1989). Multidrug resistance in mitoxantrone-selected HL-60 leukemia cells in the absence of P glycoprotein overexpression. *Cancer Res.*, **49**, 4542–4549

Heartlein, M. W., Tsuji, H. and Latt, S. A. (1987). 5-Bromodeoxyuridine-dependent increase in sister chromatid exchange formation in Bloom's syndrome is associated with reduction in topoisomerase II activity. *Exp. Cell Res.*, **169**, 245–254

Heck, M. M. S. and Earnshaw, W. C. (1986). Topoisomerase II: a specific marker for cell proliferation. *J. Cell. Biol.*, **103**, 2569–2581

Heck, M. M. S., Hittelman, W. N. and Earnshaw, W. C. (1988). Differential expression of DNA topoisomerases I and II during the eukaryotic cell cycle. *Proc. Natl Acad. Sci. USA*, **85**, 1086–1090

Heck, M. M. S., Hittelman, W. N. and Earnshaw, W. C. (1989). *In vivo* phosphorylation of the 170 kDa form of eukaryotic DNA topoisomerase II cell cycle analysis. *J. Biol. Chem.*, **264**, 15161–15164

Heller, R. A., Shelton, E. R., Dietrich, V., Elgin, S. C. and Brutlag, D. L. (1986). Multiple forms and cellular localisation of *Drosophila* DNA topoisomerase II. *J. Biol. Chem.*, **261**, 8063–8069

Hertzberg, R. P., Busby, R. W., Caranfa, M. J., Holden, K. G., Johnson, R. K., Hecht, S. M. and Kingsbury, W. D. (1990). Irreversible trapping of the DNA-topoisomerase I covalent complex. *J. Biol. Chem.*, **265**, 19287–19295

Hertzberg, R. P., Caranfa, M. J. and Hecht, W. N. (1989a). On the mechanism of topoisomerase I inhibition by camptothecin: evidence for binding to an enzyme–DNA complex. *Biochemistry*, **28**, 4629–4638

Hertzberg, R. P., Caranfa, M. J., Holden, K. G., Jakas, D. R., Gallagher, G., Mattern, M. R., Mong, S. M., Bartus, J. O., Johnson, R. K. and Kingsbury, W. D. (1989b). Modification of the hydroxy lactone ring of camptothecin: inhibition of mammalian topoisomerase I and biological activity. *J. Med. Chem.*, **32**, 715–720

Hill, B. T., Whelan, R. D., Hosking, L. K., Shellard, S. A., Hinds, M. D., Mayes, J. and Zwelling, L. A. (1991). A lack of detectable modification of topoisomerase II activity in a series of human tumour cell lines expressing only low levels of etoposide resistance. *Int. J. Cancer*, **47**, 899–902

Hinds, M., Deisseroth, K., Mayes, J., Attschuler, E., Jansen, R., Ledley, F. D. and Zwelling, L. A. (1991) Identification of a point mutation in the topoisomerase II gene from a human leukemia cell line containing an amsacrine-resistant form of topoisomerase II. *Cancer Res.*, **51**, 4729–4731

Hino, O., Ohtake, K. and Rogler, C. E. (1989). Features of two Hepatitis B virus (HBV) DNA integrations suggest mechanisms of HBV integration. *J. Virol.*, **63**, 2638–2643

Hirose, S. and Suzuki, Y. (1988). *In vitro* transcription of eukaryotic genes is affected differently by the degree of DNA supercoiling. *Proc. Natl Acad. Sci. USA*, **85**, 718–722

Hogan, A. and Faust, E. A. (1986). Non-homologous recombination in the Parvovirus chromosome. *Mol. Cell. Biol.*, **6**, 3005–3009

Holden, J. A., Rolfson, D. H. and Wittwer, C. T. (1990). Human DNA topoisomerase II: evaluation of enzyme activity in normal and neoplastic tissues. *Biochemistry*, **29**, 2127–2134

Holm, C., Covey, J. T., Kerrigan, D. and Pommier, Y. (1989a). Differential requirement of DNA replication for the cytotoxicity of DNA topoisomerase I and II inhibitors in Chinese hamster DC3F cells. *Cancer Res.*, **49**, 6365–6368

Holm, C., Goto, T., Wang, J. C. and Botstein, D. (1985). DNA topoisomerase II is required at the time of mitosis in yeast. *Cell*, **41**, 553–563

Holm, C., Stearns, T. and Botstein, D. (1989b). DNA topoisomerase II must act at mitosis to prevent non-disjunction and chromosome breakage. *Mol. Cell Biol.*, **9**, 159–168

Hong, J. H. (1989). DNA topoisomerase: the mechanism of resistance to DNA topoisomerase II inhibitor VP-16. *Hiroshima J. Med. Sci.*, **38**, 197–207

Howard, M. T., Lee, M. P., Hsieh, T. S. and Griffith, J. D. (1991). *Drosophila* topoisomerase II-DNA interactions are affected by DNA structure. *J. Mol.Biol.*, **217**, 53–62

Hsiang, Y.-H., Hertzberg, R., Hecht, S. and Liu, L. F. (1985). Camptothecin induces protein-linked DNA breaks via mammalian DNA topoisomerase I. *J. Biol. Chem.*, **260**, 14873–14878

Hsiang, Y.-H., Lihou, M. G. and Liu, L. F. (1989a). Mechanism of cell killing by camptothecin: arrest of replication forks by drug-stabilized topoisomerase I-DNA cleavable complexes. *Cancer Res.*, **49**, 5077–5082

Hsiang, Y.-H. and Liu, L. F. (1988). Identification of mammalian DNA topoisomerase I as an intracellular target of the anticancer drug camptothecin. *Cancer Res.*, **48**, 1722–1726

Hsiang, Y.-H., Liu, L. F., Wall, M. E., Wani, M. C., Nicholas, A. W., Manikumar, G., Kirschenbaum, S., Silber, R. and Potmesil, M. (1989b). DNA topoisomerase I-mediated DNA cleavage and cytotoxicity of camptothecin analogues. *Cancer Res.*, **49**, 4385–4389

Hsiang, Y.-H., Wu, H. Y. and Liu, L. F. (1988). Proliferation-dependent regulation of DNA topoisomerase II in cultured human cells. *Cancer Res.*, **48**, 3230–3235

Hsieh, T. (1990). DNA topoisomerases. *Curr. Opinion Cell Biol.*, **2**, 461–463

Huang, H. W., Juang, J. K. and Liu, H. J. (1992). The recognition of DNA cleavage sites by porcine spleen topoisomerase II. *Nucleic Acids Res.*, **20**, 467–473

Hwong, C.-L., Chen, M.-S. and Hwang, J. (1989). Phorbol ester transiently increases topoisomerase I mRNA levels in human skin fibroblasts. *J. Biol. Chem.*, **264**, 14923–14926

Hwong, C.-L., Wang, C. H., Chen, Y. J., Whang, P. J. and Hwang, J. C. (1990). Induction of topoisomerase II gene expression in human lymphocytes upon

phytohemagglutinin stimulation. *Cancer Res. (Suppl.)*, **50**, 5649–5652
Ikeda, H. (1986). Illegitimate recombination: role of type II DNA topoisomerase. *Adv. Biophys.*, **21**, 149–162
Jarman, A. P. and Higgs, D. R. (1988). Nuclear scaffold attachment sites in the human globin gene complexes. *EMBO Jl*, **7**, 3337–3344
Jaxel, C., Capranico, G., Kerrigan, D., Kohn, K. W. and Pommier, Y. (1991a). Effect of local DNA sequence on topoisomerase I cleavage in the presence or absence of camptothecin. *J. Biol. Chem.*, **266**, 20418–20423
Jaxel, C., Capranico, G., Wasserman, K., Kerrigan, D., Kohn, K. W. and Pommier, Y. (1991b). DNA sequence at sites of topoisomerase I cleavage induced by camptothecin in SV40 DNA. In Potmesil, M. and Kohn, K. W. (Eds), *DNA Topoisomerases in Cancer*. Oxford University Press, New York, pp. 182–195
Jaxel, C., Kohn, K. W. and Pommier, Y. (1988a). Topoisomerase I interaction with SV40 DNA in the presence and absence of camptothecin. *Nucleic Acids Res.*, **16**, 11157–11170
Jaxel, C., Kohn, K. W., Wani, M. C., Wall, M. E. and Pommier, Y. (1989). Structure–activity study of the actions of camptothecin derivatives on mammalian topoisomerase I. Evidence for a specific receptor site and a relation to antitumor activity. *Cancer Res.*, **49**, 1465–1469
Jaxel, C., Taudou, G., Portemer, C., Mirambeau, G., Panijel, J. and Duguet, M. (1988b). Topoisomerase inhibitors induce irreversible fragmentation of replicated DNA in concanavalin A stimulated splenocytes. *Biochemistry*, **27**, 95–99
Jensen, P. B., Sørensen, B. S., Demant, E. J. F., Sehested, M., Jensen, P. S., Vindelov, L. and Hansen, H. H. (1990). Antagonistic effect of aclarubicin on the cytotoxicity of etoposide and 4′-(9-acridinylamino)-methansulfon-*m*-anisidide in human small cell lung cancer cell lines and on topoisomerase II mediated DNA cleavage. *Cancer Res.*, **50**, 3311–3316
Jongstra-Bilen, J., Ittel, M. E., Niedergang, C., Vosberg, H. P. and Mandel, P. (1983). DNA topoisomerase I from calf thymus is inhibited *in vitro* by poly(ADP) ribosylation. *Eur. J. Biochem.*, **136**, 391–396
Juan, C.-C., Hwang, J., Liu, A. A., Whang-Peng, J., Knutsen, T., Huebner, K., Croce, C. M., Zhang, H., Wang, J. C. and Liu, L. F. (1988). Human DNA topoisomerase I is encoded by a single-copy gene that maps to chromosome 20q 12-13.2. *Proc. Natl Acad. Sci. USA*, **85**, 8910–8913
Kaiserman, H. B., Ingebritsen, T. S. and Benbow, R. M. (1988). Regulation of *Xenopus laevis* DNA topoisomerase I activity by phosphorylation *in vitro*. *Biochemistry*, **27**, 3126–3222
Kamath, N., Grabowski, D., Ford, J., Drake, F., Kerrigan, D., Pommier, Y. and Ganapathi, R. (1991). Trifluoperazine modulation of resistance to the topoisomerase II inhibitor etoposide in doxorubicin resistant L1210 murine leukemia cells. *Cancer Commun.*, **3**, 37–44
Kampinga, H. H., van den Kruk, G. and Konings, A. W. (1989). Reduced DNA break formation and cytotoxicity of the topoisomerase II drug 4′-(9′-acridinylamino)methanesulfon-*m*-anisidide when combined with hyperthermia in human and rodent cell lines. *Cancer Res.*, **49**, 1712–1717
Kann, H. E. and Kohn, K. W. (1972). Effects of DNA-reactive drugs on RNA synthesis patterns in leukemia L1210 cells. *Mol. Pharmacol.*, **8**, 551–560
Kanter, P. M. and Schwartz, H. S. (1982). A fluorescence enhancement assay for cellular DNA damage. *Mol. Pharmacol.*, **22**, 145–154
Kanzawa, F., Sugimoto, Y., Minato, K., Kasahara, K., Bungo, M., Nakagawa, K., Fujiwara, Y., Liu, L. F. and Saijo, N. (1990). Establishment of a camptothecin analogue (CPT-11)-resistant cell line of human non-small cell lung cancer: char-

acterization and mechanism of resistance. *Cancer Res.*, **50**, 5919–5924

Käs, E. and Chasin, L. A. (1987). Anchorage of the Chinese hamster dihydrofolate reductase gene to the nuclear scaffold occurs in an intra-genic region. *J. Mol. Biol.*, **198**, 677–692

Käs, E. and Laemmli, U. K. (1992). *In vivo* topoisomerase II cleavage of the *Drosophila* histone and satellite III repeats: DNA sequence and structural characteristics. *EMBO Jl*, **11**, 705–716

Kasid, U. N., Halligan, B., Liu, L. F., Dritschlo, A. and Smulson, M. (1989). Poly(ADP-ribose) mediated post-translational modification of chromatin-associated human topoisomerase I. Inhibitory effects on catalytic activity. *J. Biol. Chem.*, **264**, 18687–18692

Kaufmann, S. H. (1989). Induction of endonucleolytic DNA cleavage in human acute myelogenous leukemia cells by etoposide, camptothecin and other cytotoxic anticancer drugs: a cautionary note. *Cancer Res.*, **49**, 5870–5878

Kaufmann, S. H. (1991). Antagonism between camptothecin and topoisomerase II-directed chemotherapeutic agents in a human leukemia cell line. *Cancer Res.*, **51**, 1129–1136

Kaufmann, S., Brunet, G., Talbot, B., Lamarr, D., Dumas, C., Shaper, J. H. and Poirier, G. (1991a). Association of poly(ADP-ribose) polymerase with the nuclear matrix: the role of intermolecular disulfide bond formation, RNA retention and cell type. *Exp. Cell Res.*, **192**, 511–523

Kaufmann, S. H. and Shaper, J. (1991). Association of topoisomerase II with the hepatoma cell nuclear matrix: the role of intermolecular disulphide bond formation. *Exp. Cell Res.*, **192**, 511–523

Kaufmann, W. K., Boyer, J. C., Estabrooks, L. L. and Wilson, S. J. (1991b). Inhibition of replicon initiation in human cells following stabilisation of topoisomerase-DNA cleavable complexes. *Mol. Cell Biol.*, **11**, 3711–3718

Kawada, S., Yamashita, Y., Fujii, N. and Nakano, H. (1991). Induction of a heat-stable topoisomerase II-DNA cleavable complex by non-intercalative terpenoids, terpentecin and clerocidin. *Cancer Res.*, **51**, 2922–2925

Kessel, D., Bosmann, H. B. and Lohr, K. (1972). Camptothecin effects on DNA synthesis in murine leukemia cells. *Biochim. Biophys. Acta*, **269**, 210–216

Kessel, D. and Wheeler, C. (1984). mAMSA as a probe for transport phenomena associated with anthracycline resistance. *Biochem. Pharmacol.*, **33**, 991–994

Kiguchi, K., Constantinou, A. I. and Huberman, E. (1990). Genistein induced cell differentiation and protein-linked DNA strand breakage in human melanoma cells. *Cancer Commun.*, **2**, 271–277

Kim, R., Hirabayashi, N., Nishiyama, M., Saeki, S., Toge, T. and Okada, K. (1991). Expression of MDR1, GST-pi and topoisomerase II as an indicator of clinical response to adriamycin. *Anticancer Res.*, **11**, 429–431

Kim, R. A. and Wang, J. C. (1989a). A subthreshold level of DNA topoisomerase leads to the excision of yeast rDNA as extrachromosomal rings. *Cell*, **57**, 975–985

Kim, R. A. and Wang, J. C. (1989b). Function of DNA topoisomerases as replication swivels in *Saccharomyces cerevisiae*. *J. Mol. Biol.*, **208**, 257–267

Kingsbury, W. D., Boehm, J. C., Jakas, D. R., Holden, K. G., Hecht, S. M., Gallagher, G., Caranfa, M. J., McCabe, F., Faucette, L. F., Johnson, R. K. and Hertzberg, R. P. (1991). Synthesis of water-soluble (aminoalkyl)camptothecin analogues: inhibition of topoisomerase I and antitumor activity. *J. Med. Chem.*, **34**, 98–107

Kjeldsen, E., Bonven, B., Andoh, T., Ishii, K., Okada, K., Bolund, L. and Westergaard, O. (1988a). Characterization of a camptothecin-resistant human DNA topoisomerase I. *J. Biol. Chem.*, **263**, 3912–3916

Kjeldsen, E., Mollerup, S., Thomsen, B., Bonven, B. J., Bolund, L. and Westergaard, O. (1988b). Sequence-dependent effect of camptothecin on human topoisomerase I DNA cleavage. *J. Mol. Biol.*, **202**, 333–342

Kohn, K. W., Ewig, R. A. G., Erickson, L. A. and Zwelling, L. A. (1981). Measurement of strand breaks and crosslinks by alkaline elution. In Friedberg, E. C. and Hanawalt, P. C. (Eds), *DNA Repair: A Laboratory Manual of Research Techniques*, Marcel Dekker, New York, pp. 379–401

Kohn, K. W., Pommier, Y., Kerrigan, D., Markovits, J., and Covey, J. M. (1987). Topoisomerase II as a target of anticancer drug action in mammalian cells. *N. C. I. Monograph*, **4**, 61–71

Konopka, A. K. (1988). Compilation of DNA strand exchange sites for non-homologous recombination in somatic cells. *Nucleic Acids Res.*, **16**, 1739–1758

Kosovsky, M. J. and Soslau, G. (1991). Mitochondrial DNA topoisomerase I from human platelets. *Biochim. Biophys. Acta*, **1078**, 56–62

Kroeger, P. E. and Rowe, T. C. (1989). Interaction of topoisomerase I with the transcribed region of *Drosophila* HSP70 heat shock gene. *Nucleic Acids Res.*, **17**, 8495–8509

Krogh, S., Mortensen, U. H., Westergaard, O. and Bonven, B. J. (1991). Eukaryotic topoisomerase I-DNA interaction is stabilized by helix curvature. *Nucleic Acids Res.*, **19**, 1235–1241

Kroll, D. J. and Rowe, T. C. (1991). Phosphorylation of DNA topoisomerase II in a human tumour cell line. *J. Biol. Chem.*, **266**, 7957–7961

Kubo, M., Kano, Y., Nakamura, H., Nagata, A. and Imamoto, F. (1979). *In vivo* enhancement of general and specific transcription in *E. coli* by DNA gyrase activity. *Gene*, **7**, 153–171

Kunze, N., Yang, G., Dölberg, M., Sundarp, R., Knippers, R. and Richter, A. (1991). Structure of the human type I DNA topoisomerase gene. *J. Biol. Chem.*, **266**, 9610–9616

Kunze, N., Yang, G. C., Jiang, Z. Y., Hameister, H., Adolph, S., Wiedorn, K.-H., Richter, A. and Knippers, R. (1989). Localization of the active type I DNA topoisomerase gene on human chromosome 20q 11.2-13.1 and two pseudogenes on chromosomes 1q 23-24 and 22q 11-13.1. *Human Genet.*, **84**, 6–10

Kupfer, G., Bodley, A. L. and Liu, L. F. (1987). Involvement of intracellular ATP in cytotoxicity of topoisomerase II targetting antitumour drugs. *N. C. I. Monograph*, **4**, 37–40

Kuwahara, J., Suzuki, T., Funakoshi, K. and Sugiura, Y. (1986). Photosensitive DNA cleavage and phage inactivation by copper(II)-camptothecin. *Biochemistry*, **25**, 1216–1221

Kyprianou, N., Alexander, R. B. and Isaacs, J. (1991). Activation of programmed cell death by recombinant human tumour necrosis factor plus topoisomerase II-targeted drugs in L929 tumour cells. *J. Natl Cancer Inst.*, **83**, 346–350

Lake, R. A., Wotton, D. and Owen, M. (1990). A 3′-transcriptional enhancer regulates tissue-specific expression of the human CD2 gene. *EMBO Jl*, **9**, 3129–3136

Lassota, P., Melamed, M. R. and Darzynkiewicz, Z. (1991). Release of specific proteins from nuclei of HL60 and MOLT-4 cells by antitumour drugs having affinity for nucleic acids. *Biochem. Pharmacol.*, **41**, 1055–1065

Lawrence, T. S., Canman, C. E., Maybaum, J. and Davis, M. A. (1989). Dependence of etoposide-induced cytotoxicity and topoisomerase II-mediated DNA strand breakage on the intracellular ionic environment. *Cancer Res.*, **49**, 4775–4779

Lawrence, T. S. and Davis, M. A. (1990). The influence of Na+, K(+)-pump blockade on doxorubicin-mediated cytotoxicity and DNA strand breakage in human tumour cells. *Cancer Chemother. Pharmacol.*, **26**, 163–167

Lazarus, G. M., Henrich, J. P., Kelly, W. G., Schmitz, S. A. and Castora, F. J. (1987). Purification and characterization of a type I DNA topoisomerase from calf thymus mitochondria. *Biochemistry*, **26**, 6195–6203

Lee, M. P., Sander, M. and Hsieh, T. (1989). Nuclease protection by *Drosophila* DNA topoisomerase II. Enzyme–DNA contacts at the strong topoisomerase II cleavage sites. *J. Biol. Chem.*, **264**, 21779–21787

Lefevre, D., Riou, J. F., Ahomadegbe, J. C., Zhou, D. Y., Benard, J. and Riou, G. (1991). Study of molecular markers of resistance to *m*-AMSA in a human breast cancer cell line. Decrease of topoisomerase II and increase of both topoisomerase I and acidic glutathione *S* transferase. *Biochem. Pharmacol.*, **41**, 1967–1979

Legouy, E., Fossar, N., Lhomond, G. and Brison, O. (1989). Structure of four amplified DNA novel joints. *Somatic Cell. Mol. Genet.*, **15**, 309–320

Levy, W. B. and Fortier, C. (1989). The limits of the DNAse 1-sensitive domain of the human apolipoprotein B gene coincide with the locations of chromosomal anchorage loops and define the 5′ and 3′ boundaries of the gene. *J. Biol. Chem.*, **264**, 21196–21204

Lewin, B. (1990). Driving the cell cycle: M phase kinase, its partners and substrates. *Cell*, **61**, 743–752

Li, G. C. (1987). Heat shock proteins: role in thermotolerance, drug resistance, and relationship to DNA topoisomerases. *N. C. I. Monograph*, **4**, 99–103

Li, L. H., Fraser, T. J., Olin, E. J. and Bhuyan, B. K. (1972). Action of camptothecin on mammalian cells in culture. *Cancer Res.*, **32**, 2643–2650

Lim, M., Liu, L. F., Jacobson-Kram, D. and Williams, J. R. (1986). Induction of sister chromatid exchanges by inhibitors of topoisomerases. *Cell Biol. Toxicol.*, **2**, 485–494

Lin, J.-H. and Castora, F. J. (1991). DNA topoisomerase II from mammalian mitochondria is inhibited by the antitumour drugs mAMSA and VM-26. *Biochem. Biophys. Res. Commun.*, **176**, 690–697

Ling, Y. H., Tseng, M. and Nelson, J. A. (1991). Differentiation induction of human promyelocytic leukemia cells by 10-hydroxyl-camptothecin, a DNA topoisomerase I inhibitor. *Differentiation*, **46**, 135–142

Liu, L. F. (1983). DNA-topoisomerases—enzymes that catalyse the breaking and rejoining of DNA. *CRC Crit. Rev. Biochem.*, **15**, 1–24

Liu, L. F. (1989). DNA topoisomerase poisons as antitumour drugs. *Ann. Rev. Biochem.*, **58**, 351–375

Liu, L. F. and Miller, K. G. (1981). Eukaryotic DNA topoisomerases. Two forms of type I DNA topoisomerases from Hela cell nuclei. *Proc. Natl Acad. Sci. USA*, **78**, 3487–3491

Liu, L. F. and Wang, J. C. (1987). Supercoiling of the DNA template during transcription. *Proc. Natl Acad. Sci. USA*, **84**, 7024–7027

Ljungman, M. and Hanawalt, P. C. (1992). Localized torsional tension in the DNA of human cells. *Proc. Natl Acad. Sci. USA*, **89**, 6055–6059

Lock, R. B. and Ross, W. E. (1987). DNA topoisomerases in cancer therapy. *Anti-cancer Drug Res.*, **2**, 151–165

Lock, R. B. and Ross, W. E. (1990). Inhibition of $p34^{cdc2}$ kinase activity by etoposide or irradiation as a mechanism of G2 arrest in Chinese hamster ovary cells. *Cancer Res.*, **50**, 3761–3766

Lodge, J. K., Kazic, T. and Berg, D. E. (1989). Formation of supercoiling domains in plasmid pBR322. *J. Bacteriol.*, **171**, 2181–2187

Lönn, U., Lönn, S., Nylen, U. and Winblad, G. (1989). Altered formation of DNA in human cells treated with inhibitors of DNA topoisomerase II (etoposide and teniposide). *Cancer Res.*, **49**, 6202–6207

Lorico, A., Boiocchi, M., Rappa, G., Sen, S., Erba, E. and D'Incalci, M. (1990). Increase in topoisomerase II-mediated DNA breaks and cytotoxicity of VP-16 in human U937 lymphoma cells pretreated with low doses of methotrexate. *Int. J. Cancer*, **45**, 156–162

Lund, K., Andersen, A. H., Christiansen, K., Svejstrup, J. Q. and Westergaard, O. (1990). Minimal DNA requirement for topoisomerase II-mediated cleavage *in vitro*. *J. Biol. Chem.*, **265**, 13856–13863

Lynn, R. M., Bjornsti, M. A., Caron, P. R. and Wang, J. C. (1989). Peptide sequencing and site-directed mutagenesis identify tyrosine-727 as the active site tyrosine of *Saccharomyces cerevisiae* DNA topoisomerase I. *Proc. Natl Acad. Sci. USA*, **86**, 3559–3563

McCoubrey, W. K. and Champoux, J. J. (1986). The role of single-strand breaks in the catenation reaction catalysed by the rat type I topoisomerase. *J. Biol. Chem.*, **261**, 5130–5137

McHugh, M. M., Woynarowski, J. M., Sigmund, R. D. and Beerman, T. A. (1989). Effect of minor groove binding drugs on mammalian topo-isomerase I activity. *Biochem. Pharmacol.*, **38**, 2323–2328

McLane, K. E., Fisher, J. and Ramakrishnan, K. (1983). Reductive drug metabolism. *Drug Metab. Rev.*, **14**, 741–770

Markovits, J., Linassier, C., Fossé, P., Coupie, J., Pierre, J., Jacquemin, S. A., Saucier, J. M., Le, P. J. and Larsen, A. K. (1989). Inhibitory effects of the tyrosine kinase inhibitor genistein on mammalian DNA topoisomerase II. *Cancer Res.*, **49**, 5111–5117

Markovits, J., Pommier, Y., Kerrigan, D., Covey, J. M., Tilchen, E. J. and Kohn, W. K. (1987). Topoisomerase II-mediated DNA breaks and cytotoxicity in relation to cell proliferation and the cell cycle in NIH 3T3 fibroblasts and L1210 leukemia cells. *Cancer Res.*, **47**, 2050–2055

Marquardt, D., McCrone, S. and Center, M. S. (1990). Mechanisms of multidrug resistance in HL-60 cells: detection of resistance associated proteins with antibodies against synthetic peptides that correspond to the deduced sequence of P glycoprotein. *Cancer Res.*, **50**, 1426–1430

Marsh, W. and Center, M. S. (1987). Adriamycin resistance in HL-60 cells and accompanying modification of a surface membrane protein contained in drug sensitive cells. *Cancer Res.*, **47**, 5080–5086

Marshall, B. and Ralph, R. K. (1985). The mechanism of action of mAMSA. *Adv. Cancer Res.*, **44**, 267–293

Matsuo, K., Kohno, K., Takano, H., Sato, S., Kiue, A. and Kuwano, M. (1990). Reduction of drug accumulation and DNA topoisomerase II activity in acquired teniposide-resistant human cancer KB cell lines. *Cancer Res.*, **50**, 5819–5824

Mattern, M. R., Mong, S. M., Bartus, H. F., Mirabelli, C. K., Crooke, S. T. and Johnson, R. K. (1987). Relationships between the intracellular effects of camptothecin and the inhibition of DNA topoisomerase I in cultured L1210 cells. *Cancer Res.*, **47**, 1793–1798

Mattern, M. R., Mong, S., Mong, S. M., Bartus, J. O., Sarau, H. M., Clark, M. A., Foley, J. J. and Crooke, S. T. (1990). Transient activation of topoisomerase I in leukotriene D4 signal transduction in human cells. *Biochem. J.*, **265**, 101–107

Mattern, M. R., Tan, K. B., Zimmerman, J. P., Mong, S. M., Bartus, J. O., Hofmann, G. A., Drake, F. H., Johnson, R. K., Crooke, S. T. and Mirabelli,

C. K. (1989). Evidence for the participation of topoisomerase I and II in cadmium-induced metallothionein expression in Chinese hamster ovary cells. *Anti-cancer Drug Des.*, **4**, 107–124

Matthes, E., Langan, P., Brachwitz, H., Schroder, H. C., Maidhof, A., Weiler, B. E., Renneisen, K. and Muller, W. A. (1990). Alteration of DNA topoisomerase II activity during infection of H9 cells by human immunodeficiency virus type I *in vitro*: a target for potential therapeutic agents. *Antiviral Res.*, **13**, 273–286

Maul, G. G., French, B. T., Van, V. W. and Jimenez, S. A. (1986). Topoisomerase I identified by scleroderma 70 antisera: enrichment of topoisomerase I at the centromere in mouse mitotic cells before anaphase. *Proc. Natl Acad. Sci. USA*, **83**, 5145–5149

Maxwell, A. and Gellert, M. (1986). Mechanistic aspects of DNA topoisomerases. *Adv. Prot. Chem.*, **38**, 69–103

Meijer, C., Mulder, N. H., Timmer-Bosscha, H., Zijlstra, J. G. and de Vries, E. G. E. (1987). Role of free radicals in an adriamycin-resistant human small cell lung cancer cell line. *Cancer Res.*, **47**, 4613–4617

Meisner, H. and Czech, M. P. (1991). Phosphorylation of transcription factors and cell-cycle-dependent proteins by casein kinase II. *Curr. Opinion Cell Biol.*, **3**, 474–483

Melendy, T. and Ray, D. S. (1989). Novobiocin affinity purification of a mitochondrial type II topoisomerase from the trypanosomatid *Crithidia fasciculata*. *J. Biol. Chem.*, **264**, 1870–1876

Miller, K. G., Liu, L. F. and Englund, P. T. (1981). A homogeneous type II DNA topoisomerase from Hela cell nuclei. *J. Biol. Chem.*, **256**, 9334–9339

Minato, K., Kanzawa, F., Nishio, K., Nakagawa, K., Fujiwara, Y. and Saijo, N. (1990). Characterization of an etoposide-resistant human small-cell lung cancer line. *Cancer Chemother. Pharmacol.*, **26**, 313–317

Mirkovitch, J., Gasser, S. M. and Laemmli, U. K. (1988). Scaffold attachment of DNA loops in metaphase chromosomes. *J. Mol. Biol.*, **200**, 101–110

Mirkovitch, J., Spierer, P. and Laemmli, U. K. (1986). Genes and loops in 320,000 base pairs of the *Drosophila melanogaster* chromosome. *J. Mol. Biol.*, **190**, 255–258

Miskimins, R., Miskimins, W. K., Bernstein, H. and Shimizu, N. (1983). Epidermal growth factor-induced topoisomerase(s). *Exp. Cell Res.*, **146**, 53–62

Misra, N. C. and Roberts, D. (1975). Inhibition by 4′-demethyl-epipodophyllotoxin-9-(4,6-0-2-thenylidene-*b*-D-glucopyranoside) of human lymphoblast cultures in G2 phase of the cell cycle. *Cancer Res.*, **35**, 99–105

Mita, S., Rizzuto, R., Moraes, C. T., Shanske, S., Arnaudo, E., Fabrizi, G. M., Koga, Y., Di Mauro, S. and Schon, E. A. (1990). Recombination via flanking direct repeats is a major cause of large scale deletions of human mitochondrial DNA. *Nucleic Acids Res.*, **18**, 561–567

Mizutani, M., Ohta, T., Watanabe, H., Handa, H. and Hirose, S. (1991). Negative supercoiling of DNA facilitates an interaction between transcription factor IID and the fibroin gene promoter. *Proc. Natl Acad. Sci. USA*, **88**, 718–722

Moertel, C. G., Schutt, A. J., Reitemeier, R. J. and Hahn, R. G. (1972). Phase II study of camptothecin (NSC-100880) in the treatment of advanced gastrointestinal cancer. *Cancer Chemother. Rep.*, **56**, 95–101

Monnot, M., Mauffret, O., Simon, V., Lescot, E., Psaume, B., Saucier, J. M., Charra, M., Belehradek, J, Jr., and Fermandjian, S. (1991). DNA–drug recognition and effects on topoisomerase II-mediated cytotoxicity. A three-mode binding model for ellipticine derivatives. *J. Biol. Chem.*, **266**, 1820–1829

Moreno, S. and Nurse, P. (1990). Substrates for p34[cdc2]: *in vivo veritas? Cell*, **61**, 549–551

Muggia, F. M., Creaven, P. J., Hansen, H. H., Cohen, M. H. and Selawry, O. S. (1972). Phase I clinical trial of weekly and daily treatment with camptothecin (NSC-100880). Correlation with preclinical studies. *Cancer Chemother.*, **56**, 515–521

Muller, M. T., Pfund, W. P., Mehta, V. B. and Trask, D. K. (1985). Eukaryotic type 1 topoisomerase is enriched in the nucleolus and catalytically active on ribosomal RNA. *EMBO Jl*, **4**, 1237–1243

Naito, M., Hamada, H. and Tsuruo, T. (1988). ATP/$Mg^{2+}$-dependent binding of vincristine to the plasma membrane of multidrug-resistant K-562 cells. *J. Biol. Chem.*, **263**, 11887–11891

Nakaya, K., Chou, S., Kaneko, M. and Nakamura, Y. (1991). Topoisomerase inhibitors have potent differentiation-inducing activity for human and mouse myeloid leukemia cells. *Jpn J. Cancer Res.*, **82**, 184–191

Neil, G. L. and Homan, E. R. (1973). The effects of dose interval on the survival of L1210 leukemic mice treated with DNA synthesis inhibitors. *Cancer Res.*, **33**, 895–901

Nelson, E. M., Tewey, K. M. and Liu, L. F. (1984). Mechanism of antitumour drug action: poisoning of mammalian DNA topoisomerase II on DNA by 4′-(9-acridinylamino)methanesulfon-*m*-anisidide. *Proc. Natl Acad. Sci. USA*, **81**, 1361–1365

Nelson, W. G., Cho, K. R., Hsiang, Y.-H., Liu, L. F. and Coffey, D. S. (1987). Growth-related elevations of DNA topoisomerase II levels found in Dunning R3327 rat prostatic adenocarcinomas. *Cancer Res.*, **47**, 3246–3250

Nelson, W. G., Liu, L. F. and Coffey, D. S. (1986). Newly replicated DNA is associated with DNA topoisomerase II in cultured rat prostatic adenocarcinoma cells. *Nature*, **322**, 187–189

Nitiss, J. and Wang, J. C. (1988). DNA topoisomerase targeting antitumour drugs can be studied in yeast. *Proc. Natl Acad. Sci. USA*, **85**, 7501–7505

Noguchi, H., Reddy, G. P. U. and Pardee, A. P. (1983). Rapid incorporation of label from ribonucleoside disphosphates into DNA by a cell-free high molecular weight fraction from animal cell nuclei. *Cell*, **32**, 443–451

Nurse, P. (1990). Universal control mechanism regulating onset of M phase. *Nature*, **344**, 503–507

Nyce, J. (1989). Drug-induced DNA hypermethylation and drug resistance in human tumours. *Cancer Res.*, **49**, 5829–5836

Ohno, R., Okada, K., Masaoka, T., Kuramoto, A., Arima, T., Yoshida, Y., Ariyoshi, H., Ichimaru, M., Sakai, Y. and Oguro, M. (1990). An early phase II study of CPT-11, a new derivative of camptothecin, for the treatment of leukemia and lymphoma. *J. Clin. Oncol.*, **8**, 1907–1912

Osheroff, N. (1987). Role of the divalent cation in topoisomerase II mediated reactions. *Biochemistry*, **26**, 6402–6406

Osheroff, N., Zechiedrich, E. L. and Gale, K. C. (1991). Catalytic function of DNA topoisomerase II. *Bioessays*, **13**, 269–273

Patel, S., Austin, C. A. and Fisher, L. M. (1990). Development and properties of an etoposide-resistant human leukaemic CCRF-CEM cell line. *Anti-cancer Drug Des.*, **5**, 149–157

Patet, J., Huppert, J., Fourade, A. and Tapiero, H. (1986). Cytogenetic modifications of Friend Leukemia cells resistant to adriamycin. *Leuk. Res.*, **10**, 651–658

Pommier, Y., Capranico, G., Orr, A. and Kohn, K. W. (1991a). Local base sequence preferences for DNA cleavage by mammalian topoisomerase II in the

presence of amsacrine and teniposide. *Nucleic Acids Res.*, **19**, 5973–5980
Pommier, Y., Capranico, G., Orr, A. and Kohn, K. W. (1991b). Distribution of topoisomerase II cleavage sites in simian virus 40 DNA and the effects of drugs. *J. Mol. Biol.*, **222**, 909–924
Pommier, Y., Cockerill, P. N., Kohn, K. W. and Garrard, W. T. (1990a). Identification within the Simian virus 40 genome of a chromosomal loop attachment site that contains topoisomerase II cleavage sites. *J. Virol.*, **64**, 419–423
Pommier, Y., Jaxel, C., Kerrigan, D. and Kohn, K. W. (1991c). Structure activity relationship of topoisomerase I inhibition by camptothecin derivatives: evidence for the existence of a ternary complex. In Potmesil, M. and Kohn, K. W. (Eds), *DNA Topoisomerases in Cancer*, Oxford University Press, New York, pp. 121–132
Pommier, Y., Kerrigan, D., Covey, J. M., Kao-Shan, C.-S. and Whang-Peng, J. (1988). Sister chromatid exchanges, chromosomal aberrations and cytotoxicity produced by antitumour topoisomerase II inhibitors in sensitive (DC3F) and resistant (DC3F19-OHE) Chinese hamster cells. *Cancer Res.*, **48**, 512–516
Pommier, Y., Kerrigan, D., Hartman, K. D. and Glazer, R. I. (1990b). Phosphorylation of mammalian DNA topoisomerase I and activation by protein kinase C. *J. Biol. Chem.*, **265**, 9418–9422
Pommier, Y., Kohn, K. W., Capranico, G. and Jaxel, C. (1992). Base sequence selectivity of topoisomerase inhibitors suggests a common model for drug action. In Andoh, T., Ikeda, H. and Oguro, O. (Eds), *Molecular Biology of DNA Topoisomerases and Its Applications to Cancer Chemotherapy*, CRC Press, Boca Raton (in press)
Pommier, Y., Runger, T. M., Kerrigan, D. and Kraemer, K. H. (1991d). Relationship of DNA strand breakage produced by bromodeoxyuridine to topoisomerase II activity in Bloom-syndrome fibroblasts. *Mutation Res.*, **254**, 185–190
Pommier, Y., Zwelling, L. A., Kao-Shan, C.-S., Whang-Peng, J. and Bradley, M. O. (1985). Correlations between intercalator-induced DNA strand breaks and sister chromatid exchanges, mutations and cytotoxicity in Chinese hamster cells. *Cancer Res.*, **45**, 3143–3149
Porter, S. E. and Champoux, J. J. (1989a). The basis for camptothecin enhancement of DNA breakage by eukaryotic topoisomerase I. *Nucleic Acids Res.*, **17**, 8521–8532
Porter, S. E. and Champoux, J. J. (1989b). Mapping *in vivo* topoisomerase I sites on simian virus 40 DNA: asymmetric distribution of sites on replicating molecules. *Mol. Cell. Biol.*, **9**, 541–550
Preston, G. M. and White, B. (1990). Effects of the DNA topoisomerase II inhibitor VM-26 on transcriptional initiation *in vitro*. *Life Sci.*, **46**, 1309–1318
Ralph, R. K., Darkin-Rattray, S. and Schofield, P. (1990). Growth related protein kinases. *Bioessays*, **12**, 121–124
Ralph, R. K. and Schneider, E. (1987). DNA topoisomerases and anti-cancer drugs. In Kon, O. H., Chung, M. C.-M., Hwang, P. L. N., Leong, S.-F., Loke, K. N., Thiyagarajah, P. and Wong, P. T.-H. (Eds), *Integration and Control of Metabolic Processes: Pure and Applied Aspects*, ICSU Press, pp. 373–388
Rappa, G., Lorico, A. and Sartorelli, A. C. (1990). Induction of the differentiation of WEHI-3B $D^+$ monomyelocytic leukemia cells by inhibitors of topoisomerase II. *Cancer Res.*, **50**, 6723–6730
Razin, S. V. (1987). DNA interactions with the nuclear matrix and spatial organisation of replication and transcription. *Bioessays.*, **6**, 19–23
Razin, S. V., Petrov, P. and Hancock, R. (1991). Precise localisation of the

α-globin gene cluster within one of the 20–300-kilobase DNA fragments released by cleavage of chicken chromosomal DNA at topoisomerase II sites *in vivo*: evidence that the fragments are DNA loops or domains. *Proc. Natl Acad. Sci. USA*, **88**, 8515–8519

Reitman, M., and Felsenfeld, G. (1990). Developmental regulation of topoisomerase II sites and DNase I-hypersensitive sites in the chicken beta-globin locus. *Mol. Cell. Biol.*, **1990**, 2774–2786

Richard, R. E. and Bogenhagen, D. F. (1991). The 165kDa DNA topoisomerase I from *Xenopus laevis* oocytes is a tissue specific variant. *Develop. Biol.*, **146**, 4–11

Richter, A., Strausfeld, U. and Knippers, R. (1987). Effects of VM-26 (teniposide), a specific inhibitor of type II DNA topoisomerase, on SV40 DNA replication *in vivo*. *Nucleic Acids Res.*, **15**, 3455–3468

Riou, J.-F., Lefevre, D. and Riou, G. (1989). Stimulation of the topoisomerase II induced DNA cleavage sites in the c-*myc* protooncogene by antitumour drugs is associated with gene expression. *Biochemistry*, **28**, 9104–9110

Rius, C., Zorilla, A. R., Cabanas, A. R., Mata, C., Bernabeu, C. and Aller, P. (1991). Differentiation of human promonocytic leukemia U937 cells with DNA topoisomerase II inhibitors: induction of vimentin gene expression. *Mol. Pharmacol.*, **39**, 442–448

Robbie, M. A., Baguley, B. C., Denny, W. A., Gavin, J. B. and Wilson, W. R. (1988). Mechanism of resistance of noncycling mammalian cells to 4′-(9-acridinylamino)methanesulfon-*m*-anisidide: comparison of uptake, metabolism, and DNA breakage in log- and plateau-phase Chinese hamster fibroblast cell cultures. *Cancer Res.*, **48**, 310–319

Roberge, M., Th'ng, J., Hamaguchi, J. and Bradbury, E. M. (1990). The topoisomerase II inhibitor VM-26 induces marked changes in histone H1 kinase activity, histones H1 and H3 phosphorylation, and chromosome condensation in G2 phase in mitotic BHK cells. *J. Cell. Biol.*, **111**, 1753–1762

Robinson, M. J. and Osheroff, N. (1991). Effects of antineoplastic drugs on the post-strand-passage DNA cleavage/religation equilibrium of topoisomerase II. *Biochemistry*, **30**, 1807–1813

Robson, C. N., Hoban, P. R., Harris, A. L. and Hickson, I. D. (1987). Cross-sensitivity to topoisomerase II inhibitors in cytotoxic drug-hypersensitive Chinese hamster ovary cell lines. *Cancer Res.*, **47**, 1560–1565

Roca, J. and Mezquita, C. (1989). DNA topoisomerase II activity in non-replicating, transcriptionally inactive chicken late spermatids. *EMBO Jl*, **8**, 1855–1860

Romig, H. and Richter, A. (1990). Expression of the topoisomerase I gene in serum stimulated human fibroblasts. *Biochim. Biophys. Acta*, **1048**, 274–280

Rosenberg, B. H., Ungers, G. and Deutsch, J. F. (1976). Variation in DNA swivel enzyme activity during the mammalian cell cycle. *Nucleic Acids Res.*, **3**, 3305–3311

Rottman, M., Schroder, H., Gramson, M., Reinneisen, K., Karelec, B., Dorn, A., Freise, N. and Muller, E. G. (1987). Specific phosphorylation of protein in pore-complex-laminae from the sponge *Geordia cydonium* by the homologous aggregation factor and phorbol ester. Role of protein kinase C in the phosphorylation of DNA topoisomerase II. *EMBO Jl*, **6**, 3939–3944

Rowe, T. C., Chen, G. L., Hsiang, Y.-H. and Liu, F. (1986a). DNA damage by antitumour acridines mediated by mammalian DNA topoisomerase II. *Cancer Res.*, **46**, 2021–2026

Rowe, T., Wang, J. C. and Liu, L. F. (1986b). *In vivo* localization of DNA topoisomerase II cleavage sites on *Drosophila* heat shock chromatin. *Mol. Cell. Biol.*, **6**, 985–992

Rowley, R. and Kort, L. (1989). Novobiocin, nalidixic acid, etoposide and 4′-(9-acridinylamino)methanesulfon-*m*-anisidide effects on G2 and mitotic Chinese hamster ovary cell progression. *Cancer Res.*, **49**, 4752–4757

Rubin, E., Kharbanda, S., Gunji, H. and Kufe, D. (1991). Activation of the c-*jun* protooncogene in human myeloid leukemia cells treated with etoposide. *Mol. Pharmacol.*, **39**, 697–701

Ryan, A. J., Squires, S., Strutt, H. L. and Johnson, R. T. (1991). Camptothecin cytotoxicity in mammalian cells is associated with the induction of persistent double strand breaks in replicating DNA. *Nucleic Acids Res.*, **19**, 3295–3300

Sahyoun, N., Wolf, M., Bestermann, J., Hsieh, T. S., Sanders, M., le Vine, H., Chang, K. J. and Cuatrecasas, P. (1986). Protein kinase C phosphorylates topoisomerase II: topoisomerase activation and its possible role in phorbol ester-induced differentiation of HL-60 cells. *Proc. Natl Acad. Sci. USA*, **83**, 1603–1607

Saijo, M., Enomoto, T., Hanaoka, F. and Ui, M. (1990). Purification and characterization of type II DNA topoisomerase from mouse FM3A cells: phosphorylation of topoisomerase II and modification of its activity. *Biochemistry*, **29**, 583–590

Sander, M. and Hsieh, T. S. (1983). Double strand DNA cleavage by type II DNA topoisomerase from *Drosophila melanogaster*. *J. Biol. Chem.*, **258**, 8421–8428

Sander, M. and Hsieh, T. S. (1985). *Drosophila* topoisomerase II double-stranded DNA cleavage: analysis of DNA sequence homology at the cleavage site. *Nucleic Acids Res.*, **13**, 1057–1072

Sander, M., Nolan, J. M. and Hsieh, T. S. (1984). A protein kinase activity tightly associated with *Drosophila* type II DNA topoisomerase. *Proc. Natl Acad. Sci. USA*, **81**, 6938–6942

Schaak, J., Schedl, P. and Shenk, T. (1990a). Topoisomerase I and II cleavage of adenovirus DNA *in vivo*: both topoisomerase activities appear to be required for adenovirus DNA replication. *J. Virol.*, **64**, 78–85

Schaak, J., Schedl, P. and Shenk, T. (1990b). Transcription of adenovirus and Hela cell genes in the presence of drugs that inhibit topoisomerase I and II function. *Nucleic Acids Res.*, **18**, 1499–1508

Schaffer, R. and Traktman, P. (1987). Vaccinia virus encapsidates a novel topoisomerase with the properties of eukaryotic type I enzyme. *J. Biol. Chem.*, **262**, 9309–9316

Schneider, E., Hsiang, Y.-H. and Liu, L. F. (1990). DNA topoisomerases as anti-cancer drug targets. *Adv. Pharmacol.*, **21**, 149–183

Schneider, E., Hutchins, A.-M., Darkin, S., Lawson, P. A. and Ralph, R. K. (1988a). 4′-(9-acridinylamino)methanesulfon-*m*-anisidide and DNA topoisomerase II in a cold-sensitive cell cycle mutant of a murine mastocytoma cell line. *Biochim. Biophys. Acta*, **951**, 85–97

Schneider, E., Lawson, P. A. and Ralph, R. K. (1988b). Inhibition of protein synthesis reduces the cytotoxicity of 4′-(9-acridinylamino)-methanesulfon-*m*-anisidide without affecting DNA breakage and DNA topoisomerase II in a murine mastocytoma cell line. *Biochem. Pharmacol.*, **38**, 263–269

Schroder, H. C., Steffen, R., Wenger, R., Ugarkovic, D. and Muller, W. E. (1989). Age-dependent increase of DNA topoisomerase II activity in quail oviduct: modulation of the nuclear matrix-associated enzyme activity by protein phosphorylation and poly(ADP-ribosyl)ation. *Mutation Res.*, **219**, 283–294

Schultz, M. C., Brill, S. J., Ju, Q., Sternglanz, R. and Reeder, R. H. (1992). Topoisomerases and yeast rRNA transcription: negative supercoiling stimulates initiation and topoisomerase activity is required for elongation. *Genes Devel.*, **6** 1332–1341

Sekiguchi, J. A. M. and Kmiec, E. G. (1988). Studies on DNA topoisomerase activity during *in vitro* chromatin assembly. *Mol. Cell. Biochem.*, **83**, 195–205

Shapiro, T. A. and Englund, P. T. (1990). Selective cleavage of kinetoplast DNA minicircles promoted by anti-trypanosomal drugs. *Proc. Natl Acad. Sci. USA*, **87**, 950–954

Shapiro, T. A., Klein, V. A. and Englund, P. T. (1989). Drug promoted cleavage of kinetoplast DNA minicircles. *J. Biol. Chem.*, **264**, 4173–4178

Shen, J. W., Subjeck, J. R., Lock, R. B. and Ross, W. E. (1989). Depletion of topoisomerase II in isolated nuclei during a glucose-regulated stress response. *Mol. Cell. Biol.*, **9**, 3284–3291

Shibuya, M. L., Buddenbaum, W. E., Don, A. L., Utsumi, D., Suciu, T., Kosaka, T. and Elkind, M. (1991). Amsacrine-induced lesions in DNA and their modulation by novobiocin and 2,4-dinitrophenol. *Cancer Res.*, **51**, 573–580

Shin, C.-G. and Snapka, R. M. (1990a). Patterns of strongly protein-associated simian virus 40 DNA replication intermediates resulting from exposures to specific topoisomerase poisons. *Biochemistry*, **29**, 10934–10939

Shin, C.-G. and Snapka, R. M. (1990b). Exposure to camptothecin breaks leading and lagging strand simian virus 40 DNA replication forks. *Biochem. Biophys. Res. Commun.*, **168**, 135–140

Shuman, S. (1989). Vaccinia DNA topoisomerase I promotes illegitimate recombination in *Escherichia coli*. *Proc. Natl Acad. Sci. USA*, **86**, 3489–3493

Sinha, B. K., Katki, A. G. and Batist, G. (1987). Adriamycin-stimulated hydroxyl radical formation in human breast tumour cells. *Biochem. Pharmacol.*, **36**, 793–799

Smith, C. A. (1977). Absence of ethidium bromide induced nicking and degradation of mitochondrial DNA in mouse L cells. *Nucleic Acids Res.*, **4**, 1419–1427

Snapka, R. M. (1986). Topoisomerase inhibitors can selectively interfere with different stages of simian virus 40 DNA replication. *Mol. Cell. Biol.*, **6**, 4221–4227

Snapka, R. M., Powelson, M. A. and Strayer, J. M. (1988). Swiveling and decatenation of replicating simian virus 40 genomes *in vivo*. *Mol. Cell. Biol.*, **8**, 1515–1521

Snow, K. and Judd, W. (1991). Characterisation of adriamycin- and amsacrine-resistant human leukemic T cell lines. *Br. J. Cancer*, **63**, 17–28

Sobczak, J., Tournier, M.-F., Lotti, A.-M. and Duguet, M. (1989). Gene expression in regenerating liver in relation to cell proliferation and stress. *Eur. J. Biochem.*, **180**, 49–55

Sørensen, B. S., Jensen, P. S., Andersen, A. H., Christiansen, K., Alsner, J., Thomsen, B. and Westergaard, O. (1990). Stimulation of topoisomerase II mediated DNA cleavage at specific sequence elements by the 2-nitroimidazole Ro 15-0216. *Biochemistry*, **29**, 9507–9515

Sperry, A. O., Blasquez, V. C. and Garrard, W. T. (1989). Dysfunction of chromosomal loop attachment sites: illegitimate recombination linked to matrix association regions and topoisomerase II. *Proc. Natl Acad. Sci. USA*, **86**, 5497–5501

Spitzner, J. R. and Muller, M. T. (1988). A consensus sequence for cleavage by vertebrate DNA topoisomerase II. *Nucleic Acid Res.*, **16**, 5533–5556

Spitzner, J. R. and Muller, M. T. (1989). Application of a degenerate consensus sequence to quantify recognition sites by vertebrate DNA topoisomerase II. *J. Mol. Recog.*, **2**, 63–74

Stetina, R. and Vesela, D. (1991). The influence of DNA topoisomerase II inhibitors novobiocin and fostriecin on the induction and repair of DNA damage in

Chinese hamster ovary (CHO) cells treated with mitoxantrone. *Neoplasia*, **38**, 109–117

Stevnsner, T., Mortensen, U. H., Westergaard, O. and Bonven, B. J. (1989). Interactions between eukaryotic DNA topoisomerase I and a specific binding sequence. *J. Biol. Chem.*, **264**, 10110–10113

Stewart, A. F. and Schütz, G. (1987). Camptothecin-induced *in vivo* topoisomerase I cleavages in transcriptionally active tyrosine amino-transferase gene. *Cell*, **50**, 1109–1117

Sugimoto, Y., Tsukahara, S., Oh-hara, T., Isoe, T. and Tsuruo, T. (1990). Decreased expression of DNA topoisomerase I in camptothecin-resistant tumour cell lines as determined by a monoclonal antibody. *Cancer Res.*, **50**, 6925–6930

Sullivan, D. M., Glisson, B. S., Hodges, P. K., Smallwood-Kentro, S. and Ross, W. E. (1986). Proliferation dependence of topoisomerase II mediated drug action. *Biochemistry*, **25**, 2248–2256

Sullivan, D. M., Latham, M. D. and Ross, W. E. (1987). Proliferation-dependent topoisomerase II content as a determinant of antineoplastic drug action in human, mouse, and Chinese hamster ovary cells. *Cancer Res.*, **47**, 3973–3979

Sullivan, D. M., Latham, M. D., Rowe, T. C. and Ross, W. E. (1989). Purification and characterization of an altered topoisomerase II from a drug-resistant Chinese hamster ovary cell line. *Biochemistry*, **28**, 5680–5687

Svejstrup, J. Q., Christiansen, K., Andersen, A. H., Lund, K. and Westergaard, O. (1990). Minimal DNA duplex requirements for topoisomerase I-mediated cleavages *in vitro*. *J. Biol. Chem.*, **265**, 12529–12535

Tabuchi, H. and Hirose, S. (1988). DNA supercoiling facilitates formation of the transcription initiation complex of the fibroin gene promoter. *J. Biol. Chem.*, **263**, 15282–15287

Takano, H., Kohno, K., Ono, M., Uchida, Y. and Kuwano, M. (1991). Increased phosphorylation of DNA topoisomerase II in etoposide resistant mutants of human cancer KB cells. *Cancer Res.*, **51**, 3951–3957

Tamura, H., Ikegami, Y., Ono, K., Sekimizu, K. and Andoh, T. (1990). Acidic phospholipids directly inhibit DNA binding of mammalian DNA topoisomerase I. *FEBS Lett.*, **261**, 151–154

Tamura, H., Kohchi, C., Yamada, R., Ikeda, T., Koiwai, O., Patterson, E., Keene, J. D., Okada, K., Kjeldsen, E., Nishikawa, K. and Andoh, T. (1991). Molecular cloning of a cDNA of a camptothecin-resistant human DNA topoisomerase I and identification of mutation sites. *Nucleic Acids Res.*, **19**, 69–75

Tan, K. B., Mattern, M. R., Eng, W. K., McCabe, F. L. and Johnson, R. K. (1989). Nonproductive rearrangement of DNA topoisomerase I and II genes: correlation with resistance to topoisomerase inhibitors. *J. Natl Cancer Inst.*, **81**, 1731–1735

Tan, K. B., Per, S. R., Boyce, R. A., Mirabelli, C. K. and Crooke, S. T. (1988). Altered expression and transcription of the topoisomerase II gene in nitrogen mustard-resistant human cells. *Biochem. Pharmacol.*, **37**, 4413–4416

Tanabe, K., Ikegami, Y., Ishida, R. and Andoh, T. (1991). Inhibition of topoisomerase II by antitumor agents bis(2,6-dioxopiperazine) derivatives. *Cancer Res.*, **51**, 4903–4908

Taudou, G., Mirambeau, G., Lavenot, C., der Garabedian, A., Vermeersch, J. and Duguet, M. (1984). DNA topoisomerase activities in concanavalin A-stimulated lymphocytes. *FEBS Lett.*, **176**, 431–435

Tarr, M. and van Holden, P. D. (1990). Inhibition of transcription by adriamycin is a consequence of the loss of negative superhelicity in DNA mediated by topoisomerase II. *Mol. Cell. Biochem.*, **93**, 141–146

Thomsen, B., Bendixen, C., Lund, K., Andersen, A. H., Sørenson, B. S. and Westergaard, O. (1990). Characterisation of the interaction between topoisomerase II and DNA by transcriptional footprinting. *J. Mol. Biol.*, **215**, 237–244

Thomsen, B., Mollerup, S., Bonven, B. J., Frank, R., Blocker, H., Nielsen, O. F. and Westergaard, O. (1987). Sequence specificity of DNA topoisomerase I in the presence and absence of camptothecin. *EMBO Jl*, **6**, 1817–1823

Tobey, R. A., Deaven, L. L. and Oka, M. S. (1978). Kinetic response of cultured Chinese hamster cells to treatment with 4′[(9-acridinyl)-amino]methanesulphon-*m*-anisidide-HC1. *J. Natl Cancer Inst.*, **60**, 1147–1152

Tricoli, J. V., Sahai, B. M., McCormick, P. J., Jarlinski, S. J., Bertram, J. S. and Kowalski, D. (1985). DNA topoisomerase I and II activities during cell proliferation and the cell cycle in cultured mouse embryo fibroblast (C3H 10TF(1,2)) cells. *Exp. Cell Res.*, **158**, 1–14

Tsai-Pflugfelder, M., Liu, L. F., Liu, A. A., Tewey, K. M., Whang-Peng, J., Knutsen, T., Huebner, K., Croce, C. M. and Wang, J. C. (1988). Cloning and sequencing of cDNA encoding human DNA topoisomerase II and localization of the gene to chromosome region 17q 21–22. *Proc. Natl Acad. Sci. USA*, **85**, 7177–7181

Tsao, Y. P., Wu, H.-Y. and Liu, L. F. (1989). Transcription-driven supercoiling of DNA. Direct biochemical evidence from *in vitro* studies. *Cell*, **56**, 111–118

Tse, Y.-C., Kirkegaard, K., and Wang, J. C. (1980). Covalent bonds between protein and DNA. Formation of phosphotyrosine linkage between certain DNA topoisomerases and DNA. *J. Biol. Chem.*, **255**, 5560–5565

Tse-Dinh, Y.-C., Wong, T. W. and Goldberg, A. R. (1984). Virus and cell-encoded tyrosine protein kinases inactivate DNA topoisomerase *in vitro*. *Nature*, **312**, 785–786

Tsuruo, T., Matsuzaki, T., Matsushita, M., Saito, H. and Yokohura, T. (1988). Antitumour effect of CPT-11, a new derivative of camptothecin, against pleiotropic drug-resistant tumours *in vitro* and *in vivo*. *Cancer Chemother. Pharmacol.*, **21**, 71–74

Udvardy, A., Schedl, P., Sander, M. and Hsieh, T.-S. (1985). Novel partitioning of DNA cleavage sites for *Drosophila* topoisomerase II. *Cell*, **40**, 933–941

Udvardy, A., Schedl, P., Sander, M. and Hsieh, T. (1986). Topo II cleavage in chromatin. *J. Mol. Biol.*, **191**, 231–236

Uemura, T., Morino, K., Uzawa, S., Shiozaki, K. and Yanagida, M. (1987). Cloning and sequencing of *Schizosaccharomyces pombe* topoisomerase I gene, and effect of gene disruption. *Nucleic Acids Res.*, **15**, 9727–9739

Uemura, T. and Yanagida, M. (1984). Isolation of type I and II DNA topoisomerase mutants from fusion yeast: single and double mutants show different phenotypes in cell growth and chromatin organisation. *EMBO Jl*, **3**, 1737–1744

Utsugi, T., Mattern, M. R., Mirabelli, C. K. and Hanna, W. (1990). Potentiation of topoisomerase inhibitor-induced DNA strand breakage and cytotoxicity by tumour necrosis factor: enhancement of topoisomerase activity as a mechanism of potentiation. *Cancer Res.*, **50**, 2636–2640

Utsumi, H., Shibuya, M. L., Kosaka, T., Buddenbaum, W. E. and Elkind, M. M. (1990). Abrogation by novobiocin of cytotoxicity due to the topoisomerase II inhibitor amsacrine in Chinese hamster cells. *Cancer Res.*, **50**, 2577–2581

Van Dyke, M. and Roeder, R. G. (1987). Novobiocin interferes with the binding of transcription factors TFIIIA and TFIIIC to the promotor of class III genes. *Nucleic Acids Res.*, **15**, 4365–4374

van Kries, S. P., Buhrmester, H. and Strätling, W. H. (1991). A matrix/ scaffold attachment region binding protein: identification, purification and mode of binding. *Cell*, **64**, 123–135

Vasanthakumar, G. and Ahmed, N. K. (1986). Contribution of drug transport and reductases to daunorubicin resistance in human myelocytic cells. *Cancer Chemother. Pharmacol.*, **18**, 105–110

Vassetzky, Y., Bakayev, V., Kalandadze, A. G. and Razin, S. V. (1990). Topoisomerase I is associated with the regulatory region of transcriptionally active SV40 minichromosomes. *Mol. Cell. Biochem.*, **95**, 95–106

Vassetzky, Y. S., Razin, S. V. and Georgiev, G. P. (1989). DNA fragments which specifically bind to isolated nuclear matrix *in vitro* interact with matrix-associated DNA topoisomerase II. *Biochem. Biophys. Res. Commun.*, **159**, 1263–1268

Verheijen, R., Venrooij, W. V. and Ramaekers, F. (1988). The nuclear matrix. Structure and composition. *J. Cell Sci.*, **90**, 11–36

Vigani, A., Chiara, S., Miglietta, L., Repetto, L., Conte, P. F., Cimoli, G., Morelli, L., Billi, G., Parodi, S. and Russo, P. (1991). Effect of recombinant human tumour necrosis factor on A2774 human ovarian cancer cell line: potentiation of mitoxantrone cytotoxicity. *Gynecol. Oncol.*, **41**, 52–55

Vilcek, J. and Lee, T. H. (1991). Tumour necrosis factor. *J. Biol. Chem.*, **265**, 7313–7316

Villeponteau, B. (1989). Characterization of a topoisomerase-like activity at specific hypersensitive sites in the *Drosophila* histone gene cluster. *Biochem. Biophys. Res. Commun.*, **162**(1), 232–237

Vosberg, P.-H. (1985). DNA topoisomerase: enzymes that control DNA conformation. *Curr. Top. Microbiol. Immunol.*, **114**, 19–102

Walker, P. R., Smith, C., Youdale, T., Leblanc, J., Whitfield, J. F. and Sikorska, M. (1991). Topo II-reactive chemotherapeutic drugs induce apoptosis in thymocytes. *Cancer Res.*, **51**, 1078–1085

Wang, H. P. and Rogler, C. E. (1991). Topoisomerase I-mediated integration of hepadnavirus DNA *in vitro*. *J. Virol.*, **65**, 2381–2392

Wang, J. C. (1985). DNA topoisomerases. *Ann. Rev. Biochem.*, **54**, 665–697

Wang, J. C. (1987). Recent studies of DNA topoisomerases. *Biochim. Biophys. Acta*, **909**, 1–9

Wang, J. C. (1991). DNA topoisomerases: why so many? *J. Biol. Chem.*, **266**, 6659–6662

Wang, J. C., Caron, P. R. and Kim, R. A. (1990). The role of DNA topoisomerases in recombination and genome stability: a double-edged sword. *Cell*, **62**, 403–406

Wani, M. C., Nicholas, A. W., Manikumar, G. and Wall, M. E. (1987). Plant antitumor agents. 25. Total synthesis and antileukemic activity of ring A substituted camptothecin analogues. Structure–activity correlations. *J. Med. Chem.*, **30**, 1774–1779

Watanabe, T., Kondo, K. and Oishi, M. (1991). Induction of *in vitro* differentiation of mouse erythroleukemia cells by genistein, an inhibitor of tyrosine protein kinase. *Cancer Res.*, **51**, 764–768

Watson, J. D. and Crick, F. H. C. (1953). Genetic implications of the structure of deoxyribonucleic acid. *Nature*, **171**, 964–967

Waud, W. R., Harrison, S. D., Jr., Gilbert, K.S., Laster, W. R., Jr., and Griswald, D. (1991). Antitumour drug cross-resistance *in vivo* in a cisplatin-resistant murine P388 leukemia. *Cancer Chemother. Pharmacol.*, **27**, 456–463

Webb, M. L. and Jacob, S. T. (1988). Inhibition of RNA polymerase I-directed transcription by novobiocin. Potential use of novobiocin as a general inhibitor of eukaryotic transcription initiation. *J. Biol. Chem.*, **263**, 4745–4748

Webb, M. L., Maguire, K. A. and Jacob, S. T. (1987). Novobiocin inhibits initiation of RNA polymerase II-directed transcription of the mouse metallothionein-1 gene independent of its effects on DNA topoisomerase II. *Nucleic Acids Res.*, **15**, 8547–8560

Whalen, A. M., McConnell, M. and Fisher, P. A. (1991). Developmental regulation of *Drosophila* DNA topoisomerase II. *J. Cell Biol.*, **112**, 203–213

Wong, M-L. and Hsu, M-T. (1990). Involvement of topoisomerases in replication, transcription, and packaging of the linear adenovirus genome. *J. Virol.*, **64**, 691–699

Wu, H.-Y., Shyy, S. H., Wang, J. C. and Liu, L. F. (1988). RNA transcription generates negatively and positively supercoiled domains in the template. *Cell*, **53**, 433–440

Xu, Z.-Z., Miyahara, K., Liszczynsky, H., Seno, S. and Naora, H. (1991). Reversible conversion of cells between cancerous and normal states. *Proc. Jap. Acad.*, **67** (Ser. B), 11–15

Yamagishi, M. and Nomura, M. (1988). Deficiency in both type I and II DNA topoisomerase activities differentially affects rRNA and ribosomal protein synthesis in *Schizosaccharomyces pombe*. *Curr. Genet.*, **13**, 305–314

Yamashita, Y., Kawada, S.-Z., Fujii, N. and Nakano, H. (1991). Induction of mammalian DNA topoisomerase I and II mediated DNA cleavage by saintopin, a new antitumour agent from fungus. *Biochemistry*, **30**, 5838–5845

Yang, G., Kunze, N., Baumgärtner, B., Jiang, Z., Sapp, M., Knippers, R. and Richter, A. (1990). Molecular structures of two human DNA topoisomerase retro sequences. *Gene*, **91**, 247–253

Yang, L., Rowe, T. C., Nelson, E. M., and Liu, L. F. (1985). *In vivo* mapping of DNA topoisomerase II-specific cleavage sites on SV40 chromatin. *Cell*, **41**, 127–132

Yang, L., Wold, M. S., Li, J. J., Kelly, T. J. and Liu, L. F. (1987). Roles of DNA topoisomerases in simian virus 40 DNA replication *in vitro*. *Proc. Natl Acad. Sci. USA*, **84**, 950–954

Yang, Y. and Ames, G. F.-L. (1988). DNA gyrase binds to the family of prokaryotic repetitive extragenic palindomic sequences. *Proc. Natl Acad. Sci. USA*, **85**, 8850–8854

Zechiedrich, E. L. and Osheroff, N. (1990). Eukaryotic topoisomerases recognize nucleic acid topology by preferentially interacting with DNA crossovers. *EMBO Jl*, **9**, 4555–4562

Zhang, H., D'Arpa, P. and Liu, L. F. (1990). A model for tumour cell killing by topoisomerase poisons. *Cancer Cells*, **2**, 23–27

Zhang, H., Wang, J. C. and Liu, L. F. (1988). Involvement of DNA topoisomerase I in the transcription of human ribosomal RNA genes. *Proc. Natl Acad. Sci. USA*, **85**, 1060–1064

Zhou, B. S., Bastow, K. F. and Cheng, Y. C. (1989). Characterisation of the 3′ region of the human DNA topoisomerase I gene. *Cancer Res.*, **49**, 3922–3927

Zucker, R. M. and Elstein, K. H. (1991). A new action for topoisomerase inhibitors. *Chem.-biol. Interact.*, **79**, 31–40

Zunino, F. and Capranico, G. (1990). DNA topoisomerase II as a primary target of antitumour anthracyclines. *Anti-cancer Drug Des.*, **5**, 307–317

Zwelling, L. A. (1989). Topoisomerase II as a target of anti-leukemia drugs: a review of controversial areas. *Hematol. Pathol.*, **3**, 101–112

Zwelling, L. A., Estey, E., Silberman, L., Doyle, S. and Hittelman, W. (1987). Effect of cell proliferation and chromatin conformation on intercalator-induced, protein-associated DNA cleavage in human brain tumour cells and human fibroblasts. *Cancer Res.*, **47**, 251–257

Zwelling, L. A., Hinds, M., Chan, D., Altschuler, E., Mayes, J. and Zipf, T. F. (1990a). Phorbol ester effects on topoisomerase II activity and gene expression in HL60 human leukemia cells with different proclivities towards monocytoid differentiation. *Cancer Res.*, **50**, 7116–7122

Zwelling, L. A., Kerrigan, D., Michaels, S. and Kohn, K. W. (1982). Cooperative sequestration of mAMSA in L1210 cells. *Biochem. Pharmacol.*, **31**, 3269–3277

Zwelling, L. A., Michaels, S., Erickson, L. C., Ungerleider, R. S., Nichols, M. and Kohn, K. W. (1981). Protein associated DNA strand breaks in L1210 cells treated with the DNA intercalating agents mAMSA and adriamycin. *Biochemistry*, **20**, 6553–6563

Zwelling, L. A., Slovak, M. L., Doroshow, J. H., Hinds, M., Chan, D., Parker, E., Mayes, J., Sie, K. L., Meltzer, P. S. and Trent, J. M. (1990b). HT1080/DR4: a P-glycoprotein-negative human fibrosarcoma cell line exhibiting resistance to topoisomerase II-reactive drugs despite the presence of a drug-sensitive topoisomerase II. *J. Natl Cancer Inst.*, **82**, 1553–1561

# 2

# Cellular and Molecular Pharmacology of the Anthrapyrazole Antitumour Agents

Laurence H. Patterson and David R. Newell

## 1 Introduction

The development of the anthrapyrazole antitumour agents may yet be one of the major achievements of developmental cancer chemotherapy in the 1980s. As with many synthetic or semisynthetic DNA binding agents, the rationale for the synthesis of the anthrapyrazoles stemmed from the desire to identify a drug with equivalent antitumour activity to that of doxorubicin (Figure 2.1, **I**) yet without the cumulative dose-limiting cardiotoxicity which has consistently been a feature of anthracycline chemotherapy. As discussed later in this chapter, early clinical data suggest that at least one of the anthrapyrazoles has significant single-agent antitumour activity in patients and three compounds remain in clinical trials.

The objective of this chapter is to review the rationale for the synthesis of the anthrapyrazoles, the available knowledge on structure–activity relationships and the cellular and molecular pharmacology of this new class of drugs. In addition, known mechanisms of resistance to anthrapyrazoles and the preclinical and clinical pharmacology of these agents will be briefly discussed. Although there are still substantial gaps in our understanding of the pharmacology of the anthrapyrazoles, the considerable clinical interest in these drugs is likely to stimulate the studies necessary to fill these.

## 2 Rationale for Development of the Anthrapyrazoles

The desire to identify an agent with the activity but not the toxicity of doxorubicin has been a powerful stimulus in cancer chemotherapy. As outlined below, the anthracyclines can undergo metabolic reduction to free

I,III,IV

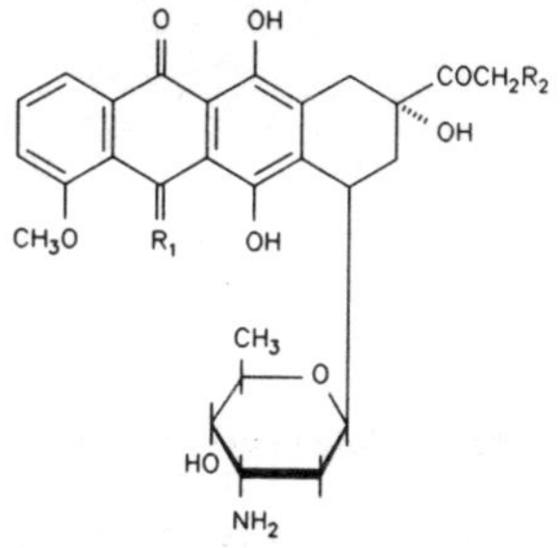


II

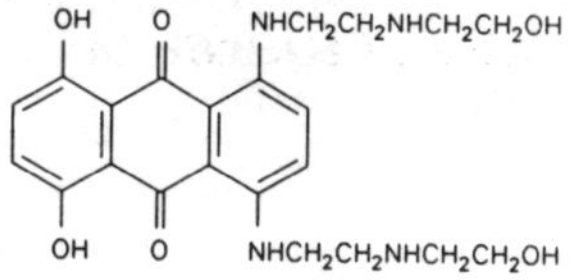


**Figure 2.1** The chemical structures of:
**I**, Doxorubicin ($R_1$ = O; $R_2$ = OH)
**II**, Mitoxantrone
**III**, 5-Iminodoxorubicin ($R_1$ = NH; $R_2$ = OH)
**IV**, 5-Iminodaunorubicin ($R_1$ = NH; $R_2$ = H)

radicals which can in turn produce reactive oxygen species. Drug and reactive oxygen free radical species have frequently been suggested as the mediators of anthracycline-induced cardiac damage and there is circumstantial evidence to support their involvement. For a recent review see Ohlson and Mushlin (1990). In the search for a non-cardiotoxic DNA binding agent a large number of natural product, semisynthetic and synthetic compounds have undergone clinical trial. Of the synthetic agents, clinical experience with mitoxantrone (Figure 2.1, **II**) is the most extensive, and the current consensus is that the incidence of cardiac toxicity is less than that seen with doxorubicin (van der Graaf and de Vries, 1990). Reduced cardiotoxicity is consistent with the greater resistance of anthraquinones, such as mitoxantrone, to both chemical and metabolic reduction (see below).

On the premise that reductive metabolism is involved in the cardiac toxicity of anthracyclines. Showalter and colleagues at Warner-Lambert/Parke-Davis Pharmaceutical Research initiated a drug discovery programme in the early 1980s with the aim of identifying the much sought after 'non-cardiotoxic doxorubicin'. The starting point for their syntheses was the observation that, for anthracyclines, conversion of the 5-carbonyl group to an imino group, as in 5-iminodoxorubicin (Figure 2.1, **III**) and 5-iminodaunorubicin (Figure 2.1, **IV**) led to molecules with reduced redox cycling characteristics (Lown *et al.*, 1979; Tong *et al.*, 1979; Acton and

V,VI,VII

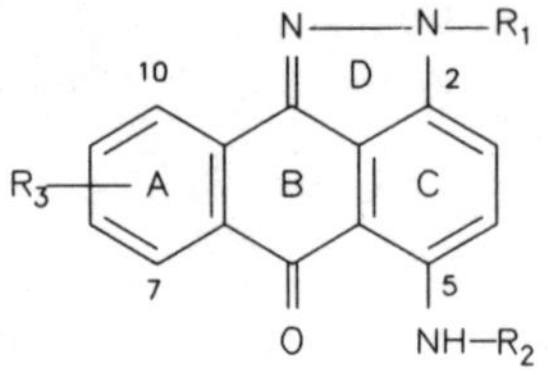

**Figure 2.2** The anthrapyrazole ring system (**V**) and:
**VI**, CI-937 $R_1 = -CH_2CH_2NHCH_2CH_2OH$
$R_2 = -CH_2CH_2NHCH_3$
$R_3 = -7,10\text{-}(OH)_2$

**VII**, CI-942 $R_1 = -CH_2CH_2NHCH_2CH_2OH$
$R_2 = -CH_2CH_2CH_2NH_2$
$R_3 = -7,10\text{-}(OH)_2$

Tong, 1981; Singh *et al.*, 1989). Studies *in vivo* with 5-iminodaunorubicin indicated a reduced cardiotoxic potential, and on the basis of these observations Showalter and co-workers identified the equivalent derivative of mitoxantrone as the target for a potentially non-cardiotoxic synthetic DNA binding agent (Showalter *et al.*, 1984, 1986). By introduction of an imino moiety, to replace one of the anthraquinone carbonyl groups, the potential for ring closure with the 2-nitrogen was created and this led to the novel anthra[1,9-*cd*]pyrazol-6(2*H*)-one, or anthrapyrazole, 4-membered ring system (Figure 2.2, **V**). As well as, it was hoped, reducing the likelihood of metabolic reduction, it was recognized that the novel structure of the anthrapyrazole nucleus might influence the activity of the compounds in both *in vitro* and *in vivo* model tumour systems, and indeed this has proved to be the case. Unfortunately, while the changes in reductive metabolism induced by converting the anthraquinone ring system to an anthrapyrazole system can be rationalized, the same cannot be said for the differences in the activities of the anthrapyrazoles and mitoxantrone in model tumour systems. Thus, at the present time the details of the comparative molecular, cellular and whole animal pharmacology of mitoxantrone and the anthrapyrazoles are insufficiently understood; nevertheless parallels will be drawn in this chapter where data are available.

## 3 Structural Requirements for Anthrapyrazole Antitumour Activity

Following the identification of the anthrapyrazole ring system as a novel lead in the development of non-cardiotoxic DNA binding agents, scientists at Warner Lambert/Parke-Davis Pharmaceutical Research, assisted by the National Cancer Institute (USA), performed an extensive structure activity analysis (Showalter *et al.*, 1984, 1987; Leopold *et al.*, 1985). Ideally, in attempting to identify structural determinants of cytotoxicity, it is preferable to primarily consider *in vitro* data. Such *in vitro* data are not complicated by whole body pharmacokinetic factors — i.e. drug absorption, distribution, systemic metabolism or excretion, which can of course be independently influenced by structural modification. Unfortunately, in the case of the anthrapyrazoles, data from prolonged-exposure *in vitro* cytotoxicity studies are not always predictive of *in vivo* antitumour activity or potency (Showalter *et al.*, 1987). Thus, 7,10-dihydroxyanthrapyrazoles were generally less active than their corresponding dihydro counterparts *in vitro* against the L1210 leukaemia, while *in vivo* against the P388 leukaemia the 7,10-dihydroxyanthrapyrazoles tended to be both more active and more potent. Table 2.1 illustrates this point for five pairs of anthrapyrazoles varying in the nature of the basic side-chain substituents at $N^2$ ($R_1$) and $N^5$ ($R_2$). In four of five cases the 7,10-dihydro compounds are 4–23-fold more potent *in vitro* than their 7,10-dihydroxy counterparts, yet *in vivo* the dihydroxyanthrapyrazoles are either more potent or/and more active (Table 2.1). The usual interpretation of such an observation would be to suggest *in vivo* metabolic activation of the dihydroxy compounds or inactivation of the dihydro molecules. However, in the case of the anthrapyrazoles, a more likely explanation lies in the oxidative instability of the dihydroxy species (Showalter *et al.*, 1987). For example, the dihydroxy compound CI-937 (Figure 2.2, **VI**) decomposed by 90% in 4 h at 37 °C in rat plasma, as measured by HPLC, a process prevented by the addition of ascorbate (Nordblom *et al.*, 1989). Similar results were obtained with the other dihydroxy compound of clinical interest, CI-942 (Figure 2.2, **VII**) (Frank *et al.*, 1989). When *in vitro* whole-cell effects are measured as the inhibition of thymidine incorporation following a 2 h exposure to anthrapyrazoles, in one series of compounds (Figure 2.3) the 7,10-dihydro compound (**VIII**) was approximately tenfold less potent than the corresponding 7,10-dihydroxy (**X**) compound (Showalter *et al.*, 1986; see below). Over this time period the oxidation of the 7,10-dihydroxy compound was presumably limited and the cytotoxic potency of the molecule revealed (Showalter *et al.*, 1986). A possible explanation for the superior *in vivo* activity of the 7,10-dihydroxy compounds is that, under *in vivo* reducing conditions, oxidation and, hence, inactivation is limited. Short-term exposure cytotoxicity data are not available for the majority of the anthrapyrazoles synthesized and, hence, the remaining discussion of structure–activity relationships will focus on *in vivo* antitumour activity data.

Table 2.2 shows the effect of selected alterations to the basic side-chain at $N^2$ ($R_1$) for three different series of 7,10-dihydroxyanthrapyrazoles where the $N^5$ ($R_2$) side-chain is held constant in each series. Significant *in vivo* activity (T/C > 200%), combined with potency (optimal dose < 100 mg/kg), is only seen when $R_1$ contains an amine group. The amine can be primary, secondary or tertiary and can be separated from $N^2$ by either two

**Table 2.1** Comparative *in vitro* cytotoxicity and *in vivo* activity of 7,10-dihydro and 7,10-dihydroxy anthrapyrazoles[a]

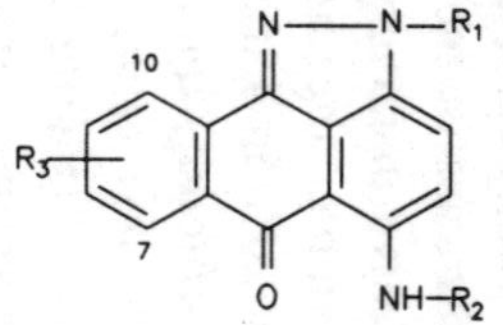


| $R_1$ | $R_2$ | $R_3$ | L1210 $IC_{50}$ (nM)[b] | P388 T/C (cures)[c] | Dose (mg/kg)[d] |
|---|---|---|---|---|---|
| $-CH_2CH_2NH_2$ | $-CH_2CH_2NHCH_2CH_2OH$ | 7,10-H | 80 | 187% | 25 |
| | | 7,10-$(OH)_2$ | 580 | 270% (5/6) | 12.5 |
| $-CH_2CH_2NHCH_2CH_2OH$ | $-CH_2CH_2NH_2$ | 7,10-H | 69 | 189% (1/6) | 25 |
| | | 7,10-$(OH)_2$ | 1600 | 197% (1/5) | 3.12 |
| $-CH_2CH_2NHCH_2CH_2OH$ | $-CH_2CH_2N(CH_3)_2$ | 7,10-H | 32 | 174% | 6.25 |
| | | 7,10-$(OH)_2$ | 230 | 250% (1/5) | 12.5 |
| $-CH_2CH_2N(CH_2CH_3)_2$ | $-CH_2CH_2NH_2$ | 7,10-H | 46 | 104% | 25 |
| | | 7,10-$(OH)_2$ | 46 | 280% (2/5) | 12.5 |
| $-CH_2CH_2N(CH_2CH_3)_2$ | $-CH_2CH_2NHCH_2CH_2OH$ | 7,10-H | 32 | 169% | 25 |
| | | 7,10-$(OH)_2$ | 130 | 264% | 12.5 |

[a] Data are adapted from Showalter *et al.* (1987).
[b] Concentration required to give 50% inhibition of cell growth following 72 h exposure
[c] Lifespan of mice bearing the P388 leukaemia treated with drug at the optimal dose/lifespan of control animals. Cures indicates number of mice surviving at day 30 out of total number treated.
[d] Optimal dose when given as two i.p. injections on days 3 and 7 after inoculation of tumour.

VIII,IX,X,XI

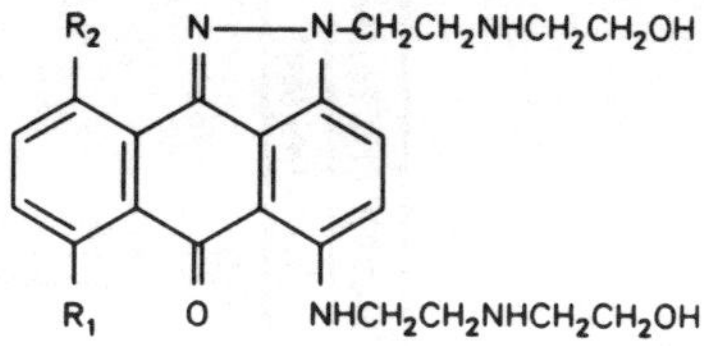


**Figure 2.3** A-ring hydroxyl substituted anthrapyrazoles:
**VIII**: $R_1 = H$; $R_2 = H$
**IX**, CI-941: $R_1 = OH$; $R_2 = H$
**X**: $R_1 = OH$; $R_2 = OH$
**XI**: $R_1 = H$; $R_2 = OH$

or three methylene groups. The effect of extending beyond three methylene groups has not been reported. Introduction of a second amine into the $R_1$ substituent was investigated in one series and activity was retained, although potency may have been reduced. Returning to the compounds with only one amine in the $R_1$ side-chain, for the tertiary amines, dimethyl and diethyl substituents were equivalent. Similarly, with a secondary amine, both methyl and hydroxyethyl substituted compounds were equiactive and equipotent. In summary, the only clear requirement for activity in the $R_1$ substituent is an amine group separated from $N^2$ by two or three methylene groups. It has been suggested (Showalter *et al.*, 1987), that for the related anthraquinones there is a more rigid requirement for only two methylene groups separating the two nitrogens. The effect of alkylamino chain length on anthrapyrazole cytotoxicity is reflected in DNA binding affinity, as discussed later.

The second structural determinant of activity to receive attention is the basic side-chain at $N^5$ ($R_2$). Shown in Table 2.3 are selected data for three series of compounds where the $N^2$ ($R_1$) substituent is held constant in each case. An amino group was included in all of the $N^2$ substituents studied, and the effect of increasing the number of methylene groups between the side-chain nitrogen and $N^5$ was investigated. Extension of the methylene chain length to four or five resulted in loss of activity, while for four pairs of compounds in which the only difference was two or three methylene groups activity was similar. Primary, secondary and tertiary aliphatic amines were again tolerated in the $R_2$ group, although the bis-hydroxyethyl substituted tertiary amines were less potent (optimal dose $\geqslant$ 100 mg/kg). Introduction of a second amine function, primary or tertiary, into $R_2$ maintained high-level activity, although potency may have been reduced slightly. In summary, the requirement of a methylene chain spacer of only two or three moieties between $N^5$ and the aliphatic nitrogen is the clearest structural determinant of activity in the $R^2$ substituent. A range of substitutions adhering to this basic structure results in compounds with good potency and curative activity against the P388 leukaemia *in vivo*.

**Table 2.2** Comparative *in vivo* antitumour activity of three series of $N^2$ ($R_1$) substituted 7,10-dihydroxyanthrapyrazoles against the P388 leukaemia[a]

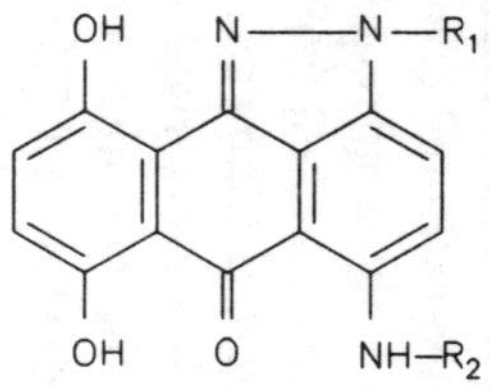


| $R_1$ | $R_2 = -(CH_2)_2NH_2$ | | $R_2 = -CH_2CH_2N(CH_2CH_3)_2$ | | $R_2 = -CH_2CH_2NHCH_2CH_2OH$ | |
|---|---|---|---|---|---|---|
| | *T/C (cures)*[b] | *Dose (mg/kg)*[c] | *T/C (cures)* | *Dose (mg/kg)* | *T/C (cures)* | *Dose (mg/kg)* |
| $-CH_3$ | 138% | 12.5 | 152% | 400 | 174% | 100 |
| $-CH_2CH_2OH$ | 196% | 100 | 190% | 400 | 271% | 200 |
| $-CH_2CH_2OCH_3$ | 127% | 50 | ND[d] | ND | 140% | 100 |
| $-CH_2CH_2NH_2$ | 221% | 12.5 | ND | ND | 270% (5/6) | 12.5 |
| $-CH_2CH_2CH_2NH_2$ | ND | ND | ND | ND | 285% (5/6) | 50 |
| $-CH_2CH_2NHCH_3$ | ND | ND | ND | ND | 324% (5/6) | 3.12 |
| $-CH_2CH_2N(CH_3)_2$ | 263% (2/5) | 12.5 | ND | ND | 242% (2/5) | 12.5 |
| $-CH_2CH_2N(CH_2CH_3)_2$ | 280% (2/5) | 12.5 | 180% | 100 | 264% | 12.5 |
| $-CH_2CH_2CH_2N(CH_3)_2$ | 230% (1/6) | 12.5 | ND | ND | 218% | 12.5 |
| $-CH_2CH(OH)CH_2N(CH_2CH_3)_2$ | 262% | 6.25 | 157% | 100 | 223% | 12.5 |
| $-CH_2CH_2NHCH_2CH_2OH$ | 197% (1/5) | 3.12 | 241% | 50 | 206% | 6.25 |
| $-CH_2CH_2NHCH_2CH_2N(CH_3)_2$ | ND | ND | ND | ND | 226% | 50 |

[a] Data are adapted from Showalter *et al.* (1987).
[b] Lifespan of mice bearing the P388 leukaemia treated with drug at the optimal dose/lifespan of control animals. Cures indicates number of mice surviving at day 30 out of total number treated.
[c] Optimal dose when given as two i.p. injections on days 3 and 7 after inoculation of tumour.
[d] ND = not determined.

**Table 2.3** Comparative *in vivo* antitumour activity of three series of $N^5$ ($R_2$) substituted 7,10-dihydroxyanthrapyrazoles against the P388 leukaemia[a]

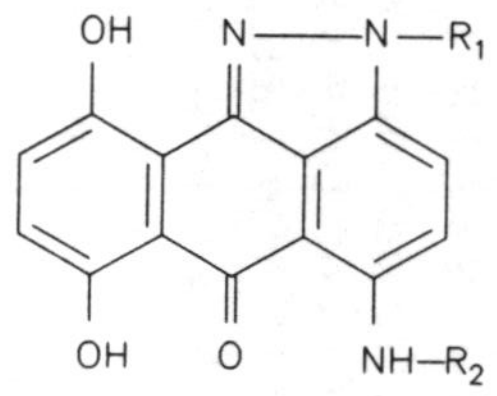


| $R_2$ | $R_1 = -(CH_2)_2NH(CH_2)_2OH$ | | $R_1 = -(CH_2)_2NH_2$ | | $R_1 = -(CH_2)_2N(CH_3)_2$ | |
|---|---|---|---|---|---|---|
| | *T/C (cures)*[b] | *Dose (mg/kg)*[c] | *T/C (cures)* | *Dose (mg/kg)* | *T/C (cures)* | *Dose (mg/kg)* |
| $-(CH_2)_2NH_2$ | 197% (1/5) | 3.12 | 221% | 12.5 | 263% (2/5) | 12.5 |
| $-(CH_2)_3NH_2$ | 294% (4/8) | 45 | ND[d] | ND | 208% | 12.5 |
| $-(CH_2)_4NH_2$ | 173% | 100 | ND | ND | ND | ND |
| $-(CH_2)_5NH_2$ | 150% | 100 | ND | ND | ND | ND |
| $-(CH_2)_2NHCH_3$ | 277% | 12.5 | ND | ND | ND | ND |
| $-(CH_2)_2NH(CH_2)_2OH$ | 206% (1/5) | 6.25 | 270% (5/6) | 12.5 | 242% (2/5) | 12.5 |
| $-(CH_2)_3NH(CH_2)_2OH$ | 254% (3/6) | 6.25 | ND | ND | ND | ND |
| $-(CH_2)_2NH(CH_2)_2NH_2$ | 277% (3/6) | 25 | ND | ND | ND | ND |
| $-(CH_2)_2NH(CH_2)_2N(CH_3)$ | 254% (4/6) | 50 | ND | ND | ND | ND |
| $-(CH_2)_2N(CH_3)_2$ | 250% (1/5) | 12.5 | ND | ND | ND | ND |
| $-(CH_2)_2N(CH_2CH_3)_2$ | 241% | 50 | ND | ND | ND | ND |
| $-(CH_2)_2N(CH_2CH_2OH)_2$ | 277% (4/6) | 200 | ND | ND | ND | ND |
| $-(CH_2)_3N(CH_2CH_2OH)_2$ | 194% | 100 | ND | ND | ND | ND |

[a] Data are adapted from Showalter *et al.* (1987).
[b] Lifespan of mice bearing the P388 leukaemia treated with drug at the optimal dose/lifespan of control animals. Cures indicates number of mice surviving at day 30 out of total number treated.
[c] Optimal dose when given as two i.p. injections on days 3 and 7 after inoculation of tumour.
[d] ND = not determined.

In addition to the studies of the structural requirements for activity discussed above, two further major aspects have been investigated. The first concerns the effect of altering the position of the second basic side-chain from $C^5$ to $C^7$. As shown in Table 2.4, in four of five cases the $C^7$ substituted compounds were less active both *in vivo* and *in vitro*. In this case the lack of A-ring hydroxyl substituents should stabilize the molecules to oxidation and, hence, the *in vitro* cytotoxicity data can be considered. The inactivity of the $C^7$ substituted compounds is encouraging, since such a major structural alteration would be expected to have profound effects on DNA interactions. Unfortunately, the only $C^7$ substituted compound where DNA interactions have been directly studied was the compound shown in Table 2.4 with a methyl group at $R_1$ and $NH(CH_2)_2NH(CH_2)_2OH$ at $R_3$ (Hartley *et al.*, 1988b). The ethidium displacement and DNA unwinding demonstrated for this compound may have been due to the molecule interacting with DNA in an inverse manner, i.e. with the C ring taking the place occupied by the A ring in the binding of the $N^2$,$C^5$ bis-alkylamino substituted molecules. Despite the DNA binding and unwinding properties of the compound, it was inactive *in vivo* and was of relatively low potency *in vitro* (Table 2.4) (Showalter *et al.*, 1987).

The final major structural determinant of the *in vivo* antitumour activity of the anthrapyrazoles to be investigated is the effect of A-ring substitution. As indicated above (Table 2.1), non-A-ring hydroxylated anthrapyrazoles (i.e. dihydro- or deshydroxyanthrapyrazoles) have in most cases less *in vivo* activity and/or potency than their 7,10-dihydroxy counterparts. To investigate the optimal A-ring hydroxyl substitution pattern, two series of 7-monohydroxy, 7,10-dihydroxy and 7,8,10-trihydroxy compounds were synthesized and compared in a range of tumour models (Leopold *et al.*, 1985). The trihydroxy compounds are very readily oxidized (see below and Showalter *et al.*, 1986) and again only *in vivo* efficacy data are useful. Table 2.5 compares the activities of the two series of compounds against four murine tumour models. As is clear, with the possible exception of the one compound which was inactive against the Colon 11a, and only marginally active against the Ridgeway osteogenic sarcoma, the other compounds are of comparable activity. Despite the good activity of the trihydroxy compounds, rapid spontaneous oxidation limits their potential as pharmaceuticals (Showalter *et al.*, 1986). The more important observation concerning A-ring hydroxyl substitution is the greatly reduced activity of a 10-monohydroxy compound in comparison with the corresponding 7-monohydroxy and the 7,10-dihydroxy molecules. Although this has only been studied in one series (Figure 2.3, **VIII–XI**), the data are unequivocal. As shown in Table 2.6, the 7-monohydroxy compound was found to be 35-fold more potent *in vitro* and both more potent (*ca* fourfold) and more active *in vivo* than the 10-monohydroxy molecule (Showalter *et al.*, 1984, 1986). As discussed below, despite careful studies on the comparative

**Table 2.4** Comparative *in vitro* cytotoxicity and *in vivo* activity of anthrapyrazoles with $C^5$ or $C^7$ aminoalkylamino substituents[a]

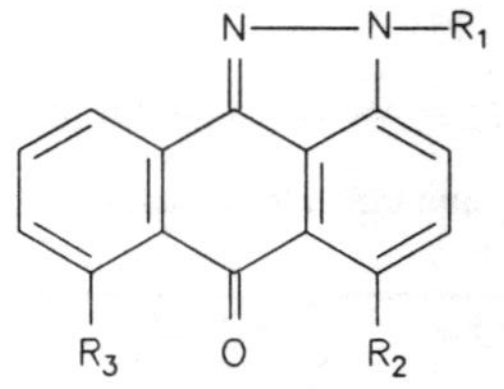

| $R_1$ | $R_2$ | $R_3$ | L1210 $IC_{50}$ (nM)[b] | P388 T/C[c] | Dose (mg/kg)[d] |
|---|---|---|---|---|---|
| $-CH_3$ | $NH-CH_2CH_2NHCH_2CH_2OH$ | $-H$ | 710 | 136% | 100 |
| | $-H$ | $NH-CH_2CH_2NHCH_2CH_2OH$ | 1200 | 118% | 200 |
| $-CH_3$ | $NH-CH_2CH_2N(CH_2CH_3)_2$ | $-H$ | 670 | 98% | 400 |
| | $-H$ | $NH-CH_2CH_2N(CH_2CH_3)_2$ | Inactive | ND[e] | ND |
| $-CH_2CH_2NHCH_2CH_2OH$ | $NH-CH_2CH_2NHCH_2CH_2OH$ | $-H$ | 74 | 199% | 50 |
| | $-H$ | $NH-CH_2CH_2NHCH_2CH_2OH$ | 390 | 133% | 12.5 |
| $-CH_2CH_2N(CH_2CH_3)_2$ | $NH-CH_2CH_2NHCH_2CH_2OH$ | $-H$ | 32 | 169% | 25 |
| | $-H$ | $NH-CH_2CH_2NHCH_2CH_2OH$ | 230 | 117% | 12.5 |
| $-CH_2CH_2N(CH_2CH_3)_2$ | $NH-CH_2CH_2N(CH_2CH_3)_2$ | $-H$ | 390 | 106% | 50 |
| | $-H$ | $NH-CH_2CH_2N(CH_2CH_3)_2$ | 600 | 103 | 50 |

[a] Data are adapted from Showalter *et al.* (1987).
[b] Concentration required to give 50% inhibition of cell growth following 72 h exposure.
[c] Lifespan of mice bearing the P388 leukaemia treated with drug at the optimal dose/lifespan of control animals.
[d] Optimal dose when given as two i.p. injections on days 3 and 7 after inoculation of tumour.
[e] ND = no data given.

**Table 2.5** Effect of A-ring hydroxyl substitution on the *in vivo* activity of anthrapyrazoles against four transplanted murine tumours[a]

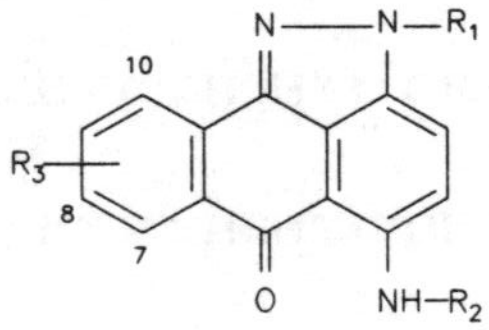


| $R_1$ | $R_2$ | $R_3$ | P388[b] | Mammary 16C[b] | Colon 11a[b] | Ridgeway OS[c] |
|---|---|---|---|---|---|---|
| $-(CH_2)_2NH(CH_2)_2OH$ | $-(CH_2)_2NH(CH_2)_2OH$ | 7-OH | ++++ | ++++ | + | AA |
| $-(CH_2)_2NH(CH_2)_2OH$ | $-(CH_2)_2NH(CH_2)_2OH$ | 7,10-$(OH)_2$ | ++++ | ++++ | – | A |
| $-(CH_2)_2NH(CH_2)_2OH$ | $-(CH_2)_2NH(CH_2)_2OH$ | 7,8,10-$(OH)_3$ | ++++ | +++ | + | AA |
| $-(CH_2)_2NH(CH_2)_2OH$ | $-(CH_2)_2NHCH_3$ | 7-OH | ++++ | +++ | + | AA |
| $-(CH_2)_2NH(CH_2)_2OH$ | $-(CH_2)_2NHCH_3$ | 7,10-$(OH)_2$ | ++++ | ++++ | ++ | AA |
| $-(CH_2)_2NH(CH_2)_2OH$ | $-(CH_2)_2NHCH_3$ | 7,8,10-$(OH)_3$ | ++++ | ++++ | ++ | AA |

[a] Data are adapted from Leopold *et al.* (1985).
[b] Data are given as the gross $\log_{10}$ cell kill following treatment at approximately equitoxic, maximum tolerated doses: ++++, >2.8; +++, 2.0–2.8; ++, 1.3–1.9; +, 0.7–1.2; –, <0.7.
[c] Activity rating derived from the day 35 treated/control tumour sizes: A = 11–42%, AA = <11%.

**Table 2.6** Effect of A-ring hydroxyl substitution pattern on the *in vitro* and *in vivo* properties of a series of anthrapyrazoles[a]

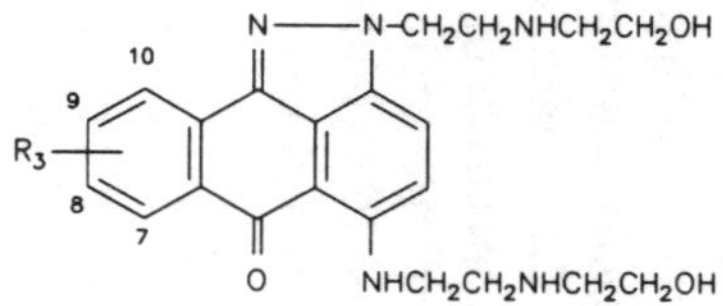


| $R_3$ | *Alkaline elution*[b] | *Nucleic acid synthesis* $IC_{50}$[c] | | *L1210* $IC_{50}$ in vitro[d] | *P388* in vivo[e] | | *Mammary 16C* in vivo[f] | |
|---|---|---|---|---|---|---|---|---|
| | | *DNA (μM)* | *RNA (μM)* | | *T/C (cures)* | *Dose* | *Growth delay* | *Dose* |
| H | 0.24 $h^{-1}$ | 2.2 | 15.5 | 74 nM | 199% | 50 | 8 days | 15 |
| 7-OH | 0.97 $h^{-1}$ | 0.3 | 4.5 | 5 nM | 248% (5/6) | 25 | 11.4 days | 12.5 |
| 10-OH | 0.09 $h^{-1}$ | ND[g] | ND | 180 nM | 221% | 100 | ND | ND |
| 7,10-OH | 1.01 $h^{-1}$ | 0.2 | 16.5 | 740 nM | 206% (1/5) | 6.25 | 8 days | 15 |
| 7,9,10-OH | 0.01 $h^{-1}$ | ND | ND | ND | 165% | 3.12 | ND | ND |
| 7,8,10-OH | 0.02 $h^{-1}$ | ND | ND | ND | –(6/6) | 25 | 8.4 days | 12.5 |

[a] Data are adapted from Showalter *et al.* (1984, 1986, 1987) and Leopold *et al.* (1985).
[b] Alkaline elution rate constant for DNA prepared from cells following a 1 h exposure to 1 μM drug in each case.
[c] Concentration required to give 50% inhibition of [$^3$H]-thymidine (DNA) or [$^3$H]-uridine (RNA) incorporation.
[d] Concentration required to give 50% inhibition of L1210 cell growth following 72 h exposure.
[e] Lifespan of mice bearing the P388 leukaemia treated with drug at the optimal dose/lifespan of control animals. Optimal dose is given as mg/kg per injection when administered as described in Showalter *et al.* (1984, 1986, 1987) and Leopold *et al.* (1985). Where more than one result is reported, the optimum value is given.
[f] Tumour growth delay in mice bearing the mammary 16C adenocarcinoma treated with the dose as given (mg/kg per injection).
[g] ND = no data given.

DNA interactions of these molecules, no clear explanation for this difference in activity has been identified (Hartley *et al.*, 1988b).

The overall features of structure–activity relationships for anthrapyrazoles can be summarized as follows. For optimal activity and potency against the P388 leukaemia *in vivo* the molecule requires (Tables 2.1–2.6):

(1) A 7-hydroxyl substituent in the A ring.

(2) Alkylamine substituents at $N^2$ and $N^5$ with two or three methylene groups between the alkyl chain amine and the nitrogen attached to the anthrapyrazole ring.

(3) The ethylamine or propylamine substituents at $N^2$ and $N^5$ can be left as primary amines or converted to either secondary or tertiary amines. Methyl, ethyl, hydroxyethyl and aminoethyl amines are, for example, all active.

Although useful, the above structure–activity analyses did not allow of the identification of a small enough subset of compounds from which clinical candidates could be chosen. Hence, a representative sample of mainly 7,10-dihydroxy anthrapyrazoles were evaluated against a broader range of murine tumours (Leopold *et al.*, 1985). Table 2.7, which summarizes certain of the data obtained, illustrates the broad spectrum of results that can be obtained even within a closely related group of compounds. All of these compounds had high-level activity against the P388 leukaemia ($\geq 4$ decades of cell kill), and most of the compounds tested were also highly active against other commonly used rapidly growing murine tumours, i.e. the L1210 leukaemia, the B16 melanoma and the M5076 sarcoma (Leopold *et al.*, 1985). In contrast, the three murine tumours employed to produce the data in Table 2.7 were relatively slow-growing and, in particular, two were adenocarcinomas. The data in Table 2.7 show that, for maximum activity against the mammary adenocarcinoma 16C, a secondary ethanolamine is required in the $R_1$ alkylamine side-chain. In contrast, the Ridgeway osteogenic sarcoma is sensitive to compounds with either a primary, secondary or tertiary amine at this position. As previously observed in the P388 model (Table 2.2), an amine in the $R_1$ side-chain is required with the $N^2$-hydroxyethyl compound being essentially inactive. To be effective against the mammary adenocarcinoma 16C, the anthrapyrazoles also appear to require a secondary or primary amine in the $R_2$ side-chain. For example, against the mammary 16C tumour a compound with an otherwise 'active' structure was completely inactivated by bis-hydroxyethylation of the terminal amine of the $N^5$-ethylamine ($R_2$) substituent, whereas the corresponding monohydroxyethyl derivative was extremely active. If anything, the converse effect was seen against the Ridgeway osteogenic sarcoma. More importantly, as shown in Table 2.7, doxorubicin was superior to mitoxantrone in this panel of tumours, yet, despite this, a number of

**Table 2.7** Comparative *in vivo* antitumour activity of 7,10-dihydroxyanthrapyrazoles against a panel of solid murine tumours[a]

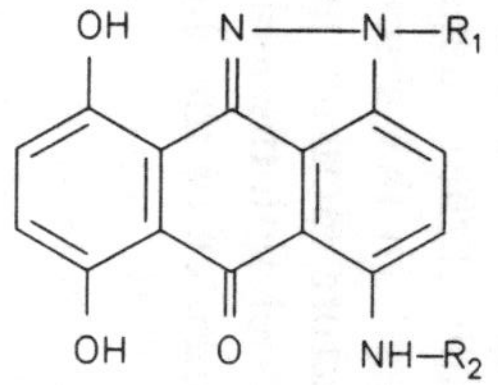


| $R_1$ | $R_2$ | *Mammary 16C*[b] | *Colon 11a*[b] | *Ridgway OS*[c] |
|---|---|---|---|---|
| $-(CH_2)_2NH(CH_2)_2OH$ | $-(CH_2)_2NH(CH_2)_2OH$ | ++++ | – | A |
| $-(CH_2)_2NH(CH_2)_2OH$ | $-(CH_2)_2N(CH_2CH_2OH)_2$ | – | NT | AA |
| $-(CH_2)_2NH(CH_2)_2OH$ | $-(CH_2)_2NH(CH_2)_2N(CH_3)_2$ | ++ | + | A |
| $-(CH_2)_2NH(CH_2)_2OH$ | $-(CH_2)_2NHCH_3$ | ++++ | ++ | AA |
| $-(CH_2)_2NH(CH_2)_2OH$ | $-(CH_2)_2NH_2$ | ++++ | – | A |
| $-(CH_2)_2NH(CH_2)_2OH$ | $-(CH_2)_3NH_2$ | ++++ | – | AA |
| $-(CH_2)_2N(CH_2CH_3)_2$ | $-(CH_2)_2NH(CH_2)_2OH$ | – | – | AA |
| $-(CH_2)_2N(CH_2CH_3)_2$ | $-(CH_2)_2NH_2$ | + | ++ | AA |
| $-(CH_2)_2N(CH_3)_2$ | $-(CH_2)_2NH(CH_2)_2OH$ | + | +++ | AA |
| $-(CH_2)_3N(CH_3)_2$ | $-(CH_2)_2NH_2$ | – | – | A |
| $-(CH_2)_2NH_2$ | $-(CH_2)_2NH(CH_2)_2OH$ | +++ | ++ | AA |
| $-(CH_2)_2NH_2$ | $-(CH_2)_2NHCH_3$ | – | + | AA |
| $-(CH_2)_3NH_2$ | $-(CH_2)_2NH(CH_2)_2OH$ | +++ | + | AA |
| $-(CH_2)_2OH$ | $-(CH_2)_2N(CH_3)_2$ | – | – | A |
| $-CH_2CH(OH)CH_2OH$ | $-(CH_2)_2NH(CH_2)_2OH$ | + | – | – |
| Reference compounds: | | | | |
| Doxorubicin | | ++++ | ++ | AA |
| Ametantrone | | + | – | – |
| Mitoxantrone | | + | – | A |

[a] Data are adapted from Leopold *et al.* (1985).
[b] Data are given as the gross $\log_{10}$ cell kill following treatment at approximately equitoxic, maximum tolerated doses: ++++, >2.8; +++, 2.0–2.8; ++, 1.3–1.9; +, 0.7–1.2; –, <0.7.
[c] Activity rating derived from the day 35 treated/control tumour sizes: A = 11–42%, AA = <11%.

anthrapyrazoles displayed antitumour activity which equalled or approached that of doxorubicin. Activity in these tumours was a major criterion for the selection of compounds for clinical trial.

Finally, in all the above discussions the protonation state of the anthrapyrazoles has been ignored. Although the exact values have not been reported, the terminal aliphatic nitrogens will have a $pK_a$ of above 9 and, hence, be protonated at physiological pH. In contrast to the alkyl side-chain nitrogen, the $N^5$ nitrogen is not considered to be basic (Showalter *et al.*, 1987). The observation of somewhat reduced activity following dialkylation of the $N^5$ position (i.e. tertiary amine) led these authors to suggest that hydrogen bonding between the proton on $N^5$ of secondary amines and the adjacent $C^6$ carbonyl is important. This suggestion is supported by the lack of activity of a compound with a nitro group at $C^5$, where again there could be no hydrogen bonding with the $C^6$ carbonyl (Leopold *et al.*, 1985). However, in lacking an $R_2$ alkylamine group, this latter compound may have been inactive for other reasons, since other series of monoalkylamine anthrapyrazoles were inactive (Table 2.2).

In conclusion, although an extremely large number of *in vitro* and *in vivo* antitumour studies have been performed, the finer details of the structural determinants of anthrapyrazole activity remain unclear. In the absence of a complete understanding of anthrapyrazole structure–activity relationships, three compounds were selected for clinical trial with the aim of covering a range of structural modifications. The structures of the three clinical candidates, CI-937 (**VI**), CI-941 (**IX**) and CI-942 (**VII**), are shown in Figures 2.2 and 2.3. Given the likely clinical impact of these compounds (see below), the lack of a full understanding of the effect of structural changes on activity, and the underlying mechanisms involved, should be rectified. Studies should be pursued using A-ring monohydroxylated compounds, since this avoids the problem of spontaneous oxidation and, hence, meaningful *in vitro* cytotoxicity data can be generated. Comparison of the molecular, cellular and whole animal pharmacology of the 7- and 10-monohydroxylated molecules in each series of compounds would be important, as the current data suggest that this apparently simple modification markedly alters the biological properties of the anthrapyrazoles. Data available on existing compounds, particularly with regard to DNA interactions and electron transfer processes, are discussed below. Prior to this the cellular pharmacology of the anthrapyrazoles will be briefly covered.

## 4 Cellular Pharmacology of the Anthrapyrazoles

Following the exposure of tumour cells to anthrapyrazoles under conditions which are not complicated by chemical instability, i.e. the use of

A-ring monohydroxylated compounds and/or short exposure to dihydroxy molecules, there is marked cytotoxicity. Cytotoxicity has been measured by either growth inhibition or clonogenic assay, and by inhibition of DNA, RNA and protein synthesis (Showalter *et al.*, 1984, 1986; Fry *et al.*, 1985). Anthrapyrazole-induced inhibition of DNA synthesis occurs at approximately one-tenth of the concentration required for the inhibition of RNA synthesis (Fry *et al.*, 1985; Table 2.6) and in this respect the anthrapyrazoles are similar to amsacrine and certain di- or trisaccharide anthracyclines (Crooke *et al.*, 1978). In contrast, both doxorubicin- and mitoxantrone-induced inhibition of DNA and RNA synthesis occurred at the same concentration, with mitoxantrone being some fourfold more potent than doxorubicin at inhibiting the synthesis of both macromolecules (Fry *et al.*, 1985).

Consistent with the more potent inhibition of DNA over RNA synthesis is the greater inhibition of DNA as compared with RNA polymerase (Fry *et al.*, 1985). Measurements of cellular ribose and deoxyribose nucleoside triphosphate pools confirmed that inhibition was not due to substrate depletion and, in addition, exogenous DNA could overcome the inhibition of both polymerases. Finally, alkaline elution studies confirmed that protein-associated DNA strand breaks were formed (Table 2.6) and, as discussed below, all these data are consistent with DNA interaction being the primary mechanism of anthrapyrazole cytotoxicity.

The final area of the cellular pharmacology of anthrapyrazoles that has received attention concerns mechanisms of resistance. In general, there is cross-resistance between doxorubicin and the anthrapyrazoles when *in vitro* studies are performed (Burchenal *et al.*, 1985; Sebolt *et al.*, 1985; Klohs *et al.*, 1986; Havelick *et al.*, 1987; Coley *et al.*, 1989; Cole, 1990). However, the cross-resistance is not always complete and its extent appears to be dependent upon the particular anthrapyrazole used. For example, Klohs *et al.* (1986) reported that resistance to five 7,10-dihydroxyanthrapyrazoles ranged from 2.8 to 298 in a P388 multidrug-resistant cell line that was 51- and 38-fold resistant to doxorubicin and daunorubicin, respectively. However, cytotoxicity was measured following 72 h exposure and, hence, the interpretation of the data may be complicated by the spontaneous oxidation of the anthrapyrazoles. Despite this problem, there was clear potentiation of cytotoxicity by verapamil and it is highly likely that the anthrapyrazoles are substrates for the P-glycoprotein drug efflux pump encoded by the *mdr*-1 gene. Using the 7-monohydroxy compound CI-941, which is less susceptible to problems of spontaneous oxidation, Coley *et al.* (1989) also reported cross-resistance to doxorubicin in two well-characterized P-glycoprotein overexpressing cell lines. In one cell line (EMT6 murine mammary tumour) the resistance factor for CI-941 (9-fold) was not of the same level as for doxorubicin (34-fold), while in the other (HT69 human small cell lung cancer) it was similar. Taken together, these data are again strong evidence that anthrapyrazoles are substrates for the P-glycoprotein pump.

In addition to accelerated drug efflux, there may be other mechanisms of resistance to anthrapyrazoles. Cole (1990) reported that three anthrapyrazoles — CI-937, CI-941 and CI-942 — were cross-resistant to doxorubicin in a small cell lung cancer line which reportedly does not overexpress P-glycoprotein (Cole *et al.*, 1989). The mechanism underlying resistance in this cell line has not been defined, although alterations in glutathione metabolism have been demonstrated (Cole *et al.*, 1990). Clearly, if a component of anthrapyrazole cytotoxicity involves the generation of reactive drug and/or oxygen free radical species (see below), glutathione-mediated resistance could be implicated. In particular, oxidation of hydroxyanthrapyrazoles could result in products likely to interact with glutathione.

The most important question concerning anthrapyrazoles and drug resistance has yet to be addressed: namely, will these agents work in tumours clinically resistant to doxorubicin? The preclinical data described above suggest not; however, ongoing clinical investigations (see below) will soon reach the stage when this question can be addressed in patients.

## 5 Molecular Pharmacology of the Anthrapyrazoles

### *Anthrapyrazole–DNA Interactions*

Anthrapyrazole–DNA equilibrium binding constants have been evaluated for a range of structurally diverse anthrapyrazoles as determined from ethidium displacement studies (Hartley *et al.*, 1988b). DNA binding constants are a measure of the overall affinity of the drug–DNA complex without specific regard for the site of interaction. Anthrapyrazoles varying in $N^2$ and $C^5$ substitutions and A-ring hydroxylation pattern all showed DNA binding constants in the range $10^6$–$10^8$ $\text{M}^{-1}$ (cf. mitoxantrone, $2.6 \times 10^8$ $\text{M}^{-1}$). As can be seen from Table 2.8, truncation or absence of either the $N^2$ or $C^5$ side-chains in a series of anthrapyrazoles with one constant side-chain of $(CH_2)_2NHCH_2CH_2OH$, as is present in mitoxantrone, and one side-chain of variable length results in a decrease in DNA binding affinity. In contrast, substitution with longer alkylamino groups or slight alteration in the terminal amino function maintains or improves binding affinity (Hartley *et al.*, 1988b). The planar nature of the anthrapyrazoles, in combination with the cationic nature of the side-chains, is predictive of intercalative binding — a process that is related to DNA unwinding (Hartley *et al.*, 1988b). The intercalative ability of deshydroxyanthrapyrazole and several A-ring hydroxy substituted anthrapyrazoles with constant side-chains of $(CH_2)_2NHCH_2CH_2OH$ groups has been determined using a topoisomerase I/ plasmid DNA unwinding assay (Hartley *et al.*, 1988b). The results (as shown in Table 2.9) demonstrate that deshydroxyanthrapyrazole produces the greatest unwinding angle and, hence, exhibits the strongest intercalative binding, which in fact is similar to that determined for ethidium. Further evidence for intercalative binding of deshydroxyanthrapyrazole comes from increased viscosity of calf thymus DNA due to helix unwinding and spectral hypochromicity and bathochromic shift as a result of charge transfer with DNA bases (Hartley *et al.*, 1988b). Substitution of the chromophore A ring with one or two hydroxy groups results in a decreased or partial intercalation, whereas trihydroxy substituted compounds do not appear to intercalate at all (Hartley *et al.*, 1988b), although the problem of stability through spontaneous oxidation of trihydroxy substituted anthrapyrazoles (Fry *et al.*, 1985; Showalter *et al.*, 1986) cannot be ignored. It is suggested that since the trihydroxyanthrapyrazoles have comparable binding constants to those of other anthrapyrazoles, then exterior binding of trihydroxy derivatives is the major mode of their DNA interaction (Hartley *et al.*, 1988b). DNA thermal denaturation studies confirmed the decrease in intercalative binding associated with hydroxylation of the anthrapyrazole A ring (Hartley *et al.*, 1988b). Experimental studies using synthetic polynucleotides indicate that the apparent binding of deshydroxyanthrapyrazole to GC base pairs is approximately three times that of AT base pairs. This finding is confirmed by thermal denaturation studies which show elevated melting temperatures of DNA with increasing GC content in the presence of deshydroxyanthrapyrazole and several A-ring hydroxylated derivatives (Hartley *et al.*, 1988b). This apparent GC preference is likely to be as a result of slow dissociation from GC sites, due to the reduced rate of opening of GC sites, compared with AT sites (Malhotra and Hopfinger, 1980). However, photosensitization studies with deshydroxyanthrapyrazole showed no sequence-specific strand breakage activity (Hartley *et al.*, 1988b) and theoretical computations on the intercalative binding of deshydroxyanthrapyrazole to double-stranded $d(GCGC)_2$ and $d(ATAT)_2$, and assuming an unwinding angle of 29°, indicate no significant base pair preference (Chen *et al.*, 1987).

Despite detailed studies relating anthrapyrazole structure to DNA binding affinity and extent of intercalation, there is no clear relationship between these equilibrium parameters and cytotoxicity (Leopold *et al.*, 1985; Showalter *et al.*, 1986). In particular, the mono- and dihydroxyanthrapyrazoles are shown to have decreased intercalation compared with the deshydroxanthrapyrazoles (Hartley *et al.*, 1988b) but are more cytotoxic both *in vitro* and *in vivo* (Leopold *et al.*, 1985; Showalter *et al.*, 1986; see above). An explanation for this may be that the rate of dissociation of anthrapyrazoles from DNA is slowed by the presence of A-ring hydroxy groups, thereby facilitating the inhibition of DNA-mediated processes. Although

**Table 2.8** Effect of anthrapyrazole side-chain alterations on DNA binding[a]

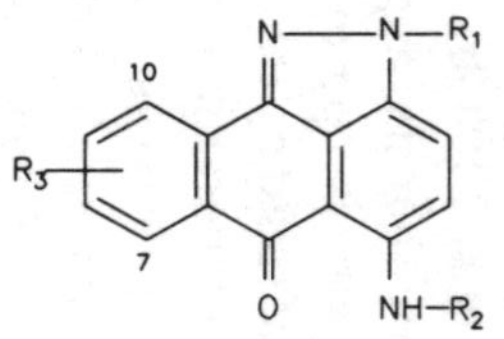

| $R_1$ | $R_2$ | $R_3$ | DNA binding constant ($\times 10^7 M^{-1}$)[b] | Unwinding angle (degrees)[c] |
|---|---|---|---|---|
| $-(CH_2)_2NH(CH_2)_2OH$ | $-(CH_2)_2NH(CH_2)_2OH$ | 7,10-H | 7.4 | 29.2 |
| $-H$ | $-(CH_2)_2NH(CH_2)_2OH$ | 7,10-H | 0.4 | 24.0 |
| $-(CH_2)_2NH(CH_2)_2OH$ | $-CH_3$ | 7,10-H | <0.2 | 22.8 |
| $-CH_2CH_2OH$ | $-(CH_2)_2NH(CH_2)_2OH$ | 7,10-$(OH)_2$ | 4.1 | 17.6 |
| $-(CH_2)_2NH(CH_2)_2OH$ | $-(CH_2)_5NH_2$ | 7,10-$(OH)_2$ | 27 | 20.5 |
| $-(CH_2)_2NH(CH_2)_2OH$ | $-(CH_2)_3NH(CH_2)_4NH(CH_2)_3NH_2$ | 7,10-$(OH)_2$ | 7.6 | 21.6 |
| Mitoxantrone | | | 26 | 17.5[d] |

[a] Data are adapted from Hartley *et al.* (1988b).
[b] Calculated from the concentration of drug required to reduce ethidium–DNA fluorescence to 50% and assuming the binding constant of ethidium to be $10^7 M^{-1}$.
[c] Based on the drug concentration required to unwind supercoiled DNA such that topoisomerase I will have no effect on topological winding number.
[d] Unwinding angles of 26.5° and 23° also reported; see Denny and Wakelin (1990).

**Table 2.9** Effect of A-ring hydroxylation on anthrapyrazole DNA binding[a]

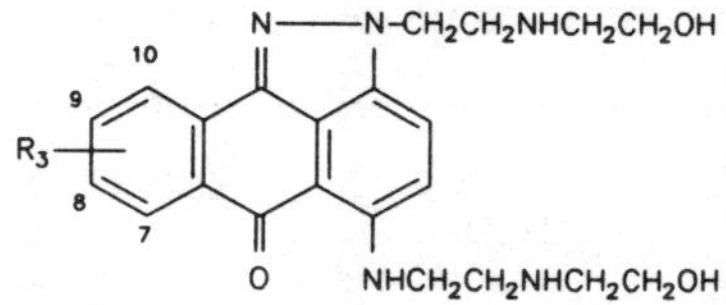


| *$R_3$* | *DNA binding constant ($\times 10^7 M^{-1}$)*[b] | *Unwinding angle (degrees)*[c] | *$\Delta T_m$(°C)*[d] |
|---|---|---|---|
| 7,10-H | 7.4 | 29.2 | 13 |
| 7-OH | 10 | 23.8 | 9 |
| 10-OH | 3.7 | 18.5 | 8 |
| 7,10-OH | 20 | 21.0 | 7 |
| 7,8,10-OH | 19 | 0 | 3 |
| 7,9,10-OH | 8.0 | 0 | ND[f] |
| Mitoxantrone | 26 | 17.5[e] | ND |

[a] Data are adapted from Hartley *et al.* (1988b).
[b] Calculated from the concentration of drug required to reduce ethidium–DNA fluorescence to 50% and assuming the binding constant of ethidium to be $10^7 M^{-1}$.
[c] Based on the drug concentration required to unwind supercoiled DNA such that topoisomerase I will have no effect on topological winding number.
[d] Difference in melting temperature mid-point ($\Delta T_m$) of calf thymus DNA in the presence and absence of drug. Data are taken from Figure 7, Hartley *et al.* (1988b).
[e] Unwinding angles of 26.5° and 23° also reported; see Denny and Wakelin (1990).
[f] ND = no data given.

there is not yet direct evidence for this, a parallel can be drawn with the anthraquinones, since kinetic studies show that there is a significant slowing of the DNA-dissociation rate of mitoxantrone compared with that of its deshydroxy analogue, ametantrone (Denny and Wakelin, 1990). Anthrapyrazole cytotoxicity may also involve DNA surface binding in combination with the self-association of these molecules (Hartley *et al.*, 1988b), which may facilitate DNA condensation and aggregation. It has been suggested that such processes contribute to the cytotoxicity of mitoxantrone (Reszka *et al.*, 1988a). The binding of anthrapyrazoles to DNA is assumed, like that for all DNA affinic agents, to be associated with the inhibition of DNA and RNA synthesis. In this respect, dihydroxyanthrapyrazoles have been shown to inhibit DNA synthesis more potently than RNA synthesis (Fry *et al.*, 1985; Table 2.6) — an observation that may be related to the slow dissociation kinetics and, hence, persistence of drug–DNA complexes (Fry *et al.*, 1985). However, differential inhibition of nucleic acid synthesis is also seen with A-ring deshydroxyanthrapyrazoles (Showalter *et al.*, 1986; Table 2.6), which, as argued above, may dissociate more rapidly from DNA.

### *Topoisomerase Inhibition*

There is no direct evidence that the anthrapyrazoles inhibit either DNA type I or type II topoisomerases. However, the intercalating ability of anthrapyrazoles and structural similarities to the known topoisomerase II inhibitors mitoxantrone and doxorubicin provide circumstantial evidence that they are likely to inhibit this enzyme. Two 7,10-dihydroxyanthrapyrazoles, CI-937 and CI-942, both caused protein-associated single- and double-strand breaks in a drug-concentration-dependent manner (Fry *et al.*, 1985; Frank *et al.*, 1989). This strand breakage, which is only revealed in cellular DNA after treatment with strong protein denaturants, is often interpreted as the end event associated with trapping of a topoisomerase–DNA cleavable complex (Lui, 1989). Effectively, the presence of intercalator drug causes an inhibition of topoisomerase-catalysed DNA strand passing. How this is related to drug-mediated lethal events in tumour cells is not clear, but in principle it appears that disturbance of topoisomerase function (for example, in DNA replication and transcription) is the essential factor and not trapping of the cleavable complex *per se* (Smith, 1990). The molecular basis of this event is not known but is likely to involve an intercalated drug–DNA–enzyme ternary complex. It has been suggested that the potent cytotoxicity associated with mitoxantrone is as a result of persistence of trapped topoisomerase II complexes (Fox and Smith, 1990). The protein-associated strand breaks mediated by 7,10-dihydroxyanthrapyrazoles (CI-937 and CI-942) are also persistent (Fry *et al.*, 1985;

Frank *et al.*, 1989), again suggesting a similar mechanism of cytotoxicity to that of mitoxantrone.

## 6 Activation of Anthrapyrazoles in Biological Systems

Binding to DNA and inhibition of DNA-associated enzymes is undoubtedly involved in the cytotoxic action of the anthrapyrazoles. In this respect, preliminary studies using confocal laser microscopy have clearly identified anthrapyrazoles associated with nuclear material in incubates of viable cells (Patterson and Smith, unpublished observations). What is less clear is the contribution of metabolic activation of the anthrapyrazoles to potentially cell-damaging species that may potentiate DNA-mediated events. Essentially, the activation of the anthrapyrazoles in biological systems is related to their potential to mediate electron transfer processes via an anthrapyrazole reactive intermediate formation of which could be catalysed by cellular enzymes, transition metals and/or visible light.

### *Metabolic Reductive Activation*

The question of whether anthrapyrazole-mediated free radical generation occurs under biologically relevant conditions is of considerable interest, since the quinoneimine component of the anthrapyrazole chromophore is considered refractory to metabolic reduction (Showalter *et al.*, 1986). This may distinguish the anthrapyrazoles from the anthraquinones and the anthracyclines, since the latter compounds are known to undergo reduction to free radical intermediates in human tissue (Basra *et al.*, 1985). At this point it is useful to summarize the principles involved in anthraquinone/anthracycline-mediated free radical generation. This process involves either (i) one-electron transfer to generate the semianthraquinone directly (reaction 2.1) or via a dismutation/comproportionation equilibrium between semianthraquinone ($AQH^{\cdot -}$) and hydroanthraquinone ($AQH_2$), as illustrated by reactions (2.1)–(2.3), or (ii) direct two-electron transfer to generate the hydroanthraquinone (reaction 2.4) with concomitant comproportionation to the semianthraquinone (reaction 2.3). Protonation of the semianthraquinones generated in these reactions will depend on their respective $pK_a$ values and the pH of the reaction medium.

$$AQ + e^- + H^+ \rightarrow AQH^{\cdot -} \quad (2.1)$$

$$AQH^{\cdot -} + e^- + H^+ \rightarrow AQH_2 \quad (2.2)$$

$$AQ + AQH_2 \rightarrow 2AQH^{\cdot -} \quad (2.3)$$

Metabolic one-electron reduction of anthraquinones and anthracyclines can occur in various subcellular compartments mediated by several NAD(P)H flavoprotein reductases, including microsomal/nuclear membrane cytochrome P450 reductase, cytosolic xanthine oxidase and mitochondrial NADH dehydrogenase (Gianni *et al.*, 1983, and references therein; Kharasch and Novak, 1983a). All these enzymes are obligate one-electron reductases (reduce in one-electron steps) and, hence, give rise to anthrasemiquinones by processes involving reactions (2.1)–(2.3). Direct two-electron reduction of mitoxantrone and daunorubicin by cytosolic DT-diaphorase has also recently been demonstrated (Fisher *et al.*, 1992), as illustrated in reaction (2.4).

$$AQ + 2e^- + 2H^+ \rightarrow AQH_2 \qquad (2.4)$$

The semianthraquinone and hydroanthraquinone species are reoxidized in the presence of dioxygen generating the parent anthraquinone, with concomitant formation of superoxide anions. This process, which involves futile consumption of NAD(P)H, is known as redox cycling (Gianni *et al.*, 1983, and references therein). The superoxide anion is not particularly reactive but can be reduced further to produce hydrogen peroxide, which is a good oxidizing agent and can give rise to toxicity at the site of formation. In addition, hydrogen peroxide is electrically neutral and thus can diffuse to other areas within the cell, causing damage distant from the site of formation. Disproportionation of hydrogen peroxide will result in formation of hydroxyl radicals which react at diffusion-controlled rates with biological macromolecules initiating cell-damaging events such as DNA strand breakage, DNA and/or DNA protein cross-linking, lipid peroxidation and protein damage (Von Sonntag, 1987). In addition, cellular micro- and macromolecules may be altered by hydroxyl-radical-mediated hydroxylation (Von Sonntag, 1987). The requirement for certain transition elements, notably iron or copper, in the overall reduction of dioxygen to the hydroxyl radical is well established (Walling, 1975).

Several studies using liver microsomal fraction as a rich source of reductase have revealed that anthrapyrazoles, including CI-941, CI-937 and CI-942, only marginally increase microsomal oxygen consumption (Fry *et al.*, 1985; Showalter *et al.*, 1986) and superoxide anion generation (Graham *et al.*, 1987). Furthermore, in the presence of liver microsomes or purified NADPH-dependent cytochrome P450 reductase, CI-941 does not generate a semianthraquinone free radical as detected by electron spin resonance spectrometry and actually inhibits microsomal NADPH oxidation (Graham *et al.*, 1987). These results indicate that the anthrapyrazoles of current clinical interest do not redox cycle significantly but instead appear to inhibit NADPH-dependent free radical processes, including lipid peroxidation (Frank and Novak, 1986; Graham *et al.*, 1987). The slight increase in oxygen consumption observed for CI-941, CI-937 and especially CI-942 (Showalter *et al.*, 1986) may be related to the hydroxy substituents on the A ring undergoing slow oxidation in the presence of oxygen. Trihydroxy substitution in the anthrapyrazole A ring results in spontaneous dioxygen consumption (Showalter *et al.*, 1986) as a result of the rapid oxidation of these catechol structures in air. The stabilization of these compounds with ascorbic acid, a mild reducing agent, would support this (Frank *et al.*, 1987, 1989; Nordblom *et al.*, 1989). This propensity for hydroxyanthrapyrazoles to undergo oxidation is discussed in more detail below.

The lack of microsomal redox cycling by anthrapyrazoles would suggest that these compounds are poor substrates for the cellular enzymes that are known to mediate anthracycline quinone reduction. In particular, CI-941 does not appear to be a substrate for NADPH-dependent cytochrome P450 reductase, although the inhibition of doxorubicin free radical formation by CI-941 would indicate that this anthrapyrazole can interact with the site responsible for doxorubicin reduction (Graham *et al.*, 1987). It is suggested from solution studies that the FMN and/or FAD components of the flavoprotein reductases are responsible for electron transfer from NAD(P)H to the anthracyclines (Kharasch and Novak, 1981) and anthraquinones (Kharasch and Novak, 1983a). It would appear that such a process does not occur with the anthrapyrazoles. Measurement of reduction potentials using polarography indicates that anthrapyrazoles are more resistant to two-electron reduction compared with doxorubicin and mitoxantrone, with addition of hydroxy groups to the A ring making the anthrapyrazole chromophore progressively harder to reduce (Showalter *et al.*, 1986). However, a truer indication of redox cycling capability is determined from one-electron reduction potentials measured at pH 7.0 ($E1/7$). Pulse radiolysis studies indicate that CI-941 has a $E1/7$ of $-538 \pm 10$ mV (Graham *et al.*, 1987), which is in fact similar to that for mitoxantrone ($E1/7$, $-527 \pm 5$ mV) (Graham *et al.*, 1987). Hence, it is intriguing that purified NADPH-dependent cytochrome P450 reductase mediates reduction of mitoxantrone (Kharasch and Novak, 1983) but not CI-941 (Graham *et al.*, 1987), and must relate to the imino functionality and possibly the pyrazole (D) ring in the anthrapyrazole chromophore. However, it must be said that mitoxantrone is also difficult to reduce, and although a mitoxantrone free radical has been reported in human liver microsomes (Basra *et al.*, 1985), more recent studies have failed to show free radical generation by this agent (but not other anthraquinones) in human-derived MCF-7 breast cancer cell fractions (Fisher *et al.*, 1989, 1990; Fisher and Patterson, 1992). CI-941 has been shown also not to redox cycle in MCF-7 cells but is considerably more cytotoxic than doxorubicin in this cell line (Fisher and Patterson, 1992). This indicates that free radical generation is not an important factor governing the cytotoxic potential of CI-941. It is tempting to conclude that the lack of reductase-mediated redox cycling and free radical generation by the clinically active anthrapyrazoles is predictive for a reduced propensity for cardiotoxicity. Indeed the quinone-imine structure of the anthrapyrazoles was, by analogy with the iminoanthracyclines, predicted to be more difficult to reduce (Peters *et al.*, 1986) and, hence, less likely to undergo redox cycling than either the anthracyclines or mitoxantrone. However, it is of concern for the continued clinical use of the anthrapyrazoles that mitoxantrone (and ametantrone) does not generate free radicals in heart tissue (Kharasch and Novak, 1983b; Patterson *et al.*, 1983; Patterson and Basra, 1985) but is clearly cardiotoxic (Stuart Harris *et al.*, 1984; Unverferth *et al.*, 1984).

### *Electron Transfer Involving Anthrapyrazole–Iron Chelates*

Several anthrapyrazoles with hydroxyl substituents in the A ring, including CI-941, bind iron(III) and iron(II) (Graham *et al.*, 1989). Hydroxyanthrapyrazole–iron chelates were associated with hydroxyl radical formation as detected by spin trapping studies using ESR. This non-enzymatic process generated DNA single-strand breaks in a supercoiled plasmid model, although no correlation with iron chelation, reactive oxygen formation and cytotoxicity to CCRF-CEM cells was observed (Graham *et al.*, 1989).

### *Metabolic Oxidative Activation*

As indicated earlier, anthrapyrazoles with hydroxyl substituents in the A ring are susceptible to oxidation. Monohydroxylated anthrapyrazoles are oxidized slowly and irreversibly, whereas di- and trihydroxy-substituted anthrapyrazoles are oxidized readily and reversibly. This phenomenon is related to the antioxidant properties of hydroxylated aromatic compounds and is consistent with anthrapyrazole-mediated inhibition of lipid peroxidation (Frank and Novak, 1986; Graham *et al.*, 1987). Oxidation of hydroxanthrapyrazoles is catalysed by horseradish peroxidase in the presence of hydrogen peroxide (Kolodziejczyk *et al.*, 1988). Although the A-ring hydroxy functionality is essential, the site of enzymic oxidation appears to be the amine function, since an anthrapyrazole with methoxy groups in place of hydroxy groups was also shown to be a substrate for peroxidase attack (Reszka *et al.*, 1988a). It is well established that arylamines undergo oxidation to reactive intermediates (Mason and Chignell, 1982). However, the relevance of oxidative activation of the anthrapyrazoles to their molecular pharmacology is yet to be determined, especially since CI-941 (a 7-hydroxyanthrapyrazole) does not damage DNA in the presence of horseradish peroxidase (Fisher and Patterson, 1991). In contrast, activation of mitoxantrone by this enzyme has been shown to result in one- and two-electron oxidation products (Kolodziejczyk *et al.*, 1986; Reszka *et al.*, 1988b) which covalently bind (Reszka *et al.* 1989) and damage DNA (Fisher and Patterson, 1991). Although not proved, it is possible that cellular peroxidases such as myeloperoxidase and prostaglandin endoperoxidase could oxidatively activate di- and trihydroxy anthrapyrazoles to cell-damaging species. The low recovery of parent compound *in vivo* of the dihydroxyanthrapyrazole CI-942 (Frank *et al.*, 1989) and the lack of dose-proportional pharmacokinetics of CI-937 (Wong *et al.*, 1989) may also be as a result of the facile oxidation of these compounds *in vivo*.

*Photoactivation*

It is clear that the chemistry of the anthrapyrazoles renders these compounds difficult to reduce enzymatically. This altered chemistry, evident in the yellow/red colour of the anthrapyrazoles compared with the dark blue associated with mitoxantrone, enables anthrapyrazoles to be more readily photosensitized and, hence, participate in energy/electron transfer reactions. Lown and others were the first to recognize that short exposure (1 min or less) of certain anthrapyrazoles to blue light resulted in their transition from the ground to the triplet state. In this way the energy/electron acceptor property (reducibility) of anthrapyrazoles can be enhanced significantly (Reszka *et al.*, 1986a–c). In early work Lown's group showed that visible light-irradiated (400–540 nm) ethanolic solutions of anthrapyrazoles resulted in singlet oxygen formation by a so-called type II (energy transfer) process (Reszka *et al.*, 1986c). Other studies in aqueous aerated systems showed that, following illumination, certain anthrapyrazoles, in the presence of NADH, consume oxygen and concomitantly generate the superoxide anion, hydrogen peroxide and hydroxyl radicals (Reszka *et al.*, 1986b). This type I (electron transfer) process can be summarized in the following scheme:

$$
\begin{array}{lllll}
AP & & \xrightarrow{h\nu} & AP^{*} & \\
AP^{*} + NADH & & \rightarrow & AP^{\cdot -} & + \; NAD^{\cdot} \; (+H^{+}) \\
AP^{\cdot -} + O_2 & & \rightarrow & AP & + \; O_2^{\cdot -} \\
O_2^{\cdot -} + O_2^{\cdot -} & + \; 2H^{+} & \rightarrow & H_2O_2 & + \; O_2
\end{array}
$$

Although feasible, there was no evidence that photosensitized anthrapyrazole generated superoxide anions from singlet oxygen or underwent obligate two-electron transfer from NADH to generate hydrogen peroxide directly (Reszka, 1986b). Anthrapyrazole-mediated photooxidation of ascorbic acid and dihydroxyphenylalanine (dopa) is also reported (Reszka *et al.*, 1986a), demonstrating the general utility of anthrapyrazole-mediated photosensitization in the presence of biologically relevant electron donors. Structure–activity studies have demonstrated that an essential feature for efficient photoactivation of the anthrapyrazoles is the absence of hydroxyl groups in the chromophore (Hartley *et al.*, 1988a). It would appear that when A-ring hydroxyl groups are present, then intramolecular hydrogen bonding with B-ring quinone and/or imine functionality facilitates radiationless deactivation of the singlet and triplet states (Hartley *et al.*, 1988a). In support of this, mitoxantrone (a dihydroxyanthraquinone) *per se* is only weakly photoactivated (Reszka *et al.*, 1988a), although illuminated deshydroxyanthrapyrazole can act as a photosensitizer for mitoxantrone by mediating transfer of an electron from NAD(P)H to generate mitoxantrone semiquinone, as detected by ESR (Reszka and Lown, 1989).

DNA is a principal target for anthrapyrazoles, owing to their DNA-intercalative and surface-binding high affinities (Fry *et al.*, 1985; Hartley *et al.*, 1988b). Photoactivation of deshydroxyanthrapyrazoles in air results in a non-sequence-specific DNA strand scission by a mechanism involving reactive oxygen (Hartley *et al.*, 1988b). In a series of deshydroxyanthrapyrazoles the extent of light-induced DNA damage was not affected by alterations in their alkylamino side-groups which nevertheless produced different affinities for DNA (Hartley *et al.*, 1988b). These data are consistent with non-bound drug generating non-specific DNA strand breaks via reactive oxygen formation. In the absence of air, photosensitized anthrapyrazoles can cleave DNA directly in a process using NADH as cofactor (Hartley *et al.*, 1988a). In the only study to date concerning whole cells, photoactivated deshydroxyanthrapyrazole was less cytotoxic in adriamycin-resistant P388 cells compared with wild-type cells, although no difference in transport of drug was observed (Kessel, 1989). This suggests that atypical resistance to photoinduced toxicity may parallel that observed for resistance to drugs such as the anthracyclines which enzymatically redox cycle to generate reactive oxygen (Sinha *et al.*, 1987). From the studies to date, it is clear that a knowledge of the photosensitization properties of anthrapyrazoles may assist rationalization of the toxicities associated with the use of these agents. In particular, the possibility of light-induced side-effects of anthrapyrazole chemotherapy must be addressed, although clinically this has not been a problem. Perhaps of greater importance is the potential use of deshydroxyanthrapyrazoles in photodynamic treatment of cancer (Reszka *et al.*, 1986a–c, 1988a).

## 7 Preclinical and Clinical Pharmacology of the Anthrapyrazoles

A detailed discussion of the preclinical and clinical pharmacology of the anthrapyrazoles is beyond the scope of this chapter, and a recent article by Judson (1991) covers this area in detail. Phase I clinical trials with all of the three clinical candidates, CI-937, CI-941 and CI-942, have now been completed and reported (Ames *et al.*, 1990; Hantel *et al.*, 1990; Allan *et al.*, 1991; Ehrlichman *et al.*, 1991; Foster *et al.*, 1992). Qualitatively and quantitatively, the clinical toxicities of the anthrapyrazoles was reasonably well predicted by preclinical studies in mice; and where discrepancies were seen (CI-942), these could be explained by species differences in pharmacokinetics. The compounds have recently become the property of Dupont Pharmaceuticals, who are currently sponsoring Phase II studies. In the future, the three clinical compounds will probably be known as DUP 937, 941 and 942. In addition, CI-941 has been referred to as biantrazole and CI-942 as either oxantrazole or, more recently, piroxantrone. Despite this confusion, one clinical fact is already clear. CI-941 is an extremely active agent for the treatment of breast cancer. In a clinical study performed under the auspices of the UK CRC Phase I/II Committee, single-agent CI-941 produced an objective response rate of 64% in women with breast cancer (Talbot *et al.*, 1991). This response rate compares very favourably with historical single-agent studies using either anthracyclines or mitoxantrone, and clinical interest in this new class of agents is considerable. Although cardiac function has been carefully monitored in the majority of clinical trials, it is still too early to comment on the clinical cardiotoxicity of the anthrapyrazoles. Minor fluctuations in left ventricular ejection fractions have been reported, but these were not of any clinical significance (Talbot *et al.*, 1991).

## 8 Conclusions

The data reviewed in this chapter indicate that an anthrapyrazole may yet become established as the long-sought non-cardiotoxic DNA affinic agent. The preclinical data indicate that anthrapyrazoles have a broader spectrum of activity than the anthraquinones, although at first sight the two drug classes appear broadly similar in structure, especially the dihydroxy derivatives. However, the hydroxyanthrapyrazoles have a chemistry that is distinct from that of the hydroxyanthraquinones, which is reflected in the ease of oxidation of dihydroxyanthrapyrazoles and the differential activities of the 7- and 10-monohydroxy compounds. Clearly, the juxtaposition of hydroxy groups to carbonyl and imino functionalities profoundly influences activity. This may relate to inherent differences in the hydrogen bonding properties of the 7-OH and 10-OH groups and/or to a possible adverse influence of the 10-OH on D-ring stability. DNA binding and inhibition of associated enzymes including (probably) topoisomerase II are functions the anthrapyrazoles share with other DNA affinic agents. However, the greater reactivity of the dihydroxyanthrapyrazoles may well be related to their enhanced antitumour activity by, for example, producing more reactive or persistent intermediates. There is a certain irony in the fact that the anthrapyrazoles were developed as compounds resistant to metabolic reduction but as a result the dihydroxyanthrapyrazoles of clinical interest appear to undergo facile oxidation.

Despite very extensive structure–activity investigations, conventional murine antitumour models were unable to identify, beyond A-ring hydroxylation, a clear single lead compound for clinical trial. The drug sponsors, Warner-Lambert/Parke-Davis Pharmaceutical Research at that stage, took the then unconventional decision of developing three anthrapyrazoles and again Phase I trials failed to clearly distinguish between the three. CI-941, as an A-ring monohydroxyanthrapyrazole, has distinct stability advantages, which greatly simplifies both mechanistic and pharmacological studies. In addition, CI-941 already has proven clinical efficacy. Whether CI-937 and/or CI-942 have similar or superior efficacy remains to be seen.

Returning to the molecular pharmacology of the anthrapyrazoles, despite quite detailed studies, DNA binding affinity *per se* is a poor indicator of anthrapyrazole cytotoxicity. This problem with equilibrium binding studies has been highlighted in the past for many other DNA affinic agents. Kinetic studies of anthrapyrazole–DNA dissociation may help to prove that persistence of DNA binding is an important determinant of activity. In addition, little attention has hitherto been paid to the role of the alkylamino side-chains beyond the recognition of the fact that these cationic functionalities help to stabilize DNA binding. It is appropriate now to establish the precise structural features of these side-chains that allow of maximum interaction with DNA and associated enzymes. The increased efficacy observed in compounds bearing primary and secondary, as opposed to tertiary, alkylamino side-chains suggests hydrogen bonding as an important factor in the stabilization of drug–DNA–enzyme ternary complexes. Furthermore, the high-level activity of ethanolamine derivatives also implicates hydrogen bonding as an important interaction. It is hoped that the ongoing clinical trials will help to provide the impetus to undertake these mechanistic studies such that the cellular and molecular pharmacology of this important new class of drugs can be more fully defined.

**References**

Acton, E. M. and Tong, G. L. (1981). Synthesis and preliminary antitumour evaluation of 5-iminodoxorubicin. *J. Med. Chem.*, **24**, 669–673

Allan, S. G., Cummings, J., Evans, S., Nicolson, M., Stewart, M. E., Cassidy, J., Soukop, M., Kaye, S. B. and Smyth, J. F. (1991). Phase I study of the anthrapyrazole biantrazole: clinical results and pharmacology. *Cancer Chemother. Pharmacol.*, **28**, 55–58

Ames, M. M., Loprinzi, C. L., Collins, J. M., van Haelst-Pisani, C., Richardson, R. L., Rubin, J. and Moertel, C. G. (1990). Phase I and clinical pharmacological evaluation of pirozantrone hydrochloride (oxantrazole). *Cancer Res.*, **50**, 3905–3909

Basra, J., Wolf, C. R., Brown, J. R. and Patterson, L. H. (1985). Evidence for human liver mediated free-radical formation by doxorubicin and mitozantrone. *Anti-cancer Drug Des.*, **1**, 45–52

Burchenal, J. H., Pancost, T. and Elslager, E. (1985). Anthrapyrazole and amsacrine analogs in mouse and human leukaemia *in vitro* and *in vivo*. *Proc. Am. Assoc. Cancer Res.*, **26**, 224

Chen, K.-X., Gresh, N. and Pullman, B. (1987). A theoretical study of the intercalative binding of the antitumour drug anthrapyrazole to double-stranded oligonucleotides. *Anti-cancer Drug Des.*, **2**, 79–84

Cole, S. P. C. (1990). Patterns of cross-resistance in a multidrug-resistant small-cell lung carcinoma cell line. *Cancer Chemother. Pharmacol.*, **26**, 250–256

Cole, S. P. C., Downes, H. F., Mirski, S. E. L. and Clements D. J. (1990). Alterations in glutathione and glutathione-related enzymes in a multidrug resistant small cell lung cancer cell line. *Mol. Pharmacol.*, **37**, 192–197

Cole, S. P. C., Downes, H. F. and Slovak, M. L. (1989). Effect of calcium antagonists on the chemosensitivity of two multidrug resistant human tumour cell lines which do not over express P-glycoprotein. *Br. J. Cancer*, **59**, 42–46

Coley, H. M., Twentyman, P. R. and Workman, P. (1989). Identification of anthracyclines and related agents that retain preferential activity over adriamycin in multidrug-resistant cell lines, and further resistance modification by verapamil and cyclosporin A. *Cancer Chemother. Pharmacol.*, **24**, 284–290

Crooke, S. T., Duvernay, V. H., Galvan, L. and Prestayko, A. W. (1978). Structure–activity relationships of anthracyclines relative to effects on macromolecular syntheses. *Mol. Pharmacol.*, **14**, 290–298

Denny, W. A. and Wakelin, L. P. G. (1990). Kinetics of the binding of mitoxantrone, ametantrone and analogues to DNA: relationship with binding mode and antitumour activity. *Anti-cancer Drug Des.*, **5**, 189–200

Erhlichman, C., Moore, M., Kerr, I. G., Wong, B., Eisenhauer, E., Zee, B. and Whitfield, L. R. (1991). Phase I pharmacokinetic and pharmacodynamic study of the new anthrapyrazole CI-937 (DUP937). *Cancer Res.*, **51**, 6317–6322

Fisher, G. R., Brown, J. R. and Patterson, L. H. (1989). Redox cycling in MCF-7 cells by antitumour agents based on mitoxantrone. *Free Radical Res. Commun.*, **7**, 221–226

Fisher, G. R., Brown, J. R. and Patterson, L. H. (1990). Involvement of hydroxyl radical formation and DNA strand breakage in the cytotoxicity of anthraquinone antitumour agents. *Free Radical Res. Commun.*, **11**, 117–125

Fisher, G. R., Gutierrez, P., Oldcorne, M. A. and Patterson, L. H. (1992) NAD(P)H (quinone acceptor) oxidoreductase (DT-diaphorase)-mediated two electron reduction of anthraquinone-based antitumour agents and generation of hydroxyl radicals. *Biochem. Pharmacol.*, **43**, 575–585

Fisher, G. R. and Patterson, L. H. (1991). DNA strand breakage by peroxidase-activated mitoxantrone. *J. Pharm. Pharmacol.*, **43**, 65–68

Fisher, G. R. and Patterson, L. H. (1992). Lack of involvement of reactive oxygen in the cytotoxicity of mitoxantrone, CI-941 and ametantrone: Comparison with doxorubicin. *Cancer Chemother. Pharmacol.*, **30**, 451–458

Foster, B. J., Newell, D. R., Graham, M. A., Gumbrell, L. A., Jenns, K. E., Kaye, S. B. and Calvert, A. H. (1992). CI-941 phase I trial: Prospective evaluation of a pharmacokinetically guided dose escalation scheme. *Eur. J. Cancer*, **28**, 463–469

Fox, M. E. and Smith, P. J. (1990). Long-term inhibition of DNA synthesis and the persistence of trapped topoisomerase II complexes in determining the toxicity of the antitumour DNA intercalators mAMSA and mitoxantrone. *Cancer Res.*, **50**, 5813–5818

Frank, P. and Novak, R. F. (1986). Effects of anthrapyrazole antineoplastic agents on lipid peroxidation. *Biochem. Biophys. Res. Commun.*, **140**, 797–807

Frank, S. K., Mathiesen, D. A., Szurszewski, M., Kuffel, M. J. and Ames, M. M. (1989). Preclinical pharmacology of the anthrapyrazole analog oxantrazole (NSC-349174, Piroxantrone). *Cancer Chemother. Pharmacol.*, **23**, 213–218

Frank, S. K., Mathiesen, D. A., Whitfield, L. R. and Ames, M. M. (1987). Reverse phase high performance liquid chromatographic assay for the experimental anticancer agent anthrapyrazole analog oxantrazole (NSC-349174). *J. Chromatogr. (Biomed. Appl.)*, **419**, 225–232

Fry, D. W., Boritzki, T. J., Besserer, J. A. and Jackson, R. C. (1985). *In vitro* DNA strand scission and inhibition of nucleic acid synthesis in L1210 leukaemia cells by a new class of DNA complexers, the anthra[1,9-*cd*]pyrazol-6(2*H*)-ones (anthrapyrazoles). *Biochem. Pharmacol.*, **34**, 3499–3508

Gianni, L., Corden, B. J. and Myers, C. E. (1983). The biochemical basis of anthracycline toxicity and antitumour activity. In Hodgson, E., Bend, J. R. and Philpot, R. M. (Eds), *Reviews in Biochemical Toxicology*, Vol. 5, Elsevier, Amsterdam, pp. 1–82

Graham, M. A., Newell, D. R., Butler, J., Hoey, B. and Patterson, L. H. (1987). The effect of the anthrapyrazole antitumour agent CI941 on rat liver microsome and cytochrome P450 reductase mediated free radical processes. Inhibition of doxorubicin activation *in vitro*. *Biochem. Pharmacol.*, **36**, 3345–3351

Graham, M. A., Newell, D. R., Patterson, L. H., Qualmann, C., Sinha, B. H. and Myers, C. E. (1989). The role of anthrapyrazole iron complexes on hydroxyl radical formation, DNA strand scission and cytotoxicity. *Br. J. Cancer*, **60**, 501

Hantel, A., Donehower, R. C., Rowinsky, E. K., Vance, E., Clarke, B. V., McGuire, W. P., Ettinger, D. S., Noe, D. A. and Grochow, L. B. (1990). Phase I and pharmacodynamics of piroxantrone (NSC 349174), a new anthrapyrazole. *Cancer Res.*, **50**, 3284–3288

Hartley, J. A., Reszka, K. and Lown, J. W. (1988a). Photosensitization by antitumour agents. 4. Anthrapyrazole-photosensitized formation of single strand breaks in DNA. *J. Free Radical Biol. Med.*, **4**, 337–343

Hartley, J. A., Reszka, K., Zuo, E. T., Wilson, W. D., Morgan, A. R. and Lown, J. W. (1988b). Characterisation of the interaction of anthrapyrazole anticancer agents with deoxyribonucleic acids: Structural requirements for DNA binding, intercalation and photosensitisation. *Mol. Pharmacol.*, **33**, 265–271

Havelick, M. J., Hamelehle, K. L. and Roberts, B. J. (1987). An *in vitro/in vivo* solid tumour model for assessing antitumour activity using murine melanoma B16 and a subline resistant to adriamycin. *Proc. Am. Assoc. Cancer Res.*, **28**, 451

Judson, I. R. (1991). Anthrapyrazoles: true successors to the anthracyclines? *Anti-cancer Drugs*, **2**, 223–231

Kessel, D. (1989). Probing modes of multi-drug resistance via photodynamic effects of anthrapyrazoles. In Tapeero, H., Robert, J. and Lampidis, T. J. (Eds),

*Anticancer Drugs*, Colloque Insern., Vol. 191, J. Libbey, Eurotext Ltd, pp. 223–232

Kharasch, E. D. and Novak, R. F. (1981). The molecular basis for the complexation of adriamycin with flavin mononucleotide and flavin dinucleotide. *Arch. Biochem. Biophys.*, **212**, 20–36

Kharasch, E. D. and Novak, R. F. (1983a). Bis(alkylamino)anthracenedione antineoplastic activation by anthracyclines. *Arch. Biochem. Biophys.*, **224**, 682–694

Kharasch, E. D. and Novak, R. F. (1983b). Inhibitory effects of anthracenedione antineoplastic agents on hepatic and cardiac lipid peroxidation. *J. Pharmacol. Exp. Ther.*, **226**, 500–506

Klohs, W. D., Steinkampf, R. W., Havelick, M. J. and Jackson, R. C. (1986). Resistance to anthrapyrazoles and anthracyclines in multidrug-resistant P388 murine leukaemia cells: Reversal by calcium blockers and calmodulin antagonists. *Cancer Res.*, **46**, 4352–4356

Kolodziejczyk, P., Reszka, K. and Lown, J. W. (1986). Horseradish peroxidase oxidation of mitoxantrone. Spectrophotometric and electron paramagnetic resonance studies. *J. Free Radical. Biol. Med.*, **2**, 25–32

Kolodziejczyk, P., Reszka, K. and Lown, J. W. (1988). Alternative to the bioreductive activation of anthracyclines: enzymatic oxidative metabolism of anthracenediones, 5-iminodaunorubicin and anthrapyrazoles. In *Oxy-radicals in Molecular Biology and Pathology*, Alan R. Liss, New York, pp. 525–539

Leopold, W. R., Mason, J. M., Plowman, J. and Jackson, R. C. (1985). Anthrapyrazoles, a new class of intercalating agents with high-level, broad spectrum activity against murine tumours. *Cancer Res.*, **45**, 5532–5539

Lown, J. L., Chen, H.-H., Plambeck, J. A. and Acton, E. M. (1979). Diminished superoxide generation by reduced 5-iminodaunorubicin relative to daunorubicin and the relationship to cardiotoxicity of the anthracycline antitumour agents. *Biochem. Pharmacol.*, **28**, 2563–2568

Lui, L. F. (1989). DNA topoisomerase poisons as antitumour drugs. *Ann. Rev. Biochem.*, **58**, 351–375

Malhotra, D. and Hopfinger, A. J. (1980). Conformational flexibility of dinucleotide dimers during unwinding from the B-form to an intercalation structure. *Nucleic Acids Res.*, **8**, 5289

Mason, R. P. and Chignell, C. F. (1982). Free radicals in pharmacology and toxicology—selected topics. *Pharmacol. Rev.*, **33**, 189–211

Nordblom, G. D., Pachla, L. A., Chang, T., Whitfield, L. R. and Showalter, H. D. H. (1989). Development of a radioimmunoassay for the anthrapyrazole chemotherapy agent CI-937 and the pharmacokinetics of CI-937 in rats. *Cancer Res.*, **49**, 5345–5341

Ohlson, R. D. and Mushlin, P. S. (1990). Doxorubicin cardiotoxicity: analysis of prevailing hypotheses. *FASEB Jl.*, **4**, 3076–3086

Patterson, L. H. and Basra, J. (1985) Lack of mitoxantrone free radicals and redox cycling in rabbit heart sarcoplasmic reticulum. *Br. J. Cancer*, **52**, 416 (abstract)

Patterson, L. H., Gandecha, B. M. and Brown, J. R. (1983). 1,4-Bis (2-(2-hydroxyethylaminoethyl))9,10 anthracenedione, an anthraquinone antitumour agent that does not cause lipid peroxidation *in vivo*: comparison with daunorubicin. *Biochem. Biophys. Res. Commun.*, **110**, 399–405

Peters, J. M., Gordon, G. R., Kashiwase, D., Lown, J. W., Yen, S.-F. and Plambeck, J. A. (1986). Redox activities of antitumour anthracyclines determined by microsomal oxygen consumption and assays of superoxide anion and hydroxyl radical generation. *Biochem. Pharmacol.*, **35**, 1309–1323

Reszka, K., Hartley, J. A., Kolodziejczyk, P. and Lown, J. W. (1989). Interaction of the peroxidase-derived metabolite of mitoxantrone with nucleic acids. Evidence for covalent binding of 14C-labelled drug. *Biochem. Pharmacol.*, **38**, 4253–4260

Reszka, K., Kolodziejczyk, P., Hartley, J. A., Wilson, W. D. and Lown, J. W. (1988a). Molecular pharmacology of anthracenedione-based anticancer agents. In Lown, W. (Ed.), *Bioactive Molecules*, Vol. 6, Elsevier, Amsterdam, pp. 401–445

Reszka, K., Kolodziejczyk, P. and Lown, J. W. (1986a). Photosensitisation by antitumour agents. 2. Anthrapyrazole-photosensitised oxidation of ascorbic acid and 3-4-dihydroxyphenylalanine. *J. Free Radical Biol. Med.*, **2**, 203–211

Reszka, K., Kolodziejczyk, P. and Lown, J. W. (1986b). Photosensitisation by antitumour agents. 3. Spectroscopic evidence for superoxide and hydroxyl radical production by anthrapyrazole-sensitised oxidation of NADH. *J. Free Radical Biol. Med.*, **2**, 267–274

Reszka, K., Kolodziejczyk, P. and Lown, J. W. (1988b). Enzymatic activation and transformation of the antitumour agent mitoxantrone. *J. Free Radical Biol. Med.*, **5**, 13–22

Reszka, K. and Lown, J. W. (1989). Photosensitisation of anticancer agents. 8. One-electron reduction of mitoxantrone: an epr and spectrophotometric study. *Photochem. Photobiol.*, **50**, 297–304

Reszka, K., Tsoungas, P. G. and Lown, J. W. (1986c). Photosensitisation by antitumour agents 1: Production of singlet oxygen during irradiation of anthrapyrazoles with visible light. *Photochem. Photobiol.*, **43**, 499–504

Sebolt, J. S., Havlick, M. J., Hamelehle, K. L., Klohs, W. D., Steinkampf, R. W. and Jackson, R. C. (1985). Establishment of adriamycin-resistant mammary carcinoma 16/C *in vitro* and its sensitivity to the anthrapyrazoles CI-942 and CI-937. *Proc. Am. Assoc. Cancer Res.*, **26**, 339

Showalter, H. D. H., Fry, D. W., Leopold, W. R., Lown, J. W., Plambeck, J. A. and Reszka, K. (1986). Design, biochemical pharmacology, electrochemistry and tumour biology of anti-tumour anthrapyrazoles. *Anti-cancer Drug Des.*, **1**, 73–85

Showalter, H. D. H., Johnson, J. L., Hoftiezer, J. M., Turner, W. R., Werbel, L. M., Leopold, W. R., Shillis, J. L., Jackson, R. C. and Elslager, E. F. (1987). Anthrapyrazole anticancer agents. Synthesis and structure–activity relationships against murine leukaemias. *J. Med. Chem.*, **30**, 121–131

Showalter, H. D. H., Johnson, J. L., Werbel, L. M., Leopold, W. R., Jackson, R. C. and Elslager, E. F. (1984). 5-[(Alkylamino)amino]-substituted anthra[1,9-*cd*]pyrazol-6(2*H*)-ones as novel anticancer agents. Synthesis and biological evaluation. *J. Med. Chem.*, **27**, 253–255

Singh, Y., Ulrich, L., Katz, D., Bowen, P. and Krishna, G. (1989). Structural requirements for anthracycline-induced cardiotoxicity and antitumour effects. *Toxicol. Appl. Pharmacol.*, **100**, 9–23

Sinha, B. K., Katki, A. G., Batist, G., Cowan, K. H. and Myers, C. E. (1987). Differential formation of hydroxyl radicals by adriamycin in sensitive and resistant human breast tumour cells: Implications for the mechanism of action. *Biochemistry*, **26**, 3776–3781

Smith, P. J. (1990). DNA topoisomerase dysfunction: A new goal for antitumour chemotherapy. *Bioessays*, **12**, 167–172

Stuart-Harris, R., Pearson, M., and Smith, I. E. (1984). Cardiotoxicity associated with mitoxantrone. *Lancet*, **2**, 219

Talbot, D. C., Smith, I. E., Mansi, J. L., Judson, I. R., Calvert, A. H. and Ashley, S. E. (1991). Anthrapyrazole CI-941: A highly active new agent in the treatment of advanced breast cancer. *J. Clin. Oncol.*, **9**, 2141–2147

Tong, G. L., Henry, D. W. and Acton, E. M. (1979). 5-Iminodaunorubicin. Reduced cardiotoxic properties in an antitumour anthracycline. *J. Med. Chem.*, **22**, 36–39

Unverferth, D. V., Bashore, T. M., Magorien, R. D., Fetters, J. K. and Neidhart, J. A. (1984). Histologic and functional characteristics of human heart after mitoxantrone therapy. *Cancer Treatment Symposia*, **3**, 47–53

van de Graaf, W. T. A. and de Vries, E. G. E. (1990). Mitoxantrone: Bluebeard for malignancies. *Anti-cancer Drugs*, **1**, 109–125

Von Sonntag, C. (1987). In *The Chemical Basis of Radiation Biology*, Taylor and Francis, London

Walling, C. (1975). Fenton's reagent revisited. *Acc. Chem. Res.*, **8**, 125

Wong, B., Nordblom, G., Chang, T. and Whitfield, L. (1989). Lack of dose proportional pharmacokinetics for CI-937, an anthrapyrazole DNA intercalator, in mice. *Res. Commun. Chem. Pathol. Pharmacol.*, **66**, 191–202

# 3
# Calicheamicin

George A. Ellestad and Wei-dong Ding

## 1 Introduction

Organic chemists have long been intrigued by the rich structural complexity of compounds obtained from natural sources. Secondary metabolites, isolated from fermentation broths of microorganisms, have been especially rewarding, not only because of their interesting chemistry but also because of their important uses in medicine and agriculture. The unprecedented enediyne antitumor antibiotics, calicheamicin $\gamma_1^I$ (**1**) (Lee *et al.*, 1987a,b, 1991), esperamicin $A_1$ (**2**) (Golik *et al.*, 1987a,b) dynemicin (**3**) (Konishi *et al.*, 1989) and neocarzinostatin (**4**) (Edo *et al.*, 1985) (Figure 3.1) are some of the most remarkable fermentation products to be isolated in many years. The seminal feature of these agents is that they bind to duplex DNA and, upon bioreductive activation, initiate oxidative strand cleavage by means of a transient diradical intermediate which abstracts proximal hydrogen atoms from the targeted deoxyribose sugars. The DNA cleavage of calicheamicin (Zein *et al.*, 1988b), esperamicin (Fry *et al.*, 1986; Long *et al.*, 1989; Sugiura *et al.*, 1989; Uesawa and Sugiura, 1991) and dynemicin (Shiraki and Sugiura, 1990; Sugiura *et al.*, 1990) is mediated by a transient *p*-benzyne, whereas that of neocarzinostatin (Myers *et al.*, 1988; Myers and Proteau, 1989; Dedon and Goldberg, 1990; Kappen *et al.*, 1991) is mediated by an unsymmetrical indenyl diradical (p. 243). DNA binding of neocarzinostatin (Goldberg, 1991) and dynemicin (Sugiura *et al.*, 1990) primarily involves intercalation of the naphthoate and the anthraquinone chromophores, respectively, between adjacent base pairs. Although less well understood, DNA association with calicheamicin and esperamicin seems to involve the oligosaccharide moieties.

Esperamicin $A_1$ (Golik *et al.*, 1987a) and calicheamicin $\gamma_1^I$ (Lee *et al.*, 1987b) are considerably more potent (10–100-fold) against a P388

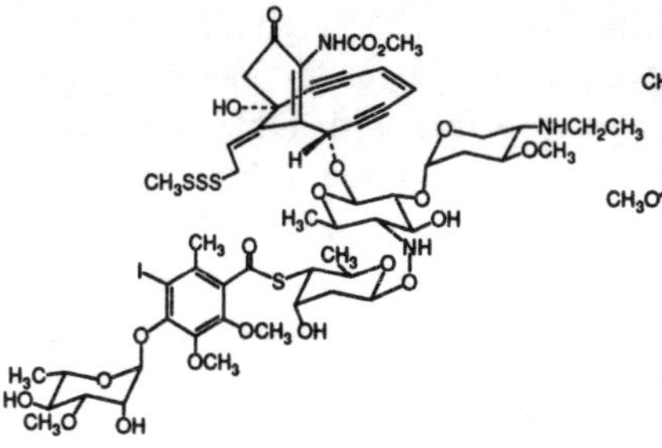


1 Calicheamicin $\gamma_1^{I}$

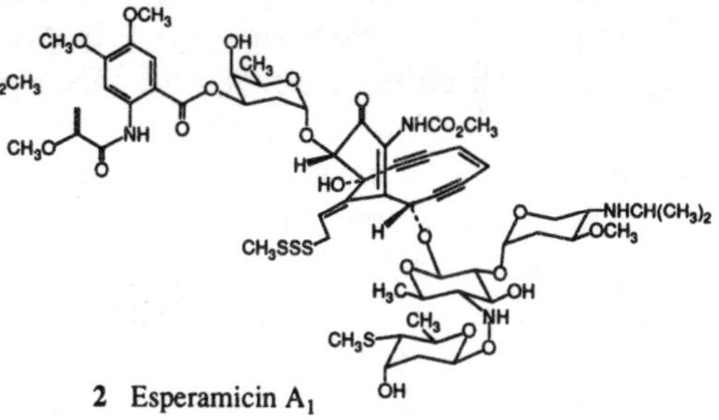


2 Esperamicin $A_1$

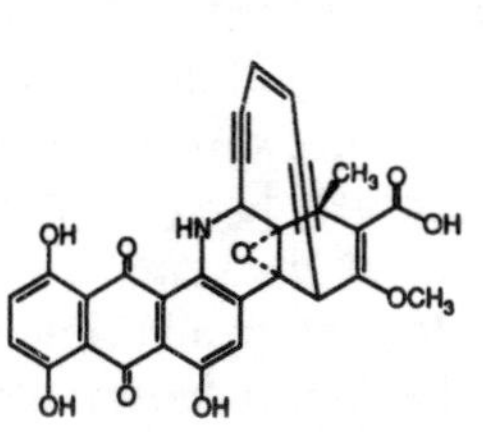


3 Dynemicin

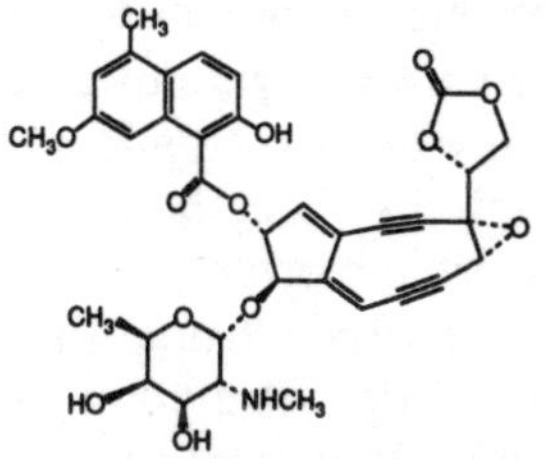


4 Neocarzinostatin chromophore

**Figure 3.1**

leukaemia model in mice than are the other two enediyne antibiotics, and are some 1000 times more active than adriamycin, with an optimal dose between 0.5 and 1.5 μg per kilogram of body weight. Whether or not the extreme cytotoxicity of these agents is due solely to their DNA cleavage properties remains to be determined. The extreme potency of calicheamicin was cause for much excitement when the activities were first obtained. However, this excitement was tempered when it was learned that this drug caused chromosomal damage in human diploid lung fibroblasts at picogram (pg) per millilitre concentrations, and mutagenesis in *Escherichia coli* (Durr and Wallace, 1988). This chapter presents an overview of the discovery and chemical properties of the calicheamicins and, primarily, our present understanding of the molecular mechanism by which calicheamicin $\gamma_1^I$ cleaves DNA. These results are based on studies carried out initially at American Cyanamid as well as more recent efforts by researchers at various academic institutions.

## 2 Isolation, Structure and Chemistry

The calicheamicins were discovered at Lederle during a screening programme using the very sensitive prophage induction assay (Elespuru and

Yarmolinsky, 1979; Elespuru and White, 1983) with a genetically engineered strain of *Escherichia coli* lysogenic for a λ-*lac* Z fusion phage. Agents which either interact with DNA or interfere with DNA synthesis cause induction of the prophage, which results in the production of β-galactosidase, the product of the *lac* Z gene. The amount of enzyme produced is determined by a colorimetric assay using a chromogenic substrate sensitive to β-galactosidase. These natural products are produced by fermentations of *Micromonospora echinospora* ssp. *calichensis* (NRRL 15839), a soil microorganism isolated from caliche or chalky soils from the state of Texas. They can be recovered from fermentations by solvent extraction of the whole broth followed by extensive chromatography on silica gel and reverse phase HPLC. Calicheamicin $\gamma_1^{Br}$ (**5**), the corresponding bromine derivative of $\gamma_1^I$, along with the corresponding isopropyl analogue, $\beta_1^{Br}$ (**6**), were the first components of the complex to be isolated from early fermentations in which the titres were extremely low (0.1 μg/ml) (Lee *et al.*, 1989). Most important for the isolation of material in quantities sufficient for structural elucidation and biological evaluation was the finding that the addition of sodium iodide to the fermentation, along with strain and media development, increased the titres to a more 'reasonable' 2–5 μg/ml (Maiese *et al.*, 1989). Calicheamicin $\gamma_1^I$, on which the structural characterization was carried out, was the major metabolite from these sodium iodide supplemented fermentations. In addition, minor amounts of analogues $\alpha_2^I$ (**7**), $\alpha_3^I$ (**8**), $\beta_1^I$ (**9**) and $\delta_1^I$ (**10**) were obtained, depending upon the fermentation conditions. Once the structure of $\gamma_1^I$ was determined, it was straightforward to go back and assign structures to the other closely related metabolites. The calicheamicin components were assigned Greek names based on their TLC mobilities and are summarized in Figure 3.2.

One other interesting point regarding the isolation of the calicheamicins, and apparently the esperamicins as well, is that these metabolites are isolated in the absence of an associated protein, in contrast to neocarzinostatin (**4**). Neocarzinostatin, however, is apparently more labile than these enediyne antibiotics, which may explain the need for a stabilizing protein (Goldberg, 1991).

As seen in structure **1**, calicheamicin $\gamma_1^I$ consists of several unusual subunits, including a glycosylated hydroxylamino sugar and a labile allylic methyltrisulfide grouping that upon reductive cleavage with thiols or phosphines initiates aromatization of the enediyne moiety, as depicted in Figure 3.3. The structural characterization relied on a combination of degradative chemistry, mass spectrometry, high-resolution NMR and the X-ray crystal structure analysis of two key hydrolysis products. The X-ray results provided unequivocal evidence for the unusual glycosylated hydroxylamino linkage as well as the orthogonal relationship of the hexasubstituted benzene ring to the carbonyl grouping (Lee *et al.*, 1987a). We were led to propose the presence of the enediyne moiety after pondering at some

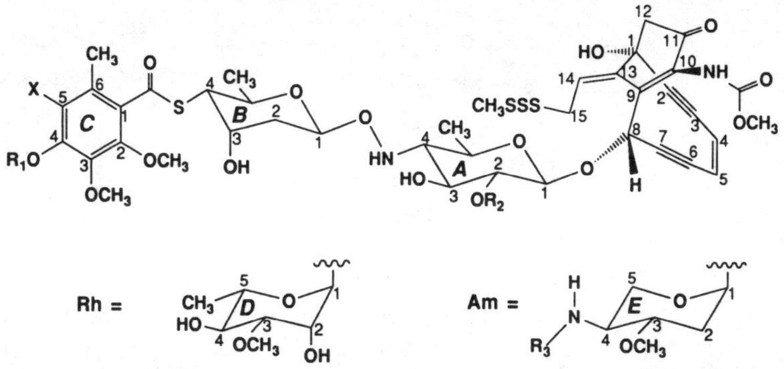


| *Calicheamicin* | *X* | $R_1$ | $R_2$ | $R_3$ |
|---|---|---|---|---|
| $\gamma_1^{Br}$ (5) | Br | Rh | Am | $CH_3CH_2$ |
| $\beta_1^{Br}$ (6) | Br | Rh | Am | $(CH_3)_2CH$ |
| $\alpha_2^I$ (7) | I | H | Am | $CH_3CH_2$ |
| $\alpha_3^I$ (8) | I | Rh | H | |
| $\beta_1^I$ (9) | I | Rh | Am | $(CH_3)_2CH$ |
| $\gamma_1^I$ (1) | I | Rh | Am | $CH_3CH_2$ |
| $\delta_1^I$ (10) | I | Rh | Am | $CH_3$ |

**Figure 3.2** Chemical structures of calicheamicins $\beta_1^{Br}$, $\gamma_1^{Br}$, $\alpha_2^I$, $\alpha_3^I$, $\beta_1^I$, $\gamma_1^I$ and $\delta_1^I$

length the facile aromatization that takes place on reductive activation of calicheamicin $\gamma_1^I$ by reducing thiols and phosphines. Further, the incorporation of two deuterium atoms at C-3 and C-6 of the aromatized end-product **13** (Figure 3.3), in the presence of $CD_2Cl_2$, was especially helpful. Therefore, it was with great satisfaction that we learned later that Bristol-Myers chemists had assigned the enediyne moiety to esperamicin $A_1$ (**2**) (Golik *et al.*, 1987b), since we anticipated that this structural proposal would be greeted with some scepticism. However, confirmation has now been obtained from Danishefsky's group which has synthesized the calicheamicin aglycone calicheamicinone (**15**), and found that its aromatized product is identical with that obtained by hydrolysis of **14** (Haseltine *et al.*, 1991).

It has been 20 years since Bergman first initiated aromatization studies on acyclic enediynes (Bergman, 1973; Lockhart and Bergman, 1981; Lockhart *et al.*, 1981). However, we were unaware of this chemistry, as well as the work of Wong and Sondheimer (1980), Darbay *et al.* (1971) and Chapman *et al.* (1976) with cyclic analogues, until the completion of the structural studies on the calicheamicins. As mentioned above, the *in vitro* triggering mechanism (and probably, the *in vivo* one) is the reductive

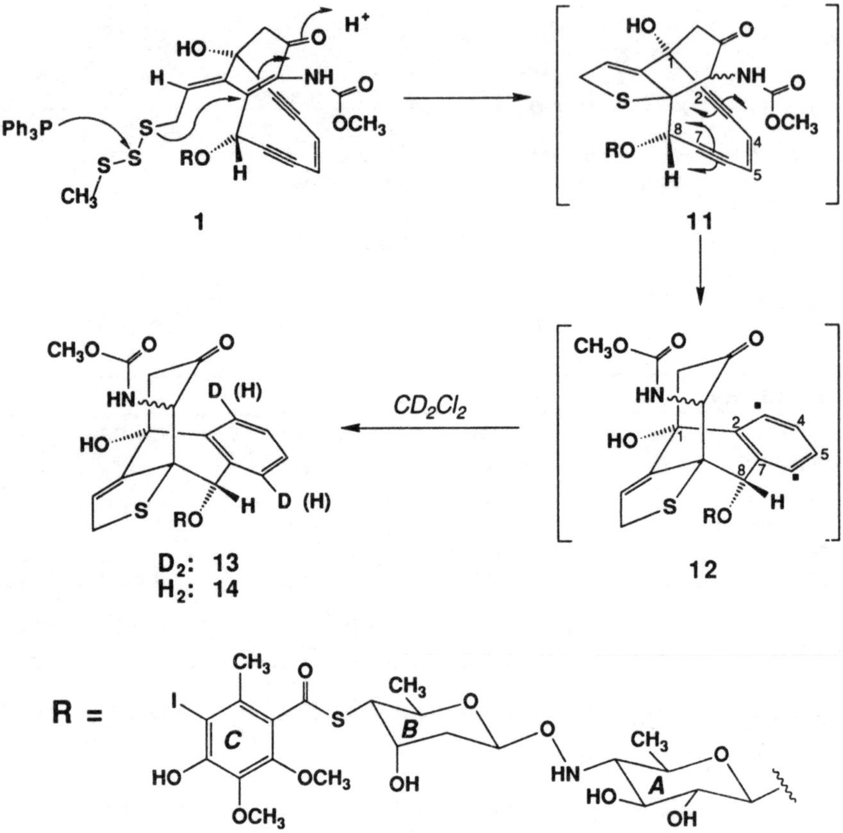


**Figure 3.3**

cleavage of the allylic trisulfide grouping. Some cleavage also occurs in the absence of thiols but is about 1000 times slower than in the presence of reducing agents. In any event, the resulting thiolate anion attacks the α,β-enone moiety via a Michael addition and creates $sp^3$ hybridization at C-9 (Figure 3.3). This relieves the strain imposed by the bridgehead double bond (Bredt's rule) and aromatization occurs. Recent studies on model enediyne-containing cyclic systems have shown that it is the difference in strain energy between the enediyne and the *p*-benzyne intermediate (**12**) that is the principal driving force for this bond reorganization (Magnus and Carter, 1988; Snyder, 1989, 1990; Magnus *et al.*, 1990), rather than the proximity of the acetylenic moieties in the ground state, as initially proposed (Nicolaou *et al.*, 1988a,b).

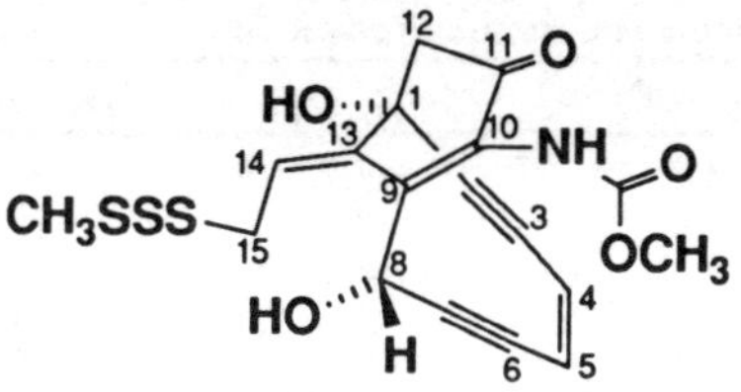


15 **Calicheamicinone**

**Figure 3.4**

## 3 Affinity of Calicheamicin $\gamma_1^I$ for DNA

Despite a number of ambiguities in the determination of the binding constant for the calicheamicin $\gamma_1^I$/DNA interaction, estimates of $10^6-10^8$ $M^{-1}$ have been reported using different DNA substrates and methodology. Drak *et al.* (1991) obtained a $K_{s1/2}$ (drug concentration at half saturation) of $1-3 \times 10^{-8}$ M for the TCCT site in the dodecamer 5′GGGTC̲CTAAATT (dodecamer-**1**), by measuring the extent of cleavage from the band intensities of resolved end-labelled cleavage products as a function of increasing drug concentration. This determination is based on the reaction below and assumes that binding is reversible and that the dissociation reaction is fast compared with cleavage:

$$\text{DNA} + \text{drug} \rightleftharpoons [\text{DNA–drug}] \rightarrow \text{cleaved DNA products}$$

Although the degree of reversibility is uncertain, $K_{s1/2}$ can be taken as an upper estimate for the equilibrium binding constant, as is the Michaelis constant for the equilibrium constant of an enzyme–substrate complex.

In another study, association constants ($K$) of approximately $10^6$ were estimated from Scatchard plots. These were obtained from optical density measurements of the drug-induced and agarose-resolved topological forms of pBR322 in the temperature range between 5 and 40 °C (Table 3.1). Sequences such as 5′TCCT, CTCT and ACTC were assumed to be preferred sites on the basis of early cleavage specificity studies. The values of $K_{obs}$, therefore, represent a weighted average of the binding/cleavage at all of these sites in the plasmid. Thus, the lower binding constants derived from these experiments compared with the $K_{s1/2}$ determined by Drak *et al.* (1991) for the single TCCT binding/cleavage site are perhaps not surprising. It was determined from a sequence search that pBR322 contained 56 such primary sites after subtraction of 16 that are less than two base pairs away from another of these preferred sequences. It was also assumed that

**Table 3.1** Binding constants and thermodynamic functions (Ding and Ellestad, 1991)

| $T(°C)$ | $K_{obs}(10^6)$ | $\Delta G^{\circ}_{obs}(kcal)$ | $\Delta H^{\circ}_{obs}(kcal)$ | $\Delta S^{\circ}_{obs}(kcal/K)$ |
|---|---|---|---|---|
| 5 | 1.2 | −7.7 | 26.6 | 0.120 |
| 10 | 1.6 | −8.0 | 20.6 | 0.985 |
| 15 | 1.8 | −8.2 | 14.5 | 0.077 |
| 20 | 2.1 | −8.5 | 8.5 | 0.057 |
| 25 | 2.5 | −8.7 | 2.4 | 0.036 |
| 30 | 2.3 | −8.8 | −3.6 | 0.016 |
| 35 | 2.2 | −8.9 | −9.7 | −0.004 |
| 40 | 1.3 | −8.8 | −15.7 | −0.023 |

cleavage occurred at each site to a similar extent, which may or may not be true, depending upon the localized geometry in the supercoiled DNA, and that the binding/cleavage was non-cooperative. Because of the supercoiled nature of the plasmid, it is possible that additional sites were recognized and cleaved that were not observed in the initial cleavage experiments with relatively short restriction fragments. Each drug binding event was taken as being productive and causing either double- or single-stranded cleavage, giving linear (form III) or nicked (form II) plasmids, respectively. The fraction of the uncleaved form I plasmid was equated to the concentration of the free DNA, and the concentration of the free drug was determined by subtraction of the combined amounts of forms II and III from the total amount of drug added. The average binding constants for the TCCT, ACCT and CTCT tracts were then obtained from Scatchard plots derived from density–concentration measurements of the various topological DNA forms. The physical chemistry implications of the variation of $K_{obs}$ as a function of temperature (Table 3.1) will be discussed below.

## 4 Plasmid DNA Cleavage Studies

An early cleavage study with supercoiled plasmid DNA revealed that comparable amounts of nicked and linear forms were produced by concentrations of calicheamicin $\gamma_1^I$ as low as 7 nM in the presence of thiols (Zein *et al.*, 1988b). These experiments also indicated a marked preference for duplex DNA over single-stranded plasmid. The high yield of linear DNA by such small amounts of drug strongly suggested a significant amount of double-strand cleavage consistent with the biradical nature of the activated drug. Drak *et al.* (1991) confirmed these results and observed a high ratio (1:2) between double- and single-strand cuts at a drug concentration as low as 1.5 nM. In addition, the same group examined the cleavage kinetics of calicheamicin $\gamma_1^I$ and the synthetic aglycon, calicheamicinone (**15**), with plasmid øx174. Fast DNA cleavage kinetics was observed for both calicheamicinone and calicheamicin $\gamma_1^I$ during the first 5 min. Thus, supercoiled DNA may be a better substrate for the aglycon than nicked DNA.

They also estimated the fraction of linear molecules that would have been obtained by accumulation of single-strand breaks, using the Freifelder–Trumbo equation. Data from calicheamicinone (0.84 mM) cleavage for the first 5 min were fitted to a linear equation and gave a ratio of 1:30 between double- and single-strand cuts per DNA molecule compared with 1:2 for the intact antibiotic at a concentration of 1.5 nM. For comparison, a ratio of 1:9 was obtained for bleomycin and 1:6–1:4 for neocarzinostatin. Thus, it is the thiobenzoate-oligosaccharide tail that converts the aglycon 'warhead' into a potent double-strand cleavage agent.

Dabrowiak and co-workers (Kishikawa *et al.*, 1991), in another cleavage rate study, developed rate laws for the interconversion of the three plasmid DNA forms in order to analyse single- versus double-strand cleavage of covalently closed circular PM2 plasmid DNA by esperamicin $A_1$, esperamicin D (**16**) (Figure 3.5) and calicheamicin $\gamma_1^I$. The hydrolytic enzyme DNase I was used for purposes of comparison, since it produces mainly single-strand breaks. Cleavage rate constants were derived from optical density data obtained from agarose gel experiments. It was assumed that the rate of nicking of each form of DNA was proportional to its concentration and to the concentration of the drug. Calculated intensities were then compared with experimental values and the deviation between the two was used to determine whether the experimental data would fit a single- or double-strand cleavage model. The cleavage data for esperamicin $A_1$ were consistent only with a single-strand model (as was DNase I), indicating that one of the radicals of the phenylene diradical is 'silent' and being quenched by solvent, DNA or the drug itself. This result stands in contrast to the data obtained for calicheamicin $\gamma_1^I$ and esperamicin D, which fit the double-strand model. Furthermore, the rate of cleavage of closed circular form I DNA by calicheamicin $\gamma_1^I$ is an order of magnitude less than that of esperamicin D and two orders of magnitude less than that of esperamicin $A_1$.

## 5 DNA Binding/Cleavage Specificity

One of the more striking features of calicheamicin $\gamma_1^I$ is the unusual specificity of the DNA binding/cleavage for regions including GC base pairs (Zein *et al.*, 1988b). This was determined by comparison of the electrophoretic mobility of the drug-induced cleavage products with the markers produced by the Maxam–Gilbert sequencing reactions. Initial studies carried out at a low drug/DNA ratio of 0.05 (M of DNA in base pairs) and with several 5′-[$^{32}$P]-end-labelled restriction fragments obtained from pUC 18 and from cloned *Streptomyces* promoter regions showed that, in the presence of thiols, oligopyrimidine/oligopurine tetramer sequences such as 5′T$\underline{C}$CT/AGGA and 5′C$\underline{T}$CT/GAGA were primary cleavage sites. How-

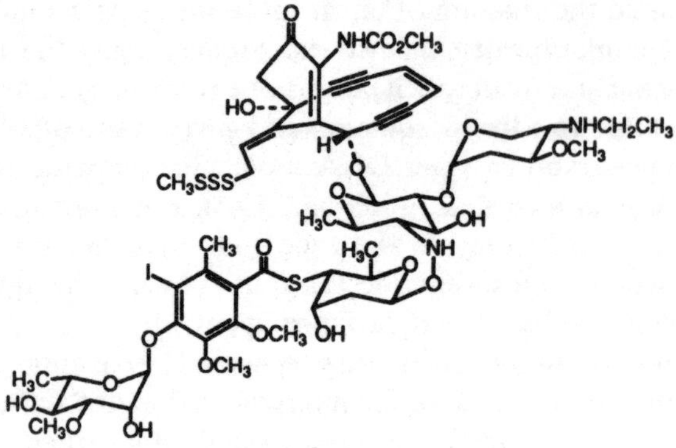

**1** Calicheamicin $\gamma_1^I$

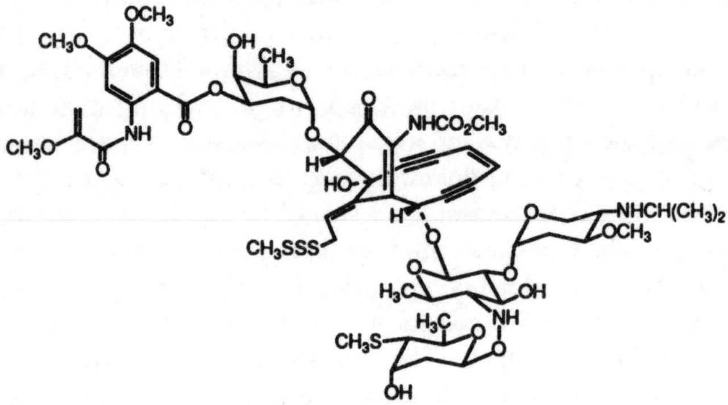

**2** Esperamicin $A_1$

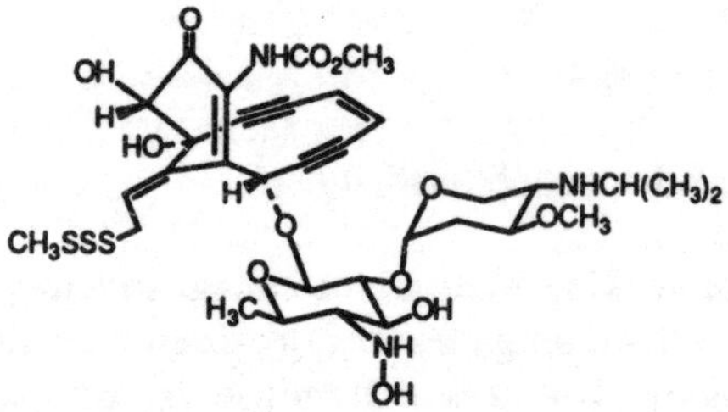

**16** Esperamicin D

**Figure 3.5**

ever, other sequences such as 5′GCCT, TCCG, TCCC, TCTC, ACCT, TCCA and their complementary sites were also cleaved in the same fashion, the extent of cleavage apparently depending on the nature of the flanking sequences. Almost always, cleavage occurred at the penultimate nucleotide, usually a pyrimidine, at the 5′ end of these tetramers. A more recent study indicated cleavage at TTCA and TTTT sites in a restriction fragment from pUC 19 (Walker *et al.*, 1992). In fact, the TTTT cleavage was preferred over an adjacent TCCC sequence. This probably reflects a wider than optimal groove width at this TCCC tract.

Cleavage on the complementary 3′–5′ strand occurred in a 3 base pair, 3′-offset from that of the 5′–3′ strand and always to a lesser extent. The difference in cleavage rate between the two strands suggested that the drug/DNA contact on the 3′–5′ strand was not as tight or that the cleavage chemistry was different from that of the TCCT-containing strand, or possibly a combination of both. The 3′-cleavage offset suggested that the DNA minor groove was the binding target for calicheamicin $\gamma_1^I$. This was confirmed by competition experiments with the minor groove-specific agent netropsin, which either prevented calicheamicin from binding to, or altered, its preferred binding/cleavage sites. In addition, the cleavage rate of DNA containing bulky carbohydrate moieties in the major groove, e.g. T4 phage DNA, was just as strong as with unblocked DNA.

Cleavage experiments with the synthetic dodecamer-**1**, 5′GGGTCC-TAAATT (Zein *et al.*, 1989b), showed sharp cleavage at the expected C residue and indicated the value of short synthetic fragments of DNA for garnering additional insight as to the spatial requirements of the binding/cleavage specificity. Creation of an AG mismatch, 5′TCAT/3′AGGA, towards the 3′ side of the TCCT site had no effect on the cleavage in both strands. However, creation of the AG mismatch 5′TACT/3′AGGA inhibited cleavage on the AGGA strand but not at the 5′ A residue on the complementary TACT tract. This suggests that the spatial requirements for hydrogen abstraction by the diradical intermediate on the AGGA-containing strand are no longer optimal in this TACT/AGGA-dodecamer. It is believed that an AG mismatch does not significantly alter the overall conformation of the DNA other than to induce a local widening at the mismatch. In contrast to an AG mismatch, creation of TC mismatches within the same sequence, 5′TCCT/3AGTA and 5′TCCT/3′ATGA, totally inhibited cleavage on either strand. It is known from X-ray analyses that the TC mismatch is one of the more unstable mismatch pairs leading to considerable broadening at this site and thus no doubt diminishing the necessary spatial complementarity between the DNA minor groove and calicheamicin $\gamma_1^I$.

Drak *et al.* (1991) also studied the calicheamicin $\gamma_1^I$ cleavage discrimination and confirmed the earlier results with dodecamer-**1**. Then they replaced the 5′TCCT/3′AGGA site with a 5′TACT/3′ATGA sequence and, surprisingly, observed cleavage at the 5′ **A** residue. In addition, they examined the cleavage pattern with the following 53 base pair duplex containing the TCCT site

5′TTTAACCGATCAGAATTCCGGTGCATGCTCCT-
AAGTGTACGCCTAAGCTTCTT

and two other previously identified binding/cleavage sites. Cleavage was observed as expected at the central TCCT tract but not at the 5′TCCG and 5′GCCT sites. Cleavage was observed, however, at the 5′TCTG tract on the complementary strand and the homopyrimidine–homopurine region to the right of the 5′GCCT tract. These authors make the point that the cleavage sites identified so far probably represent only a subset of all the possible ones.

## 6 Structural Features Important for DNA Binding and Discrimination

Although a detailed understanding of the molecular features responsible for the observed binding/cleavage discrimination is still rather obscure, some progress has been made. Comparison of the cleavage specificity of the aglycon, calicheamicinone (**15**), with that of the intact antibiotic provided evidence that the thiobenzoate oligosaccharide tail portion plays a key role in the binding and cleavage specificity (Drak *et al.*, 1991) as opposed to just binding alone, as proposed originally (Zein *et al.*, 1989b). Calicheamicinone (**15**) at concentrations 1000 times that of calicheamicin $\gamma_1^I$ were required for cleavage that was sequence-independent. Further evidence for the importance of the carbohydrate tail portion for sequence discrimination comes from cleavage experiments with a truncated calicheamicin $\gamma_1^I$ (Walker *et al.*, 1992). Selective hydrolysis of the N–O glycosidic linkage to the B sugar of calicheamicin $\gamma_1^I$ provided a truncated calicheamicin, **17** (Figure 3.6). This degradation product was shown to have greatly reduced cleavage discrimination compared with the parent drug, confirming that the distal portion of the oligosaccharide tail is important for DNA specificity. However, the ratio between nicked and linear DNA produced from supercoiled pBR 322 is about the same for both the truncated calicheamicin and calicheamicin $\gamma_1^I$, whereas calicheamicinone produced mainly nicked DNA, as mentioned earlier. Thus, the attachment of the disaccharide composed of sugars A and E to the aglycon appears to provide important rigidity at the glycosidic linkage between the aglycon and sugar A for the 'warhead' to abstract hydrogens from both strands. Earlier studies (Zein *et al.*, 1989b) showed that the $\alpha_2^I$ component (**7**), which lacks the terminal 3-methoxyrhamnose (ring D), exhibits cleavage discrimination identical with that of the parent drug but is considerably less

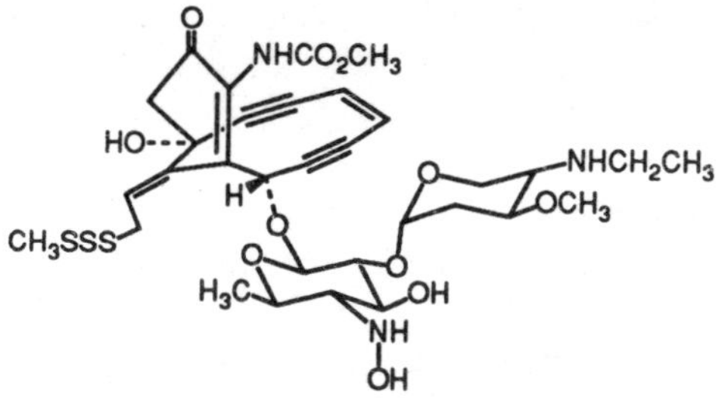


**17** Truncated calicheamicin

**Figure 3.6**

efficient in cutting. This indicates that the terminal 3-methoxyrhamnose contributes to the overall binding interaction between the drug and DNA but not to the cleavage specificity.

Removal of the ethylamino sugar (ring E) provided a derivative, $\alpha_3^I$ (**8**), which also exhibited the same cleavage specificity as the parent drug but again with considerably less cutting efficiency. Cramer and Townsend (1991) have obtained kinetic evidence that the ethylamino sugar does not play a catalytic role in the reductive cleavage of the methyl trisulfide grouping, as proposed earlier on the basis of disulfide exchange reactions in acetonitrile (Zein *et al.*, 1989b). Instead, they suggest that this sugar may contribute to non-specific, general electrostatic associations with the polyanionic DNA. Interestingly, previous studies with the positively charged minor groove binding drugs netropsin and daunomycin (although an intercalator, the amino sugar binds in the minor groove) have shown that ion pair formation does not contribute greatly to the overall binding energy. This is based on the large magnitude of the thermodynamic binding constant, $K_t^0$, which is the binding constant corrected for ionic contributions (Chaires *et al.*, 1982; Marky and Kupke, 1989; Chaires, 1990). Thus, these drug–DNA complexes are apparently stabilized by extensive non-ionic interactions and/or water release which appears to be the case with calicheamicin $\gamma_1^I$. A calculation of $K_t^0$ for calicheamicin binding/cleavage as a function of salt concentration between 0.1 and 1.0 M NaCl will be necessary to define better the influence of ion pair formation on binding affinity (Record *et al.*, 1978; Chaires *et al.*, 1982).

## 7 NMR Evidence for Solution Conformation

Most minor groove binders exist in a solution conformation that is complementary to that of the DNA minor groove (Neidle *et al.*, 1987).

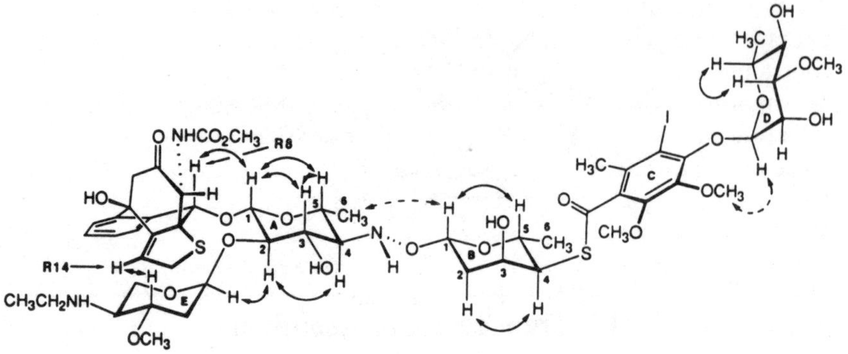


**Figure 3.7** Calicheamicin ε. Selected through-space connectivities from ROESY experiments are indicated by bold (strong NOEs) or dashed (weak NOEs) arrows (from Walker *et al.*, 1989)

Although the structure of calicheamicin $\gamma_1^I$ with its carbohydrate tail did not bear any obvious relationship to the more classical groove binders such as the polypeptide-like netropsin, the concept of preorganization for this drug was an attractive hypothesis. Walker *et al.* (1990) studied the conformational mobility of calicheamicin ε, **14** (the inactive end-product from the aromatization of parent antibiotic), by rotating frame nuclear Overhauser enhancement spectroscopy (ROESY) which provided through-space connectivities, coupling constants and NOEs for the oligosaccharide chain. The results of these experiments suggested that calicheamicin $\gamma_1^I$, which contains the same carbohydrate portion as calicheamicin ε, possesses a significant degree of structural preorganization (Figure 3.7) when not bound to DNA. Subsequent experiments (Yang *et al.*, 1991; Walker *et al.*, 1991) showed that there are substantial rotation and inversion barriers around the N–O bond which could force the molecule to adopt an overall curvature complementary to that of the DNA minor groove. Consistent with this argument is that the crystal structure of a calicheamicin $\gamma_1^I$ degradation product (Lee *et al.*, 1987a) shows that the C–N–O–C grouping adopts an eclipsed conformation which matches the preferred conformation derived from the above solution conformational analysis. The conformational rigidity of the calicheamicin molecule suggested by these NMR studies would appear to be an important component of the DNA binding/cleavage specificity of this agent.

## 8 Evidence for a Hydrophobic Contribution to the Calicheamicin–DNA Association

Because of the highly lipophilic character of calicheamicin $\gamma_1^I$, the question arises as to the nature of the attractive forces between the drug and DNA.

As mentioned above, the positively charged deoxysugar (residue E) and perhaps the glycosylated hydroxyl amino grouping could be expected to contribute to a general non-specific electrostatic component. However, hydrophobic associations are known to be important in the binding of substrates in biological systems, and we speculated that such associations could be important in the binding of calicheamicin to DNA. Initial support for a hydrophobic binding component was obtained in cleavage experiments in the presence of the lyotropic or Hofmeister series of salts. Certain salts in this series were shown to have a significant influence on the rate of plasmid DNA (pBR322) cleavage by calicheamicin $\gamma_1^I$, which was interpreted as being evidence for a favourable hydrophobic association between the drug and the DNA binding site (Ding and Ellestad, 1991). For example, increasing concentrations (0.1–0.4 M) of the strongly hydrated, anti-chaotropic and salting-out agent $Na_2SO_4$ enhanced cutting significantly, whereas chaotropic salts such as $LiClO_4$, $NaClO_4$ and LiSCN at increasing concentrations of 0.5–4 M, caused a significant decrease in cutting efficiency. This suggested that the addition of $Na_2SO_4$ to the cleavage cocktail decreases the solubility of calicheamicin by electrostriction of the solvent, thereby forcing the drug to associate more strongly with the non-polar regions of the DNA minor groove. In contrast, the weakly hydrated salts, such as $LiClO_4$ or $NaClO_4$, appeared to increase the solubility of calicheamicin in the aqueous solution by adsorbing to the non-polar surfaces of the drug. This results in a decrease in the rate of the binding/cleavage event. LiCl and NaCl had little or no effect on the cleavage rate, a result consistent with the central position of these salts in the Hofmeister series. The effect on the rate of cleavage in the presence of these salts may be a result of a change in the three-dimensional structure of the DNA. However, circular dichroism (CD) titration measurements with LiCl (a salt which has little or no effect on the cleavage rate) and $LiClO_4$ (a salt that decreases the cleavage rate) resulted in essentially identical CD changes with a strong decrease in the positive 275 nm band and relatively little effect on the negative 245 nm band. This CD change has been ascribed to an increase in the winding angle due to $Li^+$ binding in the minor groove (Kilkuskie *et al.*, 1988).

These results are clearly suggestive of a significant hydrophobic driving force for the calicheamicin $\gamma_1$–DNA association. However, it has become clear over the past several years that hydrophobic associations are characterized by a large negative heat capacity change ($\Delta C_P$) resulting in enthalpy ($\Delta H_{obs}$)–entropy ($T\Delta S_{obs}$) compensation, along with an essentially constant free energy ($\Delta G_{obs}$) change (Sturtevant, 1977; Eftink *et al.*, 1983; Baldwin, 1986; Ha *et al.*, 1989; Spolar *et al.*, 1989; Dill, 1990; Smithrud *et al.*, 1991). This thermodynamic behaviour has been observed in ligand–protein, protein folding and repressor and restriction enzyme–DNA associations, and it was found that the calicheamicin/DNA binding/cleavage behaved in a similar way to that summarized in Figure 3.8. This is observed

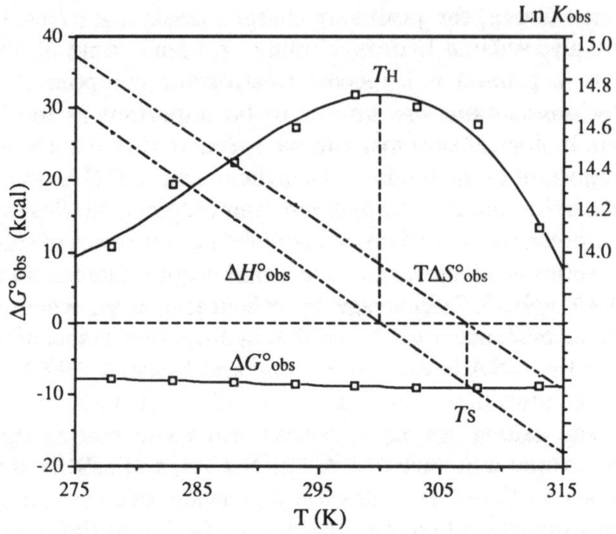


**Figure 3.8** Thermodynamics for calicheamicin–DNA binding as a function of $T$. $\Delta H^{\circ}_{obs}$ and $T\Delta S^{\circ}_{obs}$ are calculated assuming $\Delta C_p$ is $-1.21$ kcal/$T$ (From Table 3.1)

by measuring the binding constant (see Table 3.1 above) as a function of temperature over the range 5–40 °C, in which the association increases to a maximum at around 25 °C and then decreases as the temperature is raised further to 40 °C. Thus, the $\Delta H_{obs}$ for the calicheamicin $\gamma_1^I$–DNA association is positive below about 27 °C, passing through zero at the temperature where the binding is at a maximum ($T_H$), and then becoming negative at higher temperatures.

The large negative heat capacity changes observed with ligand–macromolecule binding have been explained by the removal of the non-polar surfaces of the reactants from an aqueous environment. In the present case, the extremely water-insoluble calicheamicin appears to bind preferentially to the reduced dielectric regions within certain sequences of the DNA minor groove (Breslauer *et al.*, 1987; Marky and Breslauer, 1987; Jin and Breslauer, 1988) analogous to the binding of a substrate to the hydrophobic cavity of an enzyme. A more visible manifestation of this effect was observed in NMR experiments to be discussed below, where cloudy solutions of the drug were made clear by mixing with the DNA solution (Ding and Ellestad, 1990).

A chemical precedent involving the effects of high ionic strength on the reactivity of DNA has been reported by Rokita *et al.* (1990). They observed a significant increase in the rate of photochemical reaction of acetone with DNA at high concentrations of NaCl. The role of hydrophobic interactions as a driving force in drug–DNA binding has also been discussed with DNA minor groove binding agents such as netropsin, distamycin, bisbenzimide and CC-1065, which bind to AT-rich regions of the DNA receptor (Kopka *et al.*, 1985; Breslauer *et al.*, 1987; Boger *et al.*, 1990; Hurley *et al.*, 1990). However, in contrast to calicheamicin, for which the association with DNA appears to be entropically driven between 20 and 25 °C, the binding of these more classical minor groove binding agents has been shown to be enthalpically driven in the same temperature range. Based on the non-linear van't Hoff behaviour and the temperature-dependent thermodynamic binding parameters summarized in Table 3.1 and Figure 3.8, the calicheamicin–DNA interaction appears to mimic the classical hydrophobic effect at 25 °C, i.e. a positive entropic contribution and a near-zero enthalpic contribution consistent with desolvation of the apolar surfaces (Jencks, 1969; Eftink *et al.*, 1983).

In summary, it is clear that calicheamicin $\gamma_1^I$ can recognize, bind and cleave sites within a number of different oligopyrimidine–oligopurine tracts. This supports the idea that the drug recognizes three-dimensional DNA minor groove geometry rather than specific base pair contacts. Thus, it seems reasonable to propose that it is primarily the net physical character of the calicheamicin molecule as a whole (size, shape, lipophilicity) that is recognized by a DNA binding site (Jencks, 1981; Walker *et al.*, 1992). However, specific hydrogen bond associations probably contribute to the 'fine tuning' of the preferred binding/cleavage event.

Boger *et al.* (1991) have suggested that sequence-dependent DNA conformational variability may be particularly important for the selective binding of agents such as CC-1065 that rely predominantly on van der Waals contacts and hydrophobic binding for non-covalent binding stabilization. Likewise, the binding/cleavage specificity of calicheamicin may eventually be shown to correlate with DNA structural motifs related to conformational variability/flexibility and state of hydration (Breslauer *et al.*, 1987). Recent X-ray studies with a number of decamers and dodecamers have provided evidence for a direct relationship between minor groove width and the degree of hydration (Heinemann and Alings, 1989; Marky and Kupke, 1989; Chuprina *et al.*, 1991; Grzeskowiak *et al.*, 1991; Prive *et al.*, 1991).

## 9 DNA Cleavage Chemistry

A schematic depiction of the proposed oxidative cleavage chemistry of calicheamicin $\gamma_1^I$ is given in Figure 3.9. This is obviously a very simplified picture of the sugar breakdown chemistry, especially considering the potential complexity of C4′ oxyradical reaction pathways (McGall *et al.*, 1992). It is consistent, however, with our present understanding of the calicheamicin-induced strand cleavage. The role of oxygen in the cleavage event was indicated by gel cleavage studies carried out under anaerobic conditions. When oxygen was removed from the cleavage cocktail by purging with argon, little or no strand scission was observed (Zein *et al.*, 1988b). However, in an aerobic atmosphere, excess superoxide dismutase or catalase did not alter the rate of cleavage due to calicheamicin $\gamma_1^I$. This indicates that neither superoxide radicals nor hydrogen peroxide are involved in strand cleavage. The tight, non-Gaussian cleavage pattern is consistent with the involvement of a non-diffusible carbon-centred radical species such as **12** for mediating the strand cleavage.

Presumptive evidence for the nature of the DNA cleavage chemistry was obtained by high-resolution gel electrophoresis of the calicheamicin-induced oligonucleotide fragments. Cleavage at the penultimate 5′C̲ of the 5′TC̲CT tetramer in 5′-end-labelled restriction fragments gave polynucleotide cleavage fragments the electrophoretic mobilities of which matched those of phosphate-ended control oligonucleotides produced by Maxam–Gilbert cleavage chemistry (Zein *et al.*, 1988b). This suggested that calicheamicin $\gamma_1^I$ caused the formation of fragments ending in a 3′-phosphate at the 3′ cleavage terminus. In contrast, cleavage experiments with 3′-[$^{32}$P] end-labelled restriction fragments resulted in products which migrated more slowly than the Maxam–Gilbert products as if they were two nucleotides longer. However, this was readily explained by the previous findings of Goldberg (1991) and co-workers with neocarzinostatin. Thus, calicheamicin $\gamma_1^I$, like neocarzinostatin, apparently abstracted one of the 5′ hydrogens (most likely the pro S based on models) of the targeted deoxyribose. This was proposed to give a cleavage fragment terminating in a nucleoside 5′-aldehyde as depicted in Figure 3.9. Subsequent treatment of these fragments with base then removed the altered terminal nucleoside by a β-elimination which provided a polynucleotide fragment terminating in a 5′-phosphate. The electrophoretic mobilities of these fragments now matched those of the chemically produced phosphate-ended controls.

The site of hydrogen abstraction on the complementary strand was initially less obvious. However, evidence for 4′-hydrogen atom abstraction was obtained from cleavage experiments with small 5′-end-labelled restriction fragments. Cleavage products were occasionally observed which appeared to migrate slightly faster than the Maxam–Gilbert products, similar to bleomycin-mediated cleavage in which the 4′ hydrogen of the targeted deoxyribose is abstracted generating a 4′-carbon-centred radical (Rabow *et al.*, 1990). A clearer indication of 4′-hydrogen atom abstraction on the AGGA-containing strand was obtained from cleavage experiments on synthetic dodecamers (Zein *et al.*, 1989b). 5′-End-labelled cleavage fragments clearly migrated slightly faster on sequencing gels than did the chemically produced controls consistent with the formation of 3′ termini ending in phosphoglycolates. Also similar to bleomycin–DNA cleavage chemistry was the formation of malondialdehyde in about 40% yield. This

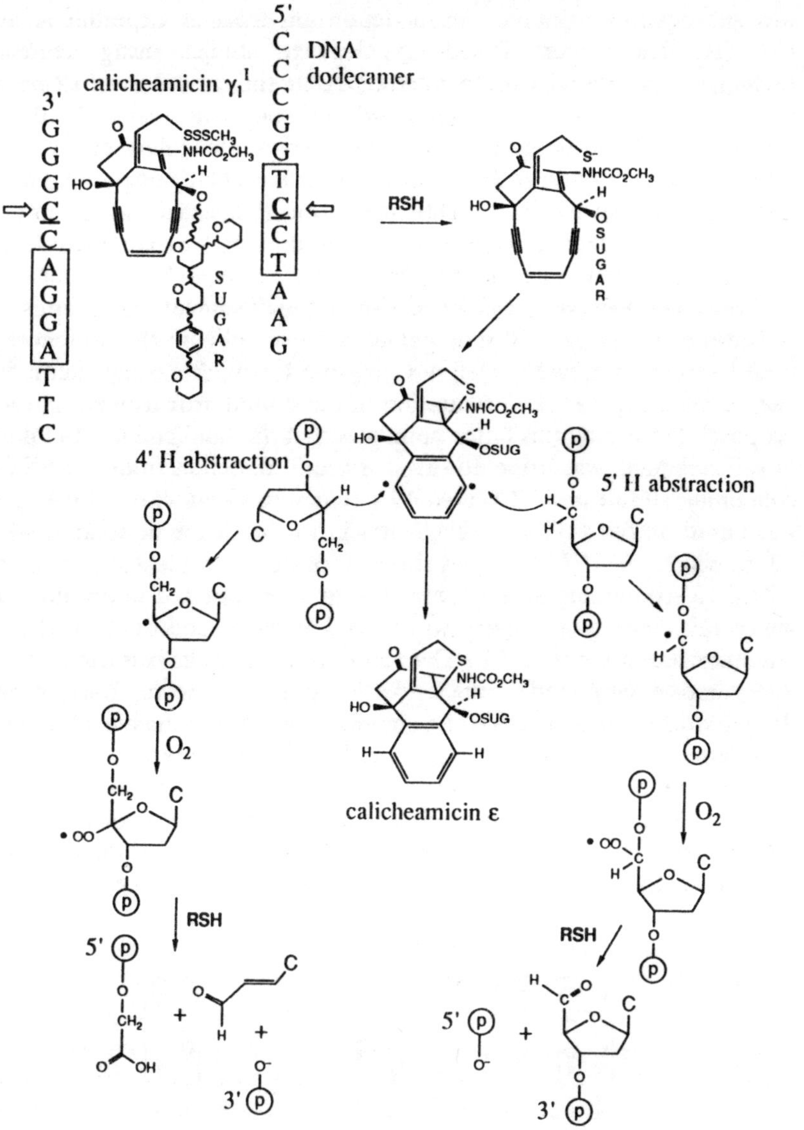


**Figure 3.9**

was based on the yield of calicheamicin ε as determined by UV analysis of the thiobarbituric acid adduct in a calicheamicin–calf thymus DNA reaction mixture (Ding and Ellestad, 1990).

Unequivocal chemical evidence confirming these early deductions about the identity of the deoxyribose hydrogens abstracted by calicheamicin $\gamma_1^I$

was subsequently obtained from deuterium transfer experiments using H-NMR. These were based on cleavage studies using deuterium-exchanged, sonicated calf thymus DNA in the presence of deuterated methyl thioglycolate, perdeuterated ethanol (for solubilization of calicheamicin) and deuterium-exchanged Tris buffer (Zein *et al.*, 1989a). The aromatized end-product, calicheamicin ε (**14**), contained only hydrogen, as shown by the H-NMR spectrum, thus indicating DNA as the sole source of hydrogen. When the experiment was conducted in the deuterated medium but without DNA, better than 95% deuterium incorporation was observed at C-3 and C-6 of calicheamicin ε.

These results suggested that synthetic duplex oligonucleotides could be used to study in greater detail not only the binding/cleavage chemistry of calicheamicin $\gamma_1^I$, but the orientation of the bound drug as well. An initial proposal concerning this latter point was that the thiobenzoate tail portion of calicheamicin was oriented towards the 5′ direction from the 5′TCCT-containing strand in a TCCT/AGGA sequence (Zein *et al.*, 1989b). This was based on the ability of calicheamicin to cleave the dodecamers-**2** and -**3**, in which the TCCT/AGGA site was located either at the 3′ end of the 5′TCCT3′-containing strand or in the middle, but not dodecamer-**4**, in which this cleavage sequence was placed at the 5′ end of the dodecamer (summarized in Figure 3.10). The fact that no cleavage was observed with dodecamer-**4** suggested to the Lederle group that the tail portion of the drug required duplex DNA for binding which was unavailable in this dodecamer. On the other hand, Schreiber and co-workers suggested that the opposite orientation was likely (Hawley *et al.*, 1989). Their suggestion was based on molecular modelling studies and X-ray crystal structures which provided evidence for binding associations between amino and aromatic halogens in crystals. They proposed that the iodo grouping of the

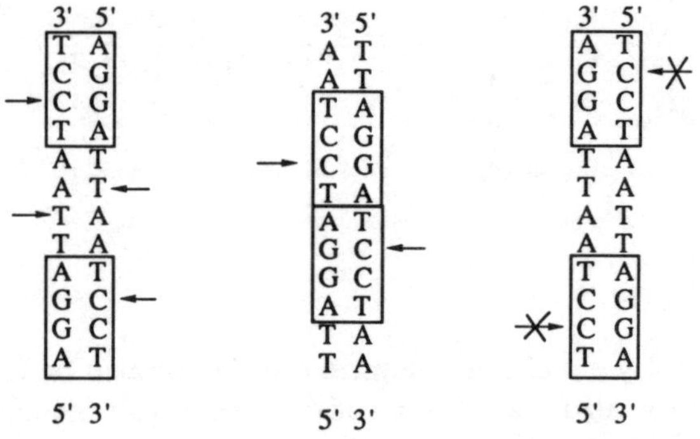


**2** **3** **4**

**Figure 3.10** Calicheamicin $\gamma_1^I$ cleavage sites for dodecamers 2, 3 and 4

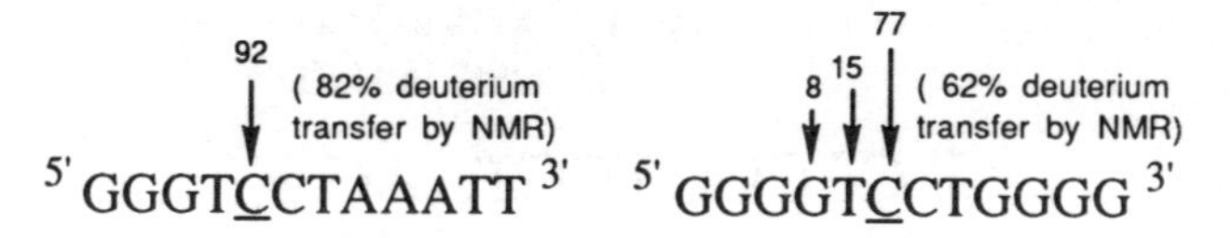


dodecamer **1** dodecamer **5**

**Figure 3.11** Summary of cleavage of dodecamers 1 and 5 as determined by high-resolution gel electrophoresis and NMR analysis of calicheamicin ε

thiobenzoate moiety formed a complex with the 2-amino grouping of the Gs of the AGGA sequence and that this could account for the observed binding/cleavage specificity.

This controversy was resolved by the above-mentioned cleavage experiments with deuterium-exchanged sonicated calf thymus DNA. Synthetic dodecamers-**1** and -**5** (Figure 3.11), specifically labelled with deuterium at the 5′ carbon of the 5′ penultimate **C** in the T**C**CT sequence, were cleaved with calicheamicin $\gamma_1^{\mathrm{I}}$ and with a drug/DNA ratio of 0.5:1 on a molar basis (DeVoss *et al.*, 1990b). The resulting distribution of the deuterium label on the aromatized calicheamicin ε end-product (either C-3 or C-6) was determined by high-resolution H-NMR. In the experiment in which only the DNA was labelled, deuterium was transferred solely to C-6 of calicheamicin ε, although the transfer was not complete. A second experiment in deuterated media provided ε containing no additional amount of deuterium. The basis for the discrepancy of deuterium transfer yields between the two dodecamers remains uncertain but it is clear that other unlabelled sites in dodecamer-**5** were providing a hydrogen source. The results of these atom-transfer experiments clearly indicate that Schreiber's orientational proposal is the correct one, given the absolute configuration of the aglycon portion of calicheamicin $\gamma_1^{\mathrm{I}}$ (Lee *et al.*, 1987a).

Subsequent transfer experiments with dodecamer-**6**, in which the TCCT/AGGA site is moved towards the 3′ end of the TCCT-containing strand (compared with dodecamer-**1**), provided a much better material balance (Figure 3.12) (Hangeland *et al.*, 1992). Cleavage experiments with this dodecamer, specifically labelled with deuterium at C-4′ of the targeted deoxyribose on the AGGA-containing strand, gave calicheamicin ε containing deuterium at C-3 in a transfer yield of 63% (compared with 25% with dodecamer-**1**) in an undeuterated medium. The atom transfer increased to 78% in the presence of deuterated media, confirming that cleavage on this strand is not as strong as on the complementary TCCT strand. These experiments indicate that abstraction on the AGGA strand takes place predominantly at the 4′ hydrogen of the targeted deoxyribose in this dodecamer, confirming the gel results mentioned earlier. Transfer experiments on dodecamer-**6** specifically labelled with deuterium on the

[5′–$d_2$] 87% in $D_2O$
79% in $H_2O$

5′ CCCGGTCCTAAG 3′
3′ GGGCCAGGATTC 5′

78% in $D_2O$
[4′–$d_1$] 63% in $H_2O$

**Figure 3.12** Deuterium labelling of calicheamicin ε with dodecamer 6 as determined by NMR

TCCT-containing strand resulted in deuterium labelling at C-6 of calicheamicin $\gamma_1^I$ in 79% yield in an undeuterated medium, consistent with the earlier results. Cleavage experiments with the targeted deoxyribose specifically labelled with deuterium at C-1′ (another possible site of hydrogen abstraction in the DNA minor grove) on the AGGA-containing strand did not result in any significant atom transfer to the calicheamicin ε (Hangeland *et al.*, 1992).

The molecular basis for this binding orientation remains uncertain. Although the nitrogen–halogen interaction based on crystal structures is a tempting one, many polyG-containing sequences are not cleaved. This is probably related to the increased minor groove width generally associated with polyGC tracts. Also, the more recent results mentioned above showing cleavage at a TTTT site, as opposed to an adjacent TCCC sequence, indicate that it is unnecessary to invoke this association. But it is conceivable that a nitrogen–halogen interaction plays a recognition role in TCCT-like tracts and not in others.

## 10 Mechanism of Trisulfide Cleavage

As mentioned earlier, Cramer and Townsend (1991) examined the possible role of the ethylamino group in general base-catalysis in the cleavage of the trisulfide and showed that it was unnecessary to attribute a general base-catalytic role to this functionality as previously proposed based on disulfide exchange experiments in acetonitrile (Ellestad *et al.*, 1989; Zein *et al.*, 1989b). This kinetic study was carried out by comparing the bimolecular rate constants for the trisulfide cleavage of calicheamicin $\gamma_1^I$, the N-acetylated (E ring) derivative and the $\alpha_3$ component (**8**) with aminoethanethiol under pseudo-first-order conditions in 30% methanol and 70% Tris buffer. Aminoethanethiol was chosen because its p$K_a$ (8.3) is similar to that of reduced glutathione (8.5), which is surmised to be the triggering thiol *in vivo* because of its high intracellular concentration (1–10 mM). The results of these experiments clearly showed that there is little difference in rate of disappearance of calicheamicin $\gamma_1^I$ and the derivative without the aminosugar or the N-acetylated analogue. Furthermore, the dependence of the overall rate on thiol concentration indicated that the bimolecular reaction between thiol and drug is rate-determining and at least 50 times slower than the rate of cycloaromatization. In contrast to these experiments in aqueous media, disulfide exchange reactions in acetonitrile did not occur with calicheamicin derivatives lacking the ethylamino sugar or with the N-acetylated derivative in the absence of triethylamine or triethylammonium acetate (Ellestad *et al.*, 1989). In this relatively apolar environment, the $pK_a$ of the amino sugar is apparently sufficiently low for it to act as a base as opposed to being protonated under physiological conditions.

Insight as to the rate-limiting step in a productive binding/cleavage event was obtained in a variable-temperature NMR study. DeVoss and coworkers (DeVoss *et al.*, 1990a) monitored the cleavage of calicheamicin $\gamma_1^I$ with $n$-$Bu_3P$ in methanol-d4 and calculated that the dihydrothiophene intermediate **11** has a significant solution half-life of 4.5 s at 37 °C. The first-order rate constant of $(5 \pm 2) \times 10^{-4}$ for the conversion of **11** to **14** was determined over three half-lives from which a $\Delta G^{\neq}$ of 19.3 ± 0.2 kcal/mol was determined. This value is similar to that obtained for the corresponding rearrangement step of neocarzinostatin (Myers and Proteau, 1989). This is interesting in that even though the hybridization at C-9 is now $sp^3$, the Bergman cyclization does not occur immediately. Thus, the formation of intermediate **11** may be the kinetically significant event in a series of transformations leading to the diradical intermediate **12** and, once formed, it is that species that actually searches and locates the preferred DNA binding site. The ensuing bond reorganization to the $p$-benzyne intermediate then initiates oxidative strand cleavage. It should be mentioned that a dimeric form of calicheamicin $\gamma_1^I$ (**18**) (Figure 3.13) exhibits the same DNA site discrimination as calicheamicin $\gamma_1^I$. Thus, it is likely that thiol reductively cleaves this dimeric calicheamicin to intermediate **11** in common with the parent drug (Zein *et al.*, 1987).

One other point is noteworthy regarding the kinetics of strand cleavage. A comparison of the extent of calicheamicin-induced damage at the C-5′ deuterated site by autoradiography on sequencing gels with 5′-end-labelled dodecamer showed little or no isotope-induced branching (on either the same or a neighbouring deoxyribose) when compared with experiments with non-deuterated dodecamer (DeVoss *et al.*, 1990b). Thus, it appears that the 5′-hydrogen abstraction step is not rate-limiting in the DNA cleavage event. Preliminary results from similar experiments with a dodecamer specifically labelled with deuterium on the AGGA-containing

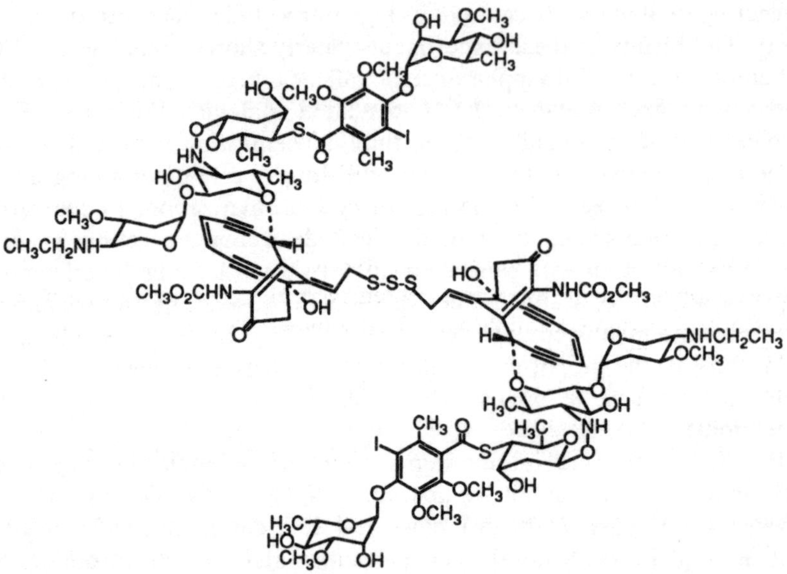

**18** Calicheamicin $\gamma_1^{\mathrm{I}}$ dimer

**Figure 3.13**

strand indicated no isotope effect here either. Similar results were obtained with C-4′ deuterium abstraction experiments with esperamicin $A_1$ (Kozarich *et al.*, 1989).

## 11 Biochemical Basis for Cytotoxicity

That DNA is a likely cellular target for calicheamicin $\gamma_1^{\mathrm{I}}$ is suggested by: (1) the extraordinary potency (pg/ml) in the biochemical prophage induction assay; (2) the chromosomal aberrations in human diploid lung fibroblast cells, again at pg/ml levels (Durr and Wallace, 1988); (3) the 50% inhibition of uptake of [$^3$H]-thymidine at a 50 pg/ml concentration of calicheamicin $\gamma_1^{\mathrm{I}}$ in HeLa cells (Zein *et al.*, 1988a); and (4) the mitochondrial DNA cleavage detected at 50 pg/ml ($10^{-11}$M) of drug in cleavage experiments with HeLa cells (Jones and Zein, 1988). Interestingly, gel analysis of the mitochondrial DNA damage from these calicheamicin $\gamma_1^{\mathrm{I}}$ treated cells showed discrete bands, instead of a generalized smear, suggestive of specific strand scission. Whether these cleavage sites match those observed in the restriction fragment experiments remains to be seen.

The susceptibility of mitochondrial DNA to calicheamicin $\gamma_1^{\mathrm{I}}$ suggests a possible explanation for the delayed lethality observed with unchallenged mice when injected with doses of this drug that prolonged life in tumour-bearing mice. A similar response has been reported with another highly potent antitumour antibiotic, CC-1065, which was found to cause ultrastructural changes in hepatic mitochondria (Reynolds *et al.*, 1986).

Further evidence suggesting that cytotoxicity is related to cellular DNA damage comes from a study by Zhao and co-workers (Zhao *et al.*, 1990). They showed that in human promyelocytic HL-60 leukaemia cells calicheamicin $\gamma_1^I$ caused a marked increase in poly(ADP-ribose)polymerase activity, apparently in response to extensive DNA damage. The increased levels of this enzyme paralleled cell death and a corresponding decrease in $NAD^+$ levels. Thus, the treated cells were apparently unable to generate sufficient energy to maintain viability. This phenomenon has been reported with other DNA-damaging agents as well (Gaal and Pearson, 1985; Durkacz *et al.*, 1980; Berger *et al.*, 1986; Stubberfield and Cohen, 1988; Marks and Fox, 1991).

Sullivan and Lyne studied the effect of calicheamicin $\gamma_1^I$ on fibroblasts derived from ataxia- telangiectasia (AT) patients. AT patients have been found to be abnormally sensitive to the therapeutic use of γ-irradiation, and cell lines from these patients are sensitive to DNA cleaving agents that act by a free radical mechanism. The AT lines used here were sensitive to both calicheamicin $\gamma_1^I$ and bleomycin relative to control lines. However, calicheamicin $\gamma_1^I$ was 40 000 times more effective on a molar basis against AT cell lines compared with bleomycin (Sullivan and Lyne, 1990). Perhaps the calicheamicin-induced nucleoside 5′-aldehyde and 3′-phosphoglycolate ends represent cytostatic lesions to AT cells. On the other hand, calicheamicin $\gamma_1^I$ did not inhibit DNA synthesis in AT-derived cell lines. Thus, calicheamicin $\gamma_1^I$ appears to act in a manner similar to ionizing radiation.

Finally, derivatives such as the *N*-acetyl calicheamicin $\gamma_1^I$ and $\alpha_3$ (**8**) analogues, which are less efficacious in producing DNA strand breaks *in vitro*, are also less active in prolonging the lifespan of mice challenged with experimental murine tumours. Taken together, the above observations point strongly to DNA damage as being primarily responsible for the potent cytotoxicity of this drug. But it is certainly possible that damage to other cellular components might also make a contribution to the overall cytotoxicity.

## 12 Summary

The discovery of the enediyne antitumour antibiotics calicheamicin $\gamma_1^I$ and esperamicin $A_1$ has triggered a great deal of research into the chemistry and biological activity of these agents in both industrial and academic laboratories. Although questions remain concerning the precise biochemical basis for the extreme cytotoxicity of these agents, it appears that their powerful DNA cleaving properties are at least partially responsible.

Structure–function studies have been hampered by the lack of material. However, strain improvement and media development efforts will no doubt make larger amounts of the calicheamicins available for such studies. A number of prototype compounds containing a portion or all of the enediyne moiety have been synthesized (Kende and Smith, 1988; Magnus and Carter, 1988; Nicolaou *et al.*, 1988a,b; Haseltine *et al.*, 1989; Mantlo and Danishefsky, 1989; Tomioka *et al.*, 1989; Doi and Takahashi, 1991) and both Nicolaou *et al.* (1990) and Halcomb *et al.* (1992) have synthesized the oligosaccharide tail fragment of calicheamicin $\gamma_1^I$. Indeed, the latter group has now coupled this fragment to an aglycon precursor, so that a total synthesis must not be far away (Halcomb *et al.*, 1992). These efforts will surely lead to a wider variety of structural analogues, hopefully with an improved therapeutic profile. An excellent review of this chemistry has appeared (Nicolaou and Dai, 1991).

Although calicheamicin $\gamma_1^I$ itself exhibits little or no margin of safety in animal tests, largely because of delayed organ toxicity (Thomas *et al.*, 1986), the preparation of monoclonal antibody conjugates to obtain agents targeted to a specific tumour cell line and with altered biodistribution appears promising (Hamann *et al.*, 1990; Hinman *et al.*, 1990; Menendez *et al.*, 1990). The synthesis of these conjugates utilizes mild and efficient disulfide exchange chemistry (Ellestad *et al.*, 1989) to attach 3-mercaptopropionyl hydrazide linkers to the drug, which can in turn be coupled to periodate-oxidized monoclonal antibodies. For example, *N*-acetylcalicheamicin $\gamma_1^I$ and the $\alpha_3$ component (**8**) have been coupled to an internalizable monoclonal antibody (7F11C7) which is specific for the human milk fat globule protein antigen present on the surface of a number of breast, ovarian, non-small-cell lung and colon tumours. Compared with the parent drug, these conjugates are 140 and 1000 times, respectively, less cytotoxic towards normal liver cells. Towards the target MX-1 breast tumour, however, they are 3 and 8 times, respectively, more cytotoxic than calicheamicin $\gamma_1^I$ itself. Most encouraging is the finding that the Lym-1 conjugate of β-mercapto, β-dimethylpropionyl hydrazide disulfide of *N*-acetyl calicheamicin $\gamma_1^I$ produced complete remissions of Raji and Sultan 'calicheamicin-resistant' tumors *in vivo* over a wide dose range. The antibody itself had little effect, and mixtures of the antibody and drug hydrazide were ineffective and toxic.

**References**

Baldwin, R. L. (1986). Temperature dependence of the hydrophobic interaction in protein folding. *Proc. Natl Acad. Sci. USA*, **83**, 8069–8072

Berger, S. J., Sudar, D. C. and Berger, N. A. (1986). Metabolic consequences of

DNA damage: DNA damage induces alterations in glucose metabolism by activation of poly(ADP-ribose)polymerase. *Biochem. Biophys. Res. Commun.*, **134**, 227–232

Bergman R. G. (1973). Reactive 1,4-dehydroaromatics. *Acc. Chem. Res.*, **6**, 25–31

Boger, D. L., Coleman, R. S., Invergo, B. J., Sakya, S. M., Ishizaki, T., Munk, S. A., Zarrinmayeh, H., Kitos, P. A. and Thompson, S. C. (1990). Synthesis and evaluation of aborted and extended CC-1065 functional analogues: (+)- and (−)-CPI-PDE-$I_1$ (+)- and (−)-CPI-$CDPI_1$, and (±)-, (+)-, and (−)-CPI-$CDPI_3$. Preparation of key partial structures and definition of an additional functional role of the CC-1065 central and right-hand subunits. *J. Am. Chem. Soc.*, **112**, 4623–4632

Boger, D. L., Zarrinmayeh, H., Munk, S. A., Kitos, P. A. and Suntornwat, O. (1991). Demonstration of a pronounced effect of noncovalent binding selectivity on the (+)-CC-1065 DNA alkylation and identification of the pharmacophore of the alkylation subunit. *Proc. Natl Acad. Sci. USA,* **88**, 1431–1435

Breslauer, K. J., Remeta, D. P., Chou, W.-y., Ferrante, R., Curry, J., Zaunczkowski, D., Snyder, J. G. and Marky, L. A. (1987). Enthalpy-entropy compensations in drug-DNA binding studies. *Proc. Natl Acad. Sci. USA*, **84**, 8922–8926

Chaires, J. B. (1990). Daunomycin binding to DNA: From the macroscopic to the microscopic. In Pullman, B. and Jortner, J. (Eds), *Molecular Basis of Specificity in Nucleic Acid–Drug Interactions*, Kluwer Academic, Dordrecht, pp. 123–136

Chaires, J. B., Dattagupta, N. and Crothers, D. M. (1982). Studies on interaction of anthracycline antibiotics and deoxyribonucleic acid: Equilibrium binding studies on interaction of daunomycin with deoxyribonucleic acid. *Biochemistry*, **21**, 3933–3940

Chapman, O. L., Chang, C. C. and Kolc, J. (1976). 9,10-Dehydroanthracene. A derivative of 1,4-dehydrobenzene. *J. Am. Chem. Soc.*, **98**, 5703–5705

Chuprina, V. P., Heinemann, U., Nurislamov, A. A., Zielenkiewicz, P., Dickerson, R. E. and Saenger, W. (1991). Molecular dynamics simulation of the hydration shell of a B-DNA decamer reveals two main types of minor-groove hydration depending on groove width. *Proc. Natl Acad. Sci. USA*, **88**, 593–597

Cramer, K. D. and Townsend, C. A. (1991) Kinetics of trisulfide cleavage in calicheamicin—assessing the role of the ethylamino group. *Tetrahedron Lett.*, **32**, 4635–4638

Darbay, N., Kim, C. U., Salaun, J. A., Shelton, K. W., Takada, S. and Masamune, S. (1971). Concerning the 1,5-didehydro[10]annulene system. *J. Chem. Soc. Chem. Commun.*, 1516–1517

Dedon, P. C. and Goldberg, I. H. (1990). Sequence-specific double-strand breakage of DNA by neocarzinostatin involves different chemical mechanisms within a staggered cleavage site. *J. Biol. Chem.*, **265**, 14713–14716

DeVoss, J. J., Hangeland, J. J. and Townsend, C. A. (1990a). Characterization of the *in vitro* cyclization chemistry of calicheamicin and its relation to DNA cleavage. *J. Am. Chem. Soc.*, **112**, 4554–4556

DeVoss, J. J., Townsend, C. A., Ding, W.-D., Morton, G. O., Ellestad, G. A., Zein, N., Tabor, A. B. and Schreiber, S. L. (1990b). Site-specific atom transfer from DNA to a bound ligand defines the geometry of a DNA-calicheamcin $\gamma_1^I$ complex. *J. Am. Chem. Soc.*, **112**, 9669–9670

Dill, A. K. (1990). Dominant forces in protein folding. *Biochemistry*, **29**, 7133–7155

Ding, W.-D. and Ellestad, G. A. (1990). Unpublished observations from Lederle Laboratories.

Ding, W.-D. and Ellestad, G. A. (1991). Evidence for hydrophobic interaction

between calicheamicin and DNA. *J. Am. Chem. Soc.*, **113**, 6617–6620
Doi, T. and Takahashi, T. (1991). Syntheses and transannular cyclizations of neocarzinostatin–chromophore and esperamicin–calicheamicin analogues. *J. Org. Chem.*, **56**, 3465–3467
Drak, J., Iwasawa, N., Danishefsky, S. and Crothers, D. M. (1991). The carbohydrate domain of calicheamicin $\gamma_1^I$ determines its sequence specificity for DNA cleavage. *Proc. Natl Acad. Sci. USA*, **88**, 7464–7468
Durkacz, B. W., Omidiji, O., Gray, D. A. and Shall, S. (1980). (ADP-ribose)$_n$ participates in DNA excision repair. *Nature*, **283**, 593–596
Durr, F. and Wallace, R. E. (1988). Unpublished results from Lederle Laboratories
Edo, K., Mizugaki, M., Koide, Y., Seto, H., Furihata, K., Otake, N. and Ishida, N. (1985). The structure of neocarzinostatin chromophore possessing a novel bicylo[7, 3, 0]dodecadiyne system. *Tetrahedron Lett.*, **26**, 331–334
Eftink, M. R., Anusiem, A. C. and Biltonen, R. L. (1983). Enthalpy–entropy compensation and heat capacity changes for protein–ligand interactions: General thermodynamic models and data for the binding of nucleotides to ribonuclease A. *Biochemistry*, **22**, 3884–3896
Elespuru, R. K. and White, R. J. (1983). Biochemical prophage induction assay: A rapid test for antitumor agents that interact with DNA. *Cancer Res.*, **43**, 2819–2830
Elespuru, R. K. and Yarmolinsky, M. B. (1979). A colorimetric assay of lysogenic induction designed for screening potential carcinogenic and carcinostatic agents. *Environ. Mutagen.*, **1**, 65–78
Ellestad, G. A., Hamann, P. R., Zein, N., Morton, G. O., Siegel, M. M., Pastel, M., Borders, D. B. and McGahren, W. J. (1989). Reactions of the trisulfide moiety in calicheamicin. *Tetrahedron Lett.*, **30**, 3033–3036
Fry, D. W., Shillis, J. L. and Leopold, W. R. (1986). Biological and biochemical activities of the novel antitumor antibiotic PD 114,759 and related derivatives. *Investigational New Drugs*, **4**, 3–10
Gaal, J. C. and Pearson, C. K. (1985). Eukaryotic nuclear ADP-ribosylation reactions. *Biochem. J.*, **230**, 1–18
Goldberg, I. H. (1991). Mechanism of neocarzinostatin action: Role of DNA microstructure in determination of chemistry of bistranded oxidative damage. *Acc. Chem. Res.*, **24**, 191–198
Golik, J., Clardy, J., Dubay, G., Groenewold, G., Kawaguchi, H., Konishi, M., Krishnan, B., Ohkuma, H., Saitoh, K. and Doyle, T. W. (1987a). Esperamicins, a novel class of potent antitumor antibiotics. 2. Structure of esperamicin X. *J. Am. Chem. Soc.*, **109**, 3461–3462
Golik, J., Dubay, G., Groenewold, G., Kawaguchi, H., Konishi, M., Krishnan, B., Ohkuma, H., Saitoh, K. and Doyle. T. W. (1987b). Esperamicins, a novel class of potent antitumor antibiotics. 3. Structures of esperamicins $A_1$, $A_2$, and $A_{1b}$. *J. Am. Chem. Soc.*, **109**, 3462–3464
Grzeskowiak, K., Yanagi, K., Prive, G. G. and Dickerson, R. E. (1991). The structure of B-helical C–G–A–T–C–G–A–T–C–G and comparison with C–C–A–A–C–G–T–T–G–G. *J. Biol. Chem.*, **266**, 8861–8883
Ha, J.-H., Spolar, R. S. and Record, M. T., Jr. (1989). Role of hydrophobic effect in stability of site-specific protein-DNA complexes. *J. Mol. Biol.*, **209**, 801–816
Halcomb, R. L., Boyer, S. H. and Danishefsky, S. J. (1992). Synthesis of the calicheamicin aryl-tetrasaccharide domain bearing a reducing terminus: Coupling of fully synthetic aglycon and carbohydrate domains by the Schmidt reaction. (We wish to thank Professor Danishefsky for a preprint of this manuscript)

Hamann, L. M., Hinman, L. M. and Upeslacis, J. (1990). 'Monoclonal antibody conjugates prepared from the calicheamicin family of highly-potent antitumor-antibiotics'. Poster and short oral presentation given at the *5th Intern. Conf. on MoAb Immunoconjugates for Cancer*, San Diego, March 1990

Hangeland, J. J., DeVoss, J. J., Heath, J. A., Townsend, C. A., Ding, W.-D., Ashcroft, J. S. and Ellestad, G. A. (1992). Specific abstraction of the 5′(*S*)- and 4′-deoxyribosyl hydrogen atoms from DNA by calicheamicin $\gamma_1^I$. *J. Am. Chem. Soc.*, **114**, 9200–9202

Haseltine, J. N., Cabal, M. P., Mantlo, N. B., Iwasawa, N., Yamashita, D. S., Coleman, R. S., Danishefsky, S. J. and Schulte, G. K. (1991). Total synthesis of calicheamicinone: New arrangements for actuation of the reductive cycloaromatization of aglycon congeners. *J. Am. Chem. Soc.*, **113**, 3850–3866

Haseltine, J. N., Danishefsky, S. J. and Schulte, G. K. (1989). Experimental modeling of the priming mechanism of the calicheamicin/esperamicin antibiotics: Actuation by the addition of intramolecular nucleophiles to the bridgehead double bond. *J. Am. Chem. Soc.*, **111**, 7638–7640

Hawley, R. C., Kiessling, L. L. and Schreiber, S. L. (1989). Model of the interactions of calicheamicin $\gamma_1^I$ with a DNA fragment from pBR 322. *Proc. Natl Acad. Sci. USA*, **86**, 1105–1109

Heinemann, U. and Alings, C. (1989). Crystallographic study of one turn of G/C-rich B-DNA. *J. Mol. Biol.*, **210**, 369–381

Hinman, L. M., Wallace, R., Hamann, P. R., Durr, F. E. and Upeslacis, J. (1990). 'Calicheamicin immunoconjugates: Influence of analog and linker modification on activity *in vivo*'. Poster given by L. Hinman at the *5th Intern. Conf. on MoAb Immunoconjugates for Cancer*, San Diego, March 1990

Hurley, L. H., Warpehoski, M. A., Lee, C.-S., McGovren, J. P., Scahill, T. A., Kelly, R. C., Mitchell, M. A., Wicnienski, N. A., Gebhard, I., Johnson, P. D. and Bradford, V. S. (1990). Sequence specificity of DNA alkylation by the unnatural enantiomer of CC-1065 and its synthetic analogues. *J. Am. Chem. Soc.*, **112**, 4633–4649

Jencks, W. P. (1969). *Catalysts in Chemistry and Enzymology*, McGraw-Hill, New York, pp. 393–396

Jencks, W. P. (1981). On the attribution and additivity of binding energies. *Proc. Natl Acad. Sci. USA*, **78**, 4046–4050

Jin, R. and Breslauer, K. J. (1988). Characterization of the minor groove environment in a drug-DNA complex: Bisbenzimide bound to the poly[d(AT)]. poly-[d(AT)] duplex. *Proc. Natl Acad. Sci. USA*, **85**, 8939–8942

Jones, T. and Zein. N. (1988). Unpublished results from Lederle Laboratories

Kappen, L. S., Goldberg, I. H., Frank, B. L., Worth, L., Jr., Christner, D. F., Kozarich, J. W. and Stubbe, J. (1991). Neocarzinostatin-induced hydrogen atom abstraction from C-4′ and C-5′ of the T residue at a d(GT) step in oligonucleotides: Shuttling between deoxyribose attack sites based on isotope selection effects. *Biochemistry*, **30**, 2034–2042

Kende, A. S. and Smith, C. A. (1988). Synthesis of a calicheamicin deoxyaglycone model by an intramolecular acetylide cyclization. *Tetrahedron. Lett.*, **29**, 4217–4220

Kilkuskie, R., Wood, N., Ringquist, S., Shinn, R. and Hanlon, S. (1988). Effects of charge modification on the helical period of duplex DNA. *Biochemistry*, **27**, 4377–4386

Kishikawa, H., Jiang, Y.-P., Goodisman, J. and Dabrowiak, J. C. (1991). Coupled kinetic analysis of cleavage of DNA by esperamicin and calicheamicin. *J. Am. Chem. Soc.*, **113**, 5434–5440

Konishi, M., Ohkuma, H., Matsumoto, K., Tsuno, T., Kamei, H., Miyaki, T., Oki, T., Kawaguchi, H., VanDuyne, G. D. and Clardy, J. (1989). Dynemicin A. A novel antibiotic with the anthraquinone and 1,5-diyne-3-ene subunit. *J. Antibiotics*, **42**, 1449–1452

Kopka, M. L., Yoon, C., Pjura, P. and Dickerson, R. E. (1985). The molecular origin of DNA-drug specificity in netropsin and distamycin. *Proc. Natl Acad. Sci USA*, **82**, 1376–1380

Kozarich, J. W., Worth, L., Jr., Frank, B. L., Christner, D. F., Vanderwall, D. E. and Stubbe, J. (1989). Sequence-specific isotope effects on the cleavage of DNA by bleomycin. *Science*, **245**, 1396–1399

Lee, M. D., Dunne, T. S., Chang, C. C., Ellestad, G. A., Siegel, M. M., Morton, G. O., McGahren, W. J. and Borders, D. B. (1987a). Calicheamicins, a novel family of antibiotics. 2. Chemistry and structure of calicheamicin $\gamma_1^I$. *J. Am. Chem. Soc.*, **109**, 3466–3468

Lee, M. D., Dunne, T. S., Siegel, M. M., Chang, C. C., Morton, G. O. and Borders, D. B. (1987b). Calicheamicins, a novel family of antibiotics. 1. Chemistry and partial structure of calicheamicin $\gamma_1^I$. *J. Am. Chem. Soc.*, **109**, 3464–3466

Lee, M. D., Ellestad, G. A. and Borders, D. B. (1991). Calicheamicins: Discovery, structure, chemistry, and interaction with DNA. *Acc. Chem. Res.*, **24**, 235–243

Lee, M. D., Manning, J. K., Williams, D. R., Kuck, N. A., Testa, R. T. and Borders, D. B. (1989). Calicheamicins, a novel family of antibiotics. 3. Isolation, purification and characterization of calicheamicins $\beta_1^{Br}$, $\gamma_1^{Br}$, $\alpha_2^I$, $\alpha_3^I$, $\beta_1^I$, $\gamma_1^I$ and $\delta_1^I$. *J. Antibiotics*, **42**, 1070–1087

Lockhart, T. P. and Bergman, R. G. (1981). Evidence for the reactive spin state of 1,4-dehydrobenzenes. *J. Am. Chem. Soc.*, **103**, 4091–4096

Lockhart, T. P., Comita, P. B. and Bergman, R. G. (1981). Kinetic evidence for the formation of discrete 1,4-dehydrobenzene intermediates. Trapping by inter- and intramolecular hydrogen atom transfer and observation of high-temperature CIDNP. *J. Am. Chem. Soc.*, **103**, 4082–4090

Long, B. H., Golik, J., Forenza, S., Ward, B., Rehfuss, R., Dabrowiak, J. C., Catino, J. J., Musial, S. T., Brookshire, K. W. and Doyle, T. W. (1989). Esperamicins, a class of potent antitumor antibiotics: mechanism of action. *Proc. Natl Acad. Sci. USA*, **86**, 2–6

McGall, G. H., Rabow, L. E., Ashley, G. W., Wu, S. H., Kozarich, J. W. and Stubbe, J. (1992). New insight into the mechanism of base propenal formation during bleomycin-mediated DNA degradation. *J. Am. Chem. Soc.*, **114**, 4958–4967

Magnus, P. and Carter, P. A. (1988). A model for the proposed mechanism of action of the potent antitumor antibiotic esperamicin $A_1$. *J. Am. Chem. Soc.*, **110**, 1626–1628

Magnus, P., Fortt, S., Pitterna, T. and Snyder, J. P. (1990). Synthetic and mechanistic studies on esperamicin $A_1$ and calicheamicin $\gamma_1^I$. Molecular strain rather than $\pi$-bond proximity determines the cycloaromatization rates of bicyclo[7.3.1] enediynes. *J. Am. Chem. Soc.*, **112**, 4986–4987

Maiese, W. M., Lechevalier, M. P., Lechevalier, H. A., Korshalla, J., Kuck, N., Fantini, A., Wildey, M. J., Thomas, J. and Greenstein, M. (1989). Calicheamicins, a novel family of antitumor antibiotics: taxonomy, fermentation and biological properties. *J. Antibiotics*, **42**, 558–563

Mantlo, N. B. and Danishefsky, S. J. (1989). A core system that simulates the cycloaromatization and DNA cleavage properties of calicheamicin-esperamicin: A correlation experiment. *J. Org. Chem.*, **54**, 2781–2783

Marks, D. I. and Fox, R. M. (1991). DNA damage, poly(ADP-ribosyl)ation and apoptotic cell death as a potential common pathway of cytotoxic drug action. *Biochem. Pharmacol.*, **42**, 1859–1867

Marky, L. A. and Breslauer, K. J. (1987). Origins of netropsin binding affinity and specificity: Correlations of thermodynamic and structural data. *Proc. Natl Acad. Sci. USA*, **84**, 4359–4363

Marky, L. A. and Kupke, D. W. (1989). Probing the hydration of the minor groove of A·T synthetic DNA polymers by volume and heat changes. *Biochemistry*, **28**, 9982–9988

Menendez, A. T., Hinman, L. M., Hamann, P. R. and Upeslacis, J. (1990). 'Reduction of non-specific toxicity of calicheamicin derivatives by conjugation to 7F11C7'. Poster given by A. T. Menendez at the *5th Intern. Conf. on MoAb Immunoconjugates for Cancer*, San Diego, March 1990

Myers, A. G. and Proteau, P. J. (1989). Evidence for spontaneous, low-temperature biradical formation from a highly reactive neocarzinostatin chromophore-thiol conjugate. *J. Am. Chem. Soc.*, **111**, 1146–1147

Myers, A. G., Proteau, P. J. and Handel, T. M. (1988). Stereochemical assignment of neocarzinostatin chromophore. Structures of neocarzinostatin chromophore-methyl thioglycolate adducts. *J. Am. Chem. Soc.*, **110**, 7212–7214

Neidle, S., Pearl, L. H. and Skelly, J. V. (1987). DNA structure and perturbation by drug binding. *Biochem. J.*, **243**, 1–13

Nicolaou, K. C. and Dai, W.-M. (1991). Chemistry and biology of the enediyne anticancer antibiotics. *Angew. Chem. Int. Ed. Engl.*, **30**, 1387–1530

Nicolaou, K. C., Groneberg, R. D., Miyazaki, T., Stylianides, N. A., Schulze, T. J. and Stahl, W. (1990). Total synthesis of the oligosaccharide fragment of calicheamicin $\gamma_1^I$. *J. Am. Chem. Soc.*, **112**, 8193–8195

Nicolaou, K. C., Ogawa, Y., Zuccarello, G. and Kataoka, H. (1988a). DNA cleavage by a synthetic mimic of the calicheamicin-esperamicin class of antibiotics. *J. Am. Chem. Soc.*, **110**, 7247–7248

Nicolaou, K. C., Zuccarello, G., Ogawa, Y., Schweiger, E. J. and Kumazawa, T. (1988b). Cyclic conjugated enediynes related to calicheamicins and esperamicins: Calculations, synthesis and properties. *J. Am. Chem. Soc.*, **110**, 4866–4868

Prive, G. G., Yanagi, K. and Dickerson, R. E. (1991). Structure of the B-DNA decamer CCAACGTTGG and comparison with the isomorphous decamers CCAAGATTGG and CCAGGCCTGG. *J. Mol. Biol.*, **217**, 177–199

Rabow, L. E., Stubbe, J. and Kozarich, J. W. (1990). Identification and quantitation of the lesion accompanying base release in bleomycin-mediated DNA degradation. *J. Am. Chem. Soc.*, **112**, 3196–3203

Record, M. T., Jr., Anderson, C. F. and Lohman, T. M. (1978). Thermodynamic analysis of ion effects on the binding and conformational equilibria of proteins and nucleic acids: the roles of ion association or release, screening, and ion effects on water activity. *Quart. Rev. Biophys.*, **11**, 103–178

Reynolds, V. L., McGovren, J. P. and Hurley, L. H. (1986). The chemistry, mechanism of action and biological properties of CC-1065, a potent antitumor antibiotic. *J. Antibiotics.*, **39**, 319–334

Rokita, S. E., Prusiewicz, S. and Romero-Fredes, L. (1990). The effect of ionic strength on the photosensitized oxidation of $d(CG)_6$. *J. Am. Chem. Soc.*, **112**, 3616–3621

Shiraki, T. and Sugiura, Y. (1990). Visible light induced DNA cleavage by the hybrid antitumor antibiotic dynemicin A. *Biochemistry*, **29**, 9795–9798

Smithrud, D. B., Wyman, T. B. and Diederich, F. (1991). Enthalpically driven cyclophane-arene inclusion complexation: Solvent-dependent calorimetric studies. *J. Am. Chem. Soc.*, **113**, 5420–5426

Snyder, J. P. (1989). The cyclization of calicheamicin-esperamicin analogues: A predictive biradicaloid transition state. *J. Am. Chem. Soc.*, **111**, 7630–7632

Snyder, J. P. (1990). Monocyclic enediyne collapse to 1,4-diyl biradicals: A pathway under strain control. *J. Am. Chem. Soc.*, **112**, 5367–5369

Spolar, R. S., Ha, J.-H., and Record, M. T., Jr. (1989). Hydrophobic effect in protein folding and other noncovalent processes involving proteins. *Proc. Natl Acad. Sci. USA*, **86**, 8382–8385

Stubberfield, C. R. and Cohen, G. M. (1988). $NAD^+$ Depletion and cytotoxicity in isolated hepatocytes. *Biochem. Pharmacol.*, **37**, 3967–3974

Sturtevant, J. M. (1977). Heat capacity and entropy changes in processes involving proteins. *Proc. Natl Acad. Sci. USA*, **74**, 2236–2240

Sugiura, Y., Shiraki, T., Konishi, M. and Oki, T. (1990). DNA intercalation and cleavage of an antitumor antibiotic dynemicin that contains anthracycline and enediyne cores. *Proc. Natl Acad. Sci. USA*, **87**, 3831–3835

Sugiura, Y., Uesawa, Y., Takahasi, Y., Kuwahara, J., Golik, J. and Doyle, T. W. (1989). Nucleotide-specific cleavage and minor-groove interaction of DNA with esperamicin antitumor antibiotics. *Proc. Natl Acad. Sci. USA*, **86**, 7672–7676

Sullivan, N. and Lyne, L. (1990). Sensitivity of fibroblasts derived from ataxia-telangiectasia patients to calicheamicin $\gamma_1^I$. *Mutation Res.*, **245**, 171–175

Thomas, J. P., Carvajal, S. G., Lindsay, H. L., Citarella, R. V., Wallace, R. E., Lee, M. D. and Durr, F. E. (1986). LL-E33288 antibiotics, a complex of novel, potent antitumor agents: Antitumor activity and tolerance in mice. *Abstr. 26th Intersci. Conf. Antimicrobial Agents Chemother.*, 1986, No. 229, p. 138

Tomioka, K., Fujita, H. and Koga, K. (1989). Synthesis of the core enediyne structure of esperamicin–calicheamicin class of antibiotics. *Tetrahedron Lett.*, **30**, 851–854

Uesawa, Y. and Sugiura, Y. (1991). Heat-induced DNA cleavage by esperamicin antitumor antibiotics. *Biochemistry*, **30**, 9242–9246

Walker, S., Landovitz, R., Ellestad, G. A., Ding, W.-D. and Kahne, D. (1992). Cleavage behavior of calicheamicin $\gamma_1^I$ and calicheamicin T. *Proc. Natl Acad. Sci. USA*, **89**, 4608–4612

Walker, S., Valentine, K. G. and Kahne, D. (1990). Sugars as DNA binders: A comment on the calicheamicin oligosaccharide. *J. Am. Chem. Soc.*, **112**, 6428–6429

Walker, S., Yang, D., Kahne, D. and Gange, D. (1991) Conformational analysis of the N–O bond in the calicheamicin oligosaccharide. *J. Am. Chem. Soc.*, **113**, 4716–4717

Wong, H. N. C. and Sondheimer, F. (1980). 5,12-dihydro-6,11-didehydronaphthacene, a derivative of 1,4-didehydronaphthalene. *Tetrahedron Lett.*, **21**, 217–220

Yang, D., Kim, S.-H. and Kahne, D. (1991). Construction of glycosidic N–O linkages in oligosaccharides. *J. Am. Chem. Soc.*, **113**, 4715–4716

Zein, N., McGahren, W. J., Morton, G. O., Ashcroft, J. and Ellestad, G. A. (1989a). Exclusive abstraction of nonexchangeable hydrogens from DNA by calicheamicin $\gamma_1^I$. *J. Am. Chem. Soc.*, **111**, 6888–6890

Zein, N., Poncin, M. and Ellestad, G. A. (1988a). Unpublished results from Lederle Laboratories

Zein, N., Poncin, M., Nilikantan, R. and Ellestad, G. A. (1989b). Calicheamicin $\gamma_1^I$ and DNA: Molecular recognition process responsible for site-specificity. *Science*, **244**, 697–699

Zein, N., Sinha, A. M. and Ellestad, G. A. (1987). Unpublished results from Lederle Laboratories

Zein, N., Sinha, A. M., McGahren, W. J. and Ellestad, G. A. (1988b). Calicheamicin $\gamma_1^I$: An antitumor antibiotic that cleaves double-stranded DNA site specifically. *Science*, **240**, 1198–1201

Zhao, B., Konno, S., Wu, J. M. and Oronsky, A. L. (1990). Modulation of nicotinamide adeninedinucleotide and poly(adenosine diphosphoribose) metabolism by calicheamicin $\gamma_1^I$ in human HL-60 cells. *Cancer Lett.*, **50**, 141–147

**Addendum**

Several papers have appeared since the submission of this manuscript that should be included in this chapter. Two groups have recently reported on DNA footprinting experiments with a synthetically prepared methyl glycoside of the thiobenzoate–carbohydrate domain of calicheamicin $\gamma_1^I$ (Aiyar *et al.*, 1992; Nicolaou *et al.*, 1992b). These results indicate protection in the general regions where calicheamicin has been shown to bind and cleave DNA. And Nicolaou has claimed that the desiodothiobenzoate–carbohydrate moiety does not footprint with DNase I in contrast to the iodo derivative. A recent cleavage study with calicheamicin $\gamma_1^I$ was observed to produce bistranded DNA damage to the virtual exclusion of single strand lesions. Hydrazine- and putrescine-sensitive lesions at the C4′ site opposite direct strand breaks at the C5′ site in a T$\underline{C}$CT sequence represent ~95% of the damage in plasmid DNA (Dedon *et al.*, 1993). A quite remarkable total synthesis of calicheamicin $\gamma_1^I$ has been reported by Nicolaou's group which confirms in every respect the structural assignment depicted in this review (Nicolaou *et al.*, 1992a). A full description of the structural elucidation of the calicheamicins has appeared (Lee *et al.*, 1992).

*Additional References*

Aiyar, J., Danishefsky, S. J. and Crothers, D. M. (1992). Interaction of the aryl tetrasaccharide domain of calicheamicin $\gamma_1^I$ with DNA: influence on aglycon and methidiumpropyl-EDTA·iron(II)-mediated DNA cleavage. *J. Am. Chem. Soc.*, **114**, 7552–7554

Dedon, P. C., Salzberg, A. A. and Xu, J. (1993). Exclusive production of bistranded DNA damage by calicheamicin. *Biochemistry*, **32**, 3618–3622

Lee, M. D., Dunne, T., Chang, C. C., Siegel, M. M., Morton, G. O., Ellestad, G. A., McGahren, W. J. and Borders, D. B. (1992). Calicheamicins, a novel family of antitumor antibiotics.4. Structure elucidation of calicheamicins $\beta_1^{Br}$, $\gamma_1^{Br}$, $\alpha_2^I$, $\alpha_3^I$, $\beta_1^I$, $\gamma_1^I$, and $\delta_1^I$. *J. Am. Chem. Soc.*, **114**, 985–997

Nicolaou, K. C., Hummel, C. W., Pitsinos, E. N., Nakada, M., Smith, A. L., Shibayama, K. and Saimoto, H. (1992a). Total synthesis of calicheamicin $\gamma_1^I$. *J. Am. Chem. Soc.*, **114**, 10082–10084

Nicolaou, K. C., Tsay, S.-C. and Joyce, G. F. (1992b). DNA carbohydrate interactions. Specific binding of the calicheamicin $\gamma_1^I$ oligosaccharide with duplex DNA. *J. Am. Chem. Soc.*, **114**, 7555–7557

# 4

# Molecular Pharmacology of Intercalator–Groove Binder Hybrid Molecules

Christian Bailly and Jean-Pierre Hénichart

## 1 Introduction

Despite many years of intense effort, there exist only a few efficient drugs against diseases arising from transcription of viral and oncogenic DNA. Therefore, the ultimate objective of various drug design programmes remains to find non-toxic molecules able to penetrate selectively into malignant cells and to bind specifically to any particular sequences of nucleic acids implicated in or directly responsible for a defined pathology.

There exist essentially two conceptual ways by which nucleic acids can be targeted. The first deals with targeting DNA and RNA by triple-helix-forming oligonucleotides and antisense oligonucleotide analogues, respectively. By virtue of tight and specific binding to their complementary nucleic acid sequences, oligonucleotides could accommodate any kind of DNA or RNA sequences with extreme fidelity and appear thus to offer considerable promise as sequence-specific inhibitors of oncogene expression. Changes in the anomeric configuration of nucleotides, replacement of the labile phosphodiester chain backbone with non-hydrolysable groups and/or combination of the oligonucleotide with a DNA intercalating or chelating agent represent elegant extensions of this approach which will probably generate new agents useful for applications *in vivo* (Tidd, 1990; Dolnick, 1991). The second approach (which has commanded our attention) concerns the use of biologically active drugs of low molecular weight a large majority of which are of natural origin and which bind to DNA with a more or less pronounced degree of sequence selectivity. Owing to their almost infinite structural diversity and their no less variable biological properties, these small molecules have for a long time been a topic of major interest. Broadly speaking, in relation to their mechanism of binding to DNA one can distinguish alkylating, intercalating and groove binding agents. Within each of these three categories of molecules structure–activity relationships have been developed in order to design more active analogues. These extensive studies have converged to show clearly that DNA sequence specificity is a major component of the efficacy of many cytotoxic drugs, several of which are of major interest in cancer chemotherapy.

Nowadays, the sequence selectivity of DNA-binding ligands can be finely appreciated at the nucleotide level largely because of the advent of nuclear magnetic resonance spectroscopy and footprinting techniques. Consequently, much of the recent focus on drug–DNA interactions has been directed towards the determination of the sequence specificity of drug binding. These studies now open the way to use well-characterized biologically active natural ligands as convenient models for the design of other DNA-binding agents. However, it must be clearly stated that to date it has not been proved that sequence specificity in binding to DNA is responsible for conferring antitumour activity on drugs. In other words, increase in DNA sequence specificity might not directly result in improved antitumour activity. There are at least three lines of evidence that bear out this idea. First, drugs such as actinomycin and daunomycin (Figure 4.1) exhibit noticeable sequence selectivity but they have therapeutic selectivity well beyond what would be expected on the basis of mere consideration of DNA binding specificity. Second, drugs having opposite biological properties can exhibit the same base or sequence selectivity. For example, the carcinogenic benzo[*a*]pyrene and the antitumour drug anthramycin both attack selectively the exocyclic amino group of guanine (Pullman, 1987). Third, the antitumour potency can vary widely within a homogeneous family of compounds showing identical DNA-binding properties. Therefore, it would appear unrealistic to design drugs with the expectation that chemotherapeutic selectivity will necessarily be improved by virtue of an increase in sequence selectivity.

Design of DNA-binding ligands demands a careful consideration of both DNA and associated cellular targets (Hurley, 1989). In fact, interaction with DNA, inhibition of replicative enzymes or DNA repair systems, poisoning effects directed towards topoisomerases, and many other processes such as free radical production or binding to membranes, are interrelated phenomena which very probably contribute to the pharmacological properties of numerous antitumour agents. Ideally, all these drug-induced effects should be considered as part of the process to create new drugs. In practice, it is obvious that any chemical strategy, efficient as it may be, cannot address all the problems simultaneously. However, efforts must be made to consider different putative targets whenever possible. It is this problem we address here.

The specific goal considered in this chapter is that it might be possible to

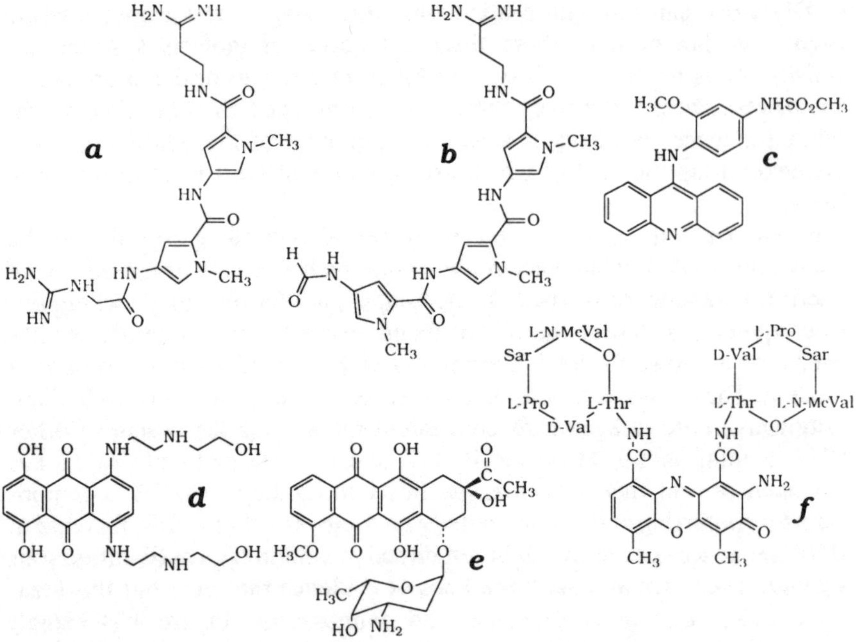


**Figure 4.1** Chemical structure of netropsin (a), distamycin (b), amsacrine (c), mitoxantrone (d), daunomycin (e) and actinomycin D (f)

consider both DNA sequence selectivity and topoisomerase interference as a basis to design new drugs. Moreover, a no less important problem (which is central to pharmacology) is that of the cellular penetration of drugs. In this respect, also, the chemical approach reported here offers interesting possibilities. This chapter is concerned with the different categories of hybrid molecules which are currently being developed, with the results which are emerging, and with their significance for the production of new antitumour drugs.

## 2 Isolexins, Lexitropsins and Combilexins

The pronounced sequence selectivity of binding to AT-rich regions of DNA by the natural antibiotic minor groove binders netropsin and distamycin (Figure 4.1) has been extensively exploited for the design of analogues that would recognize other sequences (Dervan, 1986; Bailly and Hénichart, 1991; Nielsen, 1991). In 1986 Goodsell and Dickerson published a seminal computer analysis study which suggested that appropriate chemical modifications of netropsin may provide opportunities to generate isohelical sequence-reading drug polymers, called *isolexins*, endowed with sequence-specific DNA-recognition properties. They proposed that by introducing H-bond acceptor heteroatoms into the *N*-methylpyrrole rings of netropsin, it would be possible to engage H bonds with the protruding 2-amino group of guanine. Considering this pioneering theoretical study, in the laboratories of Lown and others an impressive number of compounds have been synthesized by replacing the pyrrole units of netropsin by imidazole, thiazole, triazole, furan or recently pyrazole to generate sequence-reading oligopeptides structurally related to netropsin which were evocatively christened *lexitropsins* (Lown, 1988). By judicious substitutions with H-bond donor or acceptor heteroatoms, it was found possible to alter the high AT specificity of netropsin and distamycin so as to enhance their GC acceptance without changing their mode of binding. Efforts are still going on with the aim of designing pure GC-specific minor groove binders and drugs able to recognize a portion of DNA of defined nucleotide sequence.

The rational structural modification of natural DNA sequence-selective agents has drawn attention to the importance of several structural, stereochemical, conformational and electrostatic factors necessary for a better understanding and control of drug–DNA molecular recognition processes. It was originally believed that sequence recognition depended primarily on the formation of specific hydrogen bonds with functional groups of the DNA bases. But it now seems clear that the hydrogen-bonding capability of a ligand is a primary determinant but is not by itself sufficient to induce DNA sequence selectivity: geometrical and electrostatic factors must also be carefully considered. The distribution of the electrostatic potential in B-DNA favours the minor groove of AT sequences over the minor groove of GC sequences (Zakrzewska *et al.*, 1983a,b). The narrowness of the minor groove in AT-rich regions, compared with the markedly wider minor groove in GC-rich regions, suggests an explanation for the marked AT selectivity almost invariably observed with netropsin analogues (Gago *et al.*, 1989). Minor groove binders seem to recognize potential binding sites more by the shape of the DNA than by the specific sequence that is contained within the site. The flexibility of the DNA must also be an important feature to consider. Because of the inherently lower flexibility of GC-rich regions (Drew and Travers, 1984), the design of GC-recognition elements will require extremely accurate matching of groove and ligand surfaces, whereas a much cruder match may suffice for AT recognition (Laughton *et al.*, 1990). Therefore, a rational drug design programme should consider the notions of geometrical fitting, hydrogen bonding capability and the overall electronic properties of the interacting species.

The concept of ligands having mixed binding functions (Bailly and Hénichart, 1991) represents an alternative and complementary approach to the lexitropsin effort. The initial motivation for designing hybrid molecules derives from the observation that many minor groove binders such as netropsin, distamycin, DAPI, berenil and Hoechst 33258 bind to DNA with a powerful AT specificity (Van Dyke *et al.*, 1982; Van Dyke and Dervan, 1983; Portugal and Waring, 1987, 1988), while a marked lack of binding to AT-rich sequences is observed with many intercalators, such as ethidium (Fox and Waring, 1987) and different anthraquinones (Fox *et al.*, 1986) or benzophenanthridines (Bajaj *et al.*, 1990), some of which exhibit a pronounced GC selectivity: e.g. ellipticine (Bailly *et al.*, 1990b) and amsacrine derivatives (Bailly *et al.*, 1992b). With this in mind, an apparently simple way to associate AT and GC specificities lies in designing hybrid molecules combining a potentially intercalating heteroaromatic moiety with a peptidic or pseudopeptidic groove binding entity. The aims of such a strategy include increasing intrinsic drug potency as well as controlling the pattern of DNA sequence-specific recognition. The intricate shape of such hybrid molecules — which we name *combilexins* — would provide opportunities for binding in several distinct orientations and by several different modes along the helix.

This original approach, first introduced by Krivtsora *et al.* (1984), is simple only in appearance because, apart from the complex chemical aspect, the approach is further complicated by drug-induced DNA conformational changes. Indeed, theoretical and experimental studies have both clearly shown that binding to DNA, whatever the nature of the implied process, results in an adaptation of the DNA conformation to that of the ligand rather more than vice versa (Gilbert and Feigon, 1991). In attempting to design hybrid ligands, it must be kept in mind that intercalators and minor groove binders affect the DNA backbone conformation in very different ways (Neidle *et al.*, 1987) and that the susceptibility of DNA to such changes may vary with the nature of the target sequence. Minor groove binding results in the removal of ordered water molecules located in AT-rich regions. Only slight changes in the gross DNA structure are observed, typified by a maximal 2 Å widening of the minor groove, a bending of the helix by 8° per drug molecule and an unwinding or stiffening of the DNA helix to some extent (Kopka *et al.*, 1985a,b; Larsen *et al.*, 1989; Brown *et al.*, 1990). In contrast, in order to access their binding site intercalators provoke large-scale local structural changes such as large unwinding, stiffening and lengthening of the DNA sugar–phosphate backbone by 3.4 Å per drug molecule. Long-range structural distortions of DNA remote from the site of intercalation are also frequently observed, extending over at least 4–5 base pairs (Neidle and Abraham, 1984). Therefore, to design a minor groove binder–intercalator hybrid is a challenge. But on the basis of a large body of recent crystallographic and NMR work, it is clear that DNA is a highly polymorphic macromolecule able to adopt a remarkable range of structures. Considering the evidence for DNA bending and DNA kinking (Travers, 1991), the original definition of DNA as a simple regular helix in a classical Watson–Crick sense now appears a superseded idea. Moreover, it is known that DNA can accept two types of drugs in very close proximity (Wartell *et al.*, 1975; Patel *et al.*, 1981, 1982; Lane *et al.*, 1983; Sarker and Chen, 1989). Before discussing the DNA-binding properties of recently designed groove binder–intercalator combilexins, it is worth remembering briefly that this new topic found its origin in the study of natural drugs.

## 3 Naturally Occurring Multivalent Molecules

Nature provides numerous examples of drugs whose binding to DNA simultaneously implicates various processes. Such is the case for actinomycin D (Figure 4.1), a natural antitumour agent which consists of a phenoxazone ring disubstituted by cyclic pentapeptides. Intercalation of actinomycin occurs preferentially at some, but not at all, GpC sequences (Lane *et al.*, 1983; Van Dyke *et al.*, 1982; Van Dyke and Dervan, 1983), with the cyclic peptides fitting above and below the intercalating ring in the minor groove, each covering three base pairs. The symmetrically disposed peptides participate actively in the GC-specific recognition process through hydrogen bonding interactions between the N-3 atoms and 2-amino groups of the guanines and the amide groups of the threonine residues (Scott *et al.*, 1988a,b). Because of these sequence-specific peptide–DNA interactions, intercalation of the phenoxazone between GC base pairs is strongly preferred only if the 5′-flanking base is a pyrimidine and the 3′-flanking base is not a cytosine (Rill *et al.*, 1989).

A considerable number of DNA-binding natural products contain carbohydrate residues that are likely to serve as DNA recognition elements (Hawley *et al.*, 1989). For example, bimodal binding processes have been found in the interaction with DNA of the antitumour antibiotic daunomycin (Figure 4.1). The elucidation of the detailed interaction between this drug and a short DNA duplex by X-ray diffraction analysis (Frederick *et al.*, 1990) has demonstrated that while the chromophore intercalates at CpG steps, the glycan moiety lies in the minor groove. The sugar forms several bonds with DNA and so greatly contributes to the stability of the complex. But its binding is also important in terms of sequence specificity. Solution (Chaires *et al.*, 1987, 1990), crystallographic (Frederick *et al.*, 1990) and theoretical studies (Gresh *et al.*, 1985, 1989) have converged to show that daunomycin preferentially recognizes triplet sequences, 5′-ACG and 5′-TCG being the most preferred. The chromophore intercalates between the two CG base pairs and the sugar moiety interacts with A or T through several key hydrogen bonds (Chaires, 1990). Similar bimodal binding processes have been evidenced with many other anthracycline antibiotics, e.g. nogalamycin (Gao *et al.*, 1990; Williams *et al.*, 1990; Egli *et al.*, 1991) and arugomycin (Searle *et al.*, 1991).

Apart from the examples discussed above, there are numerous other natural antibiotics whose binding to DNA implicates intercalation and groove binding in close proximity. Such is the case with the quinoxaline antitumour antibiotics echinomycin and triostin A. These compounds share the properties of bisintercalating into DNA, placing the peptide ring in the minor groove. Here, again, specific hydrogen bonds between the peptide ring and DNA bases are prime determinants for sequence-specific recognition by the drug (Waring, 1990). Even the two natural alkylating agents of the enediyne family, dynemicin A (Sugiura *et al.*, 1990; Wender *et al.*, 1991) and neocarzinostatin (Lee and Goldberg, 1989), do not escape this rule; they bind to DNA by intercalation of their chromophore (naphthoate for neocarzinostatin and anthraquinone for dynemicin), placing their reactive enediyne-containing bicyclic core moiety in the minor groove in a suitable position for selective DNA cleavage. The list could go on. All these natural compounds can provide models for the rational design of ligands exhibiting mixed modes of binding.

In the next two sections we shall focus on two categories of minor groove binder–intercalator conjugates, namely netropsin–acridine and distamycin–ellipticine hybrid molecules, whose mechanism of action, including aspects of both cellular and molecular pharmacology, is now relatively well known. In the last section we shall briefly present a few other peptide-based hybrid molecules which are currently undergoing exploration in order to design new categories of DNA-binding entities with potential therapeutic properties.

## 4 Netropsin–Acridine Hybrid Molecules

### *Interaction with DNA*

For more than a century acridines have been of particular relevance in pharmacology. They first attracted attention because of their therapeutic action against malaria. Synthetic derivatives were continuously elaborated and extensive chemical work continues to the present day with the successful development of clinically useful antitumour drugs (Denny, 1990; Baguley, 1991). Studies on acridine derivatives have culminated with the synthesis of amsacrine (Figure 4.1), an anilino-9-aminoacridine derivative particularly active in the treatment of acute leukaemia (Cassileth and Gale, 1986). Acridines — and amsacrine, in particular — are still subject to considerable attention and more efficient analogues are in clinical trials.

On the other hand, the oligopeptides netropsin and distamycin, isolated from *Streptomyces netropsis* and *Streptomyces distacillus*, respectively, exhibit a wide spectrum of antimicrobial, antiviral and, to a lesser extent, antitumour properties. However, they are too toxic to be used in the clinic. These pseudopeptides are formed from the repetition of two and three *N*-methylpyrrole carboxamide residues that produce an arc conformation in the planar molecule (Figure 4.1). This arc matches to some extent the turn of the DNA double helix. These drugs bind avidly to AT-rich regions within the minor groove of DNA without gross helical distortion (Kopka *et al.*, 1985a,b; Coll *et al.*, 1987, 1989; Pelton and Wemmer, 1990; Boehncke *et al.*, 1991) and act to block the template function of DNA (Zimmer and Wähnert, 1986).

The first example of an acridine-linked netropsin bifunctional mixed ligand which was reported corresponded in fact to the linkage of a distamycin residue (three pyrrole rings) to a 9-amino acridine chromophore by alkyl linkers of variable length (Eliadis *et al.*, 1988). The optimum fit was obtained with a butyroyl chain between the acridine and the distamycin moiety. Although the intercalation of the acridine chromophore and the minor groove binding of the oligopyrrolecarboxamide part were demonstrated, footprinting experiments revealed identical binding site-sizes and a strict AT preference for netropsin and netropsin–acridine hybrid molecules. In parallel, we had elaborated similar hybrid molecules using a combination of the natural antitumour agents distamycin or netropsin and the anilinoaminoacridine chromophore derived from the synthetic antileukaemic drug amsacrine (Bailly *et al.*, 1989a). Apart from its evident pharmacological interest, the model of amsacrine was also attractive in our case, since it had been postulated (Denny *et al.*, 1983; Chen *et al.*, 1988) that intercalation of the acridine ring was accompanied by minor groove binding of the adjacent anilino ring. On the basis of this assumption, it seemed most interesting to tether covalently netropsin to acridine through attachment on the anilino ring. However, CPK models showed that direct linkage of netropsin to the aniline was unfavourable for DNA binding and that a linker would be required. So, we elaborated a first series of netropsin–acridine hybrid molecules in which the two parts were linked by a glycine connector. This short tether was expected to provide a frame for anchoring the minor groove moiety in proper orientation.

One of these pseudopeptides, named NetGA (Plate 4.1) was studied for its DNA-binding properties by a wide range of biochemical and physicochemical techniques. Binding data were found to be consistent with a model in which the acridine nucleus occupies an intercalation site and the netropsin residue lies in the DNA minor groove (Bailly *et al.*, 1990a). Complementary strand MPE–Fe(II) footprinting showed that NetGA binds only to AT-rich sequences such as $^{5'}$AAAT. It is interesting to note that both types of netropsin–acridine ligands (i.e. using aminoacridine and anilinoacridine) display the same selectivity: protected sites are of the strict AT type. Regions of enhanced cleavage by MPE, particularly at GC-rich sequences, were observed adjacent to and even remote from the binding site. Linear dichroism measurements revealed insertion of the acridine into the DNA helix with a tilt of about 20° relative to the plane of the DNA base pairs, higher than that measured for the acridine alone and reflecting the influence of the netropsin moiety. Nevertheless, the short spacer glycinyl, in spite of its relative rigidity, was apparently not an obstacle to binding of the two parts of the molecules. This was further supported by a molecular modelling study which showed that minor groove binding of the netropsin part and intercalation of the acridine could effectively occur simultaneously but also that this bimodal binding process would induce a local distortion of the DNA helix near the intercalation site. An energy-minimized model of the NetGA–DNA complex is presented in Plate 4.2. The hybrid molecule NetGA interacts with DNA in a well-defined geometry at specific sequences. But solely on the basis of DNA specificity the intercalating moiety of the drug would appear to be useless, since the hybrid is essentially AT-selective, just like the netropsin half of the molecule. However, the intercalating chromophore provides a better anchorage of the ligand to the DNA helix. At low drug-to-DNA ratio, the affinity of NetGA for DNA ($K_a = 9.1 \times 10^5$ M$^{-1}$) is approximately threefold higher than that of netropsin ($K_a = 2.9 \times 10^5$ M$^{-1}$) (Bailly *et al.*, 1990a). This also suggests that the two parts of the molecule are bound to DNA. The intercalating part of NetGA efficiently compensates for the missing amidine and guanidine side-chains of netropsin.

In eukaryotic cells, DNA is associated with histones in a nucleosomal structure. The highly ordered structures of DNA *in vivo* may alter the accessibility of a drug to DNA, although the drug-targeted DNA is generally believed to be in its nucleosome-free transcriptionally active form. It was recently shown that the hybrid molecule NetGA interacts with chromatin as well as with purified DNA by a bimodal process associating intercalation and groove binding (Bailly *et al.*, 1992a). When chromatin interacts with NetGA, the negative dichroism at the 430 nm band is similar to that observed for the naked DNA–NetGA complex at this band, but more negative than that of chromatin at 260 nm. This supports the view that the presence of histones in chromatin does not disturb the acridine ring intercalation, which remains mostly influenced by the netropsin moiety of the hybrid. As regards this last moiety, it appears that the presence of histones does not appreciably modify its binding to DNA. Chromatin–NetGA and DNA–NetGA complexes both exhibit positive linear dichroism at 310 nm. In other words, the minor groove is identically accessible to this ligand in chromatin and in naked DNA. This is not surprising, since 80–90% of the DNA is still accessible to the minor groove binder distamycin in chromatin fibres (Zimmer and Wähnert, 1986). The binding of netropsin and distamycin to chromatin takes place preferentially, if not exclusively, in the DNA linker regions and results in an increase in the internucleosomal distance (Koch *et al.*, 1987). Intercalators also bind preferentially to internucleosomal regions of DNA (Sen and Crothers, 1986). On the basis of these references and our results, we can assume that NetGA binds to linker DNA too.

The mode of binding of amsacrine to DNA mentioned above, i.e. binding with the aniline part located in the minor groove, is now questioned. No structural data from X-ray crystallography or NMR spectroscopy are available to define the interaction of amsacrine with its nucleic acid binding sites. Nevertheless, biochemical and spectroscopic experiments have shed some light on the mode of interaction of amsacrine-4-carboxamide derivatives with DNA (Denny and Wakelin, 1986; Wakelin *et al.*, 1990; Bailly *et al.*, 1992b). On the basis of these results and the available acridine–DNA X-ray structures, a model has been proposed in which the aniline moiety of amsacrine would occupy the major groove rather than the minor groove as discussed above. This revised interpretation was recently employed for the design of a new series of netropsin–amsacrine hybrid molecules in which the netropsin-like moiety is linked to tumour-active amsacrine-4-carboxamide derivatives. These derivatives bear the methoxy and methanesulfonamide substituents on the aniline ring which restore the oxidizable properties. Oxidative activities should contribute to an additional mechanism of toxicity. The replacement of the anilino-acridine part of NetGA by a tumour-active amsacrine analogue may lead to more effective combination molecules.

Inhibition of replication and transcription processes by drug binding to DNA and chromatin could be directly responsible for the cytotoxic effect observed. However, it is known that DNA binding is a necessary condition for cytotoxicity but cannot explain the different activities of anilino-acridine derivatives which all intercalate and exhibit similar affinities for DNA (Pommier *et al.*, 1987). There are now clear indications that topoisomerases, among other DNA-binding proteins, represent preferred targets for a large number of anticancer drugs, among which is amsacrine, acting at the DNA level. Apparently there is no correlation between the unwinding properties of intercalating agents and their effects on the catalytic activity of topoisomerase II (Pommier *et al.*, 1987) but close correlations have been found between cytotoxicity and topoisomerase II-induced DNA breakage (Schneider *et al.*, 1988). Therefore, as stated in the introduction, it seems necessary (but probably not sufficient) to target both DNA and topoisomerases to generate antitumour drugs.

### *Modulation of Topoisomerase II Activities*

Topoisomerases are enzymes which control the topology of cellular DNA by executing nicking–closing reactions. These enzymes participate in nearly all biological processes involving DNA: replication, transcription, chromosome segregation and mitosis (Wang, 1985, 1987). Moreover, they are extremely relevant in the area of human cancers and viral diseases, as topoisomerases are privileged targets for a variety of antineoplastic and antiviral agents (Drlica and Franco, 1988; Kreuzer, 1989; Liu, 1989). These enzymes now appear as one of the most promising targets for the design of more active antitumour drugs.

Topoisomerase II introduces a transient double-strand break in the DNA and forms a protein–DNA complex referred to as the cleavable complex. Certain intercalating agents, *m*-AMSA and some related acridines (Nelson *et al.*, 1984; Rowe *et al.*, 1986; Robinson and Osheroff, 1990), trap the enzyme at this stage, stabilizing the cleavable complex and thus preventing the restoration of intact DNA structure. The most satisfying model to account for the effects of intercalating agents involves making a distinction between two functional domains in the drug. With amsacrine, it is highly suspected that the acridine ring would represent the DNA binding domain and that the anilino substituent would allow the interaction with topoisomerase II (Baguley *et al.*, 1990). Similar topoisomerase II–DNA binding models, that divide the drug into two functional domains, were also proposed to explain the stabilizing effect of anthracyclines and epipodophyllotoxins (Chow *et al.*, 1989). It should be noted also that direct correlation between topoisomerase II-mediated DNA cleavage and cytotoxicity has been observed for drugs such as anthracyclines (Zunino and Capranico, 1990) and acridines including amsacrine (Rowe *et al.*, 1986; Covey *et al.*, 1988; Schneider *et al.*, 1988). In further support of this contention is the observation that (i) alteration of the topoisomerase II enzyme structure affects the cytotoxicity of amsacrine and analogues; (ii) cells resistant to amsacrine present altered topoisomerase II function (Pommier *et al.*, 1986; Per *et al.*, 1987; Finlay *et al.*, 1989). The hypothesis has been advanced that collision of DNA replication forks or transcription complexes with drug-stabilized topoisomerase–DNA adducts triggers DNA breakage and cell death (Zhang *et al.*, 1990). The detailed mechanism of interaction between the drug and the topoisomerase–DNA complex is at present unknown. It is suggested that the inhibitor binds at the interface between DNA and the active site of the enzyme, thus blocking the DNA releasing step in the topoisomerase reaction cycle by an inability to align the adjacent nucleotide residues involved in phosphodiester bond reformation (Huff and Kreuzer, 1990). In order to investigate this proposed mechanism of action, it would be of particular interest to introduce on the aniline ring of amsacrine another substituent known for its capability to interact with the topoisomerases.

In this perspective, the choice of a minor groove binding ligand is perfectly justified. Indeed, it is known that certain minor groove binders are able to impede the catalytic activity of topoisomerases I (McHugh *et*

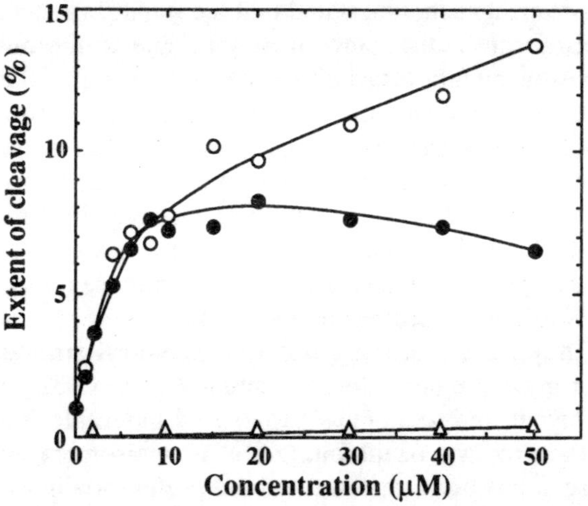


**Figure 4.2** Stimulation by NetGA (•), glycyl-anilino-9-aminoacridine (the intercalating moiety of NetGA) (Δ) and *m*-AMSA (○) of DNA topoisomerase II-mediated double-strand DNA cleavage. The extent of cleavage (%) of plasmid DNA was determined as described in Bailly *et al.* (1992a)

*al.*, 1989, 1990; Mortensen *et al.*, 1990) and II (Fesen and Pommier, 1989; Woynarowski *et al.*, 1989a,b) without stabilizing the cleavable complexes. Distamycin, netropsin and structurally related synthetic dimers were found to suppress both the catalytic activity of topoisomerase II and topoisomerase-mediated DNA–protein crosslinks in nuclei (Beerman *et al.*, 1991); consequently, it is proposed that the action of topoisomerases involves the minor groove. With this in mind, the combination of minor groove binders and intercalators could be potentially useful as a means of developing topoisomerase-mediated antitumour agents. Needless to say, all these observations prompted us to investigate the effect of NetGA on the functional activity of this enzyme (Bailly *et al.*, 1992a).

Alteration of the functional activity of topoisomerase II, detected *in vitro* by the occurrence of linear DNA manifesting stabilization of the cleavable complex, is clearly observed with NetGA (Figure 4.2). The lack of effect of the glycyl-anilino-9-aminoacridine chromophore (i.e. the intercalating moiety of the hybrid) compared with the marked effect of the hybrid suggests that the netropsin skeleton of the hybrid undergoes strong interaction with the enzyme–DNA complex. But netropsin alone does not stabilize the cleavable complex. Therefore, both moieties of the hybrid appear as determinants for the interaction with the topoisomerase II–DNA complex. If the analogy between *m*-AMSA and NetGA is correct, one would predict that the enzyme-binding domain of the hybrid is the pseudopeptidic netropsin moiety. This hypothesis is plausible because of the comparable effects of both drugs on the topoisomerase II–DNA complex. However, it cannot be stated that the stabilizing effect is specifically due to the bispyrrole entity, since it may be due to the bulky positively charged netropsin moiety attached to the anilino ring. Indeed, among a series of acridine derivatives, the one bearing the bulkiest substituent (SN 12489, an analogue bearing two consecutively linked aniline rings) was found to be the most potent topoisomerase II inhibitor (Rowe *et al.*, 1986; Pommier *et al.*, 1987). The fact that groove binders also affect topoisomerase II activity certainly needs to be considered here as well as the size of the bispyrrole moiety of NetGA. In this area, too, the design of hybrid ligands offers stimulating opportunities.

Although they have markedly different chemical structures, the AT-specific minor groove binders bisbenzimide (Hoechst 33258), DAPI, netropsin, distamycin and analogues capable of bidentate interaction with DNA share the property of inhibiting DNA topoisomerases. Meanwhile, topoisomerase II has been reported to bind preferentially to very long AT domains of the nuclear matrix, termed scaffold- or matrix-associated regions (Adachi *et al.*, 1989). Binding of histone H1 and nuclear scaffold proteins to these AT-rich regions is inhibited by distamycin (Käs *et al.*, 1989). It is likely that the inhibitory effect of the minor groove binders on topoisomerase activity is attributable to their pronounced AT selectivity. By analogy, it is tempting to postulate that the effect of the hybrid drug NetGA on topoisomerase II is also due to the preferential binding of the drug to AT-rich DNA sequences.

### *Uptake, Intracellular Distribution and Antitumour Properties*

With hybrid compounds, and, in particular, with longer ligands, the problem of cellular transport will emerge. Poor ability to penetrate into cells is one of the major problems encountered with oligonucleotide-based DNA probes. Therefore, a cellular transport programme has to be considered in addition to the DNA-binding unit strategy described so far. This problem can be addressed through the design of hybrid ligands.

Netropsin exhibits rather slow kinetics of cell penetration and only a moderate amount of drug can be recovered in the nucleus (Bailly *et al.*, 1989b). On the other hand, acridine derivatives rapidly enter cells and efficiently accumulate in the nucleus rather than in the cytoplasm (Lemay *et al.*, 1983). For this reason, the coupling of a minor groove entity to an intercalating chromophore appears extremely promising. The pattern of drug distribution between cell compartments was evaluated by the technique of electron spin resonance spectroscopy (ESR), using a spin label derivative of the hybrid NetGA in which a radical moiety had been introduced to act as a reporter group (Bailly and Hénichart, 1990). When incorporated into cells, the netropsin–acridine hybrid molecule NetGA equipped with a spin label exhibits an ESR spectrum which is determined by the position of the nitroxide as well as by the nature of the local subcellular environment. It was shown that the hybrid NetGA penetrates very rapidly into the cell. A significant ESR signal was found in both nuclear and cytoplasmic fractions after only 3 min of incubation, indicating rapid intracellular transport. After less than 10 min of incubation, NetGA was extensively accumulated in the nucleus, while the cytoplasmic fraction contained only a low amount of drug. Estimation of the total intracellular drug concentration showed that 43% of the drug was inside the cells. Nuclear accumulation is clearly a prominent feature of this hybrid ligand. A comparison of the kinetics of penetration of netropsin (Bailly *et al.*, 1989b), acridine (Lemay *et al.*, 1983) and the hybrid NetGA (Bailly and Hénichart, 1990) clearly implicated the anilinoamino-9-acridine moiety as a nuclear-binding transporter system. The acridine moiety appears as a key element in bringing as much drug as possible into close contact with nucleic acids. From these observations, we can exclude poor access into cells as a contributory cause for lack of antitumour activity.

L1210 leukaemia cells were used to evaluate the potential cytotoxic activity of NetGA. Exposure of L1210 cells to 0.5 μM NetGA for 4 days causes 50% inhibition of growth but only slightly decreases cell viability (95% of the cells remain viable in the presence of 0.5 μM NetGA). Complete cell death was obtained only at concentrations up to 25 μM. Thus, the hybrid drug NetGA is a cytostatic agent but only weakly cytotoxic. For comparison amsacrine, tested under similar conditions, induces 50% growth inhibition at a concentration of 0.075 μM. However, if amsacrine, as expected, is a powerful growth inhibitor for leukemic cells, it is also a toxic agent, since a concentration of 5 μM *m*-AMSA is sufficient to kill 100% of the cells. The hybrid NetGA is not toxic *in vivo*. By the intraperitoneal route, this agent did not evince any type of cytotoxicity in healthy mice even at a dose as high as 50 mg-kg-day. P388 leukaemia cells, inoculated intraperitoneally in DBA2 mice, were used to test the antitumour potency of NetGA *in vivo*. At 25 mg $kg^{-1}day^{-1}$ of NetGA, a noticeable increase in lifespan was observed, expressed by a T/C of 132%. With amsacrine, literature reports of T/C for P388 cells are 160% and 178%, respectively, at doses of 5.9 mg/kg (Finlay *et al.*, 1989) and 8.9 mg/kg (Baguley *et al.*, 1990). Thus, it is obvious that NetGA is less active than amsacrine both *in vitro* and *in vivo* against the selected models. It has, however, interesting biological properties. Linkage of the netropsin fragment and the anilinoaminoacridine, both individually inactive *in vivo*, affords a drug with marked cytostatic properties against L1210 cells and moderate but noticeable antitumour effect *in vivo*. Evaluation of the therapeutic potential of a drug necessitates the consideration of both activity and toxicity of the agent. NetGA exhibits moderate activity *in vivo* but is considerably less toxic (if toxic at all) compared with amsacrine *in vivo*. Therefore, we think that it could be interesting to develop other related ligands in an attempt to obtain more active and weakly cytotoxic hybrids.

To sum up, DNA sequence-specific targeting, modulation of topoisomerase II activity and rapid cellular uptake are complementary effects observed with the netropsin–acridine hybrid molecules. Each half of the molecule has a defined function: sequence specificity associated with the groove binding moiety and nuclear targeting promoted by the intercalating moiety. With judicious combination, synergistic effects could be expected. However, it is obvious that neither sequence-specific DNA binding nor alteration of topoisomerase II activities coupled with efficient intracellular transport is sufficient to generate a powerful antitumour drug. The study illustrates the difficulty of trying to correlate an extraordinary complex biological disorder such as cancer with simple physicochemical and biochemical parameters.

## *Sequence-specific DNA-cleaving Hybrid Molecules*

DNA damage is also an important mechanism of cell killing used by a variety of antitumour agents, among which glycopeptides of the bleomycin family are the best-known members (Hecht, 1986; Stubbe and Kozarich, 1987). Bleomycin, in the presence of iron and molecular oxygen, produces large amounts of highly nucleophilic oxygen radicals and induces site-specific DNA cleavage, presumably mediated by such species. In order to try to reproduce such metal-dependent DNA-cleaving processes, we elaborated simplified models of the structurally complex chelating moiety of bleomycin. A non-peptidic molecule, named Amphis (Figure 4.3), proved to mimic efficiently the metal-binding/oxygen activation domain of bleomycin (Hénichart *et al.*, 1982, 1985). Attachment of the bithiazole chromophore, i.e. the DNA-binding domain of bleomycin, to this chelating agent has led to bleomycin-like molecules (Kenani *et al.*, 1987, 1989), albeit less

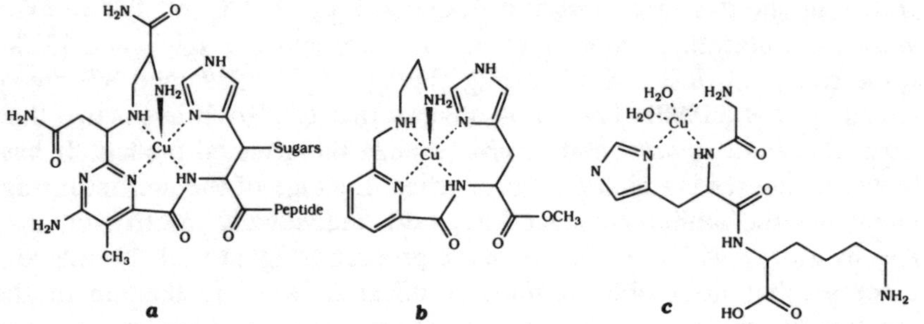


**Figure 4.3** Proposed structures of bleomycin–Cu(II) (a), Amphis–Cu(II) (b) and GHK–Cu(II) (c) complexes

effective than the natural antibiotic. Replacement of the bithiazole by a true intercalating agent, i.e. the anilinoacridine chromophore, has afforded analogues endowed with superior DNA-binding and -cleaving properties (Bailly *et al.*, 1987, 1988).

In a later effort, the non-peptidic moiety Amphis was substituted by the copper-complexing growth factor tripeptide Gly–His–Lys, GHK (Pickart and Thaler, 1973; Pickart *et al.*, 1980; Pickart and Lovejoy, 1987), with the aim of further simplifying the complexing part of bleomycin. However, it must be emphasized that the structures of the bleomycin–Cu complex and the GHK–Cu complex differ markedly. Copper coordination by bleomycin involves equatorial nitrogens from the secondary amine, pyrimidine, amide and imidazole together with an axial nitrogen from the primary amine, thus delimiting a square pyramid of nitrogen atoms (Figure 4.3). In contrast, GHK forms a tridentate complex with copper (Figure 4.3) (Freedman *et al.*, 1982; Laussac *et al.*, 1983). These distinct configurations must be carefully considered, since the structure of the ligand dictates the orientation of these complexes with respect to DNA. Indeed, in a recent informative ESR study, Chikira *et al.* (1991) have shown that the square plane of Cu–bleomycin makes an angle of about 65° with the DNA fibre axis, while the Cu–GHK complex binds with the square plane parallel to the DNA fibre axis. The cytotoxic properties of the GHK–anilinoacridine conjugate were correlated with its capacities to bind tightly to DNA and to induce strand-cleavages (Morier-Teissier *et al.*, 1989). Nevertheless, with all these different metal complex–intercalator conjugates, DNA cleavages occurred at random, i.e. without sequence selectivity. It seemed necessary to introduce a sequence-specific DNA-binding element into these molecules to improve the concept. Obviously, netropsin appeared a good candidate to confer specific recognition properties.

Accordingly the tripeptide GHK was linked to the netropsin–acridine hybrid NetGA to form a compound called GHK~NetGA which could be considered a chemical nuclease. A remarkable coincidence, which validates this hybrid model, is that the GHK–Cu complex binds to the minor groove of DNA (Chikira *et al.*, 1991), i.e. just as does netropsin. Therefore, the DNA-binding configurations of the GHK–copper complex and netropsin should be perfectly compatible and synergistic effects can be expected. A preliminary study of the interaction between this hybrid and DNA was recently reported (Bailly *et al.*, 1992d). Here, again, the drug–DNA interaction, investigated by electric linear dichroism, corresponds to a bimodal process involving intercalation and groove binding. Molecular modelling of the GHK~NetGA–DNA complex further supports this binding configuration, showing that intercalation of the acridine chromophore and simultaneous minor groove binding of the netropsin part, together with interactive anchorage of the tripeptide, are energetically feasible (Plate 4.3). Comparing the behaviour of this new molecule with that of the netropsin–acridine hybrid NetGA unsubstituted by GHK, we observed that the tripeptide does not inhibit but rather facilitates the groove binding process and allows complete intercalation of the acridine ring. The affinity constant of GHK~NetGA for calf thymus DNA is $2.2 \times 10^7$ $\text{M}^{-1}$, more than twentyfold greater than that of NetGA ($K_a = 9.1 \times 10^5$ $\text{M}^{-1}$), reinforcing the conclusion that GHK contributes positively to the DNA binding.

In the presence of the drug, the sequence $^{5'}$ATTTGATAT$^{3'}$ was protected from DNAase I cleavage. The size of the binding site revealed by footprinting proved to be twice that expected from computer modelling analysis. Moreover, cleavage initiated by addition of sodium ascorbate and hydrogen peroxide occurred on both sides of the recognized duplex sequence, although the ligand bears only one cleaving functionality. It is therefore likely that the target sequence corresponds in fact to two juxtaposed drug binding sites, to which GHK~NetGA binds in opposite orientations: first with the netropsin moiety at the sequence ATTT and the acridine ring intercalated on the 5′ side of the central GC base pair, and second with the netropsin bound to the TATA box and with acridine ring intercalated on the 3′ side of the GC base pair. However, specific binding to eight AT base pairs and one GC is still far from the 15 base pairs needed for drugs to act as specific inhibitors of gene expression.

These results establish that attachment of GHK converts a bifunctional DNA-binding ligand composed of two well-characterized antitumour moieties into an artificial endonuclease-like molecule capable, under activation, of attacking a short DNA sequence at a specifically recognized position in the minor groove. This type of small synthetic molecule — a 'designer drug' in the true sense — had its origins in a consideration of the molecular mechanism of action of a nuclease acting in the major groove. On this basis there is a brighter prospect of using chemical knowledge to design sequence-specific artificial endonucleases.

## 5 Distamycin–Ellipticine Hybrid Molecules

Recently, we examined the DNA binding ability of a distamycin-ellipticine hybrid compound, Distel, shown in Figure 4.4. This agent was built by linking the AT-specific agent distamycin to an ellipticine derivative endowed with marked GC-selective recognition properties (Bailly *et al.*, 1990b). By means of several complementary spectroscopic techniques, it was clearly demonstrated that the drug intercalates and lodges in the minor groove simultaneously (Bailly *et al.*, 1992c). The energy-minimized model of the hybrid–DNA complex shown in Plate 4.4 reveals that the recognition of the DNA by both the minor groove moiety and the intercalating chromophore requires considerable flexibility in the relative orientation of the two parts of the molecule. The propylene linker used here seems to be

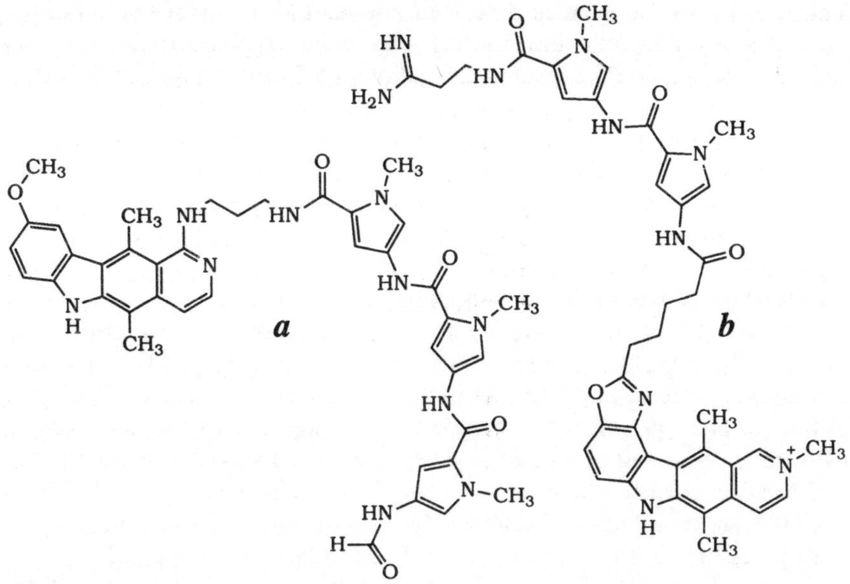

**Figure 4.4** Structures of the distamycin–ellipticine hybrid Distel (a) and the netropsin–oxazolopyridocarbazole hybrid NetOPC (b)

flexible enough to ensure simultaneous binding of both hybrid moieties. Clearly, the size and nature of the linker are crucial elements. The bimodal binding requires distortion of the DNA helix. The ellipticine chromophore forces the interbase spacing essentially to double so as to accommodate the intercalator, while the distamycin moiety induces a slight widening of the minor groove. Moreover, the DNA–hybrid ligand complex formation induces a pronounced bending towards the minor groove in the vicinity of the ellipticine binding site. These observations show that DNA can accept important constraints and adopt a particular conformation that deviates significantly from the canonical Watson–Crick B form. The conformational deformability inherent in DNA is considerably exploited by the hybrid ligand (Gilbert and Feigon, 1991). Needless to say, favourable contributions of both polar interactions (ionic contacts and hydrogen bonds) and hydrophobic effects (van der Waals contacts) substantially exceed the negative contribution from the energetic cost of conformational changes. In fact, all (or almost all) DNA-binding ligands induce, to a more or less pronounced degree, distinctive and localized sequence-dependent perturbations of base pairing and conformation which very likely play a key role in their biological functions.

The capacity of the distamycin–ellipticine hybrid to read DNA sequences was analysed by DNAase I footprinting experiments. These analyses gave conflicting results compared with those obtained with the acridine-linked netropsin molecules described above. Indeed, with the acridine–netropsin combilexins, the binding sites were confined to AT-rich sequences, while, with the distamycin–ellipticine ligand, binding sites mainly consisted of GC-rich sequences which were protected from cleavage by DNAase I. In fact, this distamycin–ellipticine hybrid compound more or less retained the specificity of the ellipticine moiety alone. Thus, in this case, as with triostin and other related structures, intercalation appears to be the crucial element. These results demonstrate that, depending on the nature of the coupled drugs and of the DNA sequences, one or other of the two binding modes can often predominate.

In a similar way, interesting features have been reported by Gosselin and co-workers, using a netropsin–oxazolopyridocarbazole (OPC derivatives are ellipticine analogues) hybrid molecule and merit mention (Plate 4.4) (Mrani *et al.*, 1991; Subra *et al.*, 1991). The presence of the intercalating OPC moiety instead of a guanidine residue does not markedly modify the AT preference of the netropsin moiety. It was shown that intercalation of the OPC chromophore and selective binding of the netropsin moiety to AT base pairs in the minor groove cannot take place simultaneously. In the presence of poly[d(AT):d(AT)], either both moieties of the Net–OPC hybrid would lie in the minor groove or the OPC would intercalate, leaving the netropsin part remaining outside the double helix. By contrast, in the presence of poly[d(GC):d(GC)], only an intercalative binding process involving the OPC part would take place. As a result, footprinting experiments failed to reveal the complementary effects expected from the coupling of a GC-selective intercalator to a powerful AT-specific groove binder. Therefore, depending on the ligands that are associated, the DNA-binding properties of the designed hybrids may differ quite significantly from one to another. Just as there is considerable structural and functional diversity within natural DNA ligands, so there will be diversity among hybrids.

Of course the model compounds NetGA and Distel described above need various improvements to increase their affinity and specificity for defined duplex sequences and to elevate their antitumour activity to a clinically useful level. An approach which we are currently examining to reach these two objectives is to combine netropsin or distamycin entities with very active amsacrine analogues. Another way might consist of combining other intercalators with different groove binding peptides so as to generate efficient synergistic effects. The next section summarizes certain recent advances in this subject which undoubtedly show promise and merit further development.

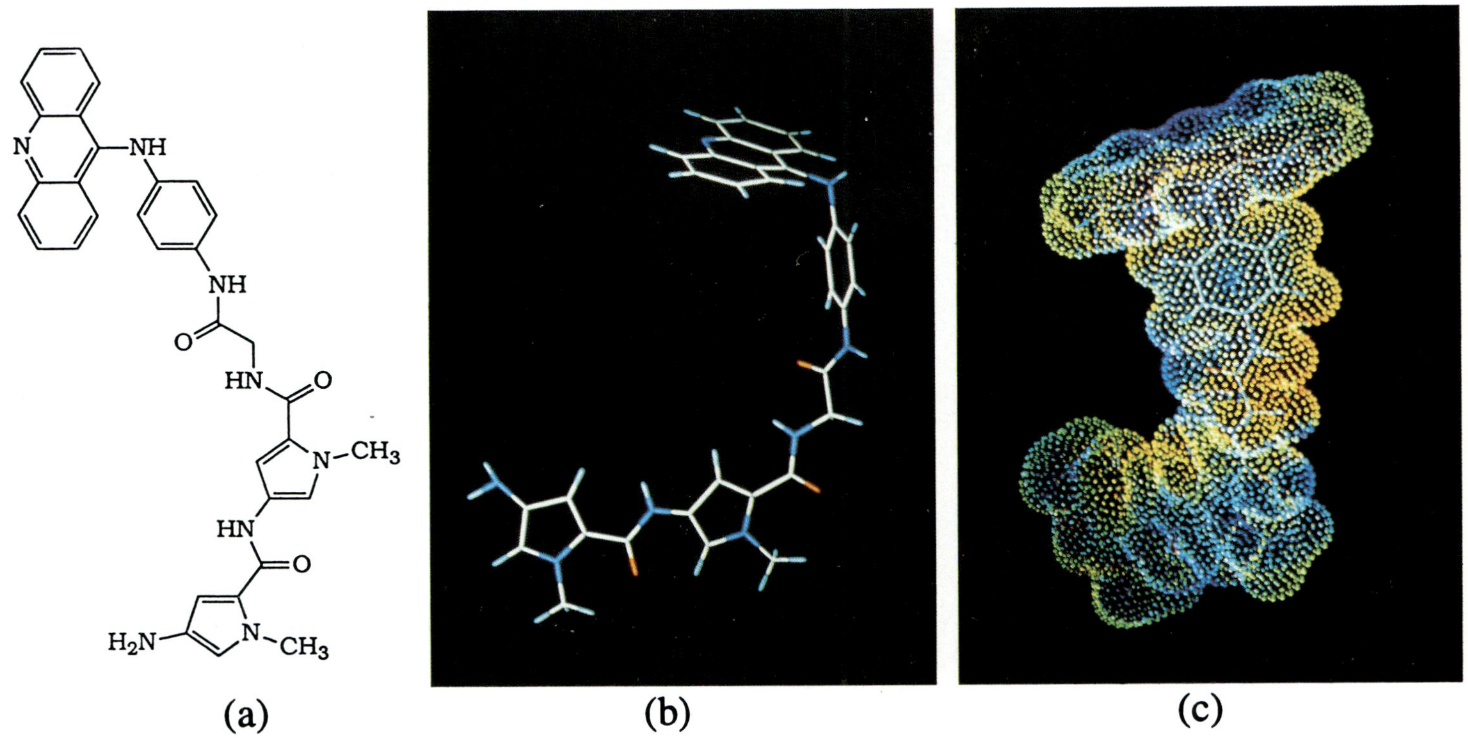


**Plate 4.1** Formula (a), energy minimized structure (b) and electrostatic potential surface (c) of the netropsin–acridine hybrid compound NetGA. Contour levels: red, $V < -4$ kcal/mol; green, $-4$ kcal/mol $< V < +4$ kcal/mol; blue, $V > +4$ kcal/mol

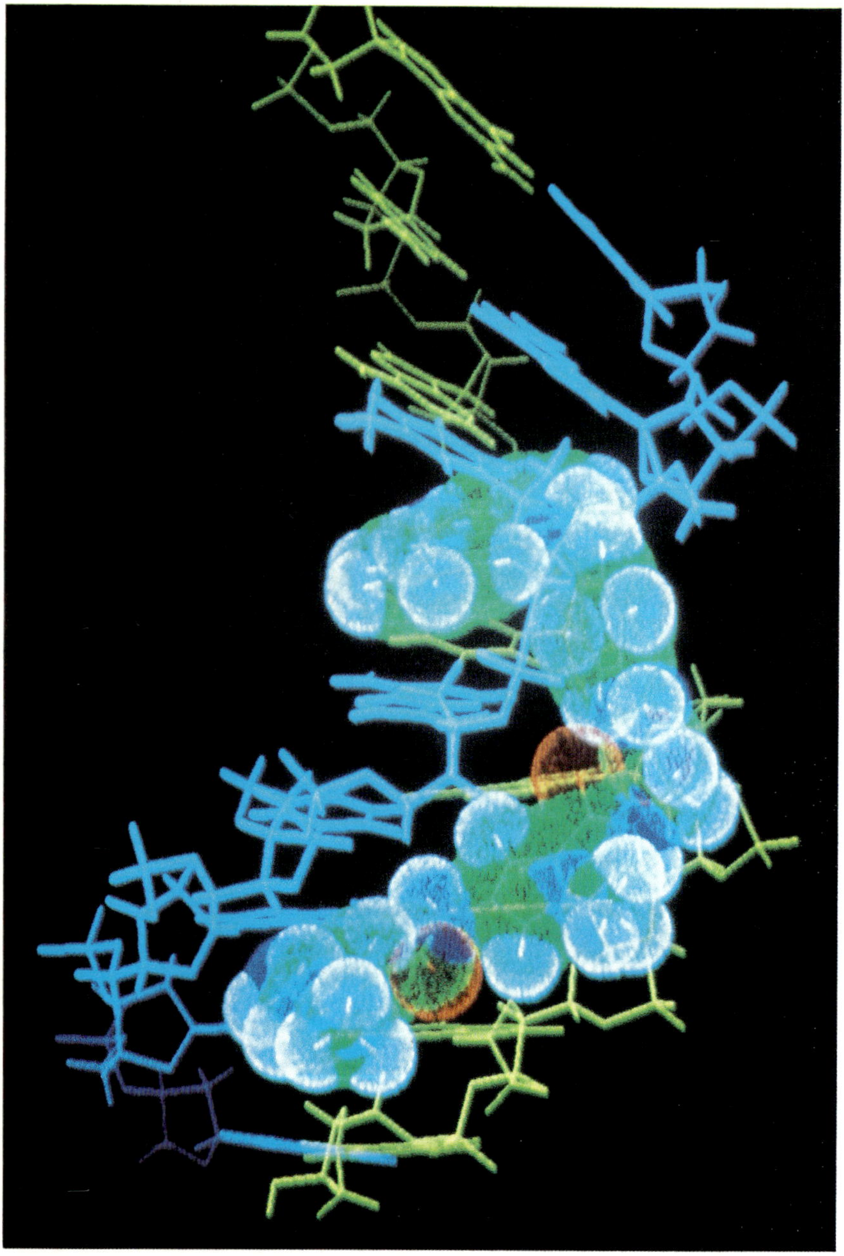

**Plate 4.2** Computer-drawn view of the NetGA–DNA complex. The complex was modelled by conformational energy minimization using the JUMNA (Junction Minimization of Nucleic Acids) program package (Lavery, 1988). Calculations were performed on a Silicon Graphics 4D/120GTXB workstation by Dr J.. S. Sun (Paris)

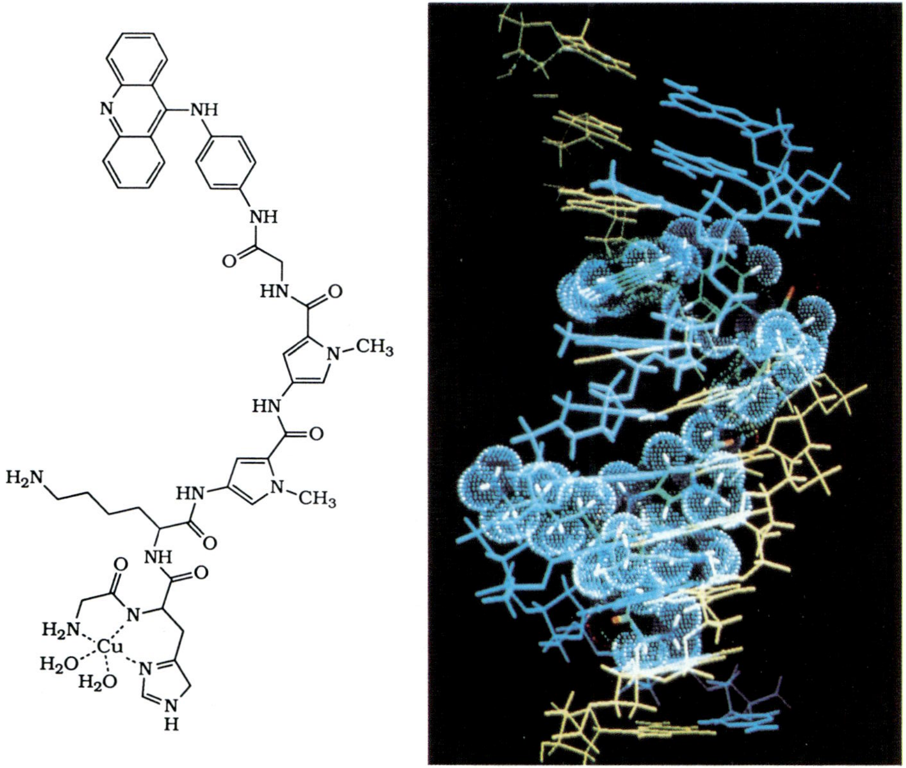

**Plate 4.3** Structure of the Cu(II)–GHK~NetGA complex and molecular graphics model of the GHK~NetGA–DNA complex. The ligand was docked into the minor groove of B-DNA to avoid steric clashes using the Insight program (Biosym Inc.). Minimization was effected by successively decreasing the number of constraints (by Dr J. S. Sun, Paris)

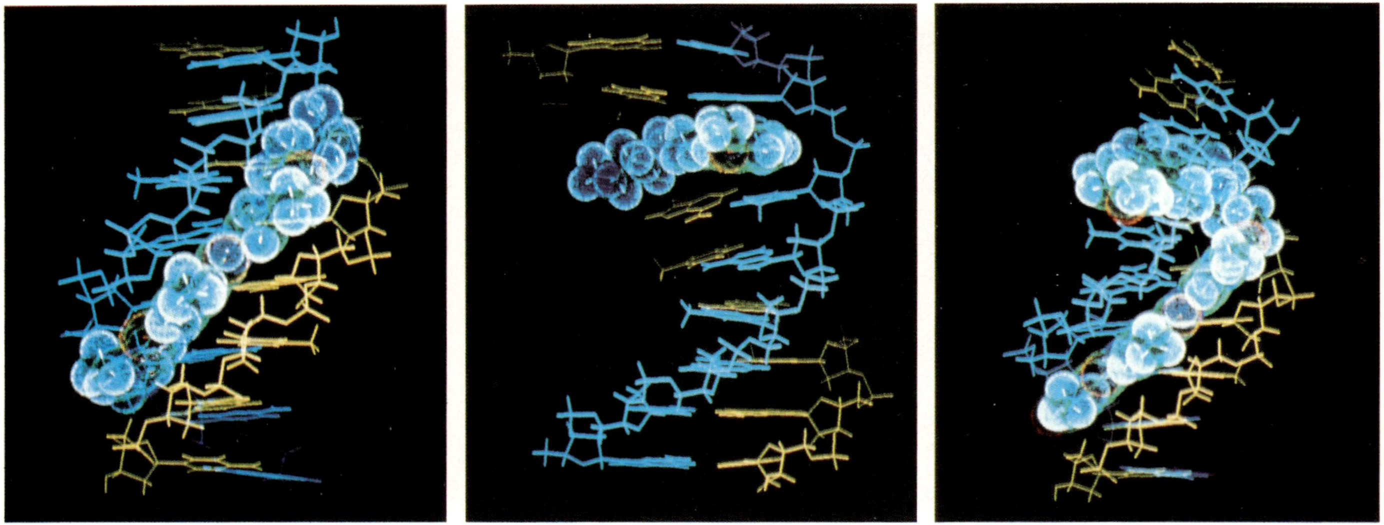

**Plate 4.4** Molecular graphics models showing (from left to right) minor groove binding of distamycin, intercalation of ellipticine and simultaneous minor groove binding and intercalation of the distamycin–ellipticine hybrid ligand (by Dr J. S. Sun)

## 6 Intercalator–Peptide Conjugates

### *Intercalation and Peptide Binding to the Minor Groove*

To illustrate this category of ligands, we must begin by referring to recent studies suggesting the tetrapeptide SPKK (serine–proline–lysine–lysine) as a new peptidic DNA-binding motif (Suzuki, 1989a,b). The SPKK peptide motif, found in the DNA-binding arms of histones H1, H2B and related proteins, is thought to form a β turn that would properly position the peptide in the minor groove of DNA. The β turn is presumably stabilized by an additional hydrogen bond between the Ser side-chain OH group and the main-chain NH group of the third Lys residue. According to the model proposed by Suzuki on the basis of statistical and modelling studies, binding of the dimer $(SPKK)_2$ to DNA would involve three H bonds between NH groups of the peptide and heteroatoms of DNA. Moreover, the stability of the peptide–DNA complex would be enhanced by electrostatic interactions between the amino terminus of lysine and the negatively charged phosphate oxygens (Churchill and Travers, 1991). Such minor groove interaction strengthens the preference for AT base pairs over GC base pairs on the same basis as mentioned above for netropsin, i.e. that the amino group of guanine would interfere sterically with the approach of the peptide to the O2 of cytidine. In support of this hypothesis, footprinting studies on the $(SPRK)_2$– and $(SPRK)_6$–DNA complexes (the tetrapeptides SPKK and SPRK bind to DNA similarly) using hydroxyl radical protection showed patterns comparable to those obtained with the Hoechst dye 33258 (Churchill and Suzuki, 1989). Therefore, the motif SPK(R)K is expected to bind to DNA in much the same way as netropsin does. Similar characteristics were proposed for a consensus peptide from non-histone chromosomal protein HMG-I (Reeves and Nissen, 1990).

This analogy between netropsin and SPKK prompted us to construct a hybrid analogue of the netropsin–acridine hybrid NetGA containing one or two SPKK motifs in place of netropsin. The bipartite compounds $(SPKK)_1$–GA and $(SPKK)_2$–GA were thus elaborated in order to evaluate their DNA-binding properties. The analogy between netropsin and the motif SPKK is, however, not so obvious as it might seem. The presence of a very basic peptide bearing as many as five free amino groups can result in a preference for external binding as opposed to intercalative binding. Electrostatic interactions may come to predominate over the expected specific hydrogen-bond formation in the minor groove of DNA, just as observed with polyamines such as spermine and spermidine (Tabor and Tabor, 1984).

A preliminary description of the $(SPKK)_{1-2}$–GA–DNA complex has been published recently (Bailly *et al.*, 1992e) and other details concerning

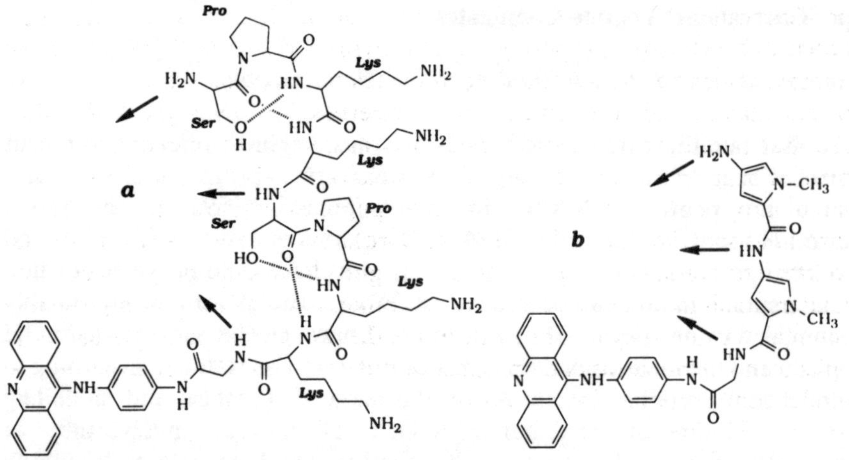


**Figure 4.5** Proposed structure of $(SPKK)_2$-GA compared with NetGA. The structure of the peptide features a combination of two intramolecular hydrogen bonds (dashed lines) stabilizing the turn structure. Both drugs have amides which can form three-centred hydrogen bonds with adjacent base pairs (indicated by arrows)

the sequence specificity of this interaction are in the course of being reported (Bailly *et al.*, in preparation). The results of spectroscopic studies revealed that the intercalating moiety provides a tight anchorage to DNA and thus greatly improves the DNA-binding affinity of the peptide motifs. When the acridine part is linked to only one SPKK motif, a perfect intercalative process is observed. By contrast, when the chromophore is coupled with the octamer $(SPKK)_2$ the parallel relationship between the acridine and the base pairs is disrupted. This effect can be attributed to the influence of the peptide chain, whose minor groove binding hinders the intercalative process. The phenomenon is reminiscent of that observed with the netropsin–acridine hybrid ligand NetGA, and thus tends to confirm the analogy proposed between netropsin and the octapeptide SPKKSPKK (Suzuki, 1989a). The hybrid molecules $(SPKK)_2$–GA and NetGA seem to share similar DNA-binding properties (Figure 4.5). This analogy is further supported by the footprinting studies. Hydroxyl radicals, generated by the Fe–EDTA complex under reducing conditions, induce random DNA breakages. In the presence of the acridine-linked octapeptide derivative, but not with the tetrapeptide analogue, DNA cleavage is drastically reduced at several loci composed almost exclusively of AT base pairs. Once again, as observed with the hybrid NetGA, no influence of the acridine part of the molecule can be detected in terms of sequence specificity of binding. The role of the chromophore is, first, to anchor to DNA the peptide which has practically no affinity alone and, second, to confer noticeable cytotoxic properties on the hybrid molecules. The potential effects of such peptide–acridine conjugates on topoisomerases remain to be elucidated, as do their antitumour properties. However, the indications are that peptide-based hybrid molecules are worthy of further development. For instance, linkage of the anilinoacridine chromophore to longer minor groove binding peptides, such as peptides derived from the helix-turn-helix motif of the 434*cro* protein (Grokhovsky *et al.*, 1991), may lead to efficient conjugates able to target longer DNA sequences selectively. Understanding the exact mechanism of binding to DNA of such peptide–intercalator mixed ligands not only offers the prospect of finding drugs with enhanced chemotherapeutic properties but also may have relevance as a model of the specific contact between proteins and DNA.

### *Intercalation and Peptide Binding to the Major Groove*

The diamino-anthracenedione derivative mitoxantrone (Figure 4.1), which stands as an explicit example of a man-made clinically used antitumour agent, is a very useful model to illustrate this category of DNA ligands. This molecule was initially elaborated with the aim of finding anthracycline analogues endowed with diminished cardiotoxicity and improved therapeutic index. Starting from antibiotics such as daunorubicin and doxorubicin, i.e. from intercalating agents bearing a glycan located in the minor groove (see above), chemists have elaborated more active, structurally related intercalating agents having aminoalkylamino side-chains whose binding properties differ markedly from those typical of the initial minor groove orientation. Indeed, computational analyses have revealed that the most favourable binding configuration of mitoxantrone locates the chromophore orthogonal to the base pair axis within the intercalation site and with the side-chains both lying in the major groove (Chen *et al.*, 1986). These results, in accordance with previous conclusions drawn from NMR (Lown and Hanstock, 1985) and computer graphics modelling (Islam *et al.*, 1985), opened the way for the design of major groove oligopeptide binders.

Independently, in the laboratories of Pullman and Palumbo, mitoxantrone constituted the starting compound for developing other peptidyl–anthraquinone conjugates endowed with recognition elements for given sequences of DNA. From recent theoretical studies (Gresh and Kahn, 1990, 1991), a bistetrapeptide derivative of mitoxantrone, Arg(L)–Gly(L)–Val(L)–Glu(L) (RGVE), is predicted to manifest a very important energetic preference favouring the major groove of $d(GGCGCC)_2$ over other

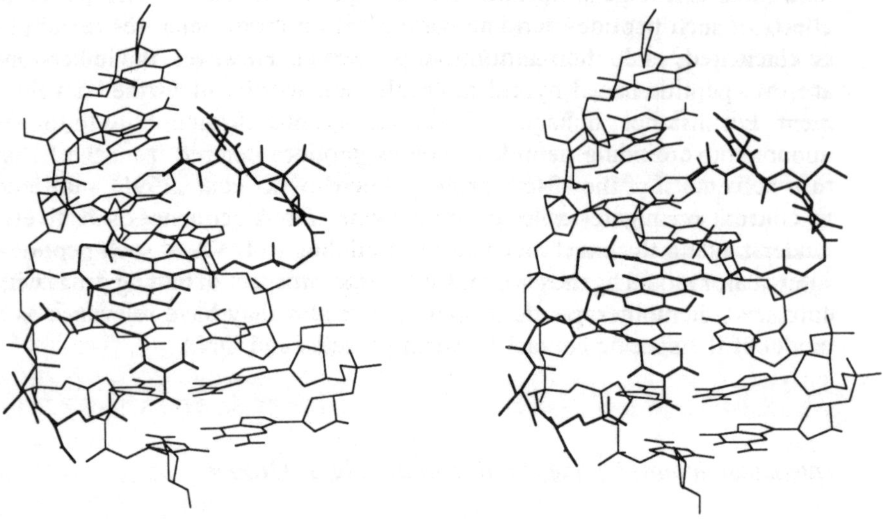

**Figure 4.6** Stereo pair drawing of the proposed major groove complex of the hybrid molecule [Arg–Gly–Val–Glu]$_2$–mitoxantrone (dark bonds) with the DNA duplex d(GGCGCC)$_2$. The computation was performed using SIBFA (Sum of Interactions Between Fragments *Ab initio* computed) and JUMNA procedures. For details see Gresh and Kahn (1990, 1991)

candidate hexameric sites (Figure 4.6). The two peptidic arms of the conjugate would adopt an antiparallel β sheet conformation stabilized by two hydrogen bonds involving the CO and NH groups of the glycine residues. The β sheet thus formed would allow the peptides to fit snugly into the major groove, just as observed for the DNA-recognition part of the *arc* repressor protein (Zagorski *et al.*, 1989; Breg *et al.*, 1989). Studies of Palumbo and co-workers showing that bis-substituted glycylanthraquinones exhibit increased affinity for GC-rich sequences (Palumbo and Gatto, 1990) provide an experimental counterpart to the computational work. However, the two approaches differ in respect of the proposed arrangement for binding of the peptides appended to the chromophore. A perpendicular mode of intercalation positioning both chains in the major groove is suggested from the energy calculations (Figure 4.6). On the other hand, a parallel intercalation process depositing one chain in the major groove and one in the minor groove is supposed by the experimental analyses (a binding arrangement similar to that described for the anthracycline antibiotic nogalamycin). However, both theoretical and experimental studies concur in that variations in the nature, sequential position and configuration of the amino acid residues can surely generate chromophore-linked oligopeptide conjugates able to recognize selectively a significant diversity of preselected sequences. Thus, the idea of connecting the mitoxantrone chromophore with longer major groove binding peptides corresponding to specific regions of repressor proteins (such as the *arc* and the *mnt* repressors) offers stimulating perspectives to decipher and ultimately to control sequence-specific DNA recognition.

In parallel with these studies, we recently reported that copper chelating GHK–anthraquinone and GGH–anthraquinone conjugates generate high levels of free radical species capable of damaging DNA (Morier-Teissier *et al.*, 1990). Indeed, the semiquinone radical produced by a chemical or enzymatic one-electron reduction of the anthraquinone chromophore is capable of transferring one electron to molecular oxygen to form the superoxide radical anion $O2^{-\cdot}$. This poorly reactive radical is in turn dismuted into hydrogen peroxide which in the presence of the reduced form of a metal ion (e.g. $Cu^{+}$ stabilized by the peptide complex) generates highly reactive hydroxyl radical $OH^{\cdot}$ by a Fenton reaction. Thus, the two moieties of these conjugates act synergistically to yield ultimately hydroxyl radical species capable of inducing DNA damage but also protein degradation and oxidation of polyunsaturated fatty acids. Metal-catalysed reduction of dioxygen to produce hydroxyl radicals or other reactive species is a fundamental process responsible for a variety of oxidative stress in biological systems.

In a manner analogous to the GHK~NetGA hybrid described above, it would thus be of particular interest to attach both a metal-chelating DNA-cleaving peptide and a sequence-specific DNA-binding peptide to the arms of the mitoxantrone chromophore so as to create a hybrid molecule able to cut DNA at specific sequences. Taken together, all these observations have recently led us to undertake the development of a series of functionalized mitoxantrone derivatives equipped with netropsin-like entities on each arm of the chromophore. These hybrid molecules associate both the AT selectivity of netropsin and the GC preference of the intercalating drug and thus might reveal a sharp sequence specificity of binding. Such complex molecules are reminiscent of the hybrid compound named 'distactin' consisting of an actinomycin-like molecule in which two distamycin moieties sandwiched a phenoxazone ring and which was shown under certain circumstances to span over ten base pairs (Krivtsora *et al.*, 1984; Dervan, 1986). With netropsin–mitoxantrone–netropsin hybrid molecules the pertinent questions are: Which part of the molecule will govern the binding to DNA? In which groove of DNA will the two netropsin moieties reside upon intercalation of the mitoxantrone part? Towards which DNA sequences will these molecules be targeted? Answers to these questions are expected to emerge from our future research in this field.

## 7 Conclusion

This discussion has covered just some of the features arising from a very limited number of recently described studies of various types of hybrid ligands. Obviously, the data on hybrid molecules are still too sparse for a true consensus picture to emerge. Additional structural, biochemical, physicochemical and theoretical studies need to be done before we may fully appreciate the relevance of this novel pharmacological approach. However, the results currently available undoubtedly reveal that this strategy warrants further development. But we must also admit that the consideration of both DNA and topoisomerases is still not sufficient to ensure that we will generate efficient anticancer drugs.

Ever since the 1950s, the basic discovery of the Watson–Crick model for the DNA double helix has inspired investigations of the interaction between DNA and biologically important planar aromatic drugs. As a result, Lerman proposed thirty years ago the intercalation model for the binding of acridines to DNA (Lerman, 1961, 1963). A large number of drugs were then studied for their ability to intercalate into DNA (Waring, 1970). Thus, the intercalation model which was originally conceived for the proflavine–DNA complex was extended to a wide range of planar chromophores. Thereafter, attempts were made to ascribe biological effects (and antitumour activity, in particular) to that process. At that time DNA was considered the principal (if not the unique) target for antitumour drugs. But nowadays nobody disputes that binding to DNA is at best a necessary feature, though certainly not a sufficient property to create anticancer drugs. However, it behoves us to keep in mind that despite this frequently reiterated dogma, one cannot disregard DNA in any rational scheme of antitumour drug discovery.

Since the 1970s, new dimensions have been added to our understanding of the control of cell proliferation as a result of advances in the molecular biology of genes. Simultaneously, a landmark discovery was the identification of the importance of topoisomerases in mediating the biological activities of a large variety of drugs (Drlica and Franco, 1988; Kreuzer, 1989; Liu, 1989). As a result, it is now commonly accepted that topoisomerases represent privileged targets for antitumour drugs and that, as a corollary, the search for topoisomerase inhibitors can provide new opportunities for the discovery of anticancer drugs. However, we may wonder if the importance of topoisomerases as a preferred target for antitumour drugs is becoming overestimated. It is worth remembering that there exist many topoisomerases having distinct functions in cells (Wang, 1991), just as a large number of topoisomerase inhibitors having different structures and biological effects have been registered. There is now strong evidence that these enzymes are directly implicated in the mechanism of action of antibacterial, antiviral and antitumour drugs. In 1968 Waring raised the question: 'How many intercalators?' In the same vein, it is tempting to raise the related question: How many topoisomerase inhibitors? and to draw a similar conclusion: if it really turns out that topoisomerase inhibition is at the bottom of the action of so many compounds, it will constitute a highly versatile process and we can look forward to more interesting explanations of how, and in what circumstances, its manifestation may lead to anticancer, antiviral or antibacterial effects. It is clear that both topoisomerases and DNA are important targets for antitumour drug action, but other targets such as membranes which are involved in tumour-cell selectivity of drugs need also to be carefully considered. There is good reason to believe that the discovery of new efficient anticancer molecules will result from the continued development and application of DNA- and topoisomerase-based hybrid molecules.

## Acknowledgement

Financial support from the 'Institut National de la Santé et de la Recherche Médicale' and the 'Association pour la Recherche contre le Cancer' is acknowledged. We would like especially to thank our colleagues cited in publications from our laboratory for their invaluable contributions over many years. Their efforts are much appreciated and their support continues to make research rewarding. Special thanks also to Dr P. Colson, Pr C. Houssier (Liège, Belgium), Pr J. W. Lown (Edmonton, Canada), Pr W. A. Denny (Auckland, New Zealand), Pr M. J. Waring (Cambridge, UK), Pr E. Bisagni (Orsay, France), Pr C. Hélène (Paris, France) for ongoing collaboration and contribution to our research. Finally, we are most grateful to Drs J. S. Sun, N. Gresh and P. H. Kahn for kindly providing molecular modelling analysis.

## References

Adachi, Y., Kas, E. and Laemmli, U. (1989). Preferential, cooperative binding of topoisomerase II to scaffold-associated regions. *EMBO Jl*, **8**, 3997–4006

Baguley, B. C. (1991). DNA intercalating anti-tumour agents. *Anti-cancer Drug Des.*, **6**, 1–35

Baguley, B. C., Holdaway, K. M. and Fray, L. M. (1990). Design of DNA intercalators to overcome topoisomerase II-mediated multidrug resistance. *J. Natl Cancer Inst.*, **82**, 399–402

Bailly, C., Bernier, J. L., Houssin, R., Helbecque, N. and Hénichart, J. P. (1987). Design of two metal-chelating, DNA-binding models: molecular combinations of bleomycin and amsacrine anti-tumour drugs. *Anti-cancer Drug Des.*, **1**, 303–312

Bailly, C., Catteau, J. P., Hénichart, J. P., Reszka, K., Shea, R. G., Krowicki, K.

and Lown, J. W. (1989a). Subcellular distribution of a nitroxide spin-labeled netropsin in living KB cells. *Biochem. Pharmacol.*, **38**, 1625–1630

Bailly, C., Collyn-d'Hooghe, M., Lantoine, D., Fournier, C., Hecquet, B., Fossé, P., Saucier, J. M., Colson, P., Houssier, C. and Hénichart, J. P. (1992a). Biological activity and molecular interaction of a netropsin–acridine hybrid ligand with chromatin and topoisomerase II. *Biochem. Pharmacol.*, **43**, 457–466

Bailly, C., Denny, W. A., Mellor, L. E., Wakelin, L. P. G. and Waring, M. J. (1992b). Sequence specificity of the binding of antitumor amsacrine-4-carboxamides to DNA studied by DNAase I footprinting. *Biochemistry*, **31**, 3514–3524

Bailly, C., Helbecque, N., Hénichart, J. P., Colson, P., Houssier, C., Rao, K. E., Shea, R. G. and Lown, J. W. (1990a). Molecular recognition between oligopeptides and nucleic acids. DNA sequence specificity and binding properties of an acridine-linked netropsin hybrid ligand. *J. Mol. Recognit.*, **3**, 26–35

Bailly, C. and Hénichart, J. P. (1990). Subcellular distribution of a nitroxide spin-labeled netropsin-acridine hybrid in living KB cells. Electron spin resonance study. *Biochem. Biophys. Res. Commun.*, **167**, 798–806

Bailly, C. and Hénichart, J. P. (1991). DNA recognition by intercalator-minor groove binder hybrid molecules. *Bioconjugate Chem.*, **2**, 379–393

Bailly, C., Kenani, A., Helbecque, N., Bernier, J. L., Houssin, R. and Hénichart, J. P. (1988). DNA-binding and DNA-cleaving properties of a synthetic model AGAGLU related to the antitumour drugs Amsa and bleomycin. *Biochem. Biophys. Res. Commun.*, **152**, 695–702

Bailly, C., OhUigin, C., Houssin, R., Colson, P., Houssier, C., Rivalle, C., Bisagni, E., Hénichart, J. P. and Waring, M. J. (1992c). DNA-binding properties of a distamycin-ellipticine hybrid molecule. *Mol. Pharmacol.*, **41**, 845–855

Bailly, C., OhUigin, C., Rivalle, C., Bisagni, E., Hénichart, J. P. and Waring, M. J. (1990b). Sequence-selective binding of an ellipticine derivative to DNA. *Nucleic Acids Res.*, **18**, 6283–6291

Bailly, C., Pommery, N., Houssin, R. and Hénichart, J. P. (1989b). Design, synthesis, DNA binding and biological activity of a series of DNA minor-groove-binding intercalating drugs. *J. Pharm. Sci.*, **78**, 910–917

Bailly, C., Sun, J. S., Colson, P., Houssier, C., Hélène, C., Waring, M. J. and Hénichart, J. P. (1992d). Design of a sequence-specific DNA-cleaving molecule which conjugates a copper-complexing peptide, a netropsin residue and an acridine chromophore. *Bioconjugate Chem.*, **3**, 100–103

Bailly, F., Bailly, C., Helbecque, N., Pommery, N., Colson, P., Houssier, C. and Hénichart, J. P. (1992e). Relationship between DNA-binding and biological activity of anilinoacridine derivatives containing the nucleic acid-binding unit SPKK. *Anti-cancer Drug Des.*, **7**, 83–100

Bailly, F., Bailly, C., Waring, M. J. and Hénichart, J. P. (1992f). Selective binding to AT sequences in DNA by an acridine-linked peptide containing the SPKK motif. *Biochem. Biophys. Res. Commun.*, **184**, 930–937

Bajaj, N. P. S., McLean, M. J., Waring, M. J. and Smekal, E. (1990). Sequence-selective, pH-dependent binding to DNA of benzophenanthridine alkaloids. *J. Mol. Recognit.*, **3**, 48–54

Beerman, T. A., Woynarowski, J. M., Sigmund, R. D., Gawron, L. S., Rao, K. E. and Lown, J. W. (1991). Netropsin and bis-netropsin analogs as inhibitors of the catalytic activity of mammalian DNA topoisomerase II and topoisomerase cleavable complexes. *Biochim. Biophys. Acta*, **1090**, 52–60

Boehncke, K., Nonella, M., Schulten, K. and Wang, A. H. J. (1991). Molecular dynamics investigation of the interaction between DNA and distamycin. *Biochemistry*, **30**, 5465–5475

Breg, J. N., Boelens, R., George, A. V. E. and Kaptein, R. (1989). Sequence-specific $^1$H NMR assignment and secondary structure of the Arc repressor of bacteriophage P22, as determined by two-dimensional $^1$H NMR spectroscopy. *Biochemistry*, **28**, 9826–9833

Brown, D. G., Sanderson, M. R., Skelly, J. V., Jenkins, T. C., Brown, T., Garman, E., Stuart, D. I. and Neidle, S. (1990). Crystal structure of a berenil–dodecanucleotide complex: the role of water in sequence-specific ligand binding. *EMBO Jl*, **9**, 1329–1334

Cassileth, P. A. and Gale, R. P. (1986). Amsacrine: a review. *Leukemia Res.*, **10**, 1257–1265

Chaires, J. B. (1990). Biophysical chemistry of the daunomycin–DNA interaction. *Biophys. Chem.*, **35**, 191–202

Chaires, J. B., Fox, K. R., Herrera, J. E., Britt, M. and Waring, M. J. (1987). Site and sequence specificity of the daunomycin–DNA interaction. *Biochemistry*, **26**, 8227–8236

Chaires, J. B., Herrera, J. E. and Waring, M. J. (1990). Preferential binding of daunomycin to 5′A/TCG and 5′A/TGC sequences revealed by footprinting titration experiments. *Biochemistry*, **29**, 6145–6153

Chen, K. X., Gresh, N. and Pullman, B. (1986). A theoretical investigation on the sequence selective binding of mitoxantrone to double-stranded tetranucleotides. *Nucleic Acids Res.*, **14**, 3799–3812

Chen, K. X., Gresh, N. and Pullman, B. (1988). Energetics and stereochemistry of DNA complexation with the antitumor AT specific intercalators tilorone and m-AMSA. *Nucleic Acids Res.*, **16**, 3061–3073

Chikira, M., Sato, T., Antholine, W. E. and Petering, D. H. (1991). Orientation of non-blue cupric complexes on DNA fibers. *J. Biol. Chem.*, **266**, 2859–2863

Chow, K. C., MacDonald, T. L. and Ross, W. E. (1989). DNA binding by epipodophyllotoxins and *N*-acyl anthracyclines: implications for mechanism of topoisomerase II inhibition. *Mol. Pharmacol.*, **34**, 467–473

Churchill, M. E. A. and Suzuki, M. (1989). 'SPKK' motifs prefer to bind to DNA at A/T-rich sites. *EMBO Jl*, **8**, 4189–4195

Churchill, M. E. A. and Travers, A. A. (1991). Protein motifs that recognize structural features of DNA. *Trends Biochem. Sci.*, **16**, 92–97

Coll, M., Aymami, J., Van der Marel, G. A., Van Boom, J. H., Rich, A. and Wang, A. H. J. (1989). Molecular structure of the netropsin–d(CGCGATATCGCG) complex: DNA conformation in an alternating AT segment. *Biochemistry*, **28**, 310–320

Coll, M., Frederick, C. A., Wang, A. H. J. and Rich, A. (1987). A bifurcated hydrogen bonded conformation in the d(A.T) base pairs of the DNA dodecamer d(CGCAATTGCG) and its complex with distamycin. *Proc. Natl Acad. Sci. USA*, **84**, 8385–8389

Covey, J. M., Kohn, K. W., Kerrigan, D., Tilchen, E. J. and Pommier, Y. (1988). Topoisomerase II-mediated DNA damage produced by 4′-(9-acridinylamino)methanesulfon-*m*-anisidide and related acridines in L1210 cells and isolated nuclei: relation to cytotoxicity. *Cancer Res.*, **48**, 860–865

Denny, W. A. (1990). DNA-intercalating ligands as anti-cancer drugs: prospects for future design. *Anti-cancer Drug Des.*, **4**, 241–263

Denny, W. A., Atwell, G. J. and Baguley, B. C. (1983). Potential antitumor agents. 39. Anilino ring geometry of amsacrine and derivatives: relationship to DNA binding and antitumor activity. *J. Med. Chem.*, **26**, 1625–1630

Denny, W. A. and Wakelin, L. P. G. (1986). Kinetic and equilibrium studies of the interaction of amsacrine and anilino ring-substituted analogues with DNA. *Cancer Res.*, **46**, 1717–1721

Dervan, P. B. (1986). Design of sequence-specific DNA-binding molecules. *Science*, **232**, 464–471

Dolnick, B. J. (1991). Antisense agents in cancer research and therapeutics. *Cancer Invest.*, **9**, 185–194

Drew, H. R. and Travers, A. A. (1984). DNA structural variations in the *E. coli tyr T* promoter. *Cell*, **37**, 491–502

Drlica, K. and Franco, R. J. (1988). Inhibitors of DNA topoisomerases. *Biochemistry*, **27**, 2253–2259

Egli, M., Williams, L. D., Frederick, C. A. and Rich, A. (1991). DNA–nogalamycin interactions. *Biochemistry*, **30**, 1364–1372

Eliadis, A., Phillips, D. R., Reiss, J. A. and Skorobogaty, A. (1988). The synthesis and DNA footprinting of acridine-linked netropsin and distamycin bifunctional mixed ligands. *J. Chem. Soc. Chem. Commun.*, 1049–1052

Fesen, M. and Pommier, Y. (1989). Mammalian topoisomerase II activity is modulated by the DNA minor groove binder distamycin in simian virus 40 DNA. *J. Biol. Chem.*, **264**, 11354–11359

Finlay, G. J., Wilson, W. R. and Baguley, B. C. (1989). Chemoprotection by 9-aminoacridine derivatives against the cytotoxicity of topoisomerase II-directed drugs. *Eur. J. Cancer Clin. Oncol.*, **25**, 1695–1701

Fox, K. R. and Waring, M. J. (1987). Footprinting at low temperature: evidence that ethidium and other simple intercalators can discriminate between different nucleotide sequences. *Nucleic Acids Res.*, **15**, 491–507

Fox, K. R., Waring, M. J., Brown, J. R. and Neidle, S. (1986). DNA sequence preferences for anticancer drug mitoxantrone and related anthraquinones revealed by DNaseI footprinting. *FEBS Lett.*, **202**, 289–294

Frederick, C. A., Williams, L. D., Ughetto, G., van der Marel, G. A., van Boom, J. H., Rich, A. and Wang, A. H. J. (1990). Structural comparison of anticancer drug–DNA complexes: adriamycin and daunomycin. *Biochemistry*, **29**, 2538–2549

Freedman, J. H., Pickart, L., Weinstein, B., Mims, W. B. and Peisach, J. (1982). Structure of the glycyl–L-histidyl–L-lysine–copper(II) complex in solution. *Biochemistry*, **21**, 4540–4544

Gago, F., Reynolds, C. A. and Richards, W. G. (1989). The binding of non-intercalative drugs to alternating DNA sequences. *Mol. Pharmacol.*, **35**, 232–241

Gao, Y. G., Liaw, Y. C., Robinson, H. and Wang, A. H. J. (1990). Binding of the antitumor drug nogalamycin and its derivatives to DNA: structural comparison. *Biochemistry*, **29**, 10307–10316

Gilbert, D. E. and Feigon, J. (1991). Structural analysis of drug–DNA interactions. *Curr. Opinion Struct. Biol.*, **1**, 439–445

Goodsell, D. and Dickerson, R. E. (1986). Isohelical analysis of DNA groove-binding drugs. *J. Mol. Biol.*, **29**, 727–733

Gresh, N. and Kahn, P. H. (1990). Theoretical design of novel, 4 base pair selective derivatives of mitoxantrone. *J. Biomol. Struct. Dyn.*, **7**, 1141–1160

Gresh, N. and Kahn, P. H. (1991). Theoretical design of a bistetrapeptide derivative of mitoxantrone targeted towards the double stranded hexanucleotide sequence $d(GGCGCC)_2$. *J. Biomol. Struct. Dyn.*, **8**, 827–846

Gresh, N., Pullman, B. and Arcamone, F. (1985). A theoretical investigation on the sequence selective binding of daunomycin to double-stranded polynucleotides. *J. Biomol. Struct. Dyn.*, **3**, 445–466

Gresh, N., Pullman, B., Arcamone, F., Menozzi, M. and Tonani, R. (1989). Joint experimental and theoretical investigation of the comparative DNA binding affini-

ties of intercalating anthracycline derivatives. *Mol. Pharmacol.*, **35**, 251–256
Grokhovsky, S. L., Surovaya, A. N., Brussov, R. V., Chernov, B. K., Sidorova, N. Yu. and Gursky, G. V. (1991). Design and synthesis of sequence-specific DNA-binding peptides. *J. Biomol. Struct. Dyn.*, **8**, 989–1025
Hawley, R. C., Kiessling, L. L. and Schreiber, S. L. (1989). Model of the interactions of calicheamicin $\gamma_1$ with a DNA fragment from pBR322. *Proc. Natl Acad. Sci. USA*, **86**, 1105–1109
Hecht, S. M. (1986). The chemistry of activated bleomycin. *Acc. Chem. Res.*, **19**, 383–391
Hénichart, J. P., Houssin, R., Bernier, J. L. and Catteau, J. P. (1982). Synthetic model of a bleomycin metal complex. *J. Chem. Soc. Chem. Commun.*, 1295–1297
Hénichart, J. P., Houssin, R., Bernier, J. L. and Catteau, J. P. (1985). Copper(II)- and iron(II)-complexes of methyl 2-(2-aminoethyl)-aminoethyl-pyridine-6-carboxyl-histidinate (AMPHIS), a peptide mimicking the metal-chelating moiety of bleomycin. An ESR investigation. *Biochem. Biophys. Res. Commun.*, **126**, 1036–1041
Huff, A. C. and Kreuzer, K. N. (1990). Evidence for a common mechanism of action for antitumor and antibacterial agents that inhibit type II topoisomerases. *J. Biol. Chem.*, **265**, 20496–20505
Hurley, L. H. (1989). DNA and associated targets for drug design. *J. Med. Chem.*, **32**, 2027–2033
Islam, S. A., Neidle, S., Grandocha, B. M., Partridge, M., Patterson, L. H. and Brown, J. R. (1985). Comparative computer graphics and solution studies of the DNA interaction of substituted anthraquinones based on doxorubicin and mitoxantrone. *J. Med. Chem.*, **28**, 857–864
Käs, E., Izaurralde, E. and Laemmli, U. K. (1989). Specific inhibition of DNA binding to nuclear scaffolds and histone H1 by distamycin. *J. Mol. Biol.*, **210**, 587–599
Kénani, A., Bailly, C., Helbecque, N., Houssin, R., Bernier, J. L. and Hénichart, J. P. (1989). Metal-complexing, DNA-binding and DNA-cleaving properties of a synthetic model AMBIGLU, a simplified model for the study of bleomycin. *Eur. J. Med. Chem.*, **24**, 371–377
Kénani, A., Lohez, M., Houssin, R., Helbecque, N., Bernier, J. L., Lemay, P. and Hénichart, J. P. (1987). Chelating, DNA-binding and DNA-cleaving properties of a synthetic model for bleomycin. *Anti-cancer Drug Des.*, **2**, 47–59
Koch, M. H. J., Sayers, Z., Vega, M. C. and Michon, A. M. (1987). The superstructure of chromatin and its condensation mechanism. IV. Enzymatic digestion, thermal denaturation, effect of netropsin and distamycin. *Eur. Biophys. J.*, **15**, 133–140
Kopka, M. L., Yoon, C., Goodsell, D., Pjura, P. and Dickerson, R. E. (1985a). The molecular origin of DNA–drug specificity in netropsin and distamycin. *Proc. Natl Acad. Sci. USA*, **82**, 1376–1380
Kopka, M. L., Yoon, C., Goodsell, D., Pjura, P. and Dickerson, R. E. (1985b). Binding of an antitumor drug to DNA: Netropsin and CGCGAATT$^{Br}$CGCG. *J. Mol. Biol.*, **183**, 553–563
Kreuzer, K. N. (1989). DNA topoisomerases as potential targets of antiviral action. *Pharmacol. Ther.*, **43**, 377–395
Krivtsora, M. A., Moroshkina, E. B. and Glibin, E. (1984). DNA interaction with low molecular weight ligands with different structures. III. Complexes of DNA with distactins. *Mol. Biol.*, **18**, 950–956
Lane, M. J., Dabrowiak, J. C. and Vournakis, J. N. (1983). Sequence specificity of

actinomycin D and netropsin binding to pBR322 DNA analyzed by protection from DNase I. *Proc. Natl Acad. Sci. USA*, **80**, 3260–3264

Larsen, T. A., Goodsell, D. S., Cascio, D., Grzeskowiak, K. and Dickerson, R. E. (1989). The structure of DAPI bound to DNA. *J. Biomol. Struct. Dyn.*, **7**, 477–491

Laughton, C. A., Jenkins, T. C., Fox, K. R. and Neidle, S. (1990). Interaction of berenil with the *tyrT* DNA sequence studied by footprinting and molecular modelling. Implications for the design of sequence-specific DNA recognition agents. *Nucleic Acids Res.*, **18**, 4479–4488

Laussac, J. P., Haran, R. and Sarkar, B. (1983). N.M.R. and e.p.r. investigation of the interaction of copper(II) and glycyl–L-histidyl–L-lysine, a growth-modulating tripeptide from plasma. *Biochem. J.*, **209**, 533–539

Lavery, R. (1988). In Olson, M. H., Sarma, M. H., Sarma, R. H. and Sundaralingam, M. (Eds), *Structure and Expression*, Vol. 3: *DNA Bending and Curvature*, Adenine Press, New York, pp. 191–211

Lee, S. H. and Goldberg, I. H. (1989). Sequence-specific, strand-selective, and directional binding of neocarzinostatin chromophore to oligodeoxyribonucleotides. *Biochemistry*, **28**, 1019–1026

Lemay, P., Bernier, J. L., Hénichart, J. P. and Catteau, J. P. (1983). Subcellular distribution of nitroxide spin-labelled 9-aminoacridine in living KB cells. *Biochem. Biophys. Res. Commun.*, **111**, 1074–1081

Lerman, L. S. (1961). Structural considerations in the interaction of DNA and acridines. *J. Mol. Biol.*, **3**, 18–30

Lerman, L. S. (1963). The structure of the DNA–acridine complex. *Proc. Natl Acad. Sci. USA*, **49**, 94–105

Liu, L. F. (1989). DNA topoisomerase poisons as antitumor drugs. *Ann. Rev. Biochem.*, **58**, 351–375

Lown, J. W. (1988). Lexitropsins: rational design of DNA sequence reading agents as novel anticancer agents and potential cellular probes. *Anti-cancer Drug Des.*, **3**, 25–40

Lown, J. W. and Hanstock, C. C. (1985). High field $^1$H NMR analysis of the 1:1 complex of the antitumour agent mitoxantrone and the DNA duplex $[d(CpGpCpG)]_2$. *J. Biomol. Struct. Dyn.*, **2**, 1097–1106

McHugh, M., Woynarowski, J. M., Sigmund, R. D. and Beerman, T. A. (1989). Effect of minor groove binding drugs on mammalian topoisomerase I activity. *Biochem. Pharmacol.*, **38**, 2323–2328

McHugh, M., Sigmund, R. D. and Beerman, T. A. (1990). Effect of minor groove binding drugs on camptothecin-induced DNA lesions in L1210 nuclei. *Biochem. Pharmacol.*, **39**, 707–714

Morier-Teissier, E., Bailly, C., Bernier, J. L., Houssin, R., Helbecque, N., Catteau, J. P., Colson, P., Houssier, C. and Hénichart, J. P. (1989). Synthesis, biological activity and DNA interaction of anilinoacridine and bithiazole peptide derivatives related to the antitumor drugs m-amsa and bleomycin. *Anti-cancer Drug Des.*, **4**, 37–52

Morier-Teissier, E., Bernier, J. L., Lohez, M., Catteau, J. P. and Hénichart, J. P. (1990). Free radical production and DNA cleavage by copper chelating peptide-anthraquinones. *Anti-cancer Drug Des.*, **5**, 291–305

Mortensen, U. H., Stevnsner, T., Krogh, S., Olesen, K., Westergaard, O. and Bonven, B. J. (1990). Distamycin inhibition of topoisomerase I–DNA interaction: a mechanistic analysis. *Nucleic Acids Res.*, **18**, 1983–1989

Mrani, D., Gosselin, G., Balzarini, J., De Clercq, E., Paoletti, C. and Imbach, J. L. (1991). Synthesis, DNA-binding and biological activity of oxazolo-

pyridocarbazole-netropsin hybrid molecules. *Eur. J. Med. Chem.*, **26**, 481–488
Neidle, S. and Abraham, Z. (1984). Structural and sequence-dependent aspects of drug intercalation into nucleic acids. *CRC Crit. Rev. Biochem.*, **17**, 73–121
Neidle, S., Pearl, L. H. and Skelly, J. V. (1987). DNA structure and perturbation by drug binding. *Biochem. J.*, **243**, 1–13
Nelson, E. M., Tewey, K. M. and Liu, L. F. (1984). Mechanism of antitumor drug action. Poisoning of mammalian DNA topoisomerase II on DNA by 4′-(9-acridinylamino)methanesulfon-*m*-anisidide. *Proc. Natl Acad. Sci. USA*, **81**, 1361–1365
Nielsen, P. E. (1991). Sequence-selective DNA recognition by synthetic ligands. *Bioconjugate Chem.*, **2**, 1–12
Palumbo, M. and Gatto, B. (1990). Aminoacyl-anthraquinones: DNA-binding and sequence specificity. In Pullman, B. and Jortner, J. (Eds), *Molecular Basis of Specificity in Nucleic Acid–Drug Interactions*, Kluwer Academic, Dordrecht, pp. 207–224
Patel, D. J., Kozlowski, S. A., Rice, J. A., Broka, C. and Itakura, K. (1981). Mutual interaction between adjacent dG.dC actinomycin binding sites and dA.dT netropsin binding sites on the self-complementary d(CGCGAATT-CGCG) duplex in solution. *Proc. Natl Acad. Sci. USA*, **78**, 7281–7284
Patel, D. J., Pardi, A. and Itakura, K. (1982). DNA conformation, dynamics, and interactions in solution. *Science*, **216**, 581–590
Pelton, J. G. and Wemmer, D. E. (1990). Binding modes of distamycin A with $d(CGCAAATTTGCG)_2$ determined by two-dimensional NMR. *J. Am. Chem. Soc.*, **112**, 1393–1399
Per, S. R., Mattern, M. R., Mirabelli, C. K., Drake, F. H., Johnson, R. K. and Crooke, S. T. (1987). Characterization of a subline of P388 leukemia resistant to amsacrine: evidence of altered topoisomerase II function. *Mol. Pharmacol.*, **32**, 17–25
Pickart, L., Freedman, J. M., Loker, W. L., Peisach, J., Perkins, C. M., Stenkamp, R. E. and Weinstein, B. (1980). Growth-modulating plasma tripeptide may function by facilitating copper-uptake into cells. *Nature*, **288**, 715–717
Pickart, L. and Lovejoy, S. (1987). Biological activity of human plasma copper binding growth factor glycyl–L-histidyl–L-lysine. *Meth. Enzymol.*, **147**, 314–338
Pickart, L. and Thaler, M. M. (1973). Tripeptide in human serum which prolongs survival on normal cells and stimulates growth in neoplastic liver. *Nature*, **243**, 85–87
Pommier, Y., Covey, J., Kerrigan, D., Mattes, W., Markovits, J. and Kohn, K. W. (1987). Role of DNA intercalation in the inhibition of purified mouse leukemia (L1210) DNA topoisomerase II by 9-aminoacridines. *Biochem. Pharmacol.*, **36**, 3477–3486
Pommier, Y., Kerrigan, D., Schwartz, R. E., Swack, J. A. and McCurdy, A. (1986). Altered DNA topoisomerase II activity in Chinese hamster cells resistant to topoisomerase II inhibitors. *Cancer Res.*, **46**, 3075–3081
Portugal, J. and Waring, M. J. (1987). Comparison of binding sites in DNA for berenil, netropsin and distamycin. A footprinting study. *Eur. J. Biochem.*, **167**, 281–289
Portugal, J. and Waring, M. J. (1988). Assignment of DNA binding sites for 4′, 6-diamidine-2-phenylindole and bisbenzimide (Hoechst 33258). A comparative footprinting study. *Biochim. Biophys. Acta*, **949**, 158–168
Pullman, B. (1987). Carcinogens, anti-tumor agents and DNA. In Chagas, C. and Pullman, B. (Eds), *Molecular Mechanisms of Carcinogenic and Antitumor Activity*, Adenine Press, New York, pp. 3–31

Reeves, R. and Nissen, M. S. (1990). The AT-DNA-binding domain of mammalian high mobility group I chromosomal proteins. A novel peptide for recognizing DNA structure. *J. Biol. Chem.*, **265**, 8573

Rill, R. L., Marsch, G. A. and Graves, D. E. (1989). 7-Azido-actinomycin D: a photoaffinity probe of the sequence specificity of DNA binding by actinomycin D. *J. Biomol. Struct. Dyn.*, **7**, 591–605

Robinson, M. J. and Osheroff, N. (1990). Stabilization of the topoisomerase II–DNA cleavable complex by antineoplastic drugs: inhibition of enzyme-mediated DNA religation by 4′-(9-acridinylamino)methane sulfon-*m*-anisidide. *Biochemistry*, **29**, 2511–2515

Rowe, T. C., Chen, G. L., Hsiang, Y. H. and Liu, L. F. (1986). DNA damage by antitumor acridines mediated by mammalian DNA topoisomerase II. *Cancer Res.*, **46**, 2021–2026

Sarker, M. and Chen, F. M. (1989). Binding of mithramycin to DNA in the presence of second drugs. *Biochemistry*, **28**, 6651–6657

Schneider, E., Darkin, S. J., Lawson, P. A., Ching, L. M., Ralph, R. K. and Baguley, B. C. (1988). Cell line selectivity and DNA breakage properties of the antitumour agent *N*-[2-(dimethylamino) ethyl]acridine-4-carboxamide: role of DNA topoisomerase II. *Eur. J. Cancer Clin. Oncol.*, **24**, 1783–1790

Scott, E. V., Jones, R. L., Banville, D. L., Zon, G., Marzilli, L. G. and Wilson, W. D. (1988a). $^{1}H$ and $^{31}P$ NMR investigations of actinomycin D binding selectivity with oligodeoxyribonucleotides containing multiple adjacent d(GC) sites. *Biochemistry*, **27**, 915–923

Scott, E. V., Zon, G., Marzilli, L. G. and Wilson, W. D. (1988b). 2D NMR investigation of the binding of the anticancer drug actinomycin D to duplexed dATCGCGAT: conformational features of the unique 2:1 adduct. *Biochemistry*, **27**, 7940–7951

Searle, M., Bicknell, W., Wakelin, L. P. G. and Denny, W. A. (1991). Anthracycline antibiotic arugomycin binds in both grooves of the DNA helix simultaneously: an NMR and molecular modelling study. *Nucleic Acids Res.*, **19**, 2897–2906

Sen, D. and Crothers, D. M. (1986). Influence of DNA-binding drugs on chromatin condensation. *Biochemistry*, **25**, 1503–1509

Stubbe, J. and Kozarich, J. W. (1987). Mechanisms of bleomycin-induced DNA degradation. *Chem. Rev.*, **87**, 1107–1136

Subra, F., Carteau, S., Pager, J., Paoletti, J., Paoletti, C., Auclair, C., Mrani, D., Gosselin, G. and Imbach, J. L. (1991). Bis(pyrrolecarboxamide) linked to intercalating chromophore oxazolopyrido carbazole (OPC): selective binding to DNA and polynucleotides. *Biochemistry*, **30**, 1642–1650

Sugiura, Y., Shiraki, T., Konishi, M. and Oki, T. (1990). DNA intercalation and cleavage of an antitumor antibiotic dynemicin that contains anthracycline and enediyne cores. *Proc. Natl Acad. Sci. USA*, **87**, 3831–3835

Suzuki, M. (1989a). SPKK, a new nucleic acid-binding unit of protein found in histone. *EMBO Jl*, **8**, 797–804

Suzuki, M. (1989b). SPXX, a frequent sequence motif in gene regulatory proteins. *J. Mol. Biol.*, **207**, 61–84

Tabor, C. W. and Tabor, H. (1984). Polyamines. *Ann. Rev. Biochem.*, **53**, 749–790

Tidd, D. M. (1990). A potential role for antisense oligonucleotide analogues in the development of oncogene targeted cancer chemotherapy. *Anticancer Res.*, **10**, 1169–1182

Travers, A. A. (1991). DNA-bending and DNA-kinking — sequence dependence and function. *Curr. Opinion Struct. Biol.*, **1**, 114–122

Van Dyke, M. W. and Dervan, P. B. (1983). Methidiumpropyl–EDTA·Fe(II) and DNAase I footprinting report different small molecule binding site sizes on DNA. *Nucleic Acids Res.*, **11**, 5555–5567

Van Dyke, M. W., Hertzberg, R. P. and Dervan, P. B. (1982). Map of distamycin, netropsin, and actinomycin binding sites on heterogeneous DNA: DNA cleavage-inhibition patterns with methidiumpropyl–EDTA·Fe(II). *Proc. Natl Acad. Sci. USA*, **79**, 5470–5474

Wakelin, L. P. G., Chetcuti, P. and Denny, W. A. (1990). Kinetic and equilibrium binding studies of amsacrine-4-carboxamides: a class of asymmetrical DNA-intercalating agents which bind by threading through the DNA helix. *J. Med. Chem.*, **33**, 2039–2044

Wang, J. C. (1985). DNA topoisomerases. *Ann. Rev. Biochem.*, **54**, 665–697

Wang, J. C. (1987). Recent studies of DNA topoisomerases. *Biochim. Biophys. Acta*, **909**, 1–9

Wang, J. C. (1991). DNA topoisomerases: why so many? *J. Biol. Chem.*, **266**, 6659–6662

Waring, M. J. (1968). Drugs which affect the structure and function of DNA. *Nature*, **219**, 1320–1325

Waring, M. J. (1970). Variation of the supercoils in closed circular DNA by binding of antibiotics and drugs: Evidence for molecular models involving intercalation. *J. Mol. Biol.*, **54**, 247–279

Waring, M. J. (1990). The molecular basis of specific recognition between echinomycin and DNA. In Pullman, B. and Jortner, J. (Eds), *Molecular Basis of Specificity in Nucleic Acid–Drug Interactions*, Kluwer Academic, Dordrecht, pp. 225–245

Wartell, R. M., Larson, J. E. and Wells, R. D. (1975). The compatibility of netropsin and actinomycin binding to natural deoxyribonucleic acid. *J. Biol. Chem.*, **250**, 2698–2702

Wender, P. A., Kelly, R. C., Beckham, S. and Miller, B. L. (1991). Studies on DNA-cleaving agents: computer modeling analysis of the mechanism of activation and cleavage of dynemicin–oligonucleotide complexes. *Proc. Natl Acad. Sci. USA*, **88**, 8835–8839

Williams, L. D., Egli, M., Gao, Q., Bash, P., Van der Marel, G. A., Van Boom, J. H., Rich, A. and Frederick, C. A. (1990). Structure of nogalamycin bound to a DNA hexamer. *Proc. Natl Acad. Sci. USA*, **87**, 2225–2229

Woynarowski, J. M., McHugh, M., Sigmund, R. D. and Beerman, T. A. (1989a). Modulation of topoisomerase II catalytic activity by DNA minor groove binding agents distamycin, Hoechst 33258, and 4′,6-diamidine-2-phenylindole. *Mol. Pharmacol.*, **35**, 177–182

Woynarowski, J. M., Sigmund, R. D. and Beerman, T. A. (1989b). DNA minor groove binding agents interfere with topoisomerase II mediated lesions induced by epipodophyllotoxin derivative VM-26 and acridine derivative m-AMSA in nuclei from L1210 cells. *Biochemistry*, **28**, 3850–3855

Zagorski, M. G., Bowie, J. U., Vershon, A. K., Sauer, R. T. and Patel, D. J. (1989). NMR studies of arc repressor mutants: proton assignments, secondary structure, and long-range contacts for the thermostable proline-8 → leucine variant of arc. *Biochemistry*, **28**, 9813–9825

Zakrzewska, K., Lavery, R. and Pullman, B. (1983a). Theoretical studies of the selective binding to DNA of two non-intercalating ligands: netropsin and SN 18071. *Nucleic Acids Res.*, **11**, 8825–8839

Zakrzewska, K. and Pullman, B. (1983b). A theoretical evaluation of the effect of netropsin binding on the reactivity of DNA towards alkylating agents. *Nucleic Acids Res.*, **11**, 8841–8845

Zhang, H., D'Arpa, P. and Liu, L. F. (1990). A model for tumor cell killing by topoisomerase poisons. *Cancer Cells*, **2**, 23–27

Zimmer, C. and Wähnert, U. (1986). Non-intercalating DNA-binding ligands: specificity of the interaction and their use as tools in biophysical, biochemical and biological investigations of the genetic material. *Prog. Biophys. Mol. Biol.*, **47**, 31–112

Zunino, F. and Capranico, G. (1990). DNA topoisomerase II as the primary target of anti-tumor anthracyclines. *Anti-cancer Drug Des.*, **5**, 307–317

# 5

# Bleomycins: Mechanism of Polynucleotide Recognition and Oxidative Degradation

Anand Natrajan and Sidney M. Hecht

## 1 Introduction

The bleomycins are a family of glycopeptide derived antibiotics originally isolated by Umezawa in 1966 from fermentation broths of *Streptomyces verticillus* (Umezawa *et al.*, 1966). The naturally occurring bleomycins (BLMs) differ structurally primarily at the C-terminus (Figure 5.1) of the glycopeptide; commercial preparations of the drug contain BLM $A_2$ and BLM $B_2$ as the principal constituents. The clinical utility of this drug for the treatment of certain kinds of tumours (Blum *et al.*, 1973) sparked tremendous interest in BLM; during the past two decades an enormous amount of effort has been spent to understand the chemistry, biochemistry and biology of bleomycin. It is now recognized that DNA serves as a target for bleomycin, which, in the presence of certain redox active metal ions and an oxygen source, oxidatively degrades the deoxyribose ring in DNA in a sequence-specific fashion. Initially, it was thought that BLM was a β-lactam type antibiotic (Takita *et al.*, 1972). However, this structure was revised in 1978 (Takita *et al.*, 1978b); the revised structure was confirmed by total synthesis in 1982 (Aoyagi *et al.*, 1982; Takita *et al.*, 1982).

Bleomycin is quite complex structurally and discussions of the molecule often focus on three key structural domains. The β-aminoalanineamide, pyrimidine and the β-hydroxyhistidine moieties constitute one domain, and are believed to be necessary for the complexation of metal ion cofactors and for oxygen activation by the metallobleomycin. The bithiazole moiety and positively charged C substituent are believed to bind to DNA. The functional role of the carbohydrate moiety is unclear at present, although it has been suggested that the carbamoyl moiety of the mannose ring may participate in metal ion complexation (Oppenheimer *et al.*,

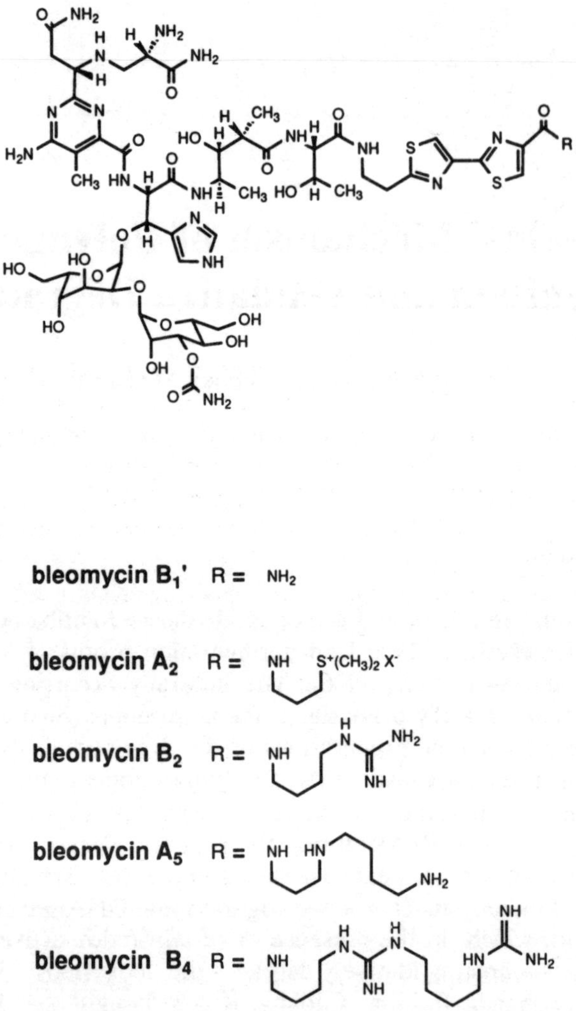


**Figure 5.1** Representative bleomycin group antibiotics

1979a). More recently it has been suggested that the carbohydrate may provide a pocket for oxygen activation (Kittaka *et al.*, 1988) and the possible involvement of the carbohydrate moiety in tumour cell recognition also seems possible.

A number of metallobleomycins, including the iron (Sausville *et al.*, 1976, 1978a,b), cobalt (Chang and Meares, 1982, 1984), manganese (Burger *et al.*, 1984; Ehrenfeld *et al.*, 1984), vanadyl (Kuwahara *et al.*, 1985) and copper (Sugiura, 1979; Ehrenfeld *et al.*, 1985, 1987) complexes have been shown to be capable of mediating DNA degradation. In 1978 Dabro-

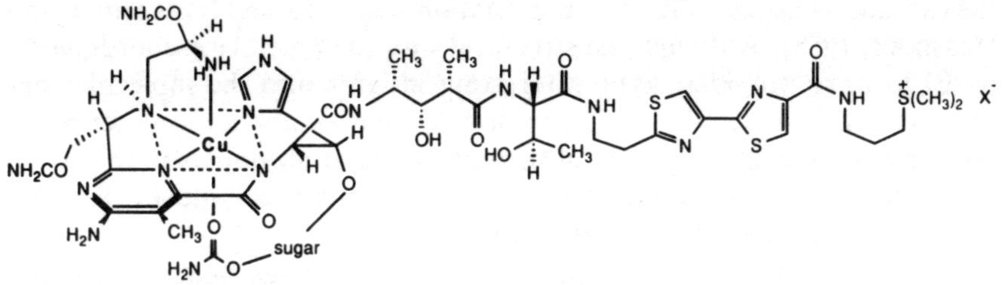

**Figure 5.2** Proposed structure for Cu(II)·BLM $A_2$

wiak *et al.* proposed a square planar structure for Cu(II)·BLM based on spectroscopic studies. The crystal structure of the copper complex of a biosynthetic intermediate of BLM, denoted P-3A, was reported in the same year (Iitaka *et al.*, 1978) and formed the basis for a proposed structure for Cu(II)·BLM (Figure 5.2) (Takita *et al.*, 1978a) and Fe(II)·BLM. In this square pyramidal structure, N-1 of the pyrimidine ring, the secondary NH of β-aminoalanineamide, the deprotonated amide nitrogen of β-hydroxyhistidine and the imidazole nitrogen constituted equatorial coordination ligands, while apical coordination was provided by the primary amine of the β-aminoalanineamide moiety and the oxygen of the carbamoyl group on mannose. The bonding between the carbonyl oxygen of the carbamoyl group and the metal ion was thought to be weak and potentially replaceable by dioxygen in Fe·BLM. The structures of other metallobleomycins, including Co·BLM (Sugiura, 1980a; Vos *et al.*, 1980; Dabrowiak and Tsukayama, 1981), Fe·BLM (Dabrowiak *et al.*, 1979; Oppenheimer *et al.*, 1979a) and Zn·BLM (Dabrowiak *et al.*, 1978; Oppenheimer *et al.*, 1979b) have been studied using NMR, EPR and optical spectroscopy. Of special interest are the NMR studies of Oppenheimer *et al.* (1979a,b) on Zn(II)·BLM and the carbon monoxide adduct of Fe(II)·BLM, two complexes in which the carbamoyl moiety of mannose was involved as a metal ligand. The Fe(II)·BLM–CO complex was found to be diamagnetic and bound to DNA, although product formation did not obtain in the absence of oxygen. The structure proposed for the Fe(II)·BLM–CO complex excluded the involvement of the amide nitrogen attached to the β-hydroxyhistidine moiety in metal ion coordination. Instead, the carbamoyl group of the mannose ring was proposed to be involved in Fe(II) coordination. Sugiura (1980b) examined the coordination geometries of a number of BLM analogues using EPR techniques. In the nitrosyl adduct of Fe(II)·BLM, the NO molecule was proposed to complete the hexacoordinate geometry around Fe, using the five ligands proposed to be involved in the coordination of Cu(II) (see above). In a more recent study, the solution structures of BLM $A_2$ and its Zn(II) complex has been examined by NMR spectroscopy and the participation of the carbamoyl group in Zn(II) coordination was proposed (Akkerman and Haasnoot, 1988). Although the structural studies of metal ion coordination by BLM have provided some interesting insights into the molecular behaviour of the drug, it is not clear whether the coordination geometries of the activated metal complexes of bleomycin are at all similar to the structures proposed thus far for metal ion–BLM binary complexes, and the inert ternary complexes studied to date.

This review is concerned with the status of studies dealing with the chemical properties of bleomycin. Although the synthesis of novel bleomycins in support of such studies has not been an easy undertaking, owing to the structural complexity of bleomycin, a number of analogues have been prepared and evaluated to provide answers to key questions concerning oxygen activation and DNA binding, as well as the basis for sequence recognition by bleomycin.

## 2 Oxygen Activation by Iron Bleomycin

While it was recognized quite early that bleomycin could inhibit DNA synthesis and cause DNA strand scission *in vivo* and *in vitro* (Suzuki *et al.*, 1968, 1969), and that the DNA cleaving properties of bleomycin could be enhanced by the addition of either a reducing agent (Nagai *et al.*, 1969b) or hydrogen peroxide (Nagai *et al.*, 1969a), the requirement for Fe and oxygen as cofactors necessary for the DNA degradation reaction was demonstrated much later (Sausville *et al.*, 1976, 1978a,b). Sausville *et al.* observed that the presence of DNA cleaving activity in bleomycin was correlated with the appearance of a 1:1 Fe(II)·BLM complex that exhibited a UV absorption maximum at 476 nm. DNA degradation was also found to occur over a wide range of pH (4.5–10) and to require oxygen. The nature of the reactive species involved in the DNA degradation reaction remained unclear, however. A number of studies suggested that superoxide or hydroxyl radicals were being generated by aerobic solutions of Fe(II)·BLM and that DNA damage was due to these oxygen radicals (Ishida and Takahashi, 1975; Lown and Sim, 1977; Sugiura and Kikuchi, 1978; Oberley and Buettner, 1979; Sugiura, 1979, 1980b). Numerous attempts were made to detect these radicals by EPR techniques using spin traps such as *N-tert*-butyl-α-phenylnitrone (BPN) (Sugiura, 1980b) and 5,5-dimethyl-1-pyrroline-1-oxide (Oberley and Buettner, 1979). Although the detection of both superoxide and hydroxyl radicals in aerobic solutions of Fe(II)·BLM was reported, no causal relationship between the production of these species and DNA degradation was demonstrated. In fact, subsequent studies showed that superoxide dismutase, catalase or low molecular weight reagents capable of sequestering reactive oxygen radicals had little effect on the ability of bleomycin to degrade DNA (Gutteridge and Shute, 1981; Rodriguez and Hecht, 1982). Burger *et al.* (1979b) reported that neither superoxide radicals nor peroxide could be detected in solutions of Fe(II)·BLM. Especially noteworthy were the results of Rodriguez and Hecht (1982), who observed that the release of [$^3$H]-thymine as a consequence of the degradation of PM-2 DNA by 30 μM Fe(II) + 5 μM BLM was not suppressed by dimethylsulfoxide (DMSO), a known hydroxyl radical scavenger. In contrast, DNA degradation by 100 μM Fe(II) + 290 μM hydrogen peroxide in a Fenton-type reaction was completely suppressed by DMSO. Similarly, the presence of 80 μM BPN, which reacts with ·OH, did not affect the release of [$^3$H]-thymine from DNA by 2.5 μM Fe(II)·BLM. The superoxide radical scavenger tetrakis(4-*N*-methylpyridyl)porphine iron(III) was also tested; it too had no effect on DNA degradation by Fe(II)·BLM.

If BLM does not produce free superoxide or hydroxyl radicals, what is the nature of the oxidizing species that causes DNA degradation? Studies of oxygen consumption by Fe(II)·BLM (Kuramochi *et al.*, 1981) and EPR studies (Burger *et al.*, 1979a,b, 1981) provided some insight into the kinetics of the activation process and led to suggestions about the structure of the 'active form' of Fe(II)·BLM. Kuramochi *et al.* (1981) found that at 0 °C, the ratio of oxygen consumed : Fe(II)·BLM in solution was 0.48 and that this ratio increased slightly to 0.59 in the presence of DNA. Remarkably, when the reductant 2-mercaptoethanol was included in the reaction mixture, the ratio almost doubled to 0.86. Since each Fe(II)·BLM consumed only about 0.5 mol of $O_2$ in the absence of reductant, it was deduced that an additional reducing equivalent was needed for the reduction of the ternary complex Fe(II)·BLM·$O_2$ and that, in the absence of an external reductant, this electron could be supplied by a second molecule of Fe·BLM. The foregoing oxygen consumption data provided some evidence that dioxygen was reduced to the level of peroxide by Fe(II)·BLM, which in the process was itself being converted to an oxygenated Fe(II)·BLM. Further, Kuramochi *et al.* (1981) also observed that the addition of hydrogen peroxide restored the DNA damaging ability of the Fe(III)·BLM produced from the aerobic oxidation of Fe(II)·BLM. Burger *et al.* (1979a, 1981) were able to characterize the reactive, oxygenated Fe·BLM species by EPR spectroscopy. The addition of $O_2$ to a complex of Fe(II)·BLM produced an EPR silent species. The rate of formation of this complex was found to be first-order with respect to the concentrations of $O_2$ and Fe(II)·BLM ($K = 6.1 \times 10^{-3}$ M/s at 2 °C). The optical and EPR spectra of this complex were similar to those of the NO and CO complexes of Fe(II)·BLM; accordingly, the oxygenated complex was assigned the structure Fe(II)·BLM·$O_2$. This ternary complex underwent a second reaction with $t_{1/2} = 6$ s at 2 °C to yield two species with distinctly different EPR spectra. The complex with $g$ values 2.26, 2.17 and 1.94 was observed to be unstable and was transformed into Fe(III)·BLM ($g = 2.45, 2.18, 1.89$) at a rate that paralleled the rate of DNA degradation. The EPR spectra also suggested that Fe(III)·BLM was formed as a primary product during the process of Fe·BLM activation. The reaction of Fe(III)·BLM with hydrogen peroxide or ethyl hydroperoxide also produced the same unstable species that decayed to Fe(III)·BLM, although the rate of production of this 'activated Fe·BLM' was slower in these cases. Since the active complex had odd electron spin, structures for activated BLM containing ferric ion bound to hydroxyl radical or superoxide ion were ruled out. Possible structures for 'activated Fe·BLM' advanced by Burger *et al.* (1981) are the following:

The first structure contains oxygen as an oxene and could result from heterolysis of the O–O bond in a peroxide bound to Fe(III)·BLM. This structure is similar to that proposed for active cytochrome P450s (Guengerich and Macdonald, 1984). The latter two structures involve either an end-on or cyclic binding of peroxide to Fe(III)·BLM. The possibility that activated BLM may be similar chemically to activated cytochrome P450 was supported by Mossbauer studies of Fe(II)·BLM (Burger *et al.* 1983). These studies indicated that iron in activated Fe·BLM is a low-spin ferric species, and that it contains two oxidizing equivalents that reside on the oxygen bound to Fe·BLM. Low-spin Fe(III)·BLM was also observed to have EPR and Mossbauer properties remarkably similar to those of low-spin ferric cytochrome P450.

The requirement of two Fe(II)·BLMs for the production of one molecule of activated bleomycin in the absence of external reductants suggested that DNA could exert an inhibitory effect on the activation reaction by binding to the Fe(II)·BLM, and thereby decreasing the concentration of free Fe(II)·BLM available for reductive activation. This decrease has been observed. In the presence of excess DNA, the ternary complex Fe(II)·BLM·$O_2$ was found to be long lived (Albertini *et al.*, 1982). Further, the rates of product release from DNA were maximal if $Fe^{2+}$ and BLM were preincubated aerobically for a short period of time prior to the addition of DNA substrate (Natrajan *et al.*, 1990a). If DNA was included in the reaction during the activation of Fe(II)·BLM, an increase in the concentration of DNA resulted in a decrease in the rate of product release from DNA. The reaction of Fe(III)·BLM with $H_2O_2$, unlike the aerobic activation reaction, was found to be slow at neutral pH (Burger *et al.*, 1981; Natrajan *et al.*, 1990a), but when the reaction was carried out at either acidic (pH 5.8) or basic pH (pH 8.4), product release from DNA was rapid and comparable to the rate observed with aerobically activated bleomycin (Natrajan *et al.*, 1990a). The most reasonable interpretation of these observations is that the normally slow reaction between Fe(III)·BLM and

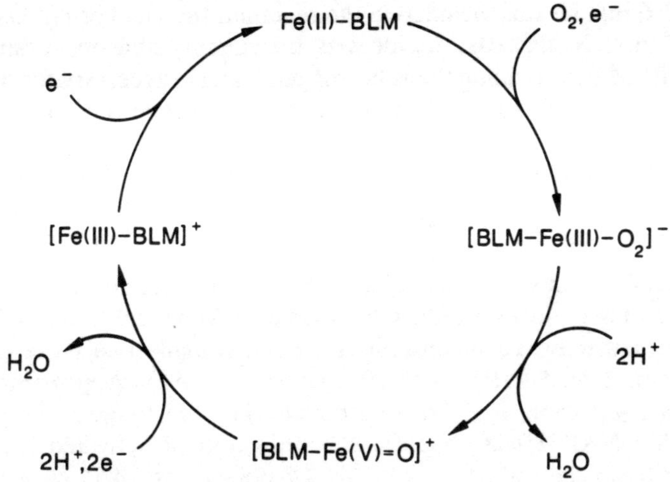


**Figure 5.3** Possible catalytic cycle for Fe•BLM

peroxide can be accelerated at both acidic and basic pHs. Production of activated bleomycin from Fe(III)•BLM + $H_2O_2$ was not inhibited by DNA, unlike the aerobic activation reaction, a circumstance that probably reflects the oxidation state of the initially formed ternary complex, as well as the possible direct activation of DNA-bound Fe(III)•BLM by hydrogen peroxide.

A possible catalytic cycle for Fe•BLM activation and its 'decay' is shown in Figure 5.3. According to this scheme, in the presence of $O_2$, Fe(II)•BLM mediates the two-electron reduction of dioxygen to peroxide, with the required reducing equivalent being derived from a second molecule of Fe(II)•BLM. In the absence of DNA or other substrates, the conversion of this BLM-bound peroxide to Fe(III)•BLM requires two additional reducing equivalents. Presumably, these two reducing equivalents are provided by two additional Fe(II)•BLMs which reduce peroxide to two water molecules. Protonation of the BLM-bound peroxide could result in O–O bond scission producing, for example, a perferryl species (cf. Figure 5.3). The stoichiometry of the foregoing processes has been verified using $^{17}$O-NMR spectroscopy (Barr *et al.*, 1990). In the absence of a DNA substrate, admixture of 20 mM Fe(II)•BLM and excess $^{17}O_2$ resulted in the formation of 9.4 mM $H_2{}^{17}O$; this observation is consistent with the formulation of the catalytic cycle of Fe(II)•BLM as involving the four-electron reduction of $O_2$ to $H_2O$, i.e.

$$4\ \mathrm{Fe(II)}\cdot\mathrm{BLM} + O_2 + 4H^+ \rightarrow 4\ \mathrm{Fe(III)}\cdot\mathrm{BLM} + 2H_2O$$

$$H_2O/\mathrm{Fe(II)}\cdot\mathrm{BLM} = 0.5$$

When 200 mM KI was included in the reaction, the yield of $H_2{}^{17}O$ doubled to 18.7 mM. In this case iodide was presumably able to substitute for Fe(II)·BLM by providing the reducing equivalents necessary for discharge of BLM-bound peroxide (or perferryl species; cf. Figure 5.3)

$$2\ Fe(II)\cdot BLM + O_2 + 2I^- + 4H^+ \rightarrow 2\ Fe(III)\cdot BLM + I_2 + 2H_2O$$

$$H_2O/Fe\cdot BLM = 1.0$$

Titration of activated bleomycin with the one-electron reducing agent iodide, and two-electron reductants such as NADH and thio NADH, also showed that activated bleomycin contained roughly two more oxidizing equivalents than Fe(III)·BLM (Burger *et al.*, 1985). Approximately 1.7 electron equivalents per activated bleomycin were found to be required when thio NADH was used for the reduction of activated bleomycin. When KI was the reductant, roughly 1.5 equivalents of $I^-$ were required for every activated bleomycin. Repetition of the same experiment by Natrajan *et al.* (1990b) afforded a value of 1.92 equivalents of $I^-$ oxidized per mole of activated BLM.

A variety of reductants such as NADPH-cytochrome P450 reductase (Kilkuskie *et al.*, 1984), glutathione (Caspary *et al.*, 1981), thiols (Antholine and Petering, 1979), Cu(II)·BLM + cysteine (Takahashi *et al.*, 1987) and liver microsomes (Ciriolo *et al.*, 1987) have been employed as sources of reducing equivalents for the Fe·BLM catalytic cycle. The last study is of special interest. Ciriolo *et al.* observed that inhibition of Fe(III)·BLM activation by excess DNA could be overcome by rat liver microsomes and NADPH. Cu(II)·BLM was observed not to be reduced by microsomes, although DNA degradation could be restored by the addition of Fe(II) to reaction mixtures containing Cu(II)·BLM. Finally, electrochemical reduction has also been employed for the generation of activated bleomycin (Melnyk *et al.*, 1981; Miyoshi *et al.*, 1988; Van Atta *et al.*, 1989). The electrochemical potential for Fe(III)·BLM was determined using cyclic voltammetry at a glassy carbon electrode (Van Atta *et al.*, 1989). The voltammogram corresponded to a one-electron Fe(III)/Fe(II) couple with $E_{1/2} = -0.08$ V vs a Ag/AgCl electrode. Activated BLM generated electrochemically by the electrolysis of solutions of Fe(III)·BLM at −0.22 V in the presence of dioxygen and various oligodeoxynucleotides produced lesions in the DNA substrates catalytically (Van Atta *et al.*, 1989). In comparison with reactions that employed chemical reductants such as ascorbate, the electrochemically driven DNA degradation reactions proceeded at a slower rate but afforded the same chemical products and very similar strand selectivity.

Scission of the O–O bond in Fe(III)·BLM-bound peroxide could, in principle, occur in two different ways (Figure 5.4). Homolytic scission of this bond would produce a ferryl species, which would be accompanied by

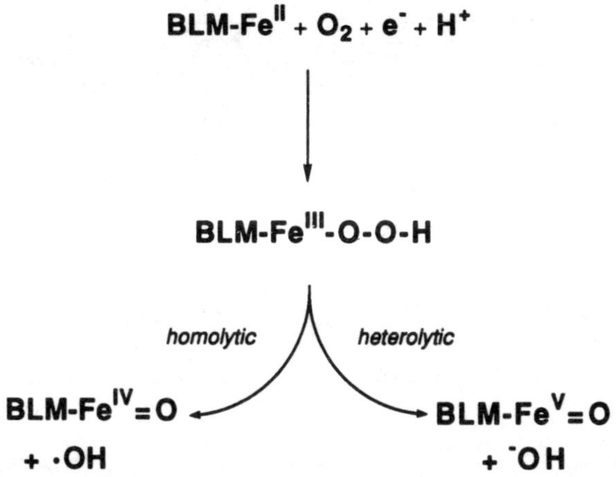


**Figure 5.4** Two possible modes of O–O bond scission in Fe(III)·BLM-bound peroxide

the formation of an equivalent of hydroxyl radical. Alternatively, O–O bond scission could proceed heterolytically to generate a perferryl species and hydroxide ion. The accumulated evidence suggests that activated bleomycin may best be represented as a perferryl species. For example, activated BLM mediated the oxygenation and oxidation of low molecular weight substrates (Murugesan *et al.*, 1982; Ehrenfeld *et al.*, 1984; Murugesan and Hecht, 1985; Heimbrook *et al.*, 1986) in reactions that are strongly reminiscent of transformations effected by activated cytochrome P450s (White and Coon, 1980; Guengerich and Macdonald, 1984). Although cytochrome P450, which is a mixed-function monooxygenase, is normally activated by NADPH-ferricytochrome oxidoreductase, oxygen transfer reagents such as iodosobenzene as well as hydrogen peroxide can substitute as oxidants. The resulting activated cytochrome P450s catalyse a variety of transformations of small substrates, such as the hydroxylation of simple aliphatic and aromatic compounds, olefin epoxidation and N-demethylation reactions (White and Coon, 1980; Guengerich and Macdonald, 1984). Analogously, the olefinic substrates *cis*-stilbene and styrene were converted to the respective epoxides upon treatment with either Fe(III)·BLM + iodosobenzene or hydrogen peroxide, or Fe(II)·BLM + $O_2$ in the presence of a reductant such as ascorbate. In analogy with cytochrome P450, only *cis*-stilbene was a good substrate, and the epoxidation reaction occurred with retention of olefin geometry to give mainly the *cis*-epoxide (Murugesan and Hecht, 1985). In addition to olefin epoxidation, bleomycin also catalysed the hydroxylation of aromatic substrates and the demethylation of *N*,*N*-dimethylaniline. The hydroxylation of *p*-deuteroanisole by Fe(III)·BLM yielded *p*-methoxyphenol; the reaction mechanism involved an NIH shift with concomitant loss of some of the

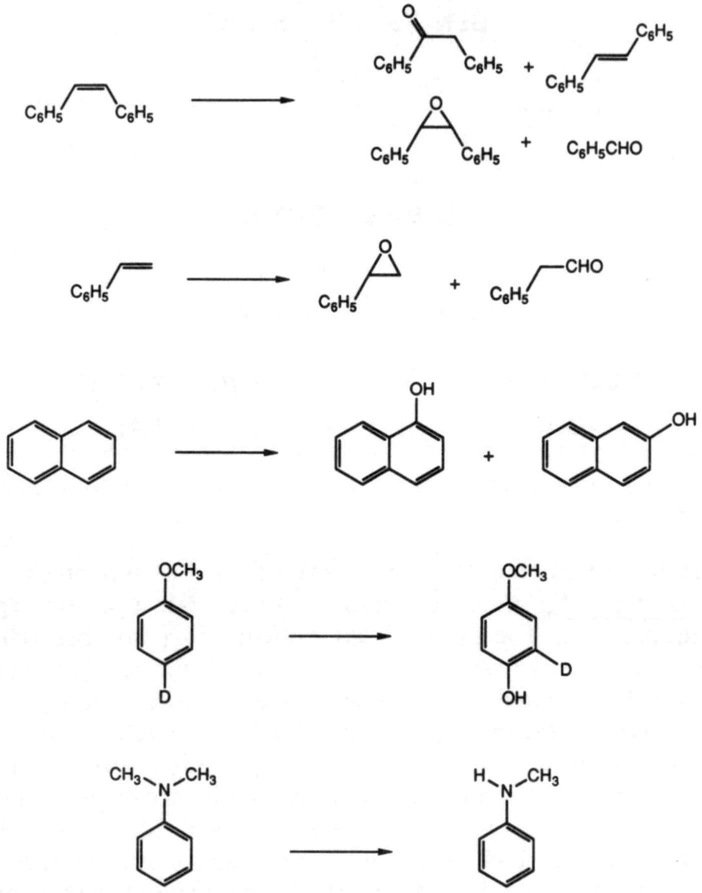


**Figure 5.5** Oxidation of small molecules by activated Fe•BLM

deuterium label (Murugesan and Hecht, 1985). Naphthalene was hydroxylated by bleomycin to afford both 1-naphthol and 2-naphthol (Natrajan *et al.*, 1990b); the reaction of *N*,*N*-dimethylaniline with activated bleomycin produced *N*-methylaniline (Murugesan and Hecht, 1985). Figure 5.5 summarizes the reactions described above.

The reactivity of the less oxidized ferryl BLM species [BLM•Fe(IV)=O] putatively generated from the reaction of Fe(III)•BLM and the fatty acid hydroperoxide 10-hydroperoxy-8,12-octadecadienoic acid has also been assessed (Padbury *et al.*, 1988; Natrajan *et al.*, 1990b; Natrajan and Hecht, 1991). The fatty acid hydroperoxide was obtained from the photooxygenation of linoleic acid (Labeque and Marnett, 1987); it can afford two different sets of products depending on the mode of scission of the peroxide O–O bond. Heterolytic scission would produce the alcohol 10-

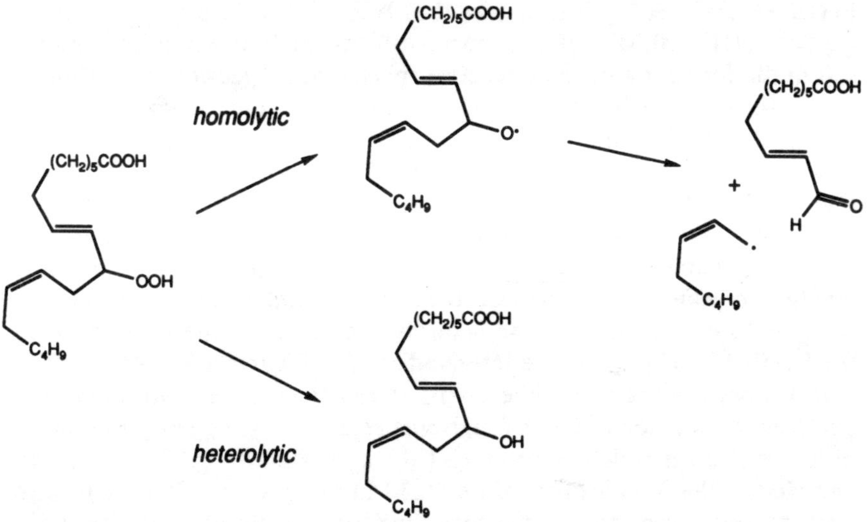


**Figure 5.6** Homolytic vs heterolytic O–O bond scission in 10-hydroperoxy-8,12-octadecadienoic acid

hydroxy-8,12-octadecadienoic acid. Homolytic scission would produce the corresponding alkoxyl radical, which would be expected to fragment to afford 10-oxo-8-decenoic acid (Figure 5.6). The appearance of either of these two products is thus diagnostic of the mechanism of scission of the peroxide bond. By the use of this fatty acid hydroperoxide, Labeque and Marnett (1987) determined that both haematin and ferrihaemoglobin catalysed homolytic scission of the O–O bond, while PGH synthase produced almost exclusively the heterolytic O–O bond scission product 10-hydroxy-8,12-octadecadienoic acid.

When Fe(III)·BLM was reacted with the fatty acid hydroperoxide, the fragmentation product 10-oxo-8-decenoic acid was the major product (Padbury *et al.*, 1988; Natrajan *et al.*, 1990b). The formation of this product via homolytic O–O bond cleavage was verified by trapping of the octenyl radical formed in parallel with 10-oxo-8-decenoic acid (cf. Figure 5.6) (Natrajan and Hecht, 1991). When solutions containing Fe(III)·BLM + 10-hydroperoxy-8,12-octadecadienoic acid were added to a supercoiled plasmid DNA, some nicking of the DNA was observed (Padbury *et al.*, 1988). However, no quantification was reported, such that the DNA strand scission observed could have been due to a small amount of putative BLM·Fe(V)=O, formed as a minor product by heterolytic cleavage of the peroxide O–O bond (Padbury *et al.*, 1988; Natrajan *et al.*, 1990b). In fact, addition of Fe(III)·BLM + 10-hydroperoxy-8,12-octadecadienoic acid to solutions of either of two oligonucleotides produced no detectable damage, in spite of the fact that both were degraded by Fe(II)·BLM + $O_2$ and by

Fe(III)·BLM + $H_2O_2$ (Natrajan *et al.*, 1990b). The conclusion that admixture of Fe(III)·BLM + 10-hydroperoxy-8,12-octadecadienoic acid did result in the formation of some reactive species was suggested by the ability of this reagent to effect the epoxidation of styrene and the demethylation of *N*,*N*-dimethylaniline.

The most reasonable interpretation of the foregoing experiments is that Fe(III)·BLM + 10-hydroperoxy-8,12-octadecadienoic acid produced BLM·Fe(IV)=O, a species incapable of effecting DNA degradation. The DNA degradation obtained by the use of Fe(II)·BLM + $H_2O_2$ must then be due to some other species; the most logical candidate would be BLM·Fe(V)=O, resulting from heterolytic cleavage of the O–O bond in the Fe·BLM-bound peroxide intermediate (cf. Figures 5.3, 5.4).

It has been noted that while Fe·BLM can function catalytically in the oxidative destruction of DNA (Sugiyama *et al.*, 1986) and the oxygenation of low molecular weight substances (Murugesan *et al.*, 1982; Murugesan and Hecht, 1985), activation of Fe·BLM in the absence of any substrate leads to loss of a portion of the functional BLM molecules (Takita *et al.*, 1978a; Burger *et al.*, 1979b, 1981; Van Atta *et al.*, 1989). For example, in the absence of DNA it was observed that following repetitive activation of BLM by the sequential addition in ten equal portions of 10 equivalents of Fe(II) to a solution of BLM the resulting bleomycin retained only 6% of its original activity (Nakamura and Peisach, 1988). In parallel with the loss of DNA cleaving activity, it was also observed that the fluorescence of the bithiazole ring system was altered. This modified BLM still bound Cu(II) and when treated with Fe(II) + $O_2$ produced a species whose EPR spectrum was identical with that of activated Fe·bleomycin. From these observations, it was suggested that the loss of DNA cleaving activity following several cycles of BLM activation resulted from the chemical modification of part of the bleomycin molecule necessary for DNA binding (Nakamura and Peisach, 1988). Owa *et al.* (1990) modelled the autoxidation of Fe·BLM by studying the oxidation of a molecule designed to mimic the metal binding domain of bleomycin. In the presence of Fe(II) + $O_2$ or Fe(III) + $H_2O_2$, this molecule fragmented by excision of the β-aminoalanineamide moiety (Figure 5.7). Since the β-aminoalanineamide moiety provides two ligands for metal ion complexation, loss of this fragment yielded a product which could not activate oxygen. This result was consistent with an earlier proposal for the mechanism of inactivation of bleomycin (Takita *et al.*, 1978a) which involved an N-dealkylation reaction analogous to the BLM-catalysed N-demethylation of *N*,*N*-dimethylaniline. These results are clearly quite different from the findings of Nakamura and Peisach (1988). Although not mutually exclusive, the possible existence of two pathways for BLM self-inactivation complicates the identification of the process primarily responsible for the inactivation of Fe·BLM. Further

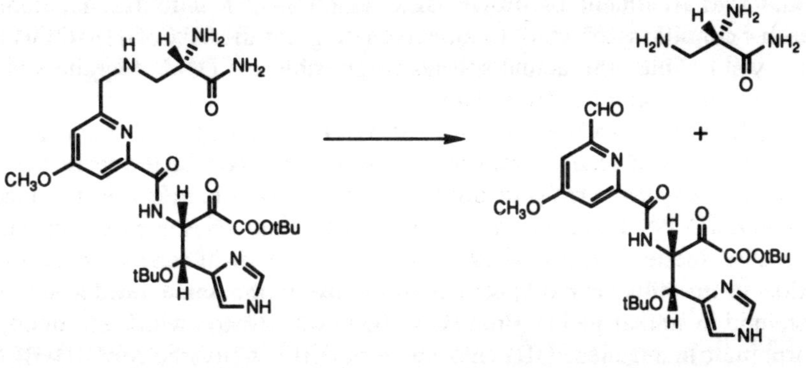


**Figure 5.7** Autoxidation of a BLM model compound in the presence of Fe(III) + $H_2O_2$ or Fe(II) + $O_2$

experimentation is needed to clarify this issue, since it is central to the design of robust BLMs capable of resisting self-inactivation.

## 3 Other Metallobleomycins

A number of other metallobleomycins, including Co·BLM (Chang and Meares, 1982, 1984), Cu·BLM (Ehrenfeld *et al.*, 1987), Mn·BLM (Burger *et al.*,1984; Ehrenfeld *et al.*, 1984; Suzuki *et al.*, 1985), VO·BLM (Kuwahara *et al.*, 1985) and Ru·BLM (Subramanian and Meares, 1985), have been shown to effect DNA strand scission. In the presence of light Co(III)·BLM caused DNA strand scission (Chang and Meares, 1982, 1984). The addition of Co(III) to aerobic solutions of bleomycin produced brown and green complexes, which were found to have water and peroxide as ligands, respectively (Chang *et al.*, 1983). Both the brown and green complexes have been reported to mediate photoinduced cleavage of DNA, but the green complex was found to be more reactive. A detailed analysis of the mechanism of DNA cleavage by Co(III)·BLM was carried out by Saito *et al.* (1989). It was shown that green Co(III)·BLM produced alkali-labile lesions in DNA oligonucleotide substrates, a transformation believed to involve hydroxylation of the C-4′ position in deoxyribose (Sugiyama *et al.*, 1985b, 1988). In parallel with this oxidation, the metallobleomycin was converted to brown Co(III)·BLM. It was suggested that photochemical activation of green BLM·Co(III)–OOH may involve scission of the O–O bond, which could lead to the observed hydroxylation of DNA and brown BLM·Co(III)–$OH_2$. Consistent with the thesis that the activity of brown Co(III)·BLM may be due to photoreduction of the Co(III)·BLM to Co(II)·BLM, followed by reoxidation with $O_2$, it was found that treatment of brown BLM·Co(III)–$OH_2$ with $NaBH_4$ under aerobic conditions effected its conversion to green BLM·Co(III)–OOH in 20% yield. Thus, the actual species responsible for DNA degradation is probably the green Co(III)·BLM.

Mn·BLM has also been shown to effect DNA cleavage and the oxidative transformation of small molecules following activation in the presence of oxygen or oxygen surrogates (Burger *et al.*, 1984; Ehrenfeld *et al.*, 1984; Suzuki *et al.*, 1985). Admixture of Mn(III)·BLM and iodosobenzene afforded a reactive species that promoted the conversion of *cis*-stilbene to *cis*-stilbene oxide, *trans*-stilbene oxide and deoxybenzoin; the same products were obtained in similar yields when *cis*-stilbene was treated with (tetraphenylporphinato)manganese (III) chloride + $C_6H_5IO$. Activated Mn(III)·BLM also effected the oxidation of styrene, cyclohexene and norbornene.

Also reported by Ehrenfeld *et al.* (1984) was the relaxation of plasmid DNA in the presence of Mn(II)·BLM + $O_2$. Burger *et al.* (1984) carried out a related study in which they observed DNA relaxation by Mn(II)·BLM only in the presence of $H_2O_2$, and found that Mn·BLM activated in this fashion would effect DNA cleavage only 1–3% as efficiently as Fe(III)·BLM + $H_2O_2$. The same type of study was also carried out by Suzuki *et al.* (1985), who found that Mn(II)·BLM could be activated for DNA relaxation in the presence of $O_2$ + 2-mercaptoethanol, $H_2O_2$ or $O_2$ + light. Reinvestigation of the conditions optimal for DNA plasmid relaxation has now established that efficient conversion can be achieved by Mn(II)·BLM + $O_2$ + Na ascorbate. Mn(II)·BLM activated by admixture with $H_2O_2$ was much less efficient in effecting the relaxation of DNA, as was Mn(II)·BLM activated in the presence of $O_2$ + 2-mercaptoethanol or $O_2$ alone (Long, E. and Hecht, S. M., unpublished observations).

Suzuki *et al.* (1985) also studied the cleavage of a [$^{32}$P] end-labelled DNA duplex by Mn·BLM. Activation of Mn(II)·BLM in the presence of $O_2$ + 2-mercaptoethanol afforded a species that produced substantial DNA damage, with lesions at positions quite similar to those observed with Fe(II)·BLM + $O_2$ + 2-mercaptoethanol. The same DNA duplex was also cleaved by Mn(II)·BLM that had been activated in the presence of $H_2O_2$, or $O_2$ + light, although less cleavage was observed in these cases.

Bleomycin forms 1:1 complexes with both Cu(I) and Cu(II). Both of these metals have an affinity for bleomycin greater than that of Fe(II) (Oppenheimer *et al.*, 1981). Cu(I)·BLM is redox-active and binds to DNA with approximately the same affinity as Fe(II)·BLM. In the presence of reducing agents such as dithiothreitol, Cu(I)·BLM + $O_2$ effected DNA degradation; for some substrate DNA duplexes the sequence selectivity of DNA cleavage differed significantly for Cu(I)·BLM and Fe(II)·BLM. DNA sites shared in common were also cleaved to a different extent in some cases. It was also shown to be possible to activate Cu(II)·BLM for

DNA cleavage in the presence of a reductant and $O_2$, provided that sufficient time was permitted for reduction of this metallobleomycin (Ehrenfeld *et al.*, 1985, 1987). Differences in the specificity of cleavage probably reflect different binding affinities of these two metalloBLMs at different DNA sequences. Although the nature of the oxidizing species responsible for these transformations is unclear, it seems likely that both Fe•BLM and Cu•BLM could serve as DNA-damaging agents *in vivo*.

In common with Fe•BLM, Cu•BLM was also found to mediate the oxidation of *cis*-stilbene in the presence of iodosobenzene. The products formed from Cu(II)•BLM + $C_6H_5IO$ included *cis*-stilbene oxide, *trans*-stilbene oxide and benzaldehyde. Interestingly, no deoxybenzoin was observed, although this compound was formed from *cis*-stilbene in the presence of Fe(III)•BLM + $C_6H_5IO$. Further, while admixture of excess Fe(II) to a reaction mixture containing 1:1 Fe(II) and BLM simply resulted in decreased product formation, excess Cu(II) substantially enhanced the amounts of products formed from *cis*-stilbene by Cu(II)•BLM + $C_6H_5IO$ (Murugesan *et al.*, 1982; Ehrenfeld *et al.*, 1987).

## 4 Interaction of Bleomycin with DNA

The degradation of DNA by bleomycin occurs at specific sequences in DNA (Asakura *et al.*, 1978; D'Andrea and Haseltine, 1978; Takeshita *et al.*, 1978). Double-stranded DNA is a better substrate than single-stranded DNA, the latter of which is cleaved poorly if at all (Shirakawa *et al.*, 1971; Takeshita *et al.*, 1978; Kross *et al.*, 1982a). Asakura *et al.* (1978) studied the degradation of SV40 DNA by bleomycin $B_2$ in the presence of oligonucleotides of varying sequence to determine whether an oligonucleotide of some particular sequence afforded more protection to SV40 DNA from BLM-induced degradation. Among the oligonucleotides studied, BLM $B_2$ bound to every purine–pyrimidine alternating polymer; poly(dG–dC) afforded the maximal level of protection to SV40 DNA. It was also observed by other workers that DNA cleavage by bleomycin occurs preferentially at 5′-GC and 5′-GT sequences, and that both single- and double-stranded lesions are produced (D'Andrea *et al.*, 1978; Takeshita *et al.*, 1978; Mirabelli *et al.*, 1982).

The propensity for bleomycin to cleave at specific sites on DNA prompted a number of studies to determine the structural origin of this specificity within the BLM molecule, and, in general, to determine the manner in which bleomycin interacts with DNA. Early studies indicated that the bithiazole ring system was necessary for DNA binding and sequence recognition, and that there may exist multiple modes of association of this moiety with DNA. More recent studies have indicated that, in addition, the N-terminal region of the BLM molecule may also participate in DNA sequence recognition and, in fact, may well be the principal determinant of this recognition. Finally, nucleotide structure has also been observed to affect substantially DNA binding and cleavage by bleomycin. The results of several key experiments and their conclusions are reviewed below.

The interaction of the bithiazole ring system with bleomycin has been examined by a number of techniques. When bleomycin binds to DNA, the fluorescence of the bithiazole moiety is quenched and, in the $^1$H-NMR spectrum, the resonances of the non-exchangeable bithiazole protons are broadened and shifted upfield (Chien *et al.*, 1977). There are at least two types of DNA-induced fluorescence quenching of BLM, one which is sensitive to the ionic strength of the medium and another which is not (Huang *et al.*, 1980). Thus, the interaction of bleomycin with DNA involves a hydrophobic component and an electrostatic component, the latter of which is affected by the ionic strength of the medium.

The magnitude of the association constant between bleomycin, and bleomycin model compounds, and DNA has been determined by fluorescence spectroscopy as well as equilibrium dialysis measurements. Kasai *et al.* (1978) measured the binding of BLM $A_2$ and its Cu(II) complex to calf thymus DNA, using fluorescence spectroscopy. The equilibrium constant, $K_b$, for the binding of metal-free BLM $A_2$ to DNA was found to be $3.0 \times 10^5$ $\mathrm{M}^{-1}$ and was determined as $2.3 \times 10^5$ $\mathrm{M}^{-1}$ in the presence of Cu(II). Earlier, Chien *et al.* (1977) had determined an apparent binding constant of $1.2 \times 10^5$ $\mathrm{M}^{-1}$ for bleomycin in 2.5 mM Tris buffer, using the same technique. They had also observed that the binding site size of bleomycin was 4–5 base pairs. Povirk *et al.* (1981) observed that the DNA binding constant of Cu(II)·BLM in 20 mM Tris, pH 8, was 20 000 $\mathrm{M}^{-1}$ but decreased to 1000 $\mathrm{M}^{-1}$ at higher ionic strength (50 mM NaCl, 25 mM Tris). Fluorescence quenching was also used to measure the binding constant of BLM and Fe(III)·BLM to DNA in 1 mM Tris, pH 8.0 (Caspary *et al.*, 1982). The results of this study indicated that each BLM molecule bound to ~5 DNA base pairs in both the presence and absence of Fe(III). Moreover, the magnitudes of the association constants were also similar, being $4.1 \times 10^5$ $\mathrm{M}^{-1}$ for BLM and $9.6 \times 10^5$ $\mathrm{M}^{-1}$ for Fe(III)·BLM. In addition to fluorescence spectroscopy, equilibrium dialysis has also been used to measure the binding constant of bleomycin to DNA (Povirk *et al.*, 1979; Roy *et al.*, 1981). The magnitude of the DNA association constant of the BLM 'fragment' tripeptide S (**3**, Figure 5.8), was found to be similar to that of bleomycin, suggesting that the source of affinity of bleomycin for DNA resided solely within the C-terminal region of the peptide (Povirk *et al.*, 1979). Equilibrium dialysis measurements (Roy *et al.*, 1981) of the binding of BLM and Cu(II)·BLM revealed the presence of two types of binding sites on DNA. The higher-affinity sites bound metal-free BLM with $K$ = $6.8 \times 10^5$ $\mathrm{M}^{-1}$ and Cu(II)·BLM with $K = 4.4 \times 10^5$ $\mathrm{M}^{-1}$. The binding site

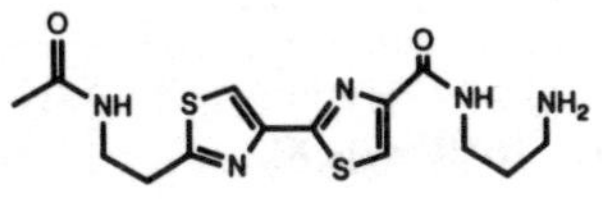

1

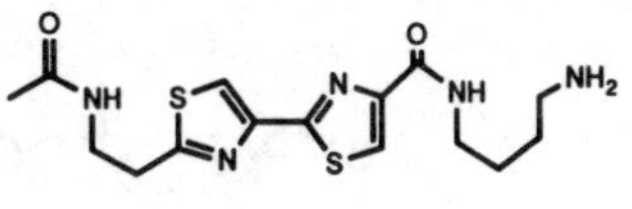

2

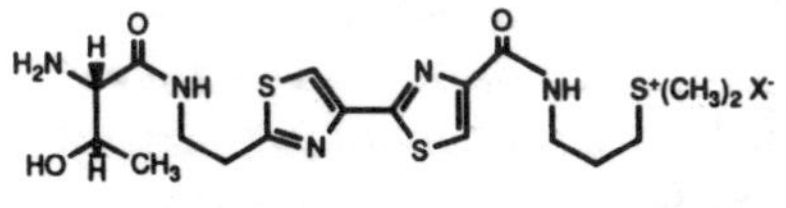

3

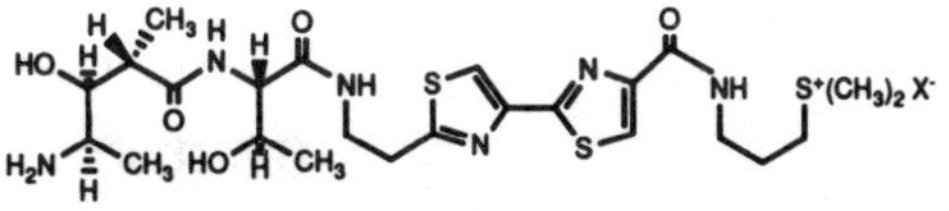

4

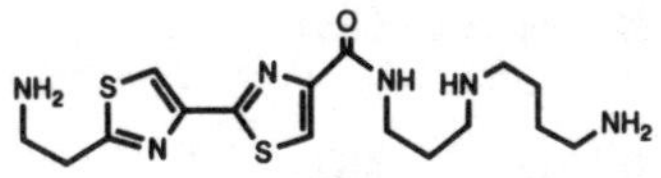

5

**Figure 5.8** Bithiazole derivatives employed to inhibit BLM-mediated DNA degradation

size for BLM was approximately 3.7 DNA base pairs; for Cu(II)•BLM the comparable value was 2.8 base pairs. Interestingly, these values were similar to those determined for Co(III)•BLM by Kuwahara and Sugiura (1988), who used DNase I footprinting to determine the number of nucleotides protected on DNA. The existence of two types of binding sites on DNA for a series of synthetic bithiazoles was also observed by Sakai *et al.* (1982). The binding constants of the compounds in Figure 5.9 were measured under two sets of conditions: 10 mM Na phosphate, 25 mM NaCl and 10 mM Na phosphate, 100 mM NaCl. Two sets of binding sites were identified, one of high affinity but low occupancy and another of low affinity and high occupancy. At low salt, the affinity of the high-occupancy site increased twofold without change in the number of binding sites, while the higher-affinity site experienced a threefold decrease in the binding constant. Occupancy numbers of roughly 19–21 base pairs per analogue molecule were observed for the high-affinity sites and 2–3 base pairs per analogue molecule for the low-affinity sites. The low-affinity sites presumably reflect ionic interactions between the analogues in Figure 5.9 and DNA, while the high-affinity sites may correspond to a hydrophobic mode of association between these analogues and DNA. The latter mode may be destabilized by loss of electrostatic shielding at low salt concentrations and consequent alteration of DNA structure (Sakai *et al.*, 1982). The existence of an ionic component in the association of bleomycin with DNA has also

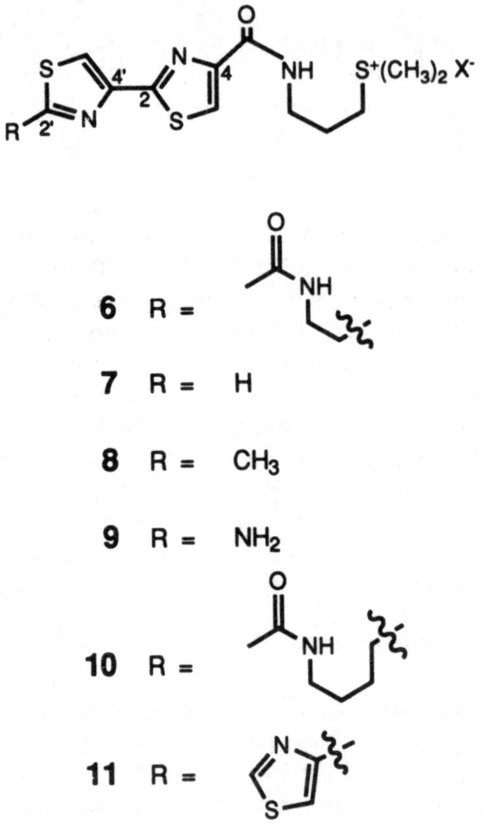


**Figure 5.9** Bithiazole derivatives used to study DNA binding and unwinding

been observed during $^1$H-NMR studies of the interaction of bleomycin $A_2$ with poly(dA–dT) (Booth *et al.*, 1983). From an analysis of the effects of temperature, pH and ionic strength of the medium, it was deduced that the cationic terminus of BLM $A_2$ interacted with the negatively charged backbone in DNA. The chemical shift of the methyl groups on the dimethylsulfonium moiety exhibited a maximal shift (0.04 ppm) at low salt and a minimal shift at high salt. Similarly, the magnitude of these chemical shifts decreased with increasing pH, indicating an ionic component to BLM binding.

The hydrophobic component of the affinity of BLM for DNA is presumably derived from the interaction of the bithiazole ring system with DNA. A number of studies with synthetic bithiazoles as well as with bleomycin suggest that the bithiazole moiety may associate with DNA via an intercalative mechanism. For instance, upon binding to DNA the fluorescence of this moiety is quenched (Chien *et al.*, 1977; Povirk *et al.*, 1981). Povirk *et al.* (1981) also observed the unwinding of supercoiled DNA upon binding of bleomycin, as well as the BLM fragment tripeptide S. Unwinding of supercoiled DNA without DNA strand scission was observed at pH 5.5 with bleomycin. Tripeptide S was effective in unwinding and then recoiling DNA at pH 5.5, as well as at pH 8.0. An unwinding angle of 12° was measured. Moreover, both BLM and tripeptide S also induced an increase in DNA length by 3.1 Å, presumably by intercalation of the planar bithiazole moiety between the bases in DNA. It has also been observed that the binding of $0.5 \times 10^{-4}$ M bleomycin to $10^{-4}$ M calf thymus DNA increased the $T_m$ of DNA by 4 °C (Hénichart *et al.*, 1985).

The unwinding of supercoiled DNA by synthetic bithiazoles (Fisher *et al.*, 1985) and bleomycin (Levy and Hecht, 1988) has also been studied, using two-dimensional agarose gel electrophoresis. All the synthetic bithiazoles listed in Figure 5.9 were effective in DNA unwinding, although bithiazoles with 2′-aromatic substituents were more effective in unwinding supercoiled DNA (Fisher *et al.*, 1985). Levy and Hecht (1988) reported an interesting new finding from their study of DNA unwinding by BLM $A_2$ and BLM demethyl $A_2$ in the presence and absence of $Cu^{2+}$ ions. The unwinding of supercoiled DNA by BLM $A_2$ was found to be greatly facilitated in the presence of $Cu^{2+}$ ions under conditions in which the metal ion alone had no effect. The addition of an equivalent of $Cu^{2+}$ ions reduced the concentration of BLM $A_2$ required for DNA unwinding by a factor of $10^2$. Thus, the minimum concentrations of BLM $A_2$ and Cu(II)·BLM $A_2$ required for the removal of one negative supercoil from ϕX174 plasmid DNA were $10^{-4}$ M and $10^{-6}$ M, respectively. Clearly, the presence of a positive charge within the N-terminal region of the BLM molecule enhanced DNA unwinding. The requirement of a positive charge at the C-terminus of the BLM molecule for effective DNA unwinding was also demonstrated by noting that Cu(II)·demethyl BLM $A_2$ was much less effective as a DNA unwinding agent. Thus, $10^{-4}$ M Cu(II)·demethyl BLM $A_2$ was able to remove only 1–2 negative supercoils from DNA. These workers also verified an earlier report of DNA unwinding by a steroidal diamine, thus affirming that DNA unwinding *per se* cannot be used to establish an intercalative associative mechanism. It is possible that DNA unwinding by bleomycin is largely ionic in nature.

In addition to the foregoing studies, a number of $^1$H-NMR studies have also been performed to define the nature of BLM–DNA interaction (Glickson *et al.*, 1981; Lin and Grollman, 1981; Sakai *et al.*, 1982; Booth *et al.*, 1983). The influence of a metal ion on BLM binding was studied by Glickson *et al.* (1981), who found that the general characteristics of the binary complexes Zn(II)·BLM $A_2$ and BLM $A_2$–DNA are retained in the Zn(II)·BLM–DNA ternary complex. Spectral perturbations resulting from the binding of BLM $A_2$ and Zn(II)·BLM $A_2$ to poly(dA–dT) were found to be remarkably similar and seemed to indicate that the metal ion had a minimal effect on the interaction of bleomycin with DNA, Similarly, perturbations at the N-terminus of the bleomycin molecule caused by Zn binding were not enhanced by nucleic acid binding. The two regions of the bleomycin molecule thus appeared to act independently, with only the bithiazole moiety contributing to the affinity of BLM for DNA.

In 1980 Chen *et al.* reported a $^1$H-NMR investigation of the binding of BLM $A_2$ to poly(dA–dT) in 100 mM sodium phosphate buffer. The chemical shifts of the bithiazole protons were measured as a function of temperature and were observed to exhibit a bell-shaped profile with a maximum shift of 0.27 ppm at 50 °C, just below the melting temperature of the DNA. At lower temperatures, the chemical shifts of the bithiazole protons were minimally perturbed. Similar observations were made by Lin and Grollman (1981) and Sakai *et al.* (1982). When the binding of the bithiazole analogue, *N*-(3-aminopropyl)-2′-(2-acetamidoethyl)-2,4′-bithiazole-4-carboxamide hydrochloride, to poly(dG–dC) was studied, it was observed that the maximum shift of 0.4 ppm of the bithiazole protons occurred at a dinucleotide/bithiazole analogue ratio of 2. On the basis of molecular modelling, Lin and Grollman (1981) proposed a binding mode of the bithiazole ring system in which the bithiazole rings intercalated into the duplex from the major groove and stacked preferentially on the purines. This preferential stacking of the thiazole rings on the purines was supported by $^1$H-NMR data which revealed significant chemical shift perturbations for the purine (0.053–0.123 ppm) rather than pyrimidine ($<$0.02 ppm) protons. This model also posited that, in the intercalated complex formed, the bithiazole protons extended into the minor groove, while the positively charged C-substituent bound to the negatively charged DNA backbone. Sakai *et al.* (1982) investigated the binding of the synthetic bithiazoles and thiazoles listed in Figure 5.9 to poly(dA–dT) and obtained some interesting results. In these synthetic bithiazoles, the presence of 2′-aliphatic substituents, of the type found in BLM, inhibited the ability of the compound to intercalate into DNA. Thus, compounds **6** and **10** in Figure 5.9 exhibited the smallest shifts of the bithiazole proton resonances upon binding to DNA, and also bound DNA with the lowest association constants. Removal of the amide-containing side-chain, as in compounds **7–9**, increased the ability of the bithiazole to intercalate into DNA. The magnitude of the shift of the bithiazole protons was maximal when a 2′-aromatic substituent was used. Structurally related monothiazoles did not bind to DNA.

A novel approach to studying the interaction of the bithiazole ring system with DNA was reported by Hénichart *et al.* (1985). By covalently linking the nitroxide 2,2′,6,6′-tetramethylpiperidine-*N*-oxyl to either of the two thiazole rings (Figure 5.10), the environment around the bithiazole ring system could be assessed when it was bound to DNA. Normally, the nitroxide exhibits a sharp three-line pattern in the EPR spectrum; this

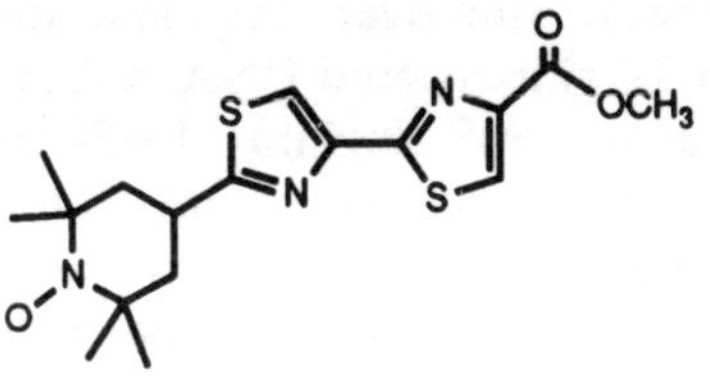

12

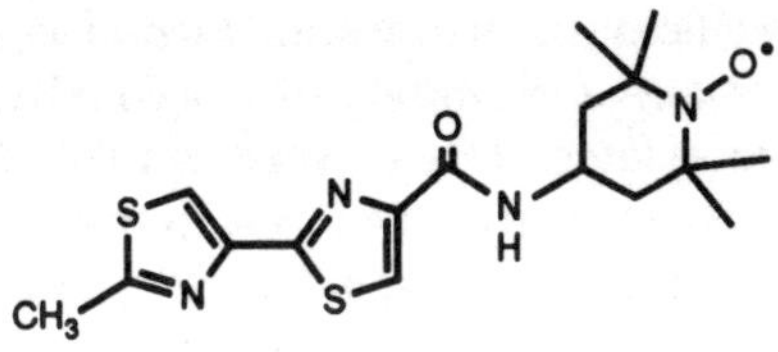

13

**Figure 5.10** Bithiazole derivatives with covalently linked nitroxides used to study DNA binding

pattern broadens and becomes asymmetric upon immobilization. Among the several compounds studied, only **12** (Figure 5.10) exhibited a modified EPR spectrum upon binding to DNA, suggesting that only one bithiazole ring was intercalated.

Regardless of the mode of association, it has been observed that various synthetic bithiazoles (Fisher *et al.*, 1985) as well as BLM 'fragments' containing the bithiazole ring system (Kross *et al.*, 1982b) are capable of inhibiting DNA binding and cleavage by bleomycin. Kross *et al.* (1982b) studied the inhibitory effect of the compounds listed in Figure 5.8 on BLM-induced DNA degradation by monitoring the efficiency of [$^3$H]-thymine release from PM-2 DNA and by evaluating the efficiency of cleavage of a 3′-[$^{32}$P] end-labelled DNA duplex. Inhibition of DNA cleavage was facilitated by the presence of a positive charge on the N-terminus as well as the C-terminus; the appropriate placement of positive charges was also shown to be important (**2** was a better inhibitor than **1**; **3** was better than **4**). Compound **5**, which has three positive charges at neutral pH, was the most effective inhibitor. Fisher *et al.* (1985) also determined that synthetic bithiazoles with 2′-aromatic substituents were more effective inhibitors of BLM-induced DNA degradation than analogues with 2′-aliphatic substituents. All of the synthetic bithiazoles inhibited DNA cleavage by bleomycin proportionately at each cleavage site and did not alter the sequence specificity of DNA cleavage by bleomycin.

While the results described above would seem to suggest that bleomycin can bind to DNA via an intercalative mechanism, evidence has also been presented which suggests that association may occur via binding in the minor groove, as opposed to intercalation (Kuwahara and Sugiura, 1988). Sugiura and Suzuki (1982) observed that the binding of distamycin to DNA altered the DNA cleavage specificity of BLM by suppressing cleavage at 5′-GT and 5′-GA sequences; in contrast, the intercalator ethidium bromide diminished cleavage, but did not alter the cleavage specificity. Modification of N-7 of guanosine in the major groove of DNA with either aflatoxin $B_1$ or dimethyl sulfate also did not alter the DNA cleavage specificity of bleomycin (Suzuki *et al.*, 1983). Moreover, the association constants for binding of Co(III)•BLM to T4 glucosylated DNA and non-glucosylated DNA were similar (Kuwahara and Sugiura, 1988) and Fe•BLM cleaved T4 DNA in the same fashion as non-glucosylated B-form DNA (Hertzberg *et al.*, 1988). Thus, DNA recognition by bleomycin does not seem to occur from the major groove in DNA. On the other hand, modification of the 2-amino group in guanosine residues in DNA by covalent attachment of anthramycin completely suppressed DNA cleavage by BLM at 5′-GC, 5′-GT as well as 5′-GA sequences in DNA (Kuwahara and Sugiura, 1988). From these results, it appeared that the guanine base in these sequences was the central recognition point for bleomycin. Binding of bleomycin to DNA was suggested to occur in the minor groove by fitting the concave side of the bithiazole ring system in the groove, and forming hydrogen bonds between the guanosine 2-amino group and the thiazole nitrogens (Kuwahara and Sugiura, 1988). Some interesting findings concerning the DNA cleavage specificities of BLMs containing altered bithiazole ring systems have been reported (Morii *et al.*, 1986, 1987). It was observed that light irradiation of Cu(II)•BLM isomerized the 2,4′-bithiazole moiety to a 4,4′-bithiazole ring system as well as to a thiazolylisothiazole-containing bleomycin analogue. The DNA cleavage specificities of these compounds were quite similar to those of BLM, with some minor differences. It is interesting to note that these two bleomycins incorporate a common N–C–C–N structural motif which was suggested to be a possible recognition feature in BLM for specific DNA sequences.

The DNA cleavage specificities of a number of BLM congeners have been evaluated. Phleomycin and tallysomycin are structurally related to bleomycin (Takita *et al.*, 1972, 1978b; Konishi *et al.*, 1977) (Figure 5.11). Phleomycin contains a thiazolinylthiazole moiety and was found not to intercalate into DNA (Povirk *et al.*, 1981). Phleomycin exhibited the same DNA cleavage specificity as bleomycin (Takeshita *et al.*, 1981; Kross *et al.*, 1982a), although it had a somewhat lesser preference for purines (Takeshita *et al.*, 1981). Tallysomycin contains an additional sugar moiety in proximity to the bithiazole ring system. Tallysomycin was found to be less effective as a DNA-damaging agent than bleomycin and exhibited altered specificity of DNA cleavage. Among the various BLM group antibiotics studied to date, tallysomycin had the greatest preference for 5′-GT sites. The substituent on the C-terminus of the BLM molecule does not appear to be a determinant of sequence recognition by bleomycin. The BLM congeners BLM $A_2$, BLM $B_2$, peplomycin, BLM $B_1'$ and BLM $B_6$ all exhibited similar DNA cleavage specificities (Takeshita *et al.*, 1981; Kross *et al.*, 1982a). The electrostatic interaction of the charged side-chain with DNA is likely to be non-specific and probably contributes only to increasing the affinity of BLM for DNA.

The DNA cleavage specificities of a number of BLM analogues formed

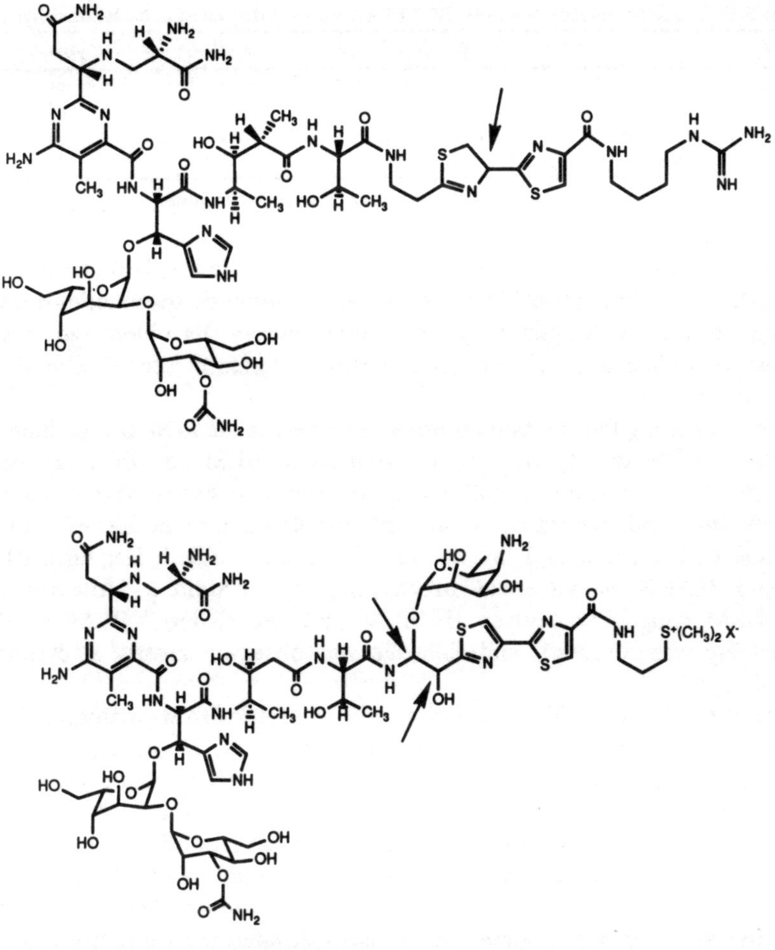

**Figure 5.11** Structures of phleomycin $D_1$ (top) and tallysomycin $S_2B$ (bottom). The arrows denote positions of unknown absolute configuration

as chemical hydrolysis products of bleomycin (Muraoka *et al.*, 1981) have been evaluated (Ehrenfeld, 1986). Deglyco BLM $A_2$, deglyco BLM demethyl $A_2$ and BLM demethyl $A_2$ exhibited the same specificity of cleavage as BLM $A_2$ itself. However, deglyco BLM $A_2$ was only about 50% as active as BLM $A_2$, while deglyco BLM demethyl $A_2$ itself was less active still. Modification of the metal binding domain by hydrolysis of some of the terminal amides decreased the activities of the resulting bleomycins but did not alter DNA cleavage specificity significantly. Epi BLM $A_2$ cleaved DNA less efficiently than BLM $A_2$ at 5′-AA, 5′-AC, 5′-AT, 5′-TC and 5′-TT sequences (Shipley and Hecht, 1988). Decarbamoyl BLM $A_2$ did not cleave DNA at 5′-AA, 5′-AC or 5′-CT sites, and exhibited diminished

**Table 5.1** Sequence selectivity of d(CGCTTTAAAGCG) degradation by BLM congeners

| *BLM* | *Specificity (%)*[a] | *Cytidine*$_3$ | *Cytidine*$_{11}$ |
|---|---|---|---|
| FE(II)·BLM $A_2$ | 78 | 15 | 85 |
| FE(II)·BLM $B_2$ | 75 | 17 | 83 |
| FE(II)·deglyco BLM $A_2$ | 98 | 79 | 21 |
| FE(II)·decarbamoyl BLM $A_2$ | 90 | 72 | 28 |

[a] Percentage of all degradation occurring at either cytidine$_3$ or cytidine$_{11}$.

cleavage at other sites. From all of the above results, it is interesting to note that most differences in DNA cleavage sequence specificities among the different BLM congeners occurred at secondary DNA cleavage sites of bleomycin. Cleavage at these sites seems particularly sensitive to BLM structure.

An interesting finding that addresses the role of the mannose carbamoyl group in BLM in directing the orientation of BLM on DNA has been reported by Sugiyama *et al.* (1986). In this study, the extent of the cleavage at cytidine$_3$ and cytidine$_{11}$ in the self-complementary dodecanucleotide 5′-CGCTTTAAAGCG-3′ by various BLMs was studied. For both BLM $A_2$ and BLM $B_2$ cleavage was directed mostly at cytidine$_{11}$. However, for the BLM congeners deglyco BLM $A_2$ and decarbamoyl BLM $A_2$ this specificity was reversed, and cleavage was observed mostly at cytidine$_3$ (Table 5.1).

Recent studies of DNA cleavage with synthetic BLM analogues have indicated that both the metal-binding domain of the bleomycin molecule and the bithiazole ring system were required for DNA sequence recognition. To test the effect of a single thiazole ring in BLM on DNA binding and sequence specificity, the analogues bleomycin monothiazoles A and B were prepared by total synthesis (Hamamichi *et al.*, 1992) (Figure 5.12). Both analogues were efficient catalysts for the oxidation of styrene and the hydroxylation of naphthalene; both also mediated the demethylation of *N,N*-dimethylaniline. However, these analogues cleaved DNA only at high concentrations and in a completely sequence-neutral fashion. Thus, oxygenation of low molecular weight substrates in reactions which do not require substrate binding are readily effected by monothiazole-containing BLMs. However, DNA degradation, which requires initial DNA binding by BLM, was diminished, owing to disruption of the bithiazole ring system, an essential DNA binding element.

To evaluate the role of the metal binding domain in BLM on DNA sequence recognition, a series of BLM analogues was prepared in which the metal binding domain was separated from the bithiazole ring system by spacers of increasing length (Carter *et al.*, 1990a; Figure 5.13). If all of these BLMs bound to DNA with the bithiazole moiety localized at a common site on DNA, analogues with spacers of increased length would enable access of the metal binding domain to sites increasingly more distant from

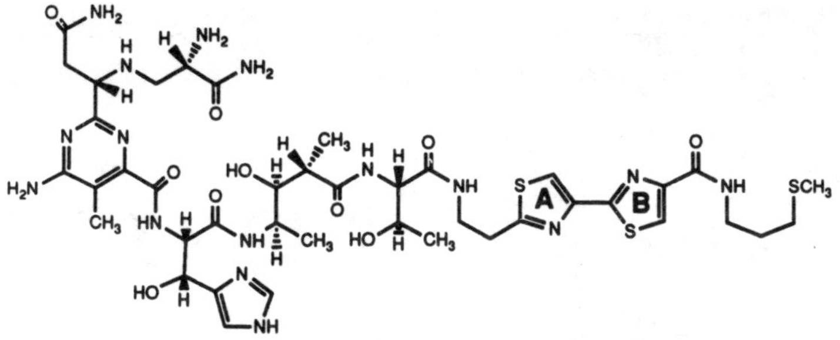

**deglycobleomycin demethyl $A_2$**

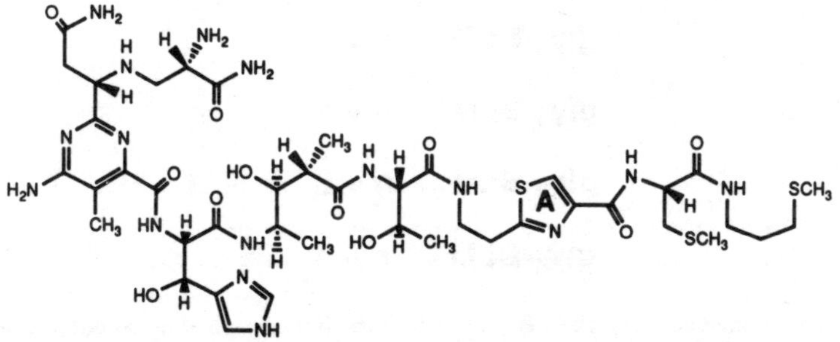

**bleomycin monothiazole A**

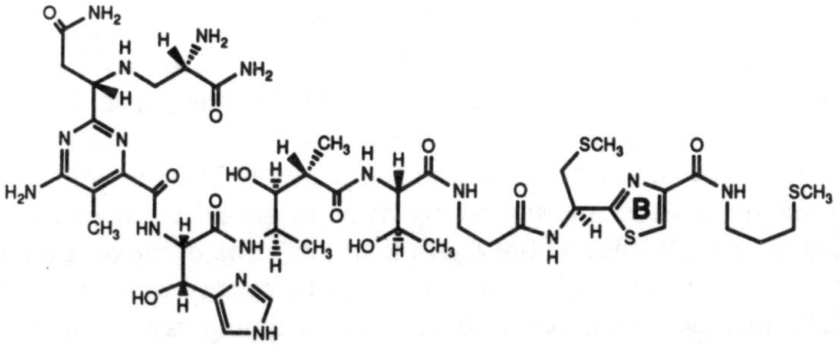

**bleomycin monothiazole B**

**Figure 5.12** Synthetic bleomycin analogues used to study the function of individual thiazole rings

the site of binding. If, on the other hand, the metal binding region of the BLM molecule were the primary determinant of DNA sequence specificity, then all analogues would bind to and cleave DNA at a common site. The latter was observed experimentally. These results strongly suggest that the N-terminal region of the BLM molecule also participates in DNA

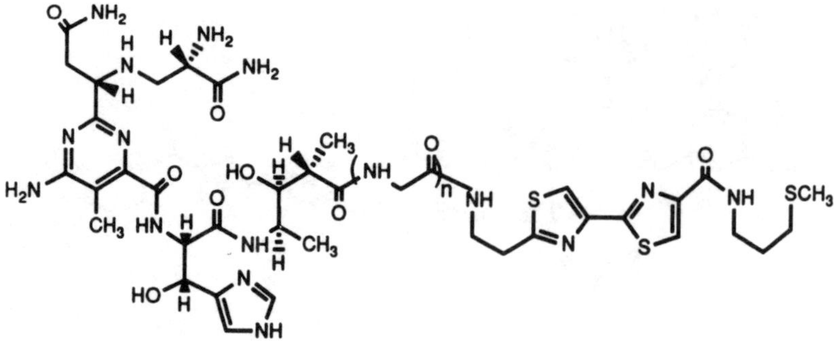

| | |
|---|---|
| **gly$_0$-BLM** | n = 0 |
| **gly$_1$-BLM** | n = 1 |
| **gly$_2$-BLM** | n = 2 |
| **gly$_4$-BLM** | n = 4 |

**Figure 5.13** Analogues of deglyco BLM in which the bithiazole and metal binding domains are separated by (oligo)glycine spacers of increasing length

sequence recognition, and is actually the dominant partner in this process.

Simplification of the structure of the metal binding region of BLM was undertaken by Ohno and co-workers (Otsuka *et al.*, 1981, 1986; Kittaka *et al.*, 1986, 1988; Ohno *et al.*, 1988). They found that the pyrimidine moiety in BLM could be replaced with a substituted pyridine moiety and that electron-releasing substituents *para* to the pyridine nitrogen enhanced the level of reactive species measurable by BPN trapping. Replacement of the carbohydrate moiety with a *t*-butyl group and deletion of the propionamide side-chain attached to the β-aminoalanineamide moiety were also studied. The BLM analogue most active in the BPN trapping assay (Figure 5.14) was assayed for its ability to degrade DNA. DNA cleavage by this analogue was found to be much less efficient than for BLM $A_2$ and the cleavage specificity of this analogue was not identical with that of BLM $A_2$ (Ohno *et al.*, 1988).

In addition to BLM structure, the structure of the polynucleotide substrate has also been found to influence strongly sequence recognition by BLM (Mascharak *et al.*, 1983; Hertzberg *et al.*, 1985, 1988; Gold *et al.*, 1988; Williams and Goldberg, 1988; Long *et al.*, 1990). The binding of *cis*-diamminedichloroplatinum, which is known to make intrastrand links in DNA by binding to N-7 of adjacent guanosine residues, alters the sequence specificity of DNA cleavage by bleomycin. The binding of plati-

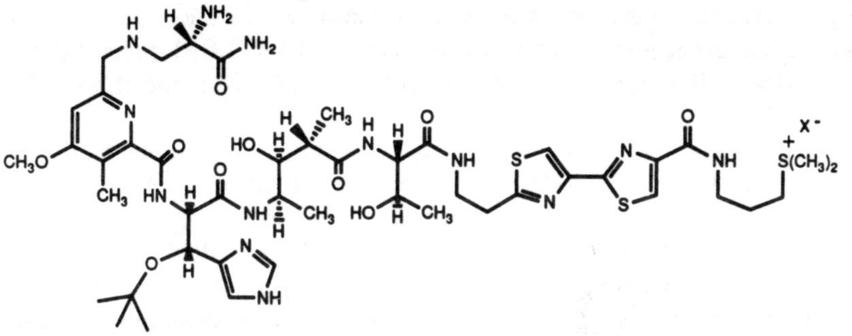


**Figure 5.14** Bleomycin analogue used to study oxygen activation and DNA degradation

num is believed to cause significant distortions in the DNA helix in the vicinity of binding. Bleomycin-induced cleavage adjacent to oligo(dG) regions is suppressed upon treatment of DNA with *cis*-platinum; instead, cleavage is directed elsewhere (Mascharak *et al.*, 1983). A detailed analysis of the effects of *cis*-platinum binding to a DNA oligonucleotide on BLM-induced cleavage has been performed by Gold *et al.* (1988), who used the oligomer 5′-CGCTTTAAAGCG-3′ annealed to its complement. The resulting unplatinated duplex contained a high-affinity bleomycin cleavage site at $G_{20}C_{21}$ in the complementary strand. Treatment of this duplex with 1 equivalent of *cis*-platinum platinated the oligomer at $G_{10}G_{11}$. Cleavage of the platinated substrate gave the same total amount of product, but much of the cleavage was redirected to the T residues, reflecting distortion of the helix.

DNA methylation has also been shown to affect BLM sequence recognition and cleavage by BLM (Hertzberg *et al.*, 1985, 1988; Long *et al.*, 1990). The effect of cytidine methylation on DNA cleavage by BLM was investigated by Hertzberg *et al.* (1985, 1988), who methylated a DNA substrate using restriction methylases. Bleomycin-mediated cleavage of a [$^{32}$P] end-labelled DNA fragment which had been methylated revealed that the sites of methylation experienced diminished cleavage. The extent of diminution varied with the BLM congener, being highest for BLM $A_2$ (78%) and lowest for BLM demethyl $A_2$ (13%) near a *Hha*I methylation site. Methylation of $N^6$ of adenine residues in the major groove of the recognition sequence TCGA with the restriction methylase *TaqI* also decreased the extent of BLM cleavage near the methylation site. The latter observation is surprising in one sense, since DNA binding by BLM is presumed not to occur from the major groove in DNA, and reinforces the conclusion that inhibition of DNA cleavage brought about by methylation probably results from an alteration of DNA conformation which is recognized by bleomycin. Cytidine methylation, for instance, is known to favour the B→Z transition in DNA by lowering the concentration of salt needed to stabilize the Z form. Thus, the degradation of cytidine-methylated duplex poly(dG–dC) was further diminished at concentrations of NaCl > 1.0 M. Strand scission of poly(dG–dC) also decreased sharply at NaCl concentrations above 2.3 M (Hertzberg *et al.*, 1988).

Long *et al.* (1990) also investigated the effect of a single methylated cytidine residue on BLM-induced DNA degradation by employing 5′-CG$^{Me}$CTTTAAAGCG-3′ as a substrate. This substrate was cleaved as efficiently as the unmethylated dodecanucleotide by BLM. However, degradation at the methylated site led to a greater proportion of alkali-labile lesions (see below). Clearly, alteration of DNA conformation perturbed BLM binding, with the consequent change in the observed chemistry.

It has also been reported that BLM mediated strand scission preferentially at DNA bulges, a bulge being an extra unpaired nucleotide on one strand of a DNA duplex (Williams and Goldberg, 1988). A series of DNA duplexes containing bulges positioned at different sites was treated with bleomycin. In each case strand scission occurred with the greatest efficiency on the strand opposite the bulge. The bulge in DNA was concluded to furnish a high-affinity binding site for BLM (Williams and Goldberg, 1988). Clearly, one implication of this work is that additional structural features in DNA may be recognized by BLM.

The foregoing results suggest that DNA binding by bleomycin is quite sensitive to DNA structure; even when binding occurred at the same site, the nature of the derived complex could be altered so as to affect the chemistry observed at that site (Long *et al.*, 1990). Some structural alterations in the BLM molecule seem to be well tolerated, in that these changes do not affect sequence recognition in a major way. At present there are still ambiguities about the mode of association of the bithiazole ring system with DNA. Even less is known about the interaction of the N-terminal region of the BLM molecule with DNA; since this domain of the BLM molecule would seem to participate in DNA sequence recognition, the details of this interaction need to be better understood.

## 5 Nucleic Acid Degradation by Bleomycin

Early experiments indicated that, upon admixture of DNA and BLM, a certain portion of the DNA was rendered acid soluble (Haidle, 1971; Haidle *et al.*, 1972). For instance, Haidle (1971) observed that when BLM was added to DNA containing both [$^{32}$P]-labelled phosphates and [$^{3}$H]-labelled thymidines, 25% of the $^{32}$P radiolabel and 80% of the tritium became soluble in acid. This was consistent with the interpretation that DNA degradation by bleomycin had resulted in both base release and fragmentation of the DNA backbone. In fact, the release of all four bases from DNA upon treatment with bleomycin, as well as the greater abundance of free pyrimidines, was also noted by other workers (Povirk *et al.*, 1978; Sausville *et al.*, 1978b). DNA degradation by bleomycin was shown to result in the formation of both single- and double-strand breaks (Suzuki *et al.*, 1969, 1970; Haidle, 1971; Povirk *et al.*, 1977; D'Andrea and Haseltine, 1978; Lloyd *et al.*, 1978). Moreover, while the DNA backbone was fragmented by bleomycin into oligonucleotides of discrete sizes, the release of nucleotides or inorganic phosphate did not occur (Haidle, 1971; Sausville *et al.*, 1978b). In addition to the four purine and pyrimidine bases, a low molecular weight aldehydic compound was also formed in the DNA degradation reaction; this material gave an intensely coloured adduct upon incubation with thiobarbituric acid (Muller *et al.*, 1972). The optical spectrum of this adduct in the 500–600 nm range was remarkably similar to the adduct obtained from the condensation of malondialdehyde and thiobarbituric acid. Formation of the adduct was increased at greater concentrations of bleomycin and $O_2$ (Kuo and Haidle, 1973; Sausville *et al.*, 1978b). In a subsequent study, Burger *et al.* (1980) determined that this aldehyde was indeed malondialdehyde but also demonstrated that BLM–DNA reaction mixtures contained malondialdehyde precursors which incorporated the nucleic acid bases; these could either condense directly with thiobarbituric acid or release malondialdehyde upon acid or base treatment. Burger *et al.* also suggested that there might exist two modes of DNA modification by bleomycin, one of which resulted in the release of free bases and the other of which led to the aldehydic species. Earlier, Povirk *et al.* (1977) had observed that the release of free bases from DNA rendered the DNA susceptible to alkali, and that the total number of DNA breaks induced by bleomycin increased upon alkali treatment.

The nature of the lesions produced in DNA as a result of its reaction with BLM was also examined (Kuo and Haidle, 1973; D'Andrea and Haseltine, 1978; Sausville *et al.*, 1978b; Takeshita *et al.*, 1978). The electrophoretic mobilities of the oligonucleotides produced by strand scission of a 5′-[$^{32}$P] end-labelled DNA fragment were found to be different from those produced by Maxam–Gilbert base-specific DNA degradation reactions (Takeshita *et al.*, 1978). This difference in mobility was attributed to structural alteration of the 3′-terminus in these lesions, since DNA containing BLM lesions did not co-migrate with DNA fragments of the same sequence having 3′-OH groups. Further, exonuclease III, which removes 3′-phosphoryl groups from DNA, failed to release inorganic phosphate from BLM-induced DNA lesions (Kuo and Haidle, 1973). Kuo and Haidle also observed that most of the 5′-termini in the oligonucleotides produced from DNA by BLM were phosphorylated. On the basis of these observations, as well as their own experimental findings, Giloni *et al.* (1981) proposed a mechanism to explain how BLM effected DNA degradation. A slightly modified version of this mechanism is shown in Figure 5.15. In this scheme, abstraction of the C-4′ H from the deoxyribose ring in DNA by a

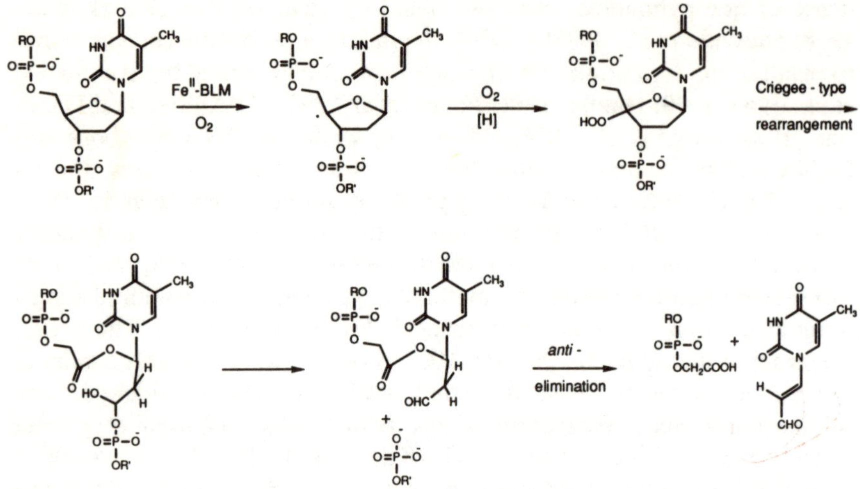


**Figure 5.15** Mechanism proposed for oxidative DNA strand scission by activated Fe•BLM

reactive metalloBLM species was envisioned to produce a C-4′ radical which combined with oxygen. The resulting peroxy radical would then be reduced to the hydroperoxide, the latter of which could undergo a Criegee-type hydroperoxide rearrangement with scission of the C-3′, C-4′ bond of deoxyribose. This would lead to the formation of a base propenal, the malondialdehyde precursor, and DNA lesions having oligonucleotide 5′-phosphate and 3′-phosphoroglycolate termini. All four base propenals were isolated from the DNA degradation mixture by Giloni *et al.* (1981); these were found not only to condense directly with thiobarbituric acid, but also to release malondialdehyde upon treatment with 88% formic acid. The presence of esterified glycolic acid residues at the 3′-ends of the DNA lesions was established by acid and enzymatic hydrolysis of the DNA lesion to release free glycolic acid, which was subsequently characterized. Consistent with the mechanism in Figure 5.15, a 1:1 correspondence was also noted between the amounts of glycolic acid recovered by hydrolysis and base propenals formed in the reaction.

The mechanism in Figure 5.15 supposes that $O_2$ is required for fragmentation of the deoxyribose ring, which results in the production of base propenal. Burger *et al.* (1982a,b) studied the effect of $O_2$ concentration on the yields of thymine and thymine propenal formed from the degradation of [$^3$H]-thymidine-labelled DNA. When DNA was in excess, increasing the $O_2$-to-Fe•BLM ratio from 6 to 30 doubled the yield of thymine propenal, without affecting thymine release. However, when DNA was limiting, 58% of the thymine was released as free base and only 2% as base propenal under anaerobic conditions. A parallel experiment conducted in the pres-

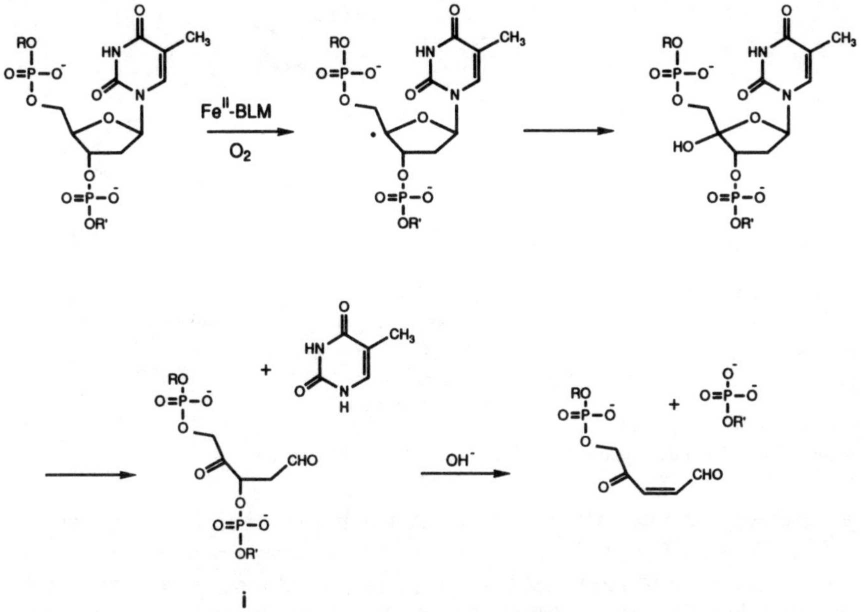


**Figure 5.16** Proposed mechanism for formation of an alkali-labile lesion by Fe•BLM

ence of $O_2$ yielded 23% free thymine and 20% propenal. These results demonstrated that while the release of free base from DNA could be effected by activated bleomycin, the fragmentation of the deoxyribose ring in DNA and the production of base propenal required $O_2$ in addition to that required for bleomycin activation. To accommodate these observations, Burger *et al.* (1982b) suggested that a common reactive intermediate was perhaps responsible for the production of both base and base propenal.

In a subsequent study on the degradation of poly(dA–dU) specifically tritiated at various positions on the deoxyribose ring, Wu *et al.* (1983) noted that the DNA substrate [$^3$H]-labelled at C-1′ of deoxyribose did not release tritium upon treatment with Fe•BLM. Instead, when C-4′ of deoxyribose was [$^3$H]-labelled, treatment with Fe•BLM resulted in the formation of $^3H_2O$. To accommodate these findings, the authors suggested that the C-4′ radical which combined with $O_2$ could also be hydroxylated by a monooxygen species. Collapse of this hydroxylated intermediate would then lead to the release of free base (Figure 5.16) and the formation of an alkali-labile lesion. Wu *et al.* (1983) also observed that DNA labelled at the 3′-position with tritium produced $^3H_2O$ upon reaction with bleomycin, even though the uracil propenal formed had the same specific activity as the DNA substrate. This result suggested that loss of this tritium as $^3H_2O$ was associated exclusively with base release, a finding consistent with the

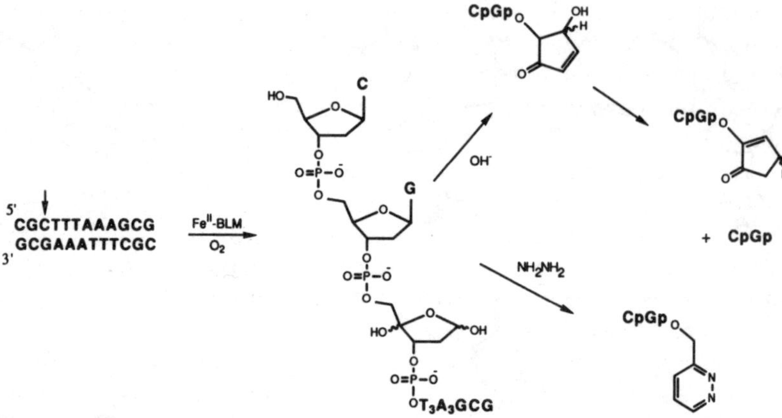


**Figure 5.17** Chemical characterization of the alkali-labile lesion

mechanistic scheme in Figure 5.16, in which C-3′ H in intermediate **i** is acidic. That the C-4′ radical may indeed be a common reactive intermediate leading to both base and base propenal was also supported by a later investigation (Wu *et al.*, 1985). During the degradation of poly(dA–[4′-$^3$H]dU) by Fe(II)•BLM, the ratio of uracil to uracil propenal could be varied between 0.03 and 7.0, depending upon the concentration of oxygen. Under these various conditions, within experimental error the same tritium selection effects were observed. More recently, Kozarich *et al.* (1989) employed gel electrophoresis to measure deuterium isotope effects for the degradation of DNA containing [4′-$^2$H]-thymidine residues. Isotope effects of 2.5–4.0 were observed for strand scission events as well as for pathways leading to the formation of alkali-labile lesions. Interestingly, the magnitude of the isotope effect was different at different cleavage sites, suggesting that not all C-4′ hydrogens at different DNA sequences were removed with the same facility by BLM.

Chemical characterization of the alkali-labile lesion formed by treatment of DNA with Fe•BLM has been accomplished (Sugiyama *et al.*, 1985b, 1988; Rabow *et al.*, 1986). Sugiyama *et al.* studied the degradation of the self-complementary dodecanucleotide dCGCTTTAAAGCG by Fe(II)•BLM. This substrate was modified oxidatively predominantly at cytidine$_3$ and cytidine$_{11}$. The release of cytosine from cytidine$_3$ resulted in the formation of an alkali-labile lesion at this position; alkali treatment of this lesion effected strand scission, and generated a 1,4-dioxo species that cyclized under the basic conditions to afford hydroxycyclopentenones (Figure 5.17). The structures of the final diastereomeric hydroxycyclopentenones were established by comparison with authentic synthetic samples of the individual diastereomers. When the alkali-labile lesion was treated with hydrazine, the pyridazine derivative in Figure 5.17 was formed in quantitative yield. This transformation was particularly important for purposes of characterization, as it preserved the same connectivity of the carbon atoms as in the alkali-labile lesion. Treatment of the alkali-labile lesion with butylamine released CpGp. Rabow *et al.* (1986) characterized the alkali-labile lesion in DNA produced from the reaction with BLM, using a different approach. This procedure involved isolation of a dinucleotide analogue containing the lesion resulting from oxidation of dCGCGCG at cytidine$_3$, followed by sodium borohydride reduction of the 1,4-dioxo species to the dinucleotide analogue containing a deoxypentitol moiety. Enzymatic digestion effected release of the sugar, which could then be characterized.

The existence of abasic lesions in DNA, formed as a result of BLM treatment, has been observed in a more direct fashion, using $^{17}O$-NMR spectroscopy (A. Natrajan and S.M. Hecht, unpublished results). Treatment of calf thymus DNA with BLM under anaerobic conditions followed by analysis of the reaction mixture in $H_2{}^{17}O$ revealed a strong resonance at ~110 ppm. This absorption was shown to contain all three oxygens in the cyclic hemiketal structure. No resonances for carbonyl oxygens were observed, suggesting that the cyclic hemiketal structure was the predominant species in aqueous solution. None the less, when the spectrum was subsequently recorded in $H_2{}^{16}O$, the resonance at 110 ppm vanished, presumably owing to label washout via the 1,4-dioxo moiety.

Recent studies by Rabow *et al.* (1990) suggest that the release of free base from DNA upon reaction with BLM may involve hydroxylation of C-4′ via the intermediacy of a C-4′ carbonium ion rather than a C-4′ radical. Sodium borohydride reduction of the alkali-labile lesion produced from the reaction of dCGCGCG with Fe(II)•BLM, followed by enzymatic digestion of the lesion, afforded free deoxypentitol which incorporated 86–98% of its oxygen at $C_4$ from solvent $H_2O$ rather than $O_2$. A mechanism to account for these observations has been proposed and involves a second, one-electron oxidation of the initially formed C-4′ radical by Fe•BLM (Figure 5.18). Both of these oxidations could be mediated by a single activated BLM. Since activated BLM is believed to contain two oxidizing equivalents above Fe(III)•BLM, this mechanistic rationale is based on the assumption that neither the perferryl nor the ferryl species exchange oxygen with solvent before reaction. At least in the case of olefin epoxidation, this appears to be true. For instance, Heimbrook *et al.* (1987) demonstrated that in the epoxidation of *cis*-stilbene by Fe(II)•BLM, the epoxide oxygen is derived mainly from the $O_2$ used to activate BLM, and not from solvent $H_2O$.

The products resulting from BLM-mediated DNA strand scission have also been characterized in a direct fashion (Uesugi *et al.*, 1984; Murugesan *et al.*, 1985; Sugiyama *et al.*, 1985a). Uesugi *et al.* (1984) studied the degradation of the hexanucleotide dCGCGCG by BLM. In addition to

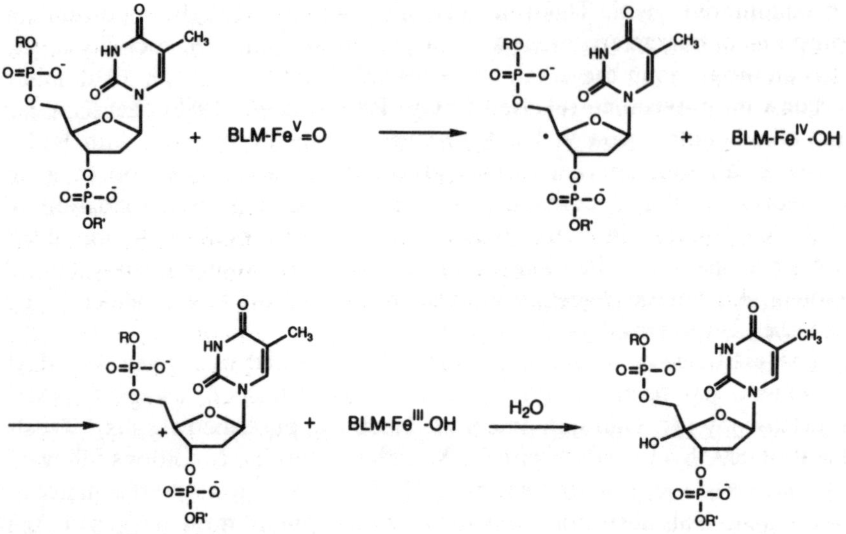

**Figure 5.18** Proposed mechanism for hydroxylation of C-4′ of deoxyribose in DNA, leading to alkali-labile lesion formation

cystosine and cytosine propenal, cleavage of this substrate produced the glycolates $CpGp_{CH_2COOH}$ and $CpGpCpGp_{CH_2COOH}$ arising from cleavage at $cytidine_3$ and $cytidine_5$, respectively, and these products were characterized in detail. Sugiyama *et al.* (1985a) described the use of the dodecanucleotide dCGCTTTAAAGCG as a highly efficient target for BLM. Strand scission of this substrate occurred mostly at $cytidine_3$ and $cytidine_{11}$ and resulted in the formation of cystosine, cytosine propenal, $CpGp_{CH_2COOH}$, 5′-dGMP, $CGCTTTAAAGp_{CH_2COOH}$ and pTTTAAAGCG. The first four products were readily separated and quantified by HPLC; the production of $CpGp_{CH_2COOH}$ and 5′-dGMP was diagnostic of cleavage at $cytidine_3$ and $cytidine_{11}$, respectively. Finally, Murugesan *et al.* (1985) observed the formation of all four nucleoside 3′-(phosphoro-2′-*O*-glycolic acid) derivatives following degradation of *E. coli* DNA by Fe(II)·BLM, and subsequent lambda exonuclease digestion to effect the release of individual nucleotides. Consistent with the observations of Giloni *et al.* (1981), all four base propenals were also observed. Examination of the double bond configuration in two of these propenals revealed that the *trans* isomers were formed exclusively as primary products of the reaction. The exclusive formation of the *trans* isomers probably resulted from a preference of the final reaction intermediate in Figure 5.15 to undergo an *anti*-elimination reaction. In fact, during studies of poly dA[2′*R* and 2′*S*-$^3$H]dU degradation by Fe(II)·BLM, Wu *et al.* (1985) observed that the 2′*R* isomer afforded uracil propenal that was devoid of radioactivity. On the other hand, degradation of the 2′*S* isomer resulted in the formation of uracil propenal that retained all of the radiolabel originally present in the deoxyuridine ring (Ajmera *et al.*, 1986).

DNA degradation by bleomycin results in the production of both single- and double-strand breaks. The formation of double-strand lesions in DNA occurs as the result of a specific event rather than the random accumulation of single-strand breaks (Povirk *et al.*, 1977; Lloyd *et al.*, 1978). Moreover, both single- and double-strand breaks in DNA were found to be produced on the same time-scale (Burger *et al.*, 1986). A model to explain the propensity of BLM to mediate double-strand breaks in DNA was proposed by Lloyd *et al.* (1978). In this model, BLM was presumed to bind to DNA as a dimer oriented in a fashion such that cleavage of DNA on both strands could be effected. At present there is no evidence to support this model. Povirk *et al.* (1977) measured the extent of single- vs. double-strand breaks in T2 DNA and Col $E_1$ DNA by velocity sedimentation and electrophoresis. They observed that the number of double-strand events in DNA was linearly related to the number of single-strand events and occurred 10% as often as single-strand breaks. Lloyd *et al.* (1978) observed that the production of double-strand breaks was suppressed by an increase in the ionic strength of the medium. Povirk *et al.* (1989) also determined the structures of the bleomycin-induced double-strand breaks in DNA. For every such break, a primary site of cleavage was identified as the pyrimidine in a 5′-G–Py sequence. The cleavage site on the opposite strand, the secondary cleavage site, was usually a site where single-strand cleavage was infrequent. When the sequence of the primary cleavage site was 5′-G–Py–Py, strand scission at the central pyrimidine was accompanied by strand breakage opposite this base on the other strand to produce blunt ends. When the sequence was 5′-G–Py–Pu, however, scission on the opposite strand occurred at the base opposite the purine to produce 5′-extensions. In general, although there was a preference for pyrimidines over purines at the secondary cleavage site, it appeared that the second cleavage event was dictated largely by the initial recognition event that produced the first break.

Several models have been proposed to explain the production of double-strand breaks in DNA as a consequence of the action of activated Fe•BLM. Povirk and co-workers (Povirk *et al.*, 1989; Steighner and Povirk, 1990) have suggested that double-strand breakage could result from (re)activation of Fe•BLM by the putative alkyl hydroperoxide DNA intermediate formed on the pathway to strand scission (cf. Figure 5.15). More recently, a related proposal has been made by the same group which suggests that the DNA hydroperoxy radical itself might (re)capture a H atom from the same Fe•BLM that abstracted this atom from DNA, presumably regenerating a perferryl species from this BLM intermediate. Both of these mechanisms are interesting and seem plausible; however, it will be necessary to evaluate both in detail in the context of additional experimental data such as rate of Fe(III)•BLM activation by $H_2O_2$, which may be too slow to support the production of both single- and double-strand lesions on the same time-scale (Burger *et al.*, 1981; Natrajan *et al.*, 1990a). An additional issue that requires resolution is the nature of the activated Fe•BLM that is present after initial H atom abstraction from deoxyribose. This would logically be thought to be a ferryl species, which has been reported to be incapable of effecting DNA degradation (Natrajan *et al.*, 1990b), and whose reconversion to the more reactive perferryl species has not been demonstrated. Further, if a reactivated Fe•BLM is envisioned to abstract a H atom directly from the opposite DNA strand, it would be helpful to analyse the distance between the potentially reactive groups as well as their relative orientations. An uncomplicated analysis of the distance between C-4′ H on one strand of DNA and C-4′ H on the other at the susceptible nucleotides involved in double-strand cleavage suggested that the two would be ~15.4 Å apart in a normal B-form duplex (R. Duff and S.M. Hecht, unpublished results) and that a single molecule of Fe•BLM would have to reorient substantially on the DNA duplex to be able to effect H atom abstraction on the second DNA strand.

Other mechanisms may exist that lead to the formation of double-strand breaks in DNA by virtue of enhanced affinity of bleomycin for a singly nicked site in DNA. To investigate this possibility, Keller and Oppenheimer (1987) evaluated the ability of BLM to interact with and cleave DNA at sites opposite to a gap on one strand. They observed that phosphorylation of either the 3′-OH or the 5′-OH in the gap enhanced the ability of bleomycin to cleave the DNA strand opposite the negatively charged gap. Phosphorylation of both the 3′- and the 5′-hydoxyls to produce a dianionic gap (similar to that produced by BLM) dramatically enhanced the ability of BLM to cleave at the gap. From these observations, the authors proposed that bleomycin-mediated double-strand scission may be a form of self-potentiation, in which the original cleavage site containing enhanced negative charge provides a high-affinity site for BLM binding and cleavage. While this proposal is consistent with the observation that the number of BLM-induced double-strand breaks in DNA decreased with increasing ionic strength of the medium (Lloyd *et al.*, 1978), it is not clear how the proportion of double- to single-strand events could remain invariant during the time-course of the DNA degradation reaction, as has been observed (Burger *et al.*, 1986).

Recently, it was reported that RNA also functions as a target for degradation by BLM (Magliozzo *et al.*, 1989; Carter *et al.*, 1990b). These results are especially interesting in the light of the finding that cell growth inhibition by BLM is not well correlated with the DNA damage induced by BLM (Berry *et al.*, 1985). The cytotoxic effects of BLM could arise from its interaction with other components of the cell; RNA is an obvious potential target, being structurally similar to DNA and more accessible, owing to its presence in the cytoplasm. Moreover, the destruction of certain critical RNAs necessary for cell function could be an important mechanism by which BLM exerts its therapeutic effects, since mechanisms for RNA repair are largely non-existent. Following a preliminary report by Magliozzo *et al.* (1989), a more thorough investigation of the cleavage of RNA by BLM was reported by Carter *et al.* (1990b). In this study, the authors observed that at micromolar concentrations, Fe(II)·BLM effected the cleavage of a *Bacillus subtilis* $tRNA^{His}$ precursor in a highly selective fashion. The cleavage of this substrate was not greatly affected by non-substrate RNAs. Further, cleavage of the $tRNA^{His}$ precursor was as efficient as DNA cleavage, and much more selective. Other RNA substrates, including a transcript encoding a large segment of the reverse transcriptase mRNA from human immunodeficiency virus-1, were also cleaved by bleomycin.

The mechanism by which bleomycin degrades RNA has been investigated (Haidle and Bearden, 1975; Krishnamoorthy *et al.*, 1988; Carter *et al.*, 1990b; Duff *et al.*, 1993). In the studies by Haidle and Krishnamoorthy, the degradation of RNA–DNA hybrids by BLM was investigated. Both studies reported that there was no cleavage of the RNA strand, although Krishnamoorthy *et al.* (1988) observed that the substrate poly(dA)·poly(rU) produced changes in the proportion of chemical products from the DNA strand, namely a greater proportion of adenine compared with adenine propenal. The ability of bleomycin to degrade sugars other than deoxyribose was demonstrated by Carter *et al.* (1990b). In this study, the ability of BLM to degrade the substrates 5′-CG*ribo*CTAGCG and 5′-CG*ara*CTAGCG was evaluated. These substrates contained either a ribose or an arabinose nucleoside at position 3, which in the normal DNA substrate is a high-affinity site for BLM. As described above, oxidative cleavage at $cytidine_3$ in $CGCT_3A_3GCG$ resulted in the production of the glycolate $CpGp_{CH_2COOH}$. Upon treatment with Fe(II)·BLM, both of the above modified substrates were degraded, and both afforded $CpGp_{CH_2COOH}$. In addition, both substrates led to increased levels of cytosine formation; recent work suggests that this resulted from a C-1′ hydroperoxide mechanism (Duff *et al.*, 1993). From these studies it appears that BLM may oxidize both the C-1′ and C-4′ positions in RNA.

## 6 Future Prospects

Despite the enormous progress that has been made, our understanding of the chemistry of bleomycin is far from complete. Only recently has it been realized that the N-terminus of BLM may also participate in DNA sequence recognition; very little is known about the details of this interaction. At present there also exists experimental evidence which suggests that the bithiazole moiety may interact with DNA by more than one mechanism. The hydrogen-bonded minor groove binding model of Kuwahara and Sugiura (1988) can be tested by systematic alteration of the structures of the participants in the proposed interaction. The chemical nature of degradation of DNA by BLM is not completely understood and the recent results on RNA cleavage have opened a new area for investigation. The finding that BLM produces double-strand breaks in DNA is intriguing; at present none of the proposed models reconciles all available experimental data. Since there appears to be a lack of correlation between cell growth inhibition caused by BLM and its DNA damaging activity, BLM may well have other targets in the cell. Clearly, the study of the mechanisms of these several processes can provide a better understanding of the way in which an important antitumour agent mediates its therapeutic effects.

**Acknowledgement**

Our studies on bleomycin at the University of Virginia were supported by Research Grants CA-27603, CA-38544 and CA-53913 from the National Cancer Institute, DHHS.

**References**

Ajmera, S., Wu, J. C., Worth, L., Jr., Rabow, L. E., Stubbe, J. and Kozarich, J. W. (1986). DNA degradation by bleomycin: evidence for 2′R-proton abstraction and for carbon-oxygen bond cleavage accompanying base propenal formation. *Biochemistry*, **25**, 6586–6592

Akkerman, M. A. J. and Haasnoot, C. A. G. (1988). Studies of the solution structure of the bleomycin-$A_2$–zinc complex by means of two-dimensional NMR spectroscopy and distance geometry calculations. *Eur. J. Biochem.*, **173**, 211–225

Albertini, J. P., Garnier-Suillerot, A. and Tosi, L. (1982). Iron bleomycin DNA system evidence of a long-lived bleomycin iron oxygen intermediate. *Biochem. Biophys. Res. Commun.*, **104**, 557–563

Antholine, W. E. and Petering, D. H. (1979). On the reaction of iron bleomycin with thiols and oxygen. *Biochem. Biophys. Res. Commun.*, **90**, 384–389

Aoyagi, Y., Katano, K., Suguna, H., Primeau, J., Chang, L.-H. and Hecht, S. M. (1982). Total synthesis of bleomycin. *J. Am. Chem. Soc.*, **104**, 5537–5538

Asakura, H., Umezawa, H. and Hori, M. (1978). DNA structures required for bleomycin binding. *J. Antibiotics*, **31**, 156–158

Barr, J. R., Van Atta, R. B., Natrajan, A. and Hecht, S. M. (1990). Iron(II) bleomycin-mediated reduction of oxygen to water: An oxygen- 17 NMR study. *J. Am. Chem. Soc.*, **112**, 4058–4060

Berry, D. E., Chang, L.-H. and Hecht, S. M. (1985). DNA damage and growth inhibition in cultured human cells by bleomycin congeners. *Biochemistry*, **24**, 3207–3214

Blum, R. H., Carter, S. K. and Agre, K. (1973). A clinical review of bleomycin—a new antineoplastic agent. *Cancer*, **31**, 903–914

Booth, T. E., Sakai, T. T. and Glickson, J. D. (1983). Interaction of bleomycin $A_2$ with poly(deoxyadenylylthymidylic acid). A proton nuclear magnetic resonance study of the influence of temperature, pH, and ionic strength. *Biochemistry*, **22**, 4211–4217

Burger, R. M., Berkowitz, A. R., Peisach, J. and Horwitz, S. B. (1980). Origin of malondialdehyde from DNA degraded by iron(II)·bleomycin. *J. Biol. Chem.*, **255**, 11832–11838

Burger, R. M., Blanchard. J. S., Horwitz, S. B. and Peisach, J. (1985). The redox state of activated bleomycin. *J. Biol. Chem.*, **260**, 15406–15409

Burger, R. M., Freedman, J. H., Horwitz, S. B. and Peisach, J. (1984). DNA degradation by manganese(II)–bleomycin plus peroxide. *Inorg. Chem.*, **23**, 2215–2217

Burger, R. M., Horwitz, S. B., Peisach, J. and Wittenberg, J. B. (1979a). Oxygenated iron bleomycin. A short-lived intermediate in the reaction of ferrous bleomycin with $O_2$. *J. Biol. Chem.*, **254**, 12299–12302

Burger, R. M., Kent, T. A., Horwitz, S. B., Munck, E. and Peisach, J. (1983). Mössbauer study of iron bleomycin and its activation intermediates. *J. Biol. Chem.*, **258**, 1559–1564

Burger, R. M., Peisach, J., Blumberg, W. E. and Horwitz, S. B. (1979b). Iron–bleomycin interactions with oxygen and oxygen analogues. *J. Biol. Chem.*, **254**, 10906–10912

Burger, R. M., Peisach. J. and Horwitz, S. B. (1981). Activated bleomycin. A transient complex of drug, iron, and oxygen that degrades DNA. *J. Biol. Chem.*, **256**, 11636–11644

Burger, R. M., Peisach, J. and Horwitz, S. B. (1982a). Effects of oxygen on the reactions of activated bleomycin. *J. Biol. Chem.*, **257**, 3372–3375

Burger, R. M., Peisach, J. and Horwitz, S. B. (1982b). Stoichiometry of DNA strand scission and aldehyde formation by bleomycin. *J. Biol. Chem.*, **257**, 8612–8614

Burger, R. M., Projan, S. J., Horwitz, S. B. and Peisach, J. (1986). The DNA cleavage mechanism of iron–bleomycin. Kinetic resolution of strand scission from base propenal release. *J. Biol. Chem.*, **261**, 15955–15959

Carter, B. J., Murty, V. S., Reddy, K. S., Wang, S.-N. and Hecht, S. M. (1990a). A role for the metal binding domain in determining the DNA sequence selectivity of iron-bleomycin. *J. Biol. Chem.*, **265**, 4193–4196

Carter, B. J., de Vroom, E., Long, E. C., van der Marel, G. A., van Boom, J. H. and Hecht, S. M. (1990b). Site-specific cleavage of RNA by Fe(II)·bleomycin. *Proc. Natl Acad. Sci. USA*, **87**, 9373–9377

Caspary, J. W., Lanzo, D. A. and Niziak, C. (1981). Intermediates in the ferrous oxidase cycle of bleomycin. *Biochemistry*, **20**, 3868–3875

Caspary, W. J., Lanzo, D. A. and Niziak, C. (1982). Effect of deoxyribonucleic acid on the production of reduced oxygen by bleomycin and iron. *Biochemistry*, **21**, 334–338

Chang, C.-H. and Meares, C. F. (1982). Light-induced nicking of deoxyribonucleic acid by cobalt(III) bleomycins. *Biochemistry*, **21**, 6332–6334

Chang, C.-H. and Meares, C. F. (1984). Cobalt-bleomycins and deoxyribonucleic acid: Sequence-dependent interactions, action spectrum for nicking, and indifference to oxygen. *Biochemistry*, **23**, 2268–2274

Chang, C.-H., Dallas, J. L. and Meares, C. F. (1983). Identification of a key structural feature of cobalt(III)-bleomycins: an exogenous ligand (e.g. hydroperoxide) bound to cobalt. *Biochem. Biophys. Res. Commun.*, **110**, 959–966

Ciriolo, M. R., Magliozzo, R. S. and Peisach, J. (1987). Microsome-stimulated activation of ferrous bleomycin in the presence of DNA. *J. Biol. Chem.*, **262**, 6290–6295

Chien, M. A., Grollman, A. P. and Horwitz, S. B. (1977). Bleomycin–DNA interactions: fluorescence and proton magnetic resonance studies. *Biochemistry*, **16**, 3641–3647

Chen, D. M., Sakai, T. T., Glickson, J. D. and Patel, D. J. (1980). Bleomycin $A_2$ complexes with poly(dA–dT): a proton nuclear magnetic resonance study of the nonexchangeable hydrogens. *Biochem. Biophys. Res. Commun.*, **92**, 197–205

Dabrowiak, J. C., Greenaway, F. T., Longo, W. E., Van Husen, M. and Crooke, S. T. (1978). A spectroscopic investigation of the metal binding site of bleomycin $A_2$. *Biochim. Biophys. Acta*, **517**, 517–526

Dabrowiak, J. C., Greenaway, F. T., Santillo, F. S. and Crooke, S. T. (1979). The iron complexes of bleomycin and tallysomycin. *Biochem. Biophys. Res. Commun.*, **91**, 721–729

Dabrowiak, J. C. and Tsukayama, M. (1981). Cobalt(III) complex of pseudotetrapeptide A of bleomycin. *J. Am. Chem. Soc.*, **103**, 7543–7550

D'Andrea, A. D. and Haseltine, W. A. (1978). Sequence specific cleavage of DNA by the antitumor antibiotics neocarzinostatin and bleomycin. *Proc. Natl Acad. Sci. USA*, **75**, 3608–3612

Duff, R. J., de Vroom, E., Geluk. A., Hecht, S. M., van der Marel, G. A. and van Boom, J. H. (1993). *J. Am. Chem. Soc.*, **115**, 3350–3351

Ehrenfeld, G. M. (1986). PhD Thesis, University of Virginia

Ehrenfeld, G. M., Murugesan, N. and Hecht, S. M. (1984). Activation of oxygen and mediation of DNA degradation by manganese–bleomycin. *Inorg. Chem.*, **23**, 1496–1498

Ehrenfeld, G. M., Rodriguez, L. O., Hecht, S. M., Chang, C., Basus, V. J. and Oppenheimer, N. J. (1985). Copper (I)–bleomycin: Structurally unique complex that mediates oxidative DNA strand scission. *Biochemistry*, **24**, 81–92

Ehrenfeld, G. M., Shipley, J. B., Heimbrook, D. C., Sugiyama, H., Long, E. C., van Boom, J. H., van der Marel, G. A., Oppenheimer, N. J. and Hecht, S. M. (1987). Copper-dependent cleavage of DNA by bleomycin. *Biochemistry*, **26**, 931–942

Fisher, L. M., Kuroda, R. and Sakai, T. T. (1985). Interaction of bleomycin $A_2$ with deoxyribonucleic acid: DNA unwinding and inhibition of bleomycin-induced DNA breakage by cationic thiazole amides related to bleomycin $A_2$. *Biochemistry*, **24**, 3199–3207

Giloni, L., Takeshita, M., Johnson, F., Iden, C. and Grollman, A. P. (1981). Bleomycin-induced strand-scission of DNA. Mechanism of deoxyribose cleavage. *J. Biol. Chem.*, **256**, 8608–8615

Glickson, J. D., Pillai, R. and Sakai, T. T. (1981). Proton NMR studies of the zinc(II)–bleomycin $A_2$–poly(dA–dT) ternary complex. *Proc. Natl Acad. Sci. USA*, **78**, 2967–2971

Gold, B., Dange, V., Moore, M. A., Eastman, A., van der Marel, G. A., van Boom, J. H. and Hecht, S. M. (1988). Alteration of bleomycin cleavage specificity in a platinated DNA oligomer of defined structure. *J. Am. Chem. Soc.*, **110**, 2347–2349

Guengerich, F. P. and Macdonald, T. L. (1984). Chemical mechanisms of catalysis by cytochromes P-450: A unified view. *Acc. Chem. Res.*, **17**, 9–16

Gutteridge, J. M. C. and Shute, D. J. (1981). Iron-dioxygen-dependent changes to the biological activities of bleomycin. *J. Inorg. Biochem.*, **15**, 349–357

Haidle, C. W. (1971). Fragmentation of deoxyribonucleic acid by bleomycin. *Mol. Pharmacol.*, **7**, 645–652

Haidle, C. W. and Bearden J., Jr. (1975). Effect of bleomycin on an RNA–DNA hybrid. *Biochem. Biophys. Res. Commun.*, **65**, 815–821

Haidle, C. W., Weiss, K. K. and Kuo, M. T. (1972). Release of free bases from deoxyribonucleic acid after reaction with bleomycin. *Mol. Pharmacol.*, **8**, 531–537

Hamamichi, N., Natrajan, A. and Hecht, S. M. (1992). On the role of individual bleomycin thiazoles in oxygen activation and DNA cleavage. *J. Am. Chem. Soc.*, **114**, 6278–6291

Heimbrook, D. C., Carr, S. A., Mentzer, M. A., Long, E. C. and Hecht, S. M. (1987). Mechanism of oxygenation of *cis*-stilbene by iron bleomycin. *Inorg. Chem.*, **26**, 3835–3836

Heimbrook, D. C., Mulholland, R. L. and Hecht, S. M. (1986). Multiple pathways in the oxidation of *cis*-stilbene by iron–bleomycin. *J. Am. Chem. Soc.*, **108**, 7839–7840

Hénichart, J.-P., Bernier, J.-L., Helbecque, N. and Houssin, R. (1985). Is the bithiazole moiety of bleomycin a classical intercalator? *Nucleic Acids Res.*, **13**, 6703–6717

Hertzberg, R. P., Caranfa, M. J. and Hecht, S. M. (1985). DNA methylation diminishes bleomycin-mediated strand scission. *Biochemistry*, **24**, 5285–5289

Hertzberg, R. P., Caranfa, M. J. and Hecht, S. M. (1988). Degradation of structurally modified DNAs by bleomycin group antibiotics. *Biochemistry*, **27**, 3164–3174

Huang, C.-H., Galvan, L. and Crooke, S. T. (1980). Interactions of bleomycin analogues with deoxyribonucleic acid and metal ions studied by fluorescence quenching. *Biochemistry*, **19**, 1761–1767

Iitaka, Y., Nakamura, H., Nakatani, T., Muraoka, Y., Fujii, A., Takita, T. and Umezawa, H. (1978). Chemistry of bleomycin. XX. The x-ray structure determination of P-3A Cu(II)-complex, a biosynthetic intermediate of bleomycin. *J. Antibiotics*, **31**, 1070–1072

Ishida, R. and Takahashi, T. (1975). Increased DNA chain breakage by combined action of bleomycin and superoxide radical. *Biochem. Biophys. Res. Commun.*, **66**, 1432–1438

Kasai, H., Naganawa, H., Takita, T. and Umezawa, H. (1978). Chemistry of bleomycin. XXII. Interaction of bleomycin with nucleic acids, preferential binding to guanine base and electrostatic effect of the terminal amine. *J. Antibiotics*, **31**, 1316–1320

Keller, J. K. and Oppenheimer, N. J. (1987). Enhanced bleomycin-mediated damage of DNA opposite charged nicks. A model for bleomycin-directed double strand scission of DNA. *J. Biol. Chem.*, **262**, 15144–15150

Kilkuskie, R. E., Macdonald, T. L. and Hecht, S. M. (1984). Bleomycin may be activated for DNA cleavage by NADPH–cytochrome P-450 reductase. *Biochemistry*, **23**, 6165–6171

Kittaka, A., Sugano, Y., Otsuka, M. and Ohno, M. (1988). Man-designed bleomycins. Synthesis of dioxygen activating molecules and a DNA cleaving molecule based on bleomycin–Fe(II)–$O_2$ complex. *Tetrahedron*, **44**, 2821–2823

Kittaka, A., Sugano, Y., Otsuka, M., Ohno, M., Sugiura, Y. and Umezawa, H. (1986). Transition-metal binding site of bleomycin. A remarkably efficient dioxygen-activating molecule based on bleomycin–Fe(II) complex. *Tetrahedron Lett.*, **27**, 3631–3634

Konishi, M., Saito, K., Numata, K., Tsuno, T., Asama, K., Tsukiura, H., Naito, T. and Kawaguchi, H. (1977). Tallysomycin, a new antitumor antibiotic complex related to bleomycin. II. Structure determination of tallysomycins A and B. *J. Antibiotics*, **30**, 789–805

Kozarich, J. W., Worth, L., Jr., Frank, B. L., Christner, D. F., Vanderwall, D. E. and Stubbe, J. (1989). Sequence-specific isotope effects on the cleavage of DNA by bleomycin. *Science*, **245**, 1396–1399

Krishnamoorthy, C. R., Vanderwall, D. E., Kozarich, J. W. and Stubbe, J. (1988). Degradation of DNA–RNA hybrids by bleomycin: evidence for DNA strand specificity and for possible structural modulation of chemical mechanism. *J. Am. Chem. Soc.*, **110**, 2008–2009

Kross, J., Henner, D. W., Hecht, S. M. and Haseltine, W. A. (1982a). Specificity of deoxyribonucleic acid cleavage by bleomycin, phleomycin, and tallysomycin. *Biochemistry*, **21**, 4310–4318

Kross, J., Henner, D. W., Hecht, S. M., Haseltine, W. A., Rodriguez, L. and Levin, M. D. (1982b). Structural basis for the deoxyribonucleic acid affinity of bleomycins. *Biochemistry*, **21**, 3711–3721

Kuo, M. T. and Haidle, C. W. (1973). Characterization of chain breakage in DNA induced by bleomycin. *Biochim. Biophys. Acta*, **335**, 109–114

Kuramochi, H., Takahashi, K., Takita, T. and Umezawa, H. (1981). An active intermediate formed in the reaction of bleomycin–Fe(II) complex with oxygen. *J. Antibiotics*, **34**, 576–582

Kuwahara, J. and Sugiura, Y. (1988). Sequence-specific recognition and cleavage of DNA by metallobleomycin: minor groove binding and possible interaction mode. *Proc. Natl Acad. Sci. USA*, **85**, 2459–2463

Kuwahara, J., Suzuki, T. and Sugiura, Y. (1985). Effective DNA cleavage by bleomycin–vanadium (IV) complex plus hydrogen peroxide. *Biochem. Biophys. Res. Commun.*, **129**, 368–374

Labeque, R. and Marnett, L. J. (1987). 10-Hydroperoxy-8,12-octadecadienoic acid. A diagnostic probe of alkoxyl radical generation in metal–hydroperoxide reactions. *J. Am. Chem. Soc.*, **109**, 2828–2829

Levy, M. J. and Hecht, S. M. (1988). Copper(II) facilitates bleomycin-mediated unwinding of plasmid DNA. *Biochemistry*, **27**, 2647–2650

Lin, S. Y. and Grollman, A. P. (1981). Interactions of a fragment of bleomycin with deoxyribodinucleotides: nuclear magnetic resonance studies. *Biochemistry*, **20**, 7589–7598

Lloyd, R. S., Haidle, C. W. and Robberson, D. L. (1978). Bleomycin-specific fragmentation of double-stranded DNA. *Biochemistry*, **17**, 1890–1896

Long, E. C., Hecht, S. M., van der Marel, G. A. and van Boom, J. H. (1990). Interaction of bleomycin with a methylated DNA oligonucleotide. *J. Am. Chem. Soc.*, **112**, 5272–5276

Lown, J. W. and Sim, S. (1977). The mechanism of the bleomycin-induced cleavage of DNA. *Biochem. Biophys. Res. Commun.*, **77**, 1150–1157

Magliozzo, R. S., Peisach, J. and Ciriolo, M. R. (1989). Transfer RNA is cleaved by activated bleomycin. *Mol. Pharmacol.*, **35**, 428–432

Mascharak, P. K., Sugiura, Y., Kuwahara, J., Suzuki, T. and Lippard, S. J. (1983). Alteration and activation of sequence-specific cleavage of DNA by bleomycin in the presence of the antitumor drug *cis*-diamminedichloroplatinum(II). *Proc. Natl Acad. Sci. USA*, **80**, 6795–6798

Melnyk, D. L. Horwitz, S. B. and Peisach, J. (1981). Redox potential of iron–bleomycin. *Biochemistry*, **20**, 5327–5331

Mirabelli, C. K., Ting, A., Huang, C.-H., Mong, S. and Crooke, S. T. (1982). Bleomycin and talisomycin sequence-specific strand scission of DNA: a mechanism of double-strand cleavage. *Cancer Res.*, **42**, 2779–2785

Miyoshi, K., Kikuchi, T., Takita, T., Murato, S. and Ishizu, K. (1988). Redox potential of the active bleomycin–iron(III) complex by cyclic voltammetry. *Inorg. Chem. Acta*, **151**, 45–47

Morii, T., Matsuura, T., Saito, I., Suzuki, T., Kuwahara, J. and Sugiura, Y. (1986). Phototransformed bleomycin antibiotics. Structure and DNA cleavage activity. *J. Am. Chem. Soc.*, **108**, 7089–7094

Morii, T., Saito, I., Matsuura, T., Kuwahara, J. and Sugiura, Y. (1987). New lumibleomycin-containing thiazolylisothiazole ring. *J. Am. Chem. Soc.*, **109**, 938–939

Muller, W. E. G., Yamazaki, Z., Brewer, H.-J. and Zahn, R. K. (1972). Action of bleomycin on DNA and RNA. *Eur. J. Biochem.*, **31**, 518–525

Muraoka, Y., Suzuki, M., Fujii, A., Umezawa, Y., Naganawa, H., Takita, T. and Umezawa, H. (1981). Chemistry of bleomycin. XXVIII. Preparation of deglycobleomycin by mild acid hydrolysis of bleomycin. *J. Antibiotics*, **34**, 353–357

Murugesan, N., Ehrenfeld, G. M. and Hecht, S. M. (1982). Oxygen transfer from bleomycin–metal complexes. *J. Biol. Chem.*, **257**, 8600–8603

Murugesan, N. and Hecht, S. M. (1985). Bleomycin as an oxene transferase. Catalytic oxygen transfer to olefins. *J. Am. Chem. Soc.*, **107**, 493–500

Murugesan, N., Xu, C., Ehrenfeld, G. M., Sugiyama, H., Kilkuskie, R. E., Rodriguez, L. O., Chang, L.-H. and Hecht, S. M. (1985). Analysis of products formed during bleomycin-mediated DNA degradation. *Biochemistry*, **24**, 5735–5744

Nagai, K., Suzuki, H., Tanaka, N. and Umezawa, H. (1969a). Decrease of melting temperature and single strand scission of DNA by bleomycin in the presence of hydrogen peroxide. *J. Antibiotics*, **22**, 624–628

Nagai, K., Suzuki, H., Tanaka, N. and Umezawa, H. (1969b). Decrease of melting temperature and single strand scission of DNA by bleomycin in the presence of 2-mercaptoethanol. *J. Antibiotics*, **22**, 569–573

Nakamura, M. and Peisach, J. (1988). Self-inactivation of iron(II)–bleomycin. *J. Antibiotics*, **41**, 638–647

Natrajan, A. and Hecht, S. M. (1991). Production of 2-octenyl radicals from the iron(III)·bleomycin-mediated fragmentation of 10-hydroperoxy-8,12-octadecadienoic acid. *J. Org. Chem.*, **56**, 5239–5241

Natrajan, A., Hecht, S. M., van der Marel, G. A. and van Boom, J. H. (1990a). A study of oxygen versus hydrogen peroxide-supported activation of iron·bleomycin. *J. Am. Chem. Soc.*, **112**, 3997–4002

Natrajan, A., Hecht, S. M., van der Marel, G. A. and van Boom, J. H. (1990b). Activation of iron (III)–bleomycin by 10-hydroperoxy-8,12-octadecadienoic acid. *J. Am. Chem. Soc.*, **112**, 4532–4538

Oberley, L. W. and Buettner, G. R. (1979). The production of hydroxyl radical by tallysomycin and copper(II). *FEBS Lett.*, **97**, 47–49

Ohno, M., Otsuka, M., Kittaka, A., Sugano, Y., Sugiura, Y., Suzuki, T., Kuwahara, J., Umezawa, K. and Umezawa, H. (1988). Man-designed bleomycins. Iron complexation and nucleotide sequence cleavage by synthetic models of bleomycin. *Int. J. Exp. Clin. Chem.*, **1**, 12–22

Oppenheimer, N. J., Chang, C., Rodriguez, L. O. and Hecht, S. M. (1981). Copper(I)·bleomycin. A structurally unique oxidation–reduction active complex. *J. Biol. Chem.*, **256**, 1514–1517

Oppenheimer, N. J., Rodriguez, L. O. and Hecht, S. M. (1979a). Structural studies of 'active complex' of bleomycins: assignment of ligands to the ferrous ion in a ferrous-bleomycin–carbon monoxide complex. *Proc. Natl Acad. Sci. USA*, **76**, 5616–5620

Oppenheimer, N. J., Rodriguez, L. O. and Hecht, S. M. (1979b). Proton nuclear magnetic resonance study of the structure of bleomycin and the zinc–bleomycin complex. *Biochemistry*, **18**, 3439–3445

Otsuka, M., Kittaka, A., Ohno, M., Suzuki, T., Kuwahara, J., Umezawa, H. and Sugiura, Y. (1986). Synthetic study towards man-designed bleomycins. Synthesis of a DNA cleaving molecule based on bleomycin. *Tetrahedron Lett.*, **27**, 3639–3642

Otsuka, M., Yoshida, M., Kobayashi, S., Ohno, M., Sugiura, T., Takita, T. and Umezawa, H. (1981). Transition-metal binding site of bleomycin. A synthetic analogue capable of binding Fe(II) to yield an oxygen-sensitive complex. *J. Am. Chem. Soc.*, **103**, 6986–6988

Owa, T., Sugiyama, T., Otsuka, M. and Ohno, M. (1990). A model study on the mechanism of the autoxidation of bleomycin. *Tetrahedron Lett.*, **31**, 6063–6066

Padbury, G., Sligar, S. G., Labeque, R. and Marnett, L. J. (1988). Ferric bleomycin catalyzed reduction of 10-hydroperoxy-8,12-octadecadienoic acid: evidence for homolytic oxygen–oxygen bond scission. *Biochemistry*, **27**, 7846–7852

Povirk, L. F., Han, Y.-H. and Steighner, R. J. (1989). Structure of bleomycin-induced DNA double-strand breaks: predominance of blunt ends and single-base 5′ extensions. *Biochemistry*, **28**, 5808–5814

Povirk, L. F., Hogan, M., Buechner, M. and Dattagupta, N. (1981). Copper(II)·bleomycin, iron(III)·bleomycin, and copper(II)·phleomycin: comparative study of deoxyribonucleic acid binding. *Biochemistry*, **20**, 665–671

Povirk, L. F., Hogan, M. and Dattagupta, N. (1979). Binding of bleomycin to DNA: intercalation of the bithiazole rings. *Biochemistry*, **18**, 96–101

Povirk, L. F., Kohnlein, W. and Hutchinson, F. (1978). Specificity of DNA base release by bleomycin. *Biochim. Biophys. Acta*, **521**, 126–133

Povirk, L. F., Wübker, W., Köhnlein, W. and Hutchinson, F. (1977). DNA double-strand breaks and alkali-labile bonds produced by bleomycin. *Nucleic Acids Res.*, **4**, 3573–3580

Rabow, L. E., McGall, G. H., Stubbe, J. and Kozarich, J. W. (1990). Identification of the source of oxygen in the alkaline-labile product accompanying cytosine release during bleomycin-mediated oxidative degradation of d(CGCGCG). *J. Am. Chem. Soc.*, **112**, 3203–3208

Rabow, L. E., Stubbe, J., Kozarich, J. W. and Gerlt, J. A. (1986). Identification of the alkaline-labile product accompanying cytosine release during bleomycin-mediated degradation of d(CGCGCG). *J. Am. Chem. Soc.*, **108**, 7130–7131

Rodriguez, L. O. and Hecht, S. M. (1982). Iron(II)–bleomycin. Biochemical and spectral properties in the presence of radical scavengers. *Biochem. Biophys. Res. Commun.*, **104**, 1470–1476

Roy, S. N., Orr, G. A., Brewer, F. and Horwitz, S. B. (1981). Chemical synthesis of radiolabeled bleomycin $A_2$ and its binding to DNA. *Cancer Res.*, **41**, 4471–4477

Saito, I., Morii, T., Sugiyama, H., Matsuura, T., Meares, C. F. and Hecht, S. M. (1989). Photoinduced DNA strand scission by cobalt bleomycin green complex. *J. Am. Chem. Soc.*, **111**, 2307–2308

Sakai, T. T., Riordan, J. M. and Glickson, J. D. (1982). Models of bleomycin interactions with poly(deoxyadenylylthymidylic acid). Fluorescence and proton nuclear magnetic resonance studies of cationic thiazole amides related to bleomycin $A_2$. *Biochemistry*, **21**, 805–816

Sausville, E. A., Peisach, J. and Horwitz, S. B. (1976). A role for ferrous ion and oxygen in the degradation of DNA by bleomycin. *Biochem. Biophys. Res. Commun.*, **73**, 814–822

Sausville, E. A., Peisach, J. and Horwitz, S. B. (1978a). Effect of chelating agents and metal ions on the degradation of DNA by bleomycin. *Biochemistry*, **17**, 2740–2745

Sausville, E. A., Stein, R. W., Peisach, J. and Horwitz, S. B. (1978b). Properties and products of the degradation of DNA by bleomycin and iron(II). *Biochemistry*, **17**, 2746–2754

Shipley, J. B. and Hecht, S. M. (1988). Bleomycin congeners exhibiting altered DNA cleavage specificity. *Chem. Res. Toxicol.*, **1**, 25–27

Shirakawa, I. M., Azegami, M., Ishii, S. and Umezawa, H. (1971). Reaction of bleomycin with DNA: strand scission of DNA in the absence of sulfhydryl or peroxide compounds. *J. Antibiotics*, **24**, 761–766

Steighner, R. J. and Povirk, L. F. (1990). Bleomycin-induced DNA lesions at mutational hot spots: implications for the mechanism of double strand cleavage. *Proc. Natl Acad. Sci. USA*, **87**, 8350–8354

Subramanian, R. and Meares, C. F. (1985). Photo-induced nicking of deoxyribonucleic acid by ruthenium(II)–bleomycin in the presence of air. *Biochem. Biophys. Res. Commun.*, **133**, 1145–1151

Sugiura, Y. (1979). The production of hydroxyl radical from copper(I) complex systems of bleomycin and tallysomycin: Comparison with copper(II) and iron(II) systems. *Biochem. Biophys. Res. Commun.*, **90**, 375–383

Sugiura, Y. (1980a). Monomeric cobalt(II)–oxygen adducts of bleomycin antibiotics in aqueous solution. A new ligand type for oxygen binding and effect of axial Lewis base. *J. Am. Chem. Soc.*, **102**, 5216–5221

Sugiura, Y. (1980b). Bleomycin–iron complexes. Electron spin resonance study, ligand effect and implication for action mechanism. *J. Am. Chem. Soc.*, **102**, 5208–5216

Sugiura, Y. and Kikuchi, T. (1978). Formation of superoxide and hydroxy radicals in iron(II)–bleomycin–oxygen system: electron spin resonance detection by spin trapping. *J. Antibiotics*, **31**, 1310–1312

Sugiura, Y. and Suzuki, T. (1982). Nucleotide sequence specificity of DNA cleavage by iron–bleomycin. Alteration on ethidium bromide-, actinomycin-, and distamycin-intercalated DNA. *J. Biol. Chem.*, **257**, 10544–10546

Sugiyama, H., Kilkuskie, R. E., Chang, L.-H., Ma, L.-T., Hecht, S. M., van der Marel, G. A. and van Boom, J. H. (1986). DNA strand scission by bleomycin: catalytic cleavage and strand selectivity. *J. Am. Chem. Soc.*, **108**, 3852–3854

Sugiyama, H., Kilkuskie, R. E. and Hecht, S. M. (1985a). An efficient, site-specific DNA target for bleomycin. *J. Am. Chem. Soc.*, **107**, 7765–7768

Sugiyama, H., Xu, C., Murugesan, N. and Hecht, S. M. (1985b). Structure of the alkali-labile product formed during iron(II)–bleomycin-mediated DNA strand scission. *J. Am. Chem. Soc.*, **107**, 4104–4105

Sugiyama, H., Xu, C., Murugesan, N., Hecht, S. M., van der Marel, G. A. and van Boom, J. H. (1988). Chemistry of the alkali-labile lesion formed from iron(II) bleomycin and d(CGCTTTAAAGCG). *Biochemistry*, **27**, 58–67

Suzuki, T., Kuwahara, J., Goto, M. and Sugiura, Y. (1985). Nucleotide sequence cleavages of manganese–bleomycin induced by reductant, hydrogen peroxide and ultraviolet light. Comparison with iron– and cobalt–bleomycins. *Biochim. Biophys. Acta*, **824**, 330–335

Suzuki, T., Kuwahara, J. and Sugiura, Y. (1983). Nucleotide sequence cleavage of guanine-modified DNA with aflatoxin $B_1$, dimethyl sulfate, and mitomycin C by bleomycin and deoxyribonuclease I. *Biochem Biophys. Res. Commun.*, **117**, 916–922

Suzuki, H., Nagai, K., Akutsu, E., Yamaki, H., Tanaka, N. and Umezawa, H. (1970). On the mechanism of action of bleomycin: strand scission of DNA caused by bleomycin and its binding to DNA *in vitro*. *J. Antibiotics*, **23**, 473–480

Suzuki, H., Nagai, K., Yamaki, H., Tanaka, N. and Umezawa, H. (1968).

Mechanism of action of bleomycin. Studies with the growing culture of bacterial and tumor cells. *J. Antibiotics*, **21**, 379–386

Suzuki, H., Nagai, K., Yamaki, H., Tanaka, N. and Umezawa, H. (1969). On the mechanism of action of bleomycin: scission of DNA strands *in vitro* and *in vivo*. *J. Antibiotics*, **22**, 446–448

Takahashi, K., Takita, T. and Umezawa, H. (1987). Activation of bleomycin–Fe(III) by bleomycin–Cu(II) and cysteine. *J. Antibiotics*, **40**, 542–546

Takeshita, M., Grollman, A. P., Ohtsubo, E. and Ohtsubo, H. (1978). Interaction of bleomycin with DNA. *Proc. Natl Acad. Sci. USA*, **75**, 5983–5987

Takeshita, M., Kappen, L. S., Grollman, A. P., Eisenberg, M. and Goldberg, I. H. (1981). Strand scission of deoxyribonucleic acid by neocarzinostatin, auromomycin, and bleomycin: studies on base release and nucleotide sequence specificity. *Biochemistry*, **20**, 7599–7606

Takita, T., Muraoka, Y., Nakatani, T., Fujii, A., Iitaka, Y. and Umezawa, H. (1978a). Chemistry of bleomycin. XXI. Metal-complex of bleomycin and its implication for the mechanism of bleomycin action. *J. Antibiotics*, **31**, 1073–1077

Takita, T., Muraoka, Y., Nakatini, T., Fujii, A., Umezawa, Y., Naganawa, H. and Umezawa, H. (1978b). Chemistry of bleomycin. XIX. Revised structures of bleomycin and phleomycin. *J. Antibiotics*, **31**, 801–804

Takita, T., Muraoka, Y., Yoshioka, T., Fujii, A., Maeda, K. and Umezawa, H. (1972). The chemistry of bleomycin. IX. The structures of bleomycin and phleomycin. *J. Antibiotics*, **25**, 755–758

Takita, T., Umezawa, Y., Saito, S., Morishima, H., Miyake, T., Kagayama, S., Umezawa, S., Muraoka, Y., Suzuki, M., Otsuka, M., Narita, M., Kobayashi, S. and Ohno, M. (1982). Total synthesis of bleomycin $A_2$. *Tetrahedron Lett.*, **23**, 521–524

Uesugi, S., Shida, T., Ikehara, M., Kobayashi, Y. and Kyogoku, Y. (1984). Identification of oligonucleotide fragments produced in a strand scission reaction of the d(CGCGCG) duplex by bleomycin. *Nucleic Acids Res.*, **12**, 1581–1592

Umezawa, H., Suhara, Y., Takita, T. and Maeda, K. (1966). Purification of bleomycins. *J. Antibiotics*, **19A**, 210–215

Van Atta, R. B., Long, E. C., Hecht, S. M., van der Marel, G. A. and van Boom, J. H. (1989). Electrochemical activation of oxygenated iron–bleomycin. *J. Am. Chem. Soc.*, **111**, 2722–2724

Vos, C. M., Westera, G. and Shipper, D. (1980). A $^{13}C$ nmr and esr study on the structure of the different forms of the cobalt–bleomycin $A_2$ complex. *J. Inorg. Biochem.*, **13**, 165–177

White, R. E. and Coon, M. J. (1980). Oxygen activation by cytochrome P-450. *Ann. Rev. Biochem.*, **49**, 315–356

Williams, L. D. and Goldberg, I. H. (1988). Selective strand scission by intercalating drugs at DNA bulges. *Biochemistry*, **27**, 3004–3011

Wu, J. C., Kozarich, J. W. and Stubbe, J. (1983). The mechanism of free base formation from DNA by bleomycin. A proposal based on site specific tritium release from poly(dA–dU). *J. Biol. Chem.*, **258**, 4694–4697

Wu, J. C., Kozarich, J. W. and Stubbe, J. (1985). Mechanism of bleomycin: evidence for a rate-determining 4′-hydrogen abstraction from poly(dA–dU) associated with the formation of both free base and base propenal. *Biochemistry*, **24**, 7562–7568

# 6

# Kinetic Analysis of Drug–Nucleic Acid Binding Modes: Absolute Rates and Effects of Salt Concentration

W. David Wilson and Farial A. Tanious

## 1 Introduction

Kinetic studies, particularly if carried out at several salt concentrations, provide information about nucleic acid complexes and how they form which is available from no other method. Kinetic experiments as a function of salt concentration, for example, can be used to determine variations in drug binding mode to different sequences of nucleic acids or between two different helical forms such as the A and B duplex states. The focus of this chapter will be on how to use kinetic results at different salt concentrations to obtain such information. Most of the kinetic results reported to date have been on duplex DNA–drug complexes; however, as a result of the interest in RNA as a potential receptor for antiviral drugs (Ratmeyer *et al.*, 1992), and therapeutic strategies involving combined addition of a drug and a third strand to a DNA duplex (Mergny *et al.*, 1992), an increasing number of studies on drug complexes with RNA and triple helical nucleic acids should appear.

Thermodynamic and kinetic studies of drug–nucleic acid complexes typically monitor some physical property, such as drug absorbance or fluorescence, that undergoes a significant change from the free to the bound state (Cantor and Schimmel, 1980; Crothers, 1971; Bloomfield *et al.*, 1974). In thermodynamic studies the property is measured in a complex solution as a function of either drug or nucleic acid concentration, and an equilibrium constant is calculated from the results. In kinetic studies the property is followed as a function of the time-course of complex formation or dissociation, and a rate constant is obtained.

Thermodynamic and dynamic experiments provide complementary

information about drug–nucleic acid complexes; however, compounds with very similar equilibrium constants can have very different rate constants for complex formation and dissociation. The rate constants have been found to be more important, at least in some systems, than the equilibrium constants in determining the biological activity of the drug (Müller and Crothers, 1968; Crothers, 1971; Gabbay *et al.*, 1976). Because changes in time are more accurately followed and digitized than changes in concentration, kinetic results of absorbance (or other physical property) versus time are often more accurately measured and are at higher resolution than equilibrium absorbance versus concentration curves. It is, thus, easier to detect different, but similar, complex states in the dynamic measurements than in the static equilibrium measurements.

Unlike thermodynamic measurements, kinetic studies provide mechanistic information about the path of drug binding to nucleic acid sites. Intermediate states, even those that exist only transiently, can be identified, and rates for DNA and/or drug conformational changes that occur upon complex formation can be detected. Because nucleic acids are polyelectrolytes, the observed rates for complex formation are dependent on salt concentration, and an analysis of this dependence can provide very useful information about the binding mechanism and, thus, about the mode of drug binding to specific nucleic acid sequences (Lohman *et al.*, 1978; Record *et al.*, 1978; Lohman, 1985; Wilson *et al.*, 1985a, 1990a–c; Tanious *et al.*, 1991, 1992a,b).

## 2 Nucleic Acid Binding Modes

From the standpoint of the physical properties of complexes, three significantly different drug binding modes have been defined with nucleic acids: (1) intercalation or classical intercalation, (2) threading intercalation, and (3) minor groove binding to DNA (Wilson, 1990). Drugs that bind by each of the three modes are shown in Figure 6.1. The classical intercalation mode was originally described in detail by Lerman (1961, 1963) for acridines and by Waring (1981, and references therein) for ethidium and related classical intercalators (Figure 6.1). In this mode a planar aromatic ring system of a drug stacks with nucleic acid base pairs, and any large or polar substituents of the drug are all in the same groove of the nucleic acid, so that there is no significant steric block to sliding the aromatic ring of the drug into the intercalation site in the duplex. In threading intercalation complexes, the aromatic system of the drug is stacked between base pairs, but two distal-substituted side-chains are on the ring system such that one must lie in each of the nucleic acid grooves in order for the intercalation complex to form (Gabbay *et al.*, 1976; Yen *et al.*, 1982; Fox and Waring, 1984; Fox *et al.*, 1985; Gandecha *et al.*, 1985; Wakelin *et al.*, 1987, 1990;

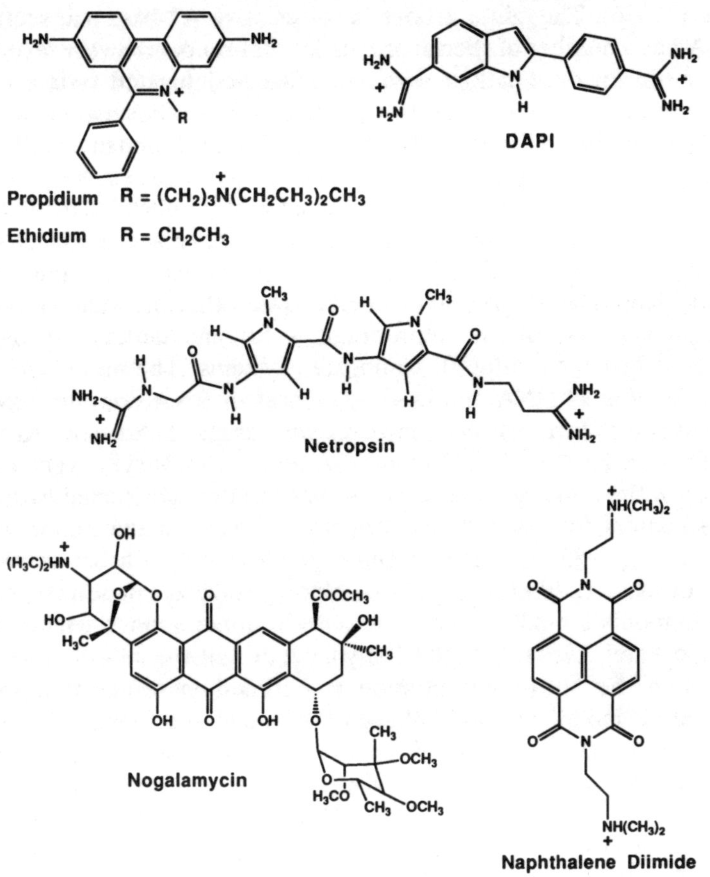


**Figure 6.1** Structures of propidium, ethidium, DAPI, netropsin, nogalamycin and naphthalene diimide

Fairley *et al.*, 1988; Tanious *et al.*, 1991, 1992a). The bulk and polarity/charge of these side-chains can present a significant kinetic block to complex formation and/or dissociation. The side-chains are typically in positions where they are not detrimental to the free energy of complex formation, and they can enhance complex stability. The primary distinction between classical and threading intercalation is in the kinetics of complex formation, and kinetic studies offer a good method for distinguishing between these two binding modes.

The B-form and A-form helical structures of DNA and RNA have two quite different minor grooves and two equally different major grooves (Saenger, 1984; Blackburn and Gait, 1990). Although some drugs can react covalently in the major groove, reversible groove-binding drug complexes have only been identified in the minor groove of DNA (Zimmer and

Wahnert, 1986). The minor groove in consecutive AT base pair sequences of DNA has a number of special properties that render it a very favourable receptor site for drug cations with an unfused-conjugated system of aromatic rings linked by alkene and/or peptide groups (Dickerson and Drew, 1981; Dervan, 1986; Zimmer and Wähnert, 1986). Pullman and Pullman (1981) have pointed out that the AT minor groove has a relatively negative molecular electrostatic potential that significantly enhances its ability to bind cationic drugs. Crystallographic studies by Dickerson and Drew of the oligomer d(CGCGAATTCGCG) (1981) and its netropsin (Figure 6.1) complex (Kopka *et al.*, 1985a–c) demonstrated that the minor groove in AT sequences was of the appropriate width and depth to provide a topological match to unfused, conjugated systems. The minor groove in GC-rich regions of DNA and in all sequences of RNA does not have the correct shape to form strong complexes with drugs of this type (Kopka *et al.*, 1985a–c). Finally, the N3 atom of A and the carboxyl oxygen of T at the floor of the minor groove serve as appropriately positioned hydrogen-bond acceptors for interactions with drugs bound in the minor groove (Kopka *et al.*, 1985a–c; Zimmer and Wähnert, 1986). Chromomycin and related derivatives bind in the DNA minor groove at GC sequences, but these compounds bind as dimers to a wide minor groove conformation (Gao and Patel, 1989a, b, 1990; Banville *et al.*, 1990).

Although the above presentation is outlined in terms of a specific drug-binding mode, we and others have found that some compounds change binding mode with nucleic acid sequence or helical type (Strickland *et al.*, 1988; Wilson *et al.*, 1989, 1990a–c; Fiel *et al.*, 1990, and references therein; Tanious *et al.*, 1992b). DAPI (Figure 6.1), for example, binds in the minor groove at AT sequences of DNA (Wilson *et al.*, 1990a), by intercalation in GC sequences of DNA (Wilson *et al.*, 1989, 1990a), and by intercalation to RNA (Tanious *et al.*, 1992b). We have suggested that intercalation and groove-binding modes should simply be viewed as two potential wells on a continuous energy surface, and that the relative depths of the wells depend primarily on the nucleic acid sequence and on the drug structure (Wilson *et al.*, 1989, 1990a–c). Kinetic studies, particularly as a function of salt concentration, provide a powerful experimental method for evaluation of drug binding modes to different nucleic acid sequences and helical types.

## 3 Ion Effects on Nucleic Acid Structure and Interactions

Even before the discovery of the double helical conformations of DNA, it was recognized that DNA properties and interactions are very sensitive to changes in salt concentration. The structures of the B-form duplex of DNA and the A-form duplex of DNA and RNA provide a means to determine the phosphate–phosphate distances in the molecules, and to quantify the effects of salt concentration. Simple Debye–Hückel analysis does not explain the effects of salt concentration on nucleic acids or other polyelectrolyte structures and interactions. However, Manning (1978, and references therein) found that a counterion condensation theory, with a critical fraction of polyelectrolyte charge neutralized by mobile bound or condensed counterions, and the remainder of the phosphate charge interacting in a Debye–Hückel screening fashion with counterions, could quantitatively explain the effect of salt concentration on nucleic acid properties and interactions. The fraction of condensed counterion is primarily dependent on nucleic acid charge density, and is relatively insensitive to changes in the salt concentration in solution.

Poisson–Boltzmann methods also predict a fraction of highly associated counterions with nucleic acids, and the predictions of the two theories are not distinguishable within experimental error (Wilson *et al.*, 1980; Klein *et al.*, 1981; Anderson and Record, 1990). Record and co-workers (1976, 1978) derived very useful closed-form equations for the thermodynamics and kinetics of DNA transitions and interactions. The basic concept is that as a charged ligand binds to a nucleic acid and/or the nucleic acid undergoes a conformational change, the charge density changes and condensed counterions (e.g. $Na^+$) are released into or absorbed from solution (Record *et al.*, 1976, 1978). For example, conformational changes, such as extension of the double-helical structure of DNA to create an intercalation site, decrease the local charge density and cause release of counterions. Thus, increasing salt concentration decreases the equilibrium constant for formation of an intercalation site or for any other process that results in counterion release. Association of nucleic acid single strands to form a duplex causes an increase in charge density and consequently counterion condensation from solution on to the duplex. Increasing the salt concentration increases the equilibrium constant (or the duplex $T_m$) for this process (Record *et al.*, 1978).

Drugs that bind strongly to nucleic acids typically have one or more cationic centres. When they form a complex with nucleic acids, the local polyanionic charge density is decreased and counterions are released. Increasing the salt concentration will decrease the observed binding constants for such cationic drugs. If the nucleic acid conformation changes during complex formation, such as with formation of an intercalation site, the counterion release will be determined by the changes in charge density induced both by the charge of the bound drug and by the conformational change (Wilson and Lopp, 1979).

It is worth emphasizing here that, because nucleic acids are such highly charged species (the DNA duplex averages one negative charge for every 1.7 Å increment along the helix axis, for example), their conformations and interactions are intrinsically dependent on strongly associated or condensed counterions. These counterions, such as $Na^+$ or $Mg^{2+}$, are not bound at specific sites on the nucleic acid, but are in a highly mobile layer in dynamic exchange with solution ions as originally described by Manning (1978). Because the counterions are not bound at specific sites and their influence on physical and chemical processes of nucleic acid complex formation is not obvious at constant salt concentration, their effects have not always been recognized. Such omission can lead to totally erroneous conclusions concerning relative binding constants, $T_m$ values or rate constants if salt concentrations and/or drug charges are not the same among the systems being compared. On the other hand, analysis of the effects of salt concentration on equilibrium and rate constants provides a very powerful means of developing a complete understanding of nucleic acid–drug complexes.

## 4 Quantitative Aspects

Record and co-workers (1976, 1978) have shown that the observed equilibrium constant ($K_{obs}$) for binding of a monocationic ligand to a nucleic acid depends on counterion concentration ($M^+$), as shown in Equation (6.1):

$$\delta \log K_{obs}/\delta \log [M^+] = -m'\Psi \tag{6.1}$$

where $m'$ is the number of charged groups on the ligand that neutralize phosphate charges on the nucleic acid, and $\Psi$, which depends on the charge density of the nucleic acid, represents the average fraction of counterion condensed per phosphate group. The constant $\Psi$ has two components, $\Psi_c$ for the condensed counterion, and $\Psi_s$, which models the counterion affected by Debye–Hückel screening interactions:

$$\Psi = \Psi_C + \Psi_S = 1 - (2\xi)^{-1} \tag{6.2}$$

where

$$\xi = e^2/\varepsilon kTb \tag{6.3}$$

$e$ is the charge on an electron, $\varepsilon$ is the dielectric constant, $k$ is Boltzmann's constant, $T$ is the temperature in K and $b$ is the average distance between phosphate charges with the nucleic acid modelled as a linear array of negative point charges (Record *et al.*, 1976, 1978; Manning, 1978). As indicated above, the $b$ value for B-form DNA is 1.7 Å and from Equation (6.2) $\Psi$ is 0.88 in water at 25 °C. This indicates that the extent of thermodynamic association of monocationic counterions with each phosphate group in the B-form duplex is an average of 0.88, and this number is insensitive to the bulk counterion concentration. Thus, a probe, drug molecule or other species external to the DNA duplex experiences a field that acts as if each phosphate group has a negative charge of 0.12 instead of the formal charge of −1. The release of associated counterion when cationic drugs bind to DNA is the basis for the salt concentration dependence of the observed equilibrium constant in Equation (6.1). It should also be kept in mind that counterion release can provide a significant part of the free energy change for binding of cationic drugs (or proteins) to nucleic acids.

Equation (6.1) predicts that when a neutral drug binds to DNA ($m' = 0$), there is no salt concentration dependence of $K_{obs}$. This is known to be incorrect for some molecules such as the neutral intercalating anticancer drug actinomycin D (Müller and Crothers, 1968; Friedman and Manning, 1984). Actinomycin has a log $K_{obs}$ versus log $[Na^+]$ slope of approximately −0.2, indicating a significant decrease in binding constant with increasing salt concentration. This finding can be explained by the counterion release generated by the conformational change to form an intercalation site when actinomycin binds to DNA. Wilson and Lopp (1979) found that for intercalators

$$\delta \log K_{obs}/\delta \log [M^+] = -P(\Psi - \Psi^*) - m'\Psi^* \qquad (6.4)$$

where $\Psi^*$ is the $\Psi$ value for DNA in an intercalation conformation and $P$ is the number of phosphate groups at the drug binding site. The first term above represents the counterion release for creation of an intercalation site in B-form DNA, and the second term is for counterion release when intercalators of positive charge $m'$ bind in the intercalation site.

Equation (6.4) correctly predicts the dependence, within experimental error, of the observed equilibrium constant on salt concentration for neutral and charged intercalators. Groove-binding drugs discovered to this time cause much smaller conformational changes in nucleic acids than intercalators (Kopka *et al.*, 1985a–c), and the influence of salt concentration on their $K_{obs}$ values is adequately described by Equation (6.1) (or with $\Psi = \Psi^*$ in Equation 6.4). It should be emphasized that, except for the rare neutral drug molecule that acts at the nucleic acid level, $K_{obs}$ values for intercalators and groove-binding drugs are very dependent on salt concentration, and this dependence must be considered when comparing drug interaction strengths under different conditions. Although, in principle, groove-binding and intercalation binding modes could be distinguished by Equations (6.1) and (6.4), in practice the difference in slope, except for neutral molecules, is too small to use with confidence in this manner.

Groove-binding molecules and intercalators with the same charge can, however, give rise to very different slopes in plots of log $k$ vs log $[M^+]$, where $k$ is the reaction rate constant, since the mechanism of complex formation is different for the two binding modes. Lohman, Record and co-workers (Lohman *et al.*, 1978; Lohman, 1985) have divided nucleic acid kinetics into two limiting types of process: (1) direct-binding processes without stable intermediates where the salt concentration effects on association are primarily controlled by Debye–Hückel screening, and (2) multiple-step reactions with at least one identifiable intermediate. A general mechanism for drug–nucleic acid interaction involves three possible major steps: (1) external electrostatic association of the drug with the nucleic acid, (2) binding of the drug to a nucleic acid binding site (this may involve conformational changes in the nucleic acid and/or drug) and (3) transfer of the drug from the initial to a more favourable binding site until thermodynamic equilibrium is reached. To evaluate sequence- and conformation-dependent binding modes in nucleic acid complexes, it is possible to use nucleic acid oligomers or polymers that have a single or multiple- identical binding sites. With such simplified models for the more complex array of sites that exist on a natural nucleic acid, the kinetic analysis is simplified and only steps (1) and (2) in the general mechanism need to be considered.

The simplest binding mechanism involves a diffusion-limited single-step reaction with no significant conformational change in the drug or nucleic acid (Lohman, 1985). One could envision some AT-specific minor groove associations, for example, that could occur by such a mechanism. The influence of salt concentration on the second-order association rate constant for such a mechanism would result from screening effects of solution counterions on the electrostatic interaction between the free nucleic acid and drug. Lohman *et al.* (1978) and Lohman (1985) have derived equations that predict the dependence of the observed association and dissociation rate constants ($k_a$ and $k_d$, respectively) on monovalent salt concentration for this mechanism:

$$\delta \log k_a/\delta \log [M^+] = -m'\Psi_s \tag{6.5}$$

$$\delta \log k_d/\delta \log [M^+] = +m'\Psi_c \tag{6.6}$$

For any single-step mechanism:

$$K_{obs} = k_a/k_d \tag{6.7}$$

and from the above equations:

$$\delta \log K_{obs}/\delta \log [M^+] = \delta \log k_a/\delta \log [M^+] - \delta \log k_d/\delta \log [M^+] \tag{6.8}$$

$$\delta \log K_{obs}/\delta \log [M^+] = -m'\Psi_s - m'\Psi_c = -m'\Psi \tag{6.9}$$

**Table 6.1** Predicted salt concentration slopes for the observed equilibrium, association and dissociation kinetic constants

| *Complex type* | $\delta \log K_{obs}/\delta \log [M^+]$ | $\delta \log k_a/\delta \log [M^+]$ | $\delta \log k_d/\delta \log [M^+]$ |
|---|---|---|---|
| Groove binding | | | |
| Monocation | −0.88 | −0.12 | 0.76 |
| Dication | −1.76 | −0.24 | 1.52 |
| Classical intercalation | | | |
| Monocation | −1.06 | −0.67 | 0.40 |
| Dication | −1.89 | −1.34 | 0.56 |
| Threading intercalation | | | |
| Monocation | −1.06 | d | d |
| Dication | −1.89 | −1.6 | 0.3 |

d = depends on whether the charged or the neutral side-chain slides through the duplex to form the complex; see text for details.

in agreement with Equation (6.1). Slopes predicted by using Equations (6.5) and (6.6) are collected in Table 6.1 for monocationic and dicationic groove-binding agents.

Some important features emerge from analysis of these equations. First, for groove binding, the association process is much less sensitive to salt concentration than is the dissociation reaction. Second, the rate constant for association decreases with increasing salt concentration, while the dissociation rate constant increases. Third, no matter what the overall magnitude of $k_a$ and $k_d$, their slopes as a function of salt concentration are characteristic of this mechanism and are fixed by Equations (6.5) or (6.6). Measurement of salt effects on kinetics is, thus, a better method for defining the mechanism and mode of binding than is direct analysis of rate constants.

Mechanisms that involve intermediate states, such as intercalation reactions, are dependent in a more complex way on salt concentration, since counterion association or release can be involved in any of the steps (Lohman *et al.*, 1978; Lohman, 1985). Wilson *et al.* (1985a) found that with simple nucleic acid model systems or simple drug molecules binding to more complex nucleic acids, intercalation can be described by a two-step mechanism that involves a pre-equilibrium externally bound complex (L–D) which intercalates as thermal motion of the duplex creates an intercalation site:

$$L + D \rightleftharpoons L{\cdot}D + n\Psi_C\, M^+ \tag{6.10}$$

$$L{\cdot}D \rightleftharpoons C + (P - n)\,(\Psi_C - \Psi^*_C)M^+ + (m' - n)\Psi_C M^+ \tag{6.11}$$

where L is the drug, D is a DNA binding site, $n$ represents the ion interactions in the pre-equilibrium complex, L–D, and C is the final complex with $m'$ ion interactions. The following equations describe the salt dependence of $k_a$ and $k_d$ for this mechanism:

$$\delta \log k_a/\delta \log [M^+] = -n\Psi \tag{6.12}$$

$$\delta \log k_d/\delta \log [M^+] = -n\Psi + P(\Psi - \Psi^*) + m'\Psi^* \tag{6.13}$$

with $P$, $\Psi$ and $\Psi^*$ defined above.

The first step in this mechanism is analogous to condensation of simple ions and simply involves exchange between the intercalator in solution and counterions in the condensation layer. In the L–D complex the intercalator is in the highly mobile condensed counterion phase around the nucleic acid polyanion and is free to diffuse along the nucleic acid in this phase. For a complex of this type $n$ is approximated by

$$n = \Psi_C \cdot m' \tag{6.14}$$

More exact treatments would require impossibly accurate experimental results to define in more detail the nature of the L·D complex. The mechanism assumes that external binding of the intercalator involves only non-specific, electrostatic effects with no specific external site binding. If such external site binding occurs, as is observed with some intercalators at low salt concentration, the mechanism would have to be expanded to include such effects.

Predicted slopes for $K_{obs}$, $k_a$ and $k_d$ with monocationic and dicationic drug–DNA complexes are compared in Table 6.1. The slopes for rate constants depend on both the drug charge and the binding mechanism, while the $K_{obs}$ slope depends primarily on drug charge. As can be seen, $k_d$ is much more sensitive to salt concentration than is $k_a$ for groove-binding drugs, while $k_a$ is more sensitive than $k_d$ for intercalators. Since the differences in slope for $k_a$ and $k_d$ are large, when $m'$ is known, the binding mode can be determined using any nucleic acid sequence or helical type. Fortunately, the $K_{obs}$ slope is relatively insensitive to binding mode (Table 6.1), and, hence, provides an experimental method for determination of $m'$, which in most cases will simply be close to the charge on the drug. Analyses of salt concentration effects on drug kinetics and equilibrium can, thus, define the drug-binding mode and whether it changes in a sequence- or helical-type dependent manner.

The threading intercalation mode is potentially more complicated, since the drug insertion into the duplex involves threading a bulky and/or polar/charged side-chain through the double helix to give the final intercalation complex. The first step in the mechanism should involve a condensation-like association of the cationic intercalator with the double helix, as with simple intercalators (Equation 6.10). For a 9,10-substituted anthracene derivative or the naphthalene diimide in Figure 6.1, a charged side-chain must pass between base pairs from one groove to the other to form the intercalation complex. When the nucleic acid is viewed as a line charge, this is not important, but for the actual duplex it is not obvious how this affects the local charge density. A counterion may condense along one groove of the duplex as the charged side-chain passes through the base pair region and be released from the opposite groove as the charged side-chain passes through the intercalation site. Alternatively, the number of associated counterions may be described in a manner much more similar to that observed for simple intercalators. In order to obtain experimental results for a fairly simple system, the naphthalene diimide shown in Figure 6.1 was prepared, to investigate the threading mechanism (Gabbay, 1978; Yen *et al.*, 1982). A number of physical techniques indicate that the diimide is an intercalator, and it can only bind strongly to nucleic acid duplexes by a threading mode to form an intercalation complex (Yen *et al.*, 1982).

The results with the diimide, which will be described in more detail below, indicate that for the dissociation step the slope of log $k_d$ against log $[M^+]$ is more similar to the value observed for monocations than for dications (Tanious *et al.*, 1991). This suggests that in the rate-determining step for dissociation of the threading intercalator, sliding one side-chain between base pairs, a single charge interaction is broken in the complex and all succeeding steps in the mechanism are significantly faster. The following mechanism accounts for all observations on threading intercalators:

$$\mathrm{T} + \mathrm{D} \rightleftharpoons \mathrm{T{\cdot}D} + n\Psi_C\mathrm{M}^+ \tag{6.15}$$

$$\mathrm{T{\cdot}D} + (n - x)\,\Psi_C\mathrm{M}^+ \rightleftharpoons \mathrm{T{\cdot}D}^* \tag{6.16}$$

$$\mathrm{T{\cdot}D}^* \rightleftharpoons \mathrm{C} + P(\Psi_C - \Psi_C{}^*)\mathrm{M}^+ + (m' - n + x)\,\Psi_C{}^*\mathrm{M}^+ \tag{6.17}$$

where T is the threading intercalator and the first step, Equation (6.15), is exactly analogous to external condensation type binding of simple intercalators described in Equation (6.10). The assumption in the second reaction of the mechanism, Equation (6.16), is that some fraction of the ion interactions are transiently released prior to threading the side-chain between base pairs to the opposite groove. The intermediate species, T·D*, can then pass one side-chain to the opposite groove to produce the final complex, C, in a reaction similar to Equation (6.11) for simple intercalators. This is a slow step that is rate-determining for this mechanism. For the naphthalene diimide in Figure 6.1, $m' = 2$, and from Equation (6.14), $n = 2{\cdot}\Psi_C$. Since the charges are symmetrical on the naphthalene diimide, $x = (n/2)$.

To a reasonable approximation, the naphthalene diimide dissociation, the reverse of Equation (6.17), is experimentally very similar to dissociation of a monocation from DNA. The experimental finding that the slope of log $k$ against $-\log\,[Na^+]$ for dissociation of several dicationic threading intercalators (Wilson *et al.*, 1990; Tanious *et al.*, 1991, 1992a), with one charge on each of the two distal side-chains, is more similar to the slope for monocations provides the basis for the mechanism in Equations (6.15)–(6.17). Threading intercalation is, thus, identified by particularly slow association rate constants, relative to other similar non-threading intercalators, and by a log $k_d$ versus log $[M^+]$ dissociation kinetics slope that is more similar to the slope for monocationic than for dicationic intercalators. Salt concentration slopes for dissociation and association rate constants for this threading intercalation mechanism are collected in Table 6.1.

The slow kinetics of association of threading intercalators can be attributed to several features that depend on the structure of the intercalator and on the nucleic acid sequence. First, the polar/charged side-chain must pass through the more non-polar region between base pairs in the duplex. Second, the steric size and rigidity of the side-chains can be important. Third, the rate at which DNA can open to form intercalation sites or larger open conformations (a bubble-type structure) to accommodate large side-chain dynamics can become dominant for large threading intercalators. A cationic alkylamine side-chain, such as on the naphthalene diimide in Figure 6.1, can assume an extended configuration that will allow it to slide through a normal intercalation site. The slow kinetics of threading could then be due to an orientation factor that requires the side-chain to be in a limited range of its total conformational space for intercalation. The resistance of the charged side-chain to enter the relatively non-polar region between base pairs could also serve to reduce the reaction rate for a threading intercalator even when it has an appropriate conformation to pass through the intercalation cavity. For threading intercalators with very large and rigid side-chains, the normal intercalation site size would not allow the side-chain to pass through to the opposite groove. In this case intercalation could only occur when slower thermal motions of the duplex produce a large opening that could arise, for example, by concerted opening of an intercalation site and breaking/breathing of one of the base pairs at the site. It should, thus, be possible to design synthetic intercalators with large and rigid side-chains producing intercalation kinetics significantly slower than for threading intercalators designed to date.

Monocationic threading intercalators must be asymmetric, since only one side-chain can be charged. Salt concentration effects on nucleic acid binding kinetics for such intercalators will depend on which side-chain passes through the helix to form the equilibrium complex. Although a neutral chain would seem to be favoured to pass through the non-polar region of the duplex, if the charged chain were less bulky and/or rigid, it could slide through the intercalation site more easily than a neutral side-chain. Highly accurate kinetic studies should be able to determine which side-chain passes through the duplex in the binding mechanism. On the basis of the mechanism above, if the neutral chain passes through, the salt effects should be similar to results for simple monocations, while if the charged side-chain passes through, the effects of salt concentration on dissociation kinetics should be more similar to effects for a neutral intercalator.

## 5 Methods

Drug cations that bind to nucleic acids generally have extended conjugated systems that undergo appreciable changes in spectral properties (electronic, fluorescence, circular dichroism and NMR) on complex formation. Induced changes in absorption and fluorescence spectra can generally be conveniently and inexpensively monitored, and rapidly digitized for computerized collection of spectral changes as a function of time (Figure 6.2). Most drug–nucleic acid reactions which have been investigated are complete in times from milliseconds to 1 min. The neutral antibiotic actinomycin and some threading intercalators can produce appreciably longer reaction times.

The vast majority of reactions of interest are therefore too fast for hand-mixing of reactants with subsequent data collection, and stopped-flow instrumentation has generally been used for mixing and data collection. For association reactions this simply involves placing the desired nucleic acid and drug solutions in separate syringes at constant temperature prior to mixing. At an appropriate trigger a syringe drive is activated, the two solutions are forced through a mixing chamber into a spectrophotometer cell, flow is stopped as the drive syringes hit a preset barrier, and computerized data collection is initiated by an electronic trigger (Figure 6.2b). Minimum mixing volumes are generally in the 50–100 μl range and the entire system is maintained at a constant temperature. The association experiment is typically carried out with a large excess of nucleic acid binding sites over drug concentration so as to give a pseudo-first-order reaction.

For dissociation reactions, the nucleic acid–drug complex is in one syringe, while an anionic detergent, such as sodium dodecyl sulfate (SDS), is in the other syringe. Mixing and data collection are as with association reactions. Müller and Crothers (1968) first discovered that detergents could be used to sequester free drug as it dissociated from its nucleic acid complex without apparently perturbing the complex itself. In any new system detergent effects can be tested by collecting dissociation results at several detergent concentrations to insure that the same rate constants are obtained in all cases. It should be noted, however, that detergents such as

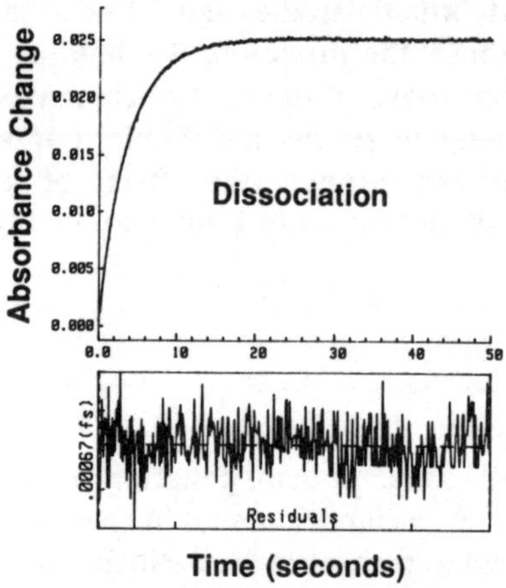


(a)

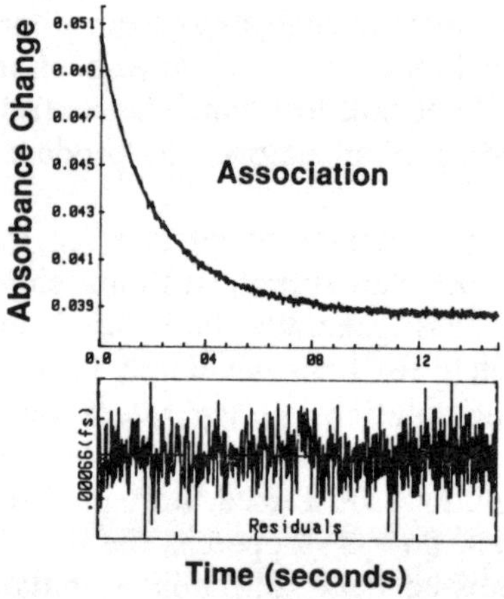


(b)

**Figure 6.2** (a) Stopped-flow kinetic trace for the SDS-driven dissociation of the naphthalene diimide (Figure 6.1) from the alternating sequence polymer polyd(G–C)$_2$ at 25 °C. The experiments were conducted in MES buffer (0.01 M 2-(*N*-morpholino)ethanesulfonic acid, 0.001 M EDTA, pH 6.2) with 0.60 M NaCl at a ratio of 1:10 naphthalene diimide to polymer base pairs. The concentration of the naphthalene diimide after mixing was $6.25 \times 10^{-6}$ M. The smooth line in the panel is the double exponential fit to the experimental data, and a residual plot for the fit is shown under the experimental plot. (b) A stopped-flow association reaction plot for the naphthalene diimide with polyd(G–C)$_2$. The total concentrations after mixing were $2.5 \times 10^{-5}$ M for polymer (base pairs) and $2.5 \times 10^{-6}$ M for the naphthalene diimide in MES buffer with 0.60 M added NaCl at 20 °C. The smooth line in the panel is the double exponential fit to the experimental data, and a residual plot for the fit is shown under the experimental plot

SDS affect the sodium ion concentration, and any change in SDS must be accompanied by a correlated reduction in NaCl concentration to keep a constant $Na^+$ concentration. The $Na^+$ for SDS must also be included in calculations of the total counterion concentration.

Dissociation can be induced by other methods, such as an increase in salt concentration, since the $K_{obs}$ value decreases as the counterion concentration increases (Equation 6.1). The nucleic acid–drug complex in one syringe is mixed with a solution of higher salt concentration in the same buffer in the second syringe. The salt concentration jump occurs in the mixing step, and the dissociation reaction occurs at fixed salt concentration after flow is stopped. The disadvantage of this method is that results can not be determined at very low salt concentration. Dissociation can also be induced by dilution of the complex. This requires two syringes of different volume in the stopped-flow instrument. The concentrated complex in a smaller syringe is mixed with a large volume of the same solvent or buffer solution delivered from a second, larger syringe. Large dilutions are generally required to induce significant dissociation, and if the volume difference in the syringes is too large, efficient mixing can become a problem. For these reasons, the detergent-driven dissociation method has been most widely used for determination of dissociation constants.

With computerized digitization and collection of results both dissociation reactions and association reactions, followed under pseudo-first-order conditions, give similar types of absorbance or fluorescence versus time profiles stored in computer memory (Figure 6.2). The data can be fitted with a non-linear least squares method by using the function

$$P = A\,\mathrm{e}^{-kt} \tag{6.18}$$

where $P$ is the spectral property being followed, $A$ is the amplitude of the spectral change, $k$ is the first-order rate constant and $t$ is the time. If there are multiple binding sites for the drug on the nucleic acid, the fitting equation is expanded:

$$P = A_1\,\mathrm{e}^{-k_1t} + A_2\,\mathrm{e}^{-k_2t} + \cdots A_n\,\mathrm{e}^{-k_nt} \tag{6.19}$$

where $A_i$ and $k_i$ are the amplitudes and the first-order rate constants for each of the binding reactions. A problem with all fitting methods, as the number of variables expands, is that multiple minima appear, and for the smooth curves of $P$ versus $t$ obtained in kinetic experiments (see Figure 6.2) there can be several minima (for example, in the sum of squared residuals (SSR) between the experimental and calculated curves) which are not distinguishable within the range of experimental error. When this occurs, two or more combinations of $A$ and $k$ values provide a fit to the experimental results that fall within the experimental error range. As an operational approach to deciding the minimum number of exponentials to use in fitting multiphasic kinetic curves, we analyse the minimized SSR, and add exponentials until this sum does not change significantly relative to the experimental random error in the dependent variable, the amplitude in these kinetic experiments. We then analyse a plot of the residuals of amplitude versus time, and again reject the fit if there is any significant systematic trend in residuals (Turner *et al.*, 1981; Krishnamoorthy *et al.*, 1986).

Depending on the level of experimental error, the multiple minima problem can become significant at a two-exponential fit (i.e. when using $A_1$, $A_2$, $k_1$ and $k_2$), and the problem becomes acute at a three-exponential fit. Without essentially perfect experimental data, there is no hope of obtaining a unique fit to more than three exponentials. Multiple binding sites will be more of a problem with natural nucleic acid samples containing complex sequences, and the problem may be reduced in complexity, even to a single binding site type, by using synthetic polymers or oligonucleotides.

Ethidium and propidium, which bind to DNA in a fairly non-specific fashion, yield kinetic curves that can be fitted within experimental error by a single-exponential function (Wilson *et al.*, 1985a,b, 1986; see also Wakelin and Waring, 1980). We have found with most other drugs that bind to nucleic acids, however, that even when studying complexes with simple polymers such as polyd(A–T)$_2$ and polyd(G–C)$_2$, fits to the experimental results require at least two exponential curves (Wilson *et al.*, 1990, 1992; Tanious *et al.*, 1991, 1992a). The two rate constants are typically not very different (less than a factor of 5), and the two processes are not resolved in lower-resolution equilibrium measurements. The two processes with similar rate constants could represent two different orientations of the bound drug, drug binding at purine–pyrimidine and pyrimidine–purine sequences in the alternating sequence polymers, major groove or minor groove orientation of side-chains in an intercalation complex, opposite orientations of a groove-binding agent in a nucleic acid groove, or some other explanation that is not obvious at present. Because of the uncertainty in origin of the two processes, the multiple minima problem and the fact that equilibrium methods with the same compounds generally provide only a single $K_{obs}$ or $T_m$ value, kinetic results have frequently been averaged according to Equation (6.20) to provide quantities that are more easily compared among compounds and to equilibrium results for the same system:

$$k = (k_1A_1 + k_2A_2)/(A_1 + A_2) \tag{6.20}$$

where the $k$ and $A$ values are as defined above.

## 6 Applications to Drug–Nucleic Acid Complexes: Classical Intercalation, Threading Intercalation and Groove Binding

As noted above, this chapter is focused on applications of effects of salt concentration on drug association and dissociation kinetics with nucleic acids, and how the results can be used to evaluate binding modes. We shall illustrate the differences in absolute association rates and the large differences in salt effects on both association and dissociation rate constants for compounds that bind by classical intercalation, threading intercalation and groove-binding modes. We shall use results from our laboratory for the intercalator, propidium (Waring, 1981, Wilson *et al.*, 1985); for the threading diimide intercalator in Figure 6.1 (Yen *et al.*, 1982; Tanious *et al.*, 1991); and for DAPI, which binds in the minor groove at AT sequences of DNA and by intercalation into GC sequences (Wilson *et al.*, 1989, 1990a–c). Studies with many other compounds in these classes suggest that they exhibit behaviour similar to that of the compounds used as examples of each binding mode.

Most of the drugs that have been studied have large $K_{obs}$ and $k_a$ values for their DNA complexes. Association reactions are, thus, quite fast and in most cases low concentrations must be employed to determine bimolecular association rate constants. Such low concentrations are feasible for compounds with large extinction coefficients or large fluorescence quantum yields that undergo appreciable changes from the free to the bound state. When large changes in some spectral property do not occur on complex formation, it becomes quite difficult to measure $k_a$ values as a function of salt concentration. Typically dissociation reactions are first-order and much slower than the process of association. Dissociation reactions can, thus, be monitored at higher concentrations, and $k_d$ values can be accurately determined for compounds with a broad range of $K_{obs}$ values. Stopped-flow kinetic plots for association and dissociation reactions for naphthalene diimide binding to polyd(G–C)$_2$ are shown as examples in Figure 6.2. Non-linear least squares curves, fitted to the experimental results as described in the section on Methods, are shown in the figure, along with residual plots.

## 7 Association Reactions

The association results from Figure 6.2 were obtained under pseudo-first-order conditions with DNA present in excess. The experiment was repeated at several DNA concentrations and the results, together with those for the other two example compounds, are plotted in Figure 6.3 as $(k_a)_{app}$, the apparent pseudo-first-order rate constant, versus the DNA concentration, according to Equation (6.21):

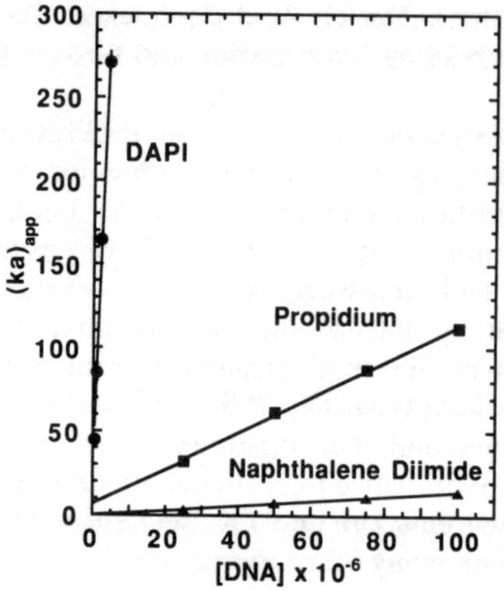


**Figure 6.3** Plots of $(k_a)_{app}$ as a function of polymer concentration in base pairs for (●) DAPI with polyd(A–T)$_2$, (■) propidium with CT–DNA and (▲) naphthalene diimide with polyd(G–C)$_2$. The slopes of the lines represent the observed second-order association rate constants. Experiments were conducted in MES buffer with 0.2 M $Na^+$ in the manner described in Figure 6.2(b)

$$(k_a)_{app} = k_a\,[\text{DNA}] + k_d \qquad (6.21)$$

The second-order association rate constant, $k_a$, under a particular set of solution conditions, is obtained from the slope of this plot. Propidium yields very similar association plots with natural DNA and with the polymers polyd(A–T)$_2$ and polyd(G–C)$_2$. All three of the example molecules are dications, so that differences among their association rates are not due to charge differences.

The compounds for which results are shown in Figure 6.3 typify three different modes of binding. The differences in slopes are very large in Figure 6.3 and serve as a general characterization of the binding modes. DAPI is a groove-binding agent at AT sequences of DNA, and associates with polyd(A–T)$_2$ at close to the diffusion-controlled limit. Propidium is a classical intercalator, and associates much more slowly with DNA than the groove-binding compound. The naphthalene diimide is a threading intercalator, and associates with DNA even more slowly than propidium. The three dications are representative of other compounds in their respective binding classes. A number of AT-specific minor groove binding drugs display $k_a$ values similar to the value for DAPI under the same conditions. Similarly, other simple intercalators display association reactions in the same range as propidium under the same conditions (Tanious *et al.*, 1992a), and other simple, threading intercalators yield association rate constants similar to the naphthalene diimide value (Tanious *et al.*, 1992a).

Association constants for the groove-binding agents are $\sim 10^8\ \text{M}^{-1}\text{s}^{-1}$, for simple intercalators are $2–3 \times 10^6\ \text{M}^{-1}\text{s}^{-1}$ and for simple threading intercalators are $1–2 \times 10^5\ \text{M}^{-1}\text{s}^{-1}$ at 20 °C in 0.2 M $Na^+$. More bulky, rigid and complex threading intercalators, such as the natural product nogalamycin (Figure 6.1), can have significantly lower $k_a$ values.

The association constants observed for these relatively simple compounds are, thus, more characteristic of the *binding mode* than of the specific drug and DNA binding site. With groove-binding molecules, which require little change in DNA or drug conformation, a rate approaching diffusion control is obtained. Classical intercalators first form a loose external complex with the anionic duplex, and then diffuse along the double helix until they undergo an appropriate collision with a thermally opened intercalation site. Provided that there is no significant block to inserting the aromatic ring system of the drug between base pairs at the intercalation site, the process should be essentially independent of the drug molecule except for charge. Threading intercalators bind according to a very similar mechanism, except that, as discussed above, they experience an additional barrier to the intercalation step, and associate by a factor of at least 10 more slowly than classical intercalators. It should be emphasized that the naphthalene diimide (Figure 6.1) and related compounds with simple alkylamine threading substituents are the simplest threading intercalators. Compounds with larger and/or more rigid threading substituents should associate with nucleic acids much more slowly in a compound-specific manner.

Results such as those shown in Figure 6.3 can also be obtained at additional salt concentrations, when the experiments fall in the stopped-flow range, and the slopes of log $k_a$ versus log $[Na^+]$ plots can be compared with predictions in Table 6.1. Results are shown in Figure 6.4 for the minor-groove association of DAPI with polyd$(A–T)_2$, and the slope for this plot is 0.24, in good agreement with the prediction of a very small concentration dependence for groove-binding dications (Table 6.1). The association plot shown for propidium in Figure 6.4 has a slope of 1.4, in good agreement with the prediction of a large salt concentration dependence for dicationic intercalators (Table 6.1).

The slope for the diimide threading intercalator in Figure 6.4 is greater than the propidium slope (1.8 versus 1.4), although the difference is close to the limit imposed by experimental error. It appears for this and other simple threading molecules that dissociation reactions occur with reduced slopes (Table 6.1), and that a directly related increased slope is observed for the association reaction as predicted by Equation (6.8). Such simple behaviour has also been observed for an anthraquinone threading intercalator (Tanious *et al.*, 1992a); however, more complex threading molecules which bind by more elaborate mechanisms may exhibit more complex effects of salt concentration on the association reaction.

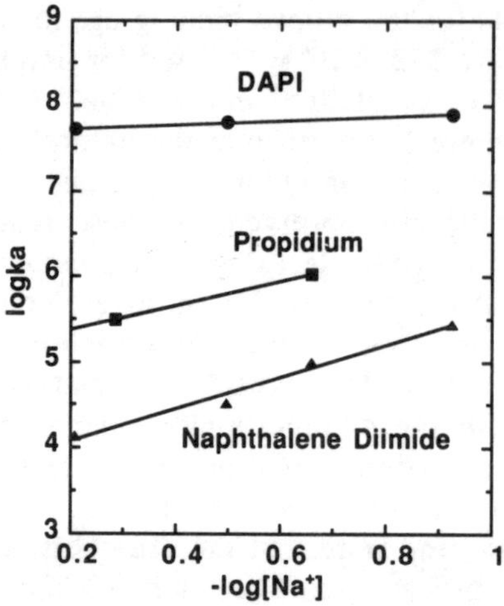


**Figure 6.4** Plots of log $k_a$ against $-\log [Na^+]$ for association of naphthalene diimide with polyd(G–C)$_2$ (○), DAPI with poly d(A–T)$_2$ (●) and propidium with calf thymus DNA (■). The $k_a$ values for DAPI and naphthalene diimide are the average values from Equation (6.20) for a double exponential fit. The propidium results are fitted by a single exponential curve

## 8 Dissociation Reactions

A typical trace of absorbance versus time is shown for SDS-driven dissociation of a naphthalene diimide–polyd(G–C)$_2$ complex in Figure 6.2. As indicated above, the dissociation reaction for such a polymer–drug complex is first-order, and can be followed at higher concentration in the stopped-flow time range than a bimolecular association reaction. For these reasons, it is generally easier to obtain high-quality experimental results for dissociation reactions, and more dissociation rate constants than association rate constants have been published for nucleic acid complexes. The association rate constants described above fall into three groups that are characterized by the general binding mode for compounds in the group. There is no such systematic grouping of dissociation constants according to binding mode.

Dissociation constants for the three compounds of Figure 6.3 are plotted as a function of salt concentration in Figure 6.5. Although all compounds are dications, the $k_d$ values vary by more than a factor of 100 under these conditions, and the slopes of the plots are also quite different. Propidium, which yields very similar $k_d$ values and slopes for dissociation from complexes with polyd(A–T)$_2$ and polyd(G–C)$_2$, yields slopes for dissociation from the two polymer complexes of 0.7–0.8, slightly higher than the value of 0.56 predicted for dicationic intercalators in Table 6.1, but well below the value of 1.52 predicted for dicationic groove-binding agents. Combina-

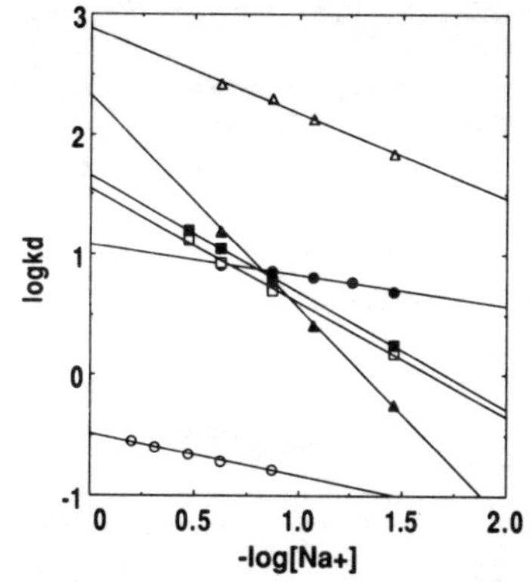


**Figure 6.5** Plots of log $k_d$ against $-\log\,[Na^+]$ for dissociation of the naphthalene diimide from polyd$(A-T)_2$ (●) or polyd$(G-C)_2$ (○); DAPI from polyd$(A-T)_2$ (▲) or polyd$(G-C)_2$ (△); propidium from polyd$(A-T)_2$ (■) or polyd$(G-C)_2$ (□). Experiments were conducted in MES buffer at different ionic strengths in the manner described in Figure 6.2(a). The kinetic constants for naphthalene diimide and DAPI are the averaged values from the double-exponential fits, while the propidium results were fitted with a single exponential curve

tion of $k_a$ and $k_d$ values according to Equation (6.7) under a fixed set of conditions allows the calculation of an equilibrium constant, and values determined in this way are in good agreement with direct measurements of the equilibrium constant for propidium (Wilson *et al.*, 1985a). These results agree with many other physical studies on propidium, ethidium and related drugs which indicate fairly non-specific intercalation binding for compounds of this type.

The dramatically different $k_d$ values and slopes for DAPI in Figure 6.5 indicate that this compound engages in much more complex interactions with DNA than does propidium. It is well known that DAPI binds strongly in the minor groove at AT sequences in DNA (Zimmer and Wähnert, 1986; Larsen *et al.*, 1989), and the association rates and slopes in Figure 6.4 agree with a groove-binding mode for DAPI at AT sequences. The slope for the AT complex of DAPI in Figure 6.5 is 1.8, somewhat higher than the predicted value of 1.52 (Table 6.1), but clearly not in the intercalation range of 0.56 for dications. The dissociation rate constant for the GC complex of DAPI is much higher than for the AT complex, and the slope is significantly lower. The DAPI–GC slope is similar to the propidium slope, and a number of other physical measurements also indicate that DAPI intercalates into GC-rich sequences of DNA (Wilson *et al.*, 1989, 1990a,b). This compound represents a member of the growing class of nucleic acid interactive drugs whose binding mode varies with sequence (Wilson *et al.*, 1990c, 1992).

The sequence-dependence of the DNA binding mode for DAPI and related compounds raises the intriguing question of how the binding mode

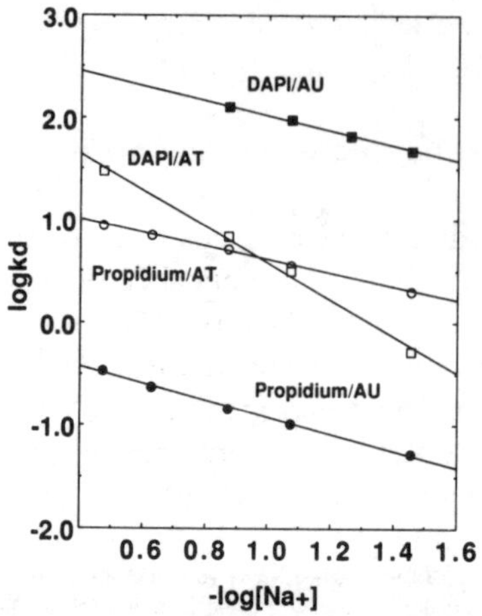


**Figure 6.6** Plots of log $k_d$ against $-\log\ [Na^+]$ for dissociation of propidium from polydA·polydT (○) or polyA·polyU (●), and DAPI from polydA·polydT (□) or polyA·polyU (■). Experiments were conducted in MES buffer at different ionic strengths in the manner described in Figure 6.2(a). The kinetic constants for DAPI are the averaged values from double-exponential fits, while the propidium results were fitted with a single exponential curve

depends on helical type in corresponding sequences. To answer this question, kinetic experiments were conducted on DAPI complexes with the non-alternating DNA sequence polydA·polydT, and the corresponding RNA sequence polyA·polyU (Tanious *et al.*, 1992b). Rate constants for dissociation of DAPI from the RNA and DNA complexes are plotted as a function of salt concentration in Figure 6.6, and are compared with results for propidium, which intercalates into both DNA and RNA. The slope for propidium with the DNA polymer agrees with predictions for an intercalator (Table 6.1), and the slope for DAPI is as predicted for a groove-binding mode as expected for an AT DNA sequence. The propidium slope for dissociation from the RNA polymer is slightly greater than that from the DNA polymer, as expected for intercalation with the higher-charge-density RNA A-form helix. The $k_d$ values for propidium are lower for RNA than for the DNA sequence, in agreement with binding studies (Bresloff and Crothers, 1981) and $T_m$ experiments (Ratmeyer *et al.*, 1992) that indicate a significantly stronger interaction of ethidium and propidium with polyA·U than with polydA·dT (Tanious *et al.*, 1992b).

DAPI yields significantly different results with the two polymers. It dissociates faster from RNA than from DNA polymers, and the slope for the RNA complex is in the intercalation, not the groove-binding, region. Other results also indicate that DAPI binds to RNA by intercalation (Tanious *et al.*, 1992b). These findings indicate that DAPI should actually be viewed primarily as an intercalator, since it appears to intercalate into all DNA and RNA sequences except for its very strong minor-groove mode of binding to DNA sequences of three or more consecutive AT base pairs.

The results from kinetic studies on the modes of binding of DAPI and related compounds have been supported by other physical methods, and it is clear that kinetic studies provide a sensitive method for evaluating drug binding modes with any sequence or helical type of nucleic acid polymer or oligomers. The sequence- and helical-type dependence of DAPI binding modes agrees with our emerging knowledge of the physical properties of grooves in nucleic acids. The minor groove at AT sequences in DNA has a negative electrostatic potential, appropriate hydrogen-bond accepting groups and groove topology suitable to interact strongly with cations such as DAPI (Dickerson and Drew, 1981; Pullman and Pullman, 1981; Kopka *et al.*, 1985a–c; Dervan, 1986; Zimmer and Wähnert, 1986). These characteristics are diminished in the grooves of DNA at regions composed of GC base pairs and in the minor groove of all sequences of RNA. Compounds such as DAPI switch to an intercalation mode in GC regions of DNA and in RNA, while dications such as netropsin, which do not form strong intercalation complexes, bind very weakly through electrostatic external interactions at GC regions of DNA or RNA (Wilson *et al.*, 1989, 1990a–c; Tanious *et al.*, 1992b).

The naphthalene diimide yields the highest slope in Figure 6.3, but the lowest slope in Figure 6.5. This type of behaviour is characteristic of a threading intercalation mode for a large number of threading intercalators that we have now investigated (Tanious *et al.*, 1992a). The slopes for the naphthalene diimide in Figure 6.5 are in the range expected for a monocationic simple intercalator, and, as indicated above, suggest that the rate-determining step in dissociation involves release of a single charged side-chain to slide between base pairs from one groove to the other (reverse of threading).

The dissociation of the naphthalene diimide complex with the GC DNA polymer is much slower than with the AT polymer (Figure 6.5). An attractive explanation for this difference could involve slower dynamics for the GC polymer to release the threaded diimide; however, this does not agree with results for other threading intercalators (Tanious *et al.*, 1992b; Yao and Wilson, manuscript in preparation). The naphthalene diimide displays an equilibrium GC-binding specificity, and its slower dissociation from the GC complex appears to be due to more favourable interactions with GC than with AT base pairs. Thus, the higher energy of activation for dissociation of the diimide–GC complex arises from the need to break additional favourable contacts and not from significantly slower dynamics of the complex. Other threading intercalators, which manifest AT specific binding interactions, dissociate more slowly from AT DNA polymer complexes than from GC complexes (Yao and Wilson, manuscript in preparation).

## 9 Mechanism of Nucleic Acid–Drug Interactions

The studies described in this review allow us to define the fundamental mechanisms by which drugs bind to nucleic acid sequences. These conclusions hold for binding of drugs to synthetic polymers or oligomers, or for initial interactions of drug with a natural nucleic acid. Binding to natural nucleic acids can involve additional mechanistic steps such as transfer of bound drugs from less to more energetically favourable sites as equilibrium is approached.

At the most general level, it is now apparent that association reactions fall into three rate groups that depend primarily on the mode of binding of the drug to the nucleic acid. Dissociation rate constants do not show any such systematic behaviour, and appear to depend primarily on the number and strength of interactions in the drug–nucleic acid complex that must be broken for dissociation to occur. In Figure 6.5, for example, the $k_d$ values for a classical intercalator, a threading intercalator and a groove-binding compound are essentially the same, near 0.1 M $Na^+$. Influences of binding mode are no doubt important in dissociation reactions, but are masked by the strong effects of specific interactions that must occur for dissociation.

Activation energies for association reactions appear to be determined by the need for the nucleic acid and drug molecules to assume the necessary conformation and orientation for complex formation. For groove-binding complexes the conformational rearrangements required are generally minor and complex formation occurs at near the diffusion-controlled limit. Classical intercalation involves the initial formation of an external, electrostatic mobile complex from which the ligand can diffuse to a thermally opened site on the nucleic acid, and insert to form the intercalation complex. The energy of activation for this process appears to be primarily determined by the energetics of opening the pre-intercalation site in the nucleic acid duplex. The site can arise through concerted thermal motions that involve torsional angles in the duplex backbone (Berman and Neidle, 1980). As the intercalator diffuses in the anionic field surrounding the nucleic acid, it also rotates so that, as the transient pre-intercalation site is approached, it can assume an orientation satisfactory for insertion to form the intercalation complex.

Threading intercalators undergo very similar mechanistic steps up to the point of insertion into the pre-intercalation site. At this point formation of the complex can be blocked by one of the bulky substituents at opposite sides of the aromatic ring system of the intercalator. For intercalation of the threading molecule to occur requires additional larger opening of the pre-intercalation site or rotation of the intercalator substituent(s) to a configuration that will allow it to slide through the normal-sized pre-intercalation site. Clearly, some combination of these effects could also lead to the threaded final intercalation complex, but in any combination a higher energy of activation is demanded than for classical intercalation. The scaling of rates for groove-binding, classical intercalation and threading intercalation reactions is readily seen in Figure 6.3.

If the side-chains of the threading intercalator are more rigid, as in nogalamycin, then side-chain rotation can not occur to yield an intercalation complex, and larger dynamic motions of the nucleic acid represent the only mechanism for complex formation. It will be interesting to synthesize aromatic systems that have large, rigid substituents that do not block stacking of the aromatic system with base pairs in the final complex. Such compounds may have significant biological activity, and will certainly provide interesting and important information about large-scale dynamic motions of nucleic acid duplexes that can not be obtained by any other method. Kinetic studies have much more to tell us about nucleic acids and their complexes.

**Acknowledgements**

This work was supported by NIH Grants AI-27196 and AI-33363. We gratefully acknowledge a number of our collaborators for compounds, information, suggestions and ideas, without which the kinetic studies from our laboratory would not have been able to progress as they have: Drs David Boykin, Lucjan Strekowski, Stephen Neidle, Terence C. Jenkins, Luigi Marzilli, and Shau-Fong Yen. Discussions with Drs J. B. Chaires, W. Denny, and L. Wakelin have also been of valuable assistance as we formulated our ideas on mechanisms of drug–nucleic acid interactions.

**References**

Anderson, C. F. and Record, M. T. (1990). *Ann. Rev. Biophys. Biophys. Chem.*, **19**, 423

Banville, D. L., Keniry, M., Kam, M. and Shafer, R. H. (1990). *Biochemistry*, **29**, 6521

Berman, H. M. and Neidle, S. (1980). In Sarma, R. H. (Ed.), *Nucleic Acid Geometry and Dynamics*, Pergamon Press, New York, p. 325

Blackburn, G. M. and Gait, M. (1990). In Blackburn, G. M. and Gait, M. (Eds), *Nucleic Acids in Chemistry and Biology*, IRL Press, Oxford, Chapter 2

Bloomfield, V. A., Crothers, D. M. and Tinoco, I., Jr. (1974). *Physical Chemistry of Nucleic Acids*, Harper, New York, Chapter 7

Bresloff, J. L. and Crothers, D. M. (1981). *Biopolymers*, **20**, 3547

Cantor, C. R. and Schimmel, P. R. (1980). In *Biophysical Chemistry*, Freeman, San Francisco

Crothers, D. M. (1971). In Hahn, F. E. (Ed.), *Progress in Molecular and Subcellular Biology: Kinetics of Binding Drugs to DNA*, Vol. II, Springer-Verlag, Berlin, New York, p. 10

Dervan, P. B. (1986). *Science*, **232**, 464

Dickerson, R. E. and Drew, H. R. (1981). *J. Mol. Biol.*, **149**, 761

Fairley, T., Molock, F., Boykin, D. W. and Wilson, W. D. (1988). *Biopolymer*, **27**, 1433

Fiel, R. J., Jenkins, B. G. and Alderfer, J. L. (1990). In Pullman, B. and Jortner, J. (Eds), *Molecular Basis of Specificity in Nucleic Acid–Drug Interactions*, Kluwer Academic, Dordrecht, p. 385

Fox, K. R., Brassett, C. and Waring, M. J. (1985). *Biochim. Biophys. Acta*, **840**, 383

Fox, K. R. and Waring, M. J. (1984). *Biochim. Biophys. Acta*, **802**, 162

Friedman, R. A. G. and Manning, G. S. (1984). *Biopolymers*, **23**, 2671

Gabbay, E. J., Grier, D., Fingerle, R. E., Reimer, R., Levy, R., Pearce, S. W. and Wilson, W. D. (1976). *Biochemistry*, **15**, 2062

Gandecha, B. M., Brown, J. R. and Crampton, M. R. (1985). *Biochem. Pharmacol.*, **34**, 733

Gao, X. and Patel, D. J. (1989a). *Biochemistry*, **28**, 751

Gao, X. and Patel, D. J. (1989b). *Quart. Rev. Biophys.*, **22**, 93

Gao, X. and Patel, D. J. (1990). *Biochemistry*, **29**, 10940

Kopka, M. L., Pjura, P., Yoon, C., Goodsell, D. and Dickerson, R. E. (1985a). In Clementi, E., Corongiu, G., Sarma, M. H. and Sarma, R. (Eds), *Structure and Motion: Membranes, Nucleic Acids and Proteins*, Adenine Press, New York, pp. 461–483

Kopka, M. L., Yoon, C., Goodsell, D., Pjura, P. and Dickerson, R. E. (1985b). *Proc. Natl Acad. Sci. USA*, **82**, 1376

Kopka, M. L., Yoon, C., Goodsell, D., Pjura, P. and Dickerson, R. E. (1985c). *J. Mol. Biol.*, **183**, 553

Klein, B. K., Anderson, C. F. and Record, M. T. (1981). *Biopolymers*, **20**, 2263

Krishnamoorthy, C. R., Yen, S. F., Smith, J. C., Lown, J. W. and Wilson, W. D. (1986). *Biochemistry*, **25**, 5933

Larsen, T. A., Goodsell, D. S., Cascio, D., Grzeskowiak, K. and Dickerson, R. E. (1989). *J. Biomol. Struct. Dyn.*, **7**, 477

Lerman, L. S. (1961). *J. Mol. Biol.*, **3**, 18

Lerman, L. S. (1963). *Proc. Natl Acad. Sci. USA*, **49**, 94

Lohman, T. M. (1985). *CRC Crit. Rev. Biochem.*, **19**, 191

Lohman, T. M., DeHaseth, P. L., and Record, M. T. (1978). *Biophys. Chem.*, **8**, 281

Manning, G. S. (1978). *Quart. Rev. Biophys.*, **11**, 179

Mergny, J. L., Duval-Valentin, G., Nguyen, C. H., Perrouault, L., Faucon, B., Rougée, M., Montenay-Garestier, T., Bisagni, E. and Hélène, C. (1992). *Science*, **256**, 1681

Müller, W. and Crothers, D. M. (1968). *J. Mol. Biol.*, **35**, 251

Pullman, A. and Pullman, B. (1981). *Quart. Rev. Biophys.*, **14**, 289

Ratmeyer, L. S., Vinayak, R., Zon, G. and Wilson, W. D. (1992). *J. Med. Chem.*, **35**, 966

Record, M. T., Jr., Anderson, C. F. and Lohman, T. M. (1978). *Quart. Rev. Biophys.*, **11**, 103

Record, M. T., Lohman, T. M. and DeHaseth, P. (1976). *J. Mol. Biol.*, **107**, 145

Saenger, W. (1984). In *Principles of Nucleic Acid Structure*, Springer-Verlag, New York

Strickland, J. A., Marzilli, L. G., Gay, K. M. and Wilson, W. D. (1988). *Biochemistry*, **27**, 8870

Tanious, F. A., Jenkins, T. C., Neidle, S. and Wilson, W. D. (1992a). *Biochemistry*, **31**, 11632

Tanious, F. A., Veal, J. M., Buczak, H., Ratmeyer, L. S. and Wilson, W. D. (1992b). *Biochemistry*, **31**, 3103

Tanious, F. A., Yen, S. F. and Wilson, W. D. (1991). *Biochemistry*, **30**, 1813
Turner, B. W., Pettigrew, D. W. and Ackers, G. K. (1981). *Meth. Enzymol.*, **71**, 596
Wakelin, L. P. G., Atwell, G. J., Rewcastle, G. W. and Denny, W. A. (1987). *J. Med. Chem.*, **30**, 855
Wakelin, L. P. G., Chetcuti, P. and Denny, W. A. (1990). *J. Med. Chem.*, **33**, 2039
Wakelin, L. P. G. and Waring, M. J. (1980). *J. Mol. Biol.*, **144**, 183–214
Waring, M. J. (1981). In Gale, E. F., Cundiffe, E., Reynolds, P. E., Richmond, M. H. and Waring, M. J., *The Molecular Basis of Antibiotic Action*, 2nd edn, Wiley, New York, p. 287
Wilson, R. W., Rau, D. C. and Bloomfield, V. A. (1980). *Biophys. J.*, **30**, 317
Wilson, W. D. (1990). In Blackburn, G. M. and Gait, M. (Eds), *Nucleic Acids in Chemistry and Biology*, IRL Press, Oxford, Chapter 8
Wilson, W. D., Krishnamoorthy, C. R., Wang, Y. H. and Smith, J. C. (1985a). *Biopolymers*, **24**, 1941
Wilson, W. D. and Lopp, I. G. (1979). *Biopolymers*, **18**, 3025
Wilson, W. D., Tanious, F. A., Barton, H. J., Jones, R. L., Fox, K., Wydra, R. L. and Strekowski, L. (1990a). *Biochemistry*, **29**, 8452
Wilson, W. D., Tanious, F. A., Barton, H. J., Strekowski, L., Boykin, D. W. and Jones, R. L. (1989). *J. Am. Chem. Soc.*, **111**, 5008
Wilson, W. D., Tanious, F. A., Barton, H. J., Wydra, R. L., Jones, R. L., Boykin, D. W. and Strekowski, L. (1990b). *Anti-cancer Drug Des.*, **5**, 31
Wilson, W. D., Tanious, F. A., Buczak, H., Ratmeyer, L. S., Venkatramanan, M. K., Kumar, A., Boykin, D. W. and Munson, B. R. (1992). In Sarma, R. H. and Sarma, M. H. (Eds), *Structure and Function*, Vol. 1: *Nucleic Acids*, Adenine Press, New York, p. 83
Wilson, W. D., Tanious, F. A., Buczak, H., Venkatramanan, M. K., Das, B. P. and Boykin, D. W. (1990c). In Pullman, B. and Jortner, J. (Eds), *Molecular Basis of Specificity in Nucleic Acid–Drug Interactions*, Kluwer Academic, Dordrecht, p. 331
Wilson, W. D., Wang, Y. H., Krishnamoorthy, C. R. and Smith, J. C. (1985b). *Biochemistry*, **24**, 3991
Wilson, W. D., Wang, Y. H., Krishnamoorthy, C. R. and Smith, J. C. (1986). *Chem.-biol. Interact.*, **58**, 41
Yen, S. F., Gabbay, E. J. and Wilson, W. D. (1982). *Biochemistry*, **21**, 2070
Zimmer, Ch. and Wähnert, U. (1986). *Prog. Biophys. Mol. Biol.*, **47**, 31

# 7
# Acridine-based Anticancer Drugs

William A. Denny and Bruce C. Baguley

## 1 Introduction

We have previously (Denny *et al.*, 1983) outlined the considerable importance of acridines in clinical medicine, dating from the early years of the twentieth century, primarily as antibacterial and antimalarial agents. We also noted the changing of this emphasis, with the more recent development of acridine-based compounds as anticancer drugs, e.g. amsacrine (*m*-AMSA, NSC 249992) (**1**) and nitracrine (**2**). This trend has continued since 1983, as can be seen in this chapter. The primary mechanism of action of amsacrine, inhibition of the religation reaction of DNA topoisomerase II, has been elucidated and shown to be common to many DNA-intercalating agents, as described elsewhere in this volume. In addition to its clinical antitumour properties, amsacrine has considerable importance as a biochemical reagent for studying topoisomerase II. Further work on the acridines has uncovered a number of derivatives with varied and promising activities. An analogue of amsacrine (CI-921; NSC 343499) (**3**) has reached clinical trial, while another acridine derivative (acridinecarboxamide; DACA, NSC 601136) (**4**) is about to begin trials. The nitroacridines (e.g. nitracrine) have been shown to have an important new type of biological activity (hypoxia-selective cytotoxicity).

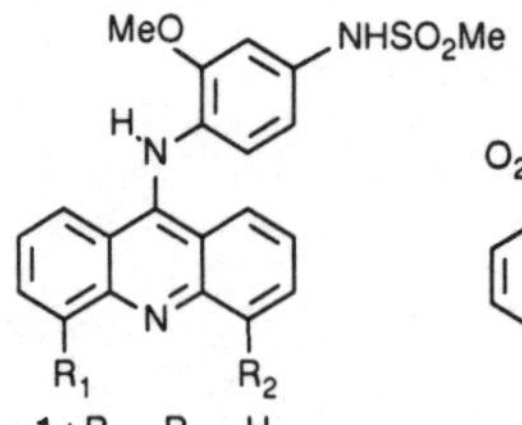


**1** : $R_1 = R_2 = H$
**3** : $R_1 = Me$; $R_2 = CONHMe$

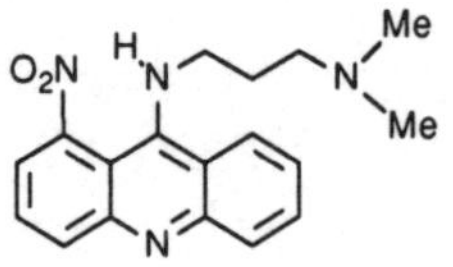


**2**

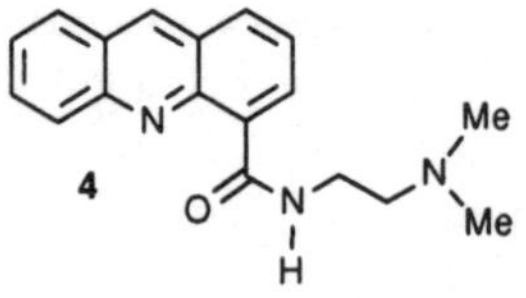


**4**

## 2 9-Anilinoacridines

### *Introduction*

The 9-anilinoacridines evolved from the bisquaternary ammonium heterocycles (Denny *et al.*, 1979a), prepared to test early theories that geometry and charge separation were critical properties governing compounds designed to bind to the minor groove of the DNA double helix. It was suggested that the more active members of the series (e.g. quinolinium compounds such as **5**) bound to DNA initially in the minor groove, a step which preceded final intercalation of the chromophore (Cain *et al.*, 1969). The first assumption has been proved correct by spectrophotometric and viscometric studies (Braithwaite and Baguley, 1980) and by NMR analysis (Leupin *et al.*, 1986). However, a series of acridinium analogues of **5** (e.g. **6**), which were also highly active (Cain *et al.*, 1971), do intercalate into DNA (Braithwaite and Baguley, 1980). Work on a series of simpler analogues produced 9-(4-aminoanilino)acridine (**7**), the corresponding methanesulfonamide (**8**) (Atwell *et al.*, 1972), and its more potent 3′-methoxy derivative (**1**; amsacrine) (Cain *et al.*, 1974).

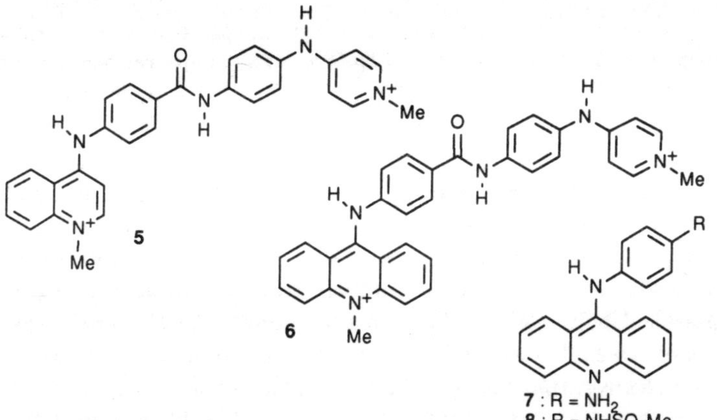

**5**

**6**

**7** : R = $NH_2$
**8** : R = $NHSO_2Me$

### *Interaction with DNA*

Many 9-anilinoacridine derivatives, including amsacrine, bind reversibly to DNA by intercalation of the acridine chromophore between adjacent base pairs. The driving force for this binding is primarily enthalpy, coming from stacking interactions between the acridine nucleus and the DNA bases (Wadkins and Graves, 1989, 1991). The association constant for amsacrine binding to calf thymus DNA at 0.01 ionic strength is $1.5 \times 10^5$ $\mathrm{M}^{-1}$

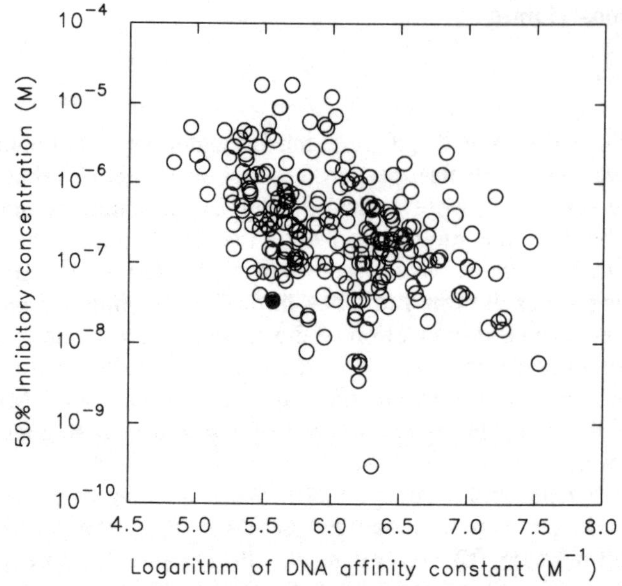


**Figure 7.1** Relationship between DNA binding affinity (poly[dA–dT] at 0.01 ionic strength) and *in vitro* growth inhibitory activity against cultured murine leukaemia L1210 cells (concentrations required for 50% growth inhibition) for various series of substituted 9-anilinoacridine derivatives. Amsacrine is depicted by the solid symbol. Although there is a significant positive relationship between potency *in vitro* and DNA binding, activity can vary by more than 1000-fold for any particular DNA binding affinity. Data from Baguley and Nash (1981), Atwell *et al.* (1984b, 1986a), Denny *et al.* (1984) and references cited therein

(Wilson *et al.*, 1981). By analogy with the crystal structure determined for 9-aminoacridine binding to a dinucleotide (Sakore *et al.*, 1979), amsacrine was postulated (Wilson *et al.*, 1981) to bind with the anilino ring lodged in the minor groove, a conformation supported by later energy calculations (Chen *et al.*, 1988a). In this model, the 1′-substituent points tangentially away from the helix. As well as modulating DNA binding affinity by electron transmission effects (Baguley *et al.*, 1981), it can contribute to electron transfer complexes (Baguley, 1990) and can presumably interact with other macromolecules, including proteins, which are associated with the DNA. Changes in the pattern of substitution of 9-anilinoacridine give rise to wide variation in both DNA binding affinity and cytotoxicity towards cultured leukaemia cells (Figure 7.1). Much of this variation is associated with changes in the 1′-substituent, giving rise to the concept of a ternary complex of DNA, drug and protein (Baguley and Nash, 1981). The combination of an appropriate 1′-substituent with a 3′-methoxy group provides an electron-rich centre which might enhance the stability of such a ternary complex (Wilson *et al.*, 1981).

*Amsacrine*

Introduction

Amsacrine was discovered in the course of testing 9-anilinoacridines for antitumour activity, using the transplantable L1210 murine leukaemia (Cain *et al.*, 1972). Following further testing, amsacrine was selected for clinical studies by the National Cancer Institute, USA (Cain and Atwell, 1974). It began Phase I/II clinical trials in 1976, showing encouraging results against both leukaemias and lymphomas (Grove *et al.*, 1982), thus becoming the first totally synthetic DNA intercalating agent to show useful clinical activity (Denny, 1989; Baguley, 1991). Amsacrine is now used in combination with antimetabolites for acute leukaemia (Zittoun, 1985; Jehn and Heinemann, 1991; Miller *et al.*, 1991).

Biological Properties

A consequence of the intercalative binding of amsacrine to DNA is the physical unwinding of the DNA double helix in order to accommodate the inserted ligand (Neidle and Abraham, 1984). Amsacrine and related compounds inhibit DNA synthesis, and this was initially considered to be their primary mechanism of cytotoxicity and anticancer activity. However, amsacrine caused DNA strand breakage at low concentration (Zwelling *et al.*, 1981). Its facile oxidation, giving it the ability to degrade DNA after oxidation by molecular oxygen, suggested an oxygen-dependent mechanism (Shoemaker *et al.*, 1982) in the presence of Cu(II) ions. Amsacrine reacts to form Cu(I) species (Wong *et al.*, 1984a), which under suitable conditions produce single-strand DNA breaks (Wong *et al.*, 1984b), probably by a free radical mechanism. However, oxygen is not required for toxicity in cultured cells (Wilson and Whitmore, 1981; Robbie *et al.*, 1990).

The major mechanism for the cellular cytotoxicity of amsacrine appears to be its ability to form a ternary complex with the enzyme DNA topoisomerase II and DNA (see Chapter 1), altering the position of equilibrium and trapping a reaction intermediate termed the 'cleavable complex' (Nelson *et al.*, 1984; Rowe *et al.*, 1986; Liu, 1989). This mechanism, first described for amsacrine, has now been shown to be general for DNA-intercalating ligands. The cytotoxic activities of amsacrine derivatives correlate more closely with their ability to stabilize formation of the topoisomerase II cleavable complex than with their DNA binding ability (Covey *et al.*, 1988), providing an explanation for the failure of the DNA binding affinity alone to predict the activities of amsacrine analogues.

Amsacrine, like etoposide and doxorubicin, is more toxic to tumour cell lines in logarithmic phase cultures than in plateau-phase cultures (Finlay *et al.*, 1987; Robbie *et al.*, 1988). Results of cytokinetic studies have shown that the length of the $G_1$ phase transit time is directly related to drug sensitivity for all three drugs, arguing for a common mechanism of resistance. It has been suggested that the degradation of topoisomerase II during $G_1$ phase (Heck *et al.*, 1988) is responsible for this effect (Holdaway *et al.*, 1992).

Metabolism and Pharmacology

In mice, amsacrine displays dose-dependent pharmacokinetics and the majority of injected drug is recovered in the faeces, suggesting a major metabolic pathway in the liver (Kestell *et al.*, 1990). The biliary oxidation of amsacrine was first demonstrated in rats, where >50% of the dose was excreted as the 6′-glutathione conjugate (**11**), resulting from two-electron oxidation to the quinoneimine (AQDI; **10**) followed by 1,4-addition of glutathione (Shoemaker *et al.*, 1982; Gaudich and Przybylski, 1983) (Scheme 7.1). Later studies (Robbie *et al.*, 1990) showed that the 5′-glutathione (**12**), resulting from inverse 1,4-addition, was also formed. This oxidation was later shown to be a two-step process, proceeding through the one-electron radical anion species (**9**), which is detectable by pulse radiolysis (Anderson *et al.*, 1988). Amsacrine has a redox potential of +280 mV, as measured by cyclic voltammetry (Jurlina *et al.*, 1987).

The quinonediimine (**10**) is a black crystalline solid which can be produced electrochemically by cyclic voltammetry, chemically (Shoemaker *et al.*, 1982) by activated $MnO_2$, biologically by liver microsome preparations (Shoemaker *et al.*, 1984), and even by leaving dilute aqueous solutions exposed to the air. It undergoes quantitative conversion to amsacrine with

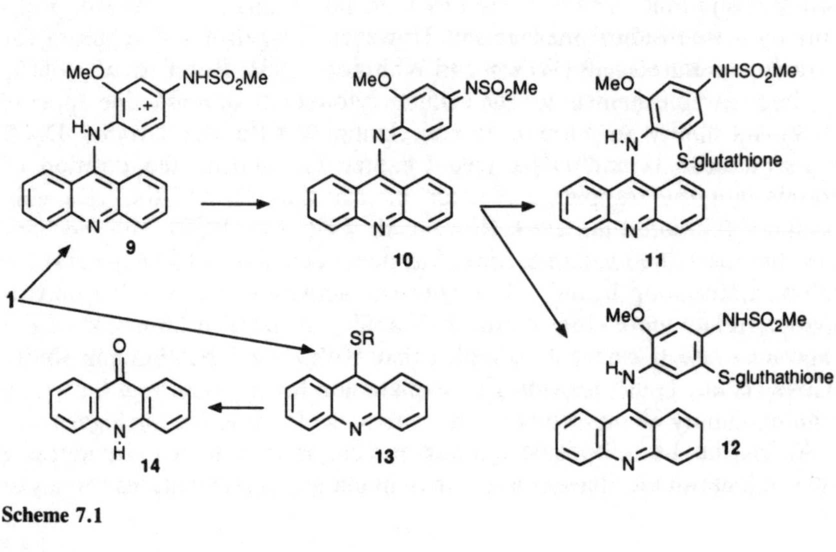


**Scheme 7.1**

mild reducing agents such as ascorbic acid, and 1,4-additions with nucleophiles such as amines and thiols (Shoemaker *et al.*, 1982; Lee *et al.*, 1988). The quinonediimine (**10**) has been suggested as the active species responsible for the cytotoxicity of amsacrine (Shoemaker *et al.*, 1984), but recent studies (Robbie *et al.*, 1990) showing its slow rate of formation in AA8 cells, suggest that this is unlikely.

Amsacrine, like other 9-anilinoacridines, is unstable in the presence of thiols, undergoing displacement of the side-chain to give unstable 9-thio species (**13**), and finally acridone (**14**) (Cain *et al.*, 1976) (Scheme 7.1). The reaction is general-acid-catalysed, with the rate-determining step being the expulsion of the anilino leaving group (Khan and Malspeis, 1982). However, while the relative rates of thiolysis of amsacrine derivatives are inversely proportional to their potency (Denny *et al.*, 1979b), this property has little bearing on their antitumour activity.

## *CI-921*

### Introduction

The limitation of amsacrine's clinical activity to leukaemia raised the question of whether the clinical results reflected the use of murine leukaemia as a screening system for detecting the antitumour activity of 9-anilinoacridine derivatives. This led to a search among the numerous analogues available for a 'second-generation' compound with a broader spectrum of action. Initial selection was on physicochemical properties, particularly better aqueous solubility and lower $pK_a$ (to improve drug distribution) than amsacrine, while retaining high DNA binding. Since structure–activity relationships suggested that the existing side-chain was close to optimal (Baguley *et al.*, 1981), changes were sought in the acridine chromophore. Previous work with carboxamide derivatives (Cain *et al.*, 1977) had shown the suitability of these substituents, and a large quantitative structure–activity relationship study (Denny *et al.*, 1982) had focused attention on the 4- and 3-positions as being the most suitable for modification. Thus, the primary focus of these studies centred on 3- and 4-carboxamides (Denny *et al.*, 1984).

The antitumour activity of these derivatives was assayed using mouse solid tumours rather than leukaemias, with the Lewis lung carcinoma chosen as a model, since it was comparatively resistant to amsacrine and to the available clinical topoisomerase II-directed agents (Goldin *et al.*, 1981). A study of the activity of amsacrine analogues against Lewis lung tumours growing in the lung showed that the analogues were ranked in a different order as compared with antileukaemic activity (Baguley *et al.*, 1983). A large series of mono- and disubstituted amsacrine derivatives were tested, and the highest activity was found for the 4-methyl-5-(*N*-methyl)carboxamide (**3**) (Baguley *et al.*, 1984). This compound, 9-[(2-methoxy-4-methylsulfonylamino)-phenylamino]-*N*,5-dimethyl-4-acridine-carboxamide; isethionate (CI-921, NSC 343499) was selected for further study. The experimental activity of CI-921 was compared with that of amsacrine in a multicentre trial, and shown to be more active than amsacrine against a variety of tumours (Leopold *et al.*, 1987). CI-921 was also found to be more effective than amsacrine when administered orally. A large-scale synthesis of CI-921 using the best available methods was developed (Brennan *et al.*, 1989) and the compound was advanced to clinical trial. In two phase II trials, two drug-induced remissions were observed in non-small-cell lung cancer plus two in a breast cancer study, while no remissions were obtained in colorectal cancer or gastric cancer (Hardy *et al.*, 1988; Sklarin *et al.*, 1990; Harvey *et al.*, 1991). These results suggest that CI-921 has clinical activity, and further studies are proceeding.

### Biological Properties

The cytotoxic action of CI-921 appears to be mediated, as with amsacrine and drugs such as etoposide, by interaction with the enzyme DNA topoisomerase II. CI-921 induces both DNA breakage and the formation of DNA–protein cross-links (Covey *et al.*, 1988; Schneider *et al.*, 1988) and, as a consequence, causes arrest of cultured cells in the $G_2$ phase of the cell division cycle (Baguley *et al.*, 1986; Traganos *et al.*, 1987), chromosomal aberrations at mitosis and the appearance of micronuclei in cells following cell division (Ferguson *et al.*, 1988). CI-921 is 2–4-fold more potent than amsacrine against a variety of human and mouse cell lines in both continuous exposure and clonogenic assays (Finlay and Baguley, 1984; Finlay *et al.*, 1986; Baguley and Wilson, 1987; Baguley and Finlay, 1988a,b; Finlay *et al.*, 1990). In contrast, CI-921 and amsacrine are almost equitoxic against bone marrow cells from both humans and mice (Ching *et al.*, 1990), arguing for greater antitumour selectivity towards at least some tumour cells.

### Metabolism and Pharmacology

The pharmacokinetics of CI-921 have been studied in mice (Kestell *et al.*, 1990) and in patients (Hardy *et al.*, 1988; Paxton *et al.*, 1988), using high-performance liquid chromatography (Jurlina and Paxton, 1985) to assay drug. One unusual feature of the plasma pharmacokinetics of CI-921 is the high protein binding, providing a free drug fraction for CI-921 in mice of 0.006, as compared with 0.07 for amsacrine. For a given free drug fraction in plasma, the uptake of CI-921 into tumours is more efficient than for amsacrine (Paxton *et al.*, 1986; Kestell *et al.*, 1990). CI-921 is readily oxidized to the corresponding quinonediimine (**15**) (Jurlina *et al.*, 1987), and in the liver this product reacts with glutathione to give the 5′- and 6′-gluthathione conjugates (**16**, **17**) which are removed in the bile (Robertson *et al.*, 1988). The pattern of biliary metabolites of CI-921 is different in the rat and mouse (Robertson *et al.*, 1992). CI-921 also reacts directly with plasma proteins with displacement of the side-chain to produce C-9 thiol adducts (e.g. **18**) (Kestell *et al.*, 1989; Robertson *et al.*, 1992), and a 4-hydroxymethyl metabolite (**19**) has also been isolated (Robertson *et al.*, 1992).

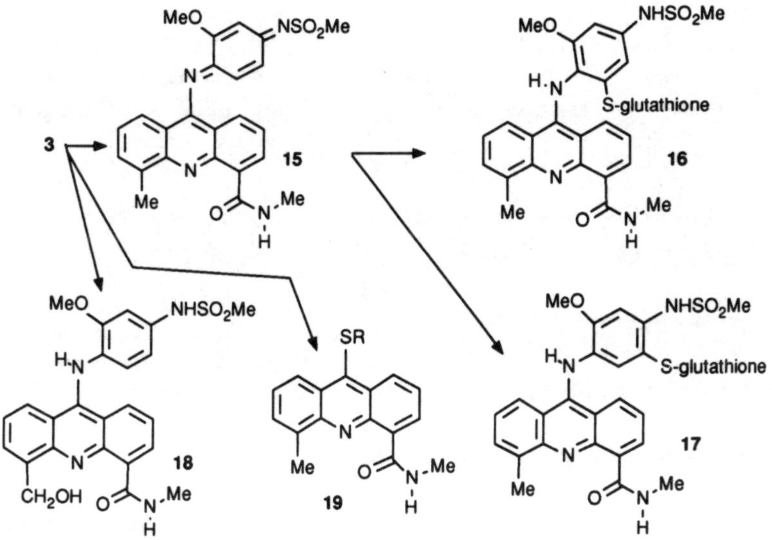


**Scheme 7.2**

*3′-NR₁R₂ Derivatives of Amsacrine*

Limited structure–activity relationships for 3′-substitution of amsacrine (Cain *et al.*, 1975) suggested the value of small, strongly-electron-donating groups, a conclusion also supported by the global QSAR study (Denny *et al.*, 1982). Studies of a series of 3′-$NR_1R_2$ analogues of amsacrine, where $R_1$ and $R_2$ were H or lower alkyl, showed that both the 3′-NHMe and 3′-$NMe_2$ derivatives (**20**, **21**) had excellent antileukaemic activity *in vivo* (Atwell *et al.*, 1984b), with **20** also being considerably more water-soluble than amsacrine. A series of acridine-substituted 3′-NHMe compounds were shown to be much stronger DNA binders than their amsacrine counterparts, and to have excellent activity *in vivo* against Lewis lung carcinoma (Atwell *et al.*, 1986a). The best compounds were comparable to

CI-921 against this solid tumour, but the structure–activity relationships for acridine substitution were different from those of the 3′-OMe series. A series of analogues of the 3′-$NMe_2$ derivative (**21**) were also studied (Atwell *et al.*, 1987a). While generally possessing much weaker DNA binding and lower potency *in vivo* than amsacrine or 3′-NHMe derivatives, the 3′-$NMe_2$ compounds showed exceptional solid tumour activity, with some compounds (e.g. **22**) effecting 100% cures against Lewis lung. Like amsacrine, both 3′-NHMe and 3′-$NMe_2$ compounds underwent facile oxidation. The 3′-$NMe_2$ quinoneimine (**23**) corresponding to **21** could be isolated, but the oxidation products of **20** proved unstable (Atwell *et al.*, 1987a).

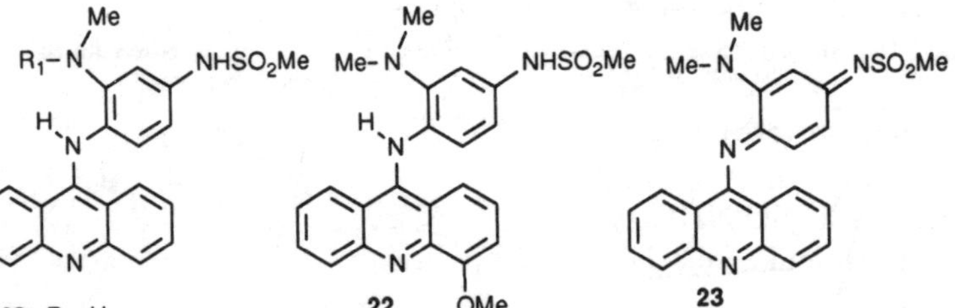


**20** : R = H
**21** : R = Me

**22**

**23**

*Other 9-Anilinoacridines*

1′-Carbamate Analogues of Amsacrine

A consequence of the hypothesis that amsacrine forms a ternary complex with DNA and topoisomerase II is that topoisomerase II molecules of differing structure may require different 1′-substituents for optimal interaction. An amsacrine-resistant subline of P388 leukaemia (P388/AMSA), in contrast to the parental line, expresses relatively higher amounts of the topoisomerase IIβ than of topoisomerase IIα isoenzyme (Chung *et al.*, 1989). However, this line responds to the compounds such as the 1′-carbamate (**24**) both *in vitro* and *in vivo*, suggesting that resistance can be overcome by redesign of the protein-binding domain (Baguley *et al.*, 1990). A series of acridine-substituted 3′-$NHCH_3$-1′-carbamates (e.g. **25**) also show excellent activity against the Lewis lung carcinoma (Rewcastle *et al.*, 1987).

Tetracyclic Chromophores

Tetracyclic derivatives of amsacrine have been evaluated for antitumour activity and for their ability to inhibit topoisomerase II (Yamato *et al.*, 1989). Indeno- (**26**), benzofuro- (**27**) and benzothienoquinoline (**28**) ana-

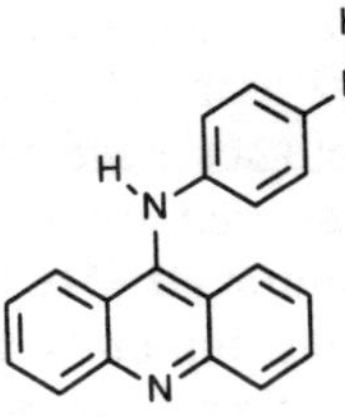

**24**

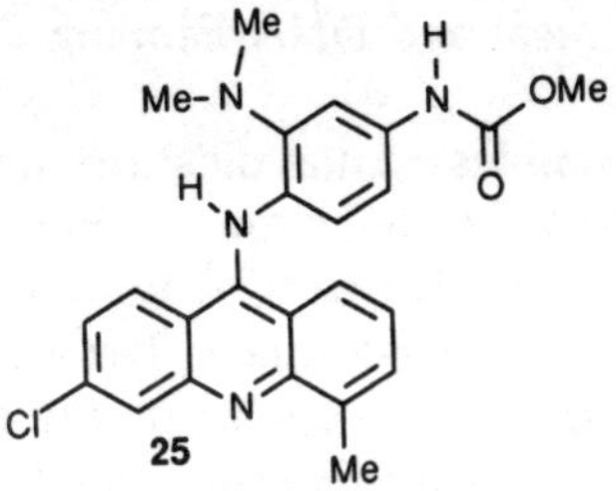

**25**

logues showed antileukaemic activities *in vitro* and *in vivo* comparable to those of amsacrine. This result is somewhat surprising, given that the benz[*b*]acridine analogue (**29**) is inactive (Denny and Baguley, 1987). All the active tetracyclic quinoline analogues were intercalating agents, and induced topoisomerase II-mediated DNA cleavage (Yamato *et al.*, 1989).

Amsacrine Isosteres

A series of 1′-$SO_2NHR$ isosteres (e.g. **30**) of amsacrine were inactive against P388 leukaemia *in vivo* (Ebeid *et al.*, 1990; El-Moghazy Aly and Safwat, 1990).

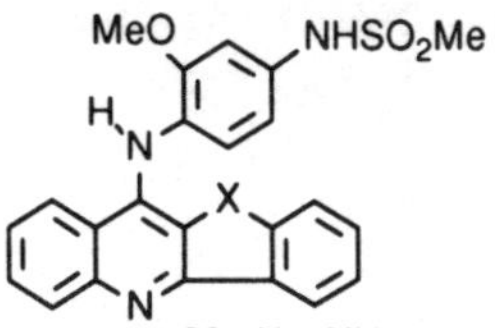

**26** : X = NH
**27** : X = O
**28** : X = S

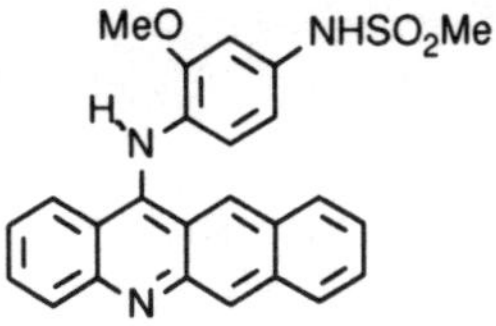

**29**

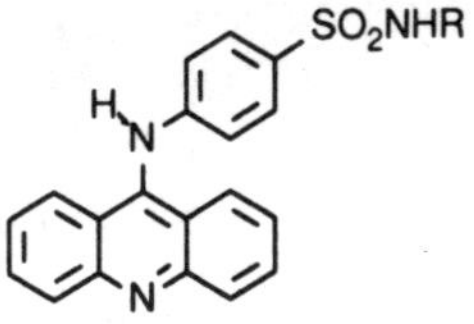

**30**

## 3 Acridinecarboxamides

### *9-Aminoacridinecarboxamides*

Introduction

9-Aminoacridine-4-carboxamides (exemplified by the parent, **31**) were developed in an attempt to increase the DNA-binding properties of acridines, particularly for GC-rich DNA, by modification of their charge-transfer properties (Atwell *et al.*, 1984a). Many derivatives showed potent cytotoxicity *in vitro* and antileukaemic activity *in vivo*, and well-defined structure–activity relationships for alterations in both the acridine chromophore and the carboxamide side-chain.

## Physicochemical and DNA Binding Properties

The carboxamides exist as dications under physiological conditions (Denny *et al.*, 1987b), although the acridine p$K_a$ is lowered from the 9.99 of 9-aminoacridine (**32**) to 8.30 for the parent compound (**31**) by addition of the carboxamide side-chain. The 4-carboxamide group in the free base of (**31**) is planar, and lies coplanar with the acridine ring, as shown by X-ray crystallography (Hudson *et al.*, 1987). Stabilization of this form is achieved by a hydrogen bond between the acridine ring N10 acceptor and the amide NH donor, with the side-chain fully extended. The length of the C4–C11 bond linking the side-chain to the chromophore is *ca* 1.5 Å, suggesting little double bond character (Hudson *et al.*, 1987). The minimum-energy conformation computed for the dications is also essentially planar, but has a hydrogen bond between the protonated N10 and the amide carbonyl oxygen. These conformations are calculated by molecular mechanics to be the lowest-energy ones, but the barrier to rotation is so low (*ca* 4 kcal/mol) that side-chain rotation about the carboxamide C4–C11 bond would be virtually unhindered. Thus, models of the binding of these ligands to DNA can feasibly have the side-chain rotated at least 30° out of plane with the chromophore.

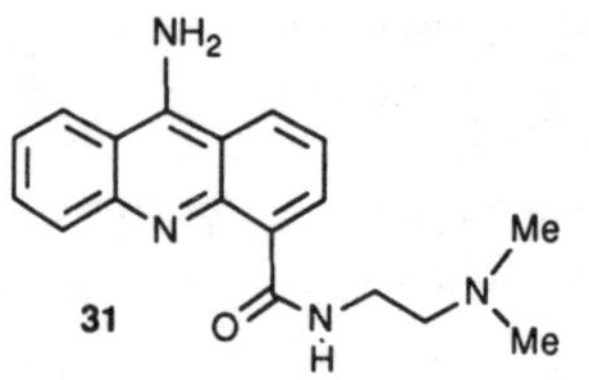


**31**

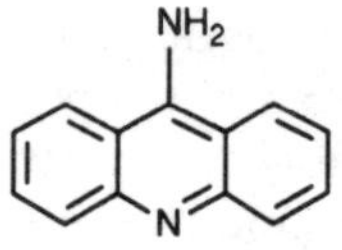


**32**

The parent compound (**31**) binds to DNA by intercalation, with an unwinding angle (16°) identical with that of 9-aminoacridine, but with a much higher association constant and a preference for binding to GC base pairs (Atwell *et al.*, 1984a). The GC-specificity is structure-dependent, being restricted to compounds bearing a 4-CONH$(CH_2)_2$N$R_1R_2$ side-chain, where $R_1$ and $R_2$ are groups which permit the nitrogen to be protonated at neutral pH (Wakelin *et al.*, 1987).

The 9-aminoacridine-4-carboxamides dissociate from calf thymus DNA by a complex mechanism involving at least three intermediate forms, but derivatives bearing the side-chains specified above also possess a fourth binding mode of greater kinetic stability. This manifests itself by a fourth, long-lived dissociation transient observed in stopped-flow studies (Wakelin *et al.*, 1987), and has been interpreted in terms of a molecular model for binding of the 9-aminoacridine-4-carboxamides to DNA in which the acridine chromophore intercalates from the narrow groove, with the major axis of the acridine lying at an angle to the major base pair axis. In such an orientation, the amide NH and protonated $N^{+}H(Me)_2$ groups of the side-chain are positioned to become hydrogen bond donors to a single DNA acceptor molecule, the O-2 of a cytosine base adjacent to the binding site (Wakelin *et al.*, 1987). This model has been challenged in favour of one where the side-chains lie in the major groove (Chen *et al.*, 1988b), but molecular mechanics calculations for related phenylquinolinecarboxamides favour minor groove binding of the side-chain (McKenna *et al.*, 1989).

## Structure–Activity Relationships

The 9-aminoacridine-4-carboxamides are potent cytotoxins, with the parent (**31**) showing an $IC_{50}$ against L1210 cells in culture of 15 nM (Atwell *et al.*, 1984a). Structure–activity relationships for ring substitution have been established (Rewcastle *et al.*, 1986; Denny *et al.*, 1987b), with substitution in the 5-position most favourable to both cytotoxicity *in vitro* (the 5-methyl derivative has an $IC_{50}$ of below 1 nM) and antileukaemic activity *in vivo*. The 5-$SO_2$Me derivative (**33**) also showed good activity against the Lewis lung carcinoma *in vivo*; the enhanced solid tumour activity of this compound was attributed to its very weak acridine $pK_a$ (5.15), which allows it to exist primarily as the monocation, thus permitting more efficient distribution to remote tumour sites (Denny *et al.*, 1987b). In addition, there is an absolute requirement for a side-chain bearing a cationic centre at the 4-position. Significant attenuation of the $pK_a$ of the side-chain nitrogen, or alteration of its position with respect to the chromophore (by variation of either the link group to the chromophore or the length of the linking chain) abolishes activity; thus, the 4-butylcarboxamide (**34**), *N*-methylcarboxamide (**35**) and sulfonamide (**36**) compounds are all inactive (Atwell *et al.*, 1984a). These are the same structure–activity requirements for the fourth, long-lived transient in the DNA dissociation kinetics (Wakelin *et al.*, 1987; see above), suggesting some biological importance for this stable binding mode. The parent compound (**31**) is an extremely potent inducer of DNA strand breaks (Denny *et al.*, 1986), suggesting that these compounds may act similarly to other DNA-intercalating agents on topoisomerase II. Among a small series of analogues of (**31**), DNA breakage ability also correlated best with slow drug–DNA dissociation kinetics.

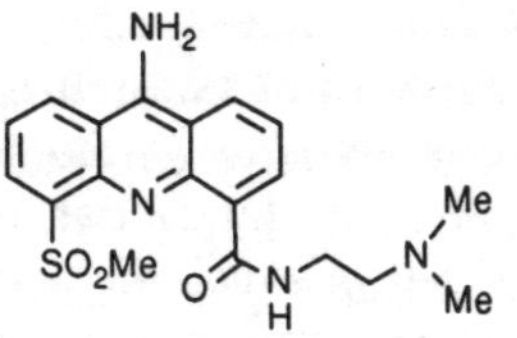


**33**

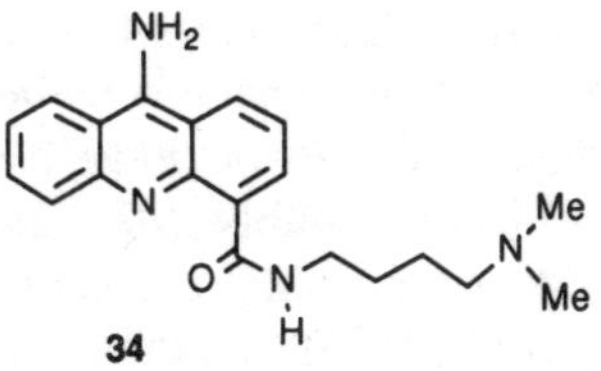


**34**

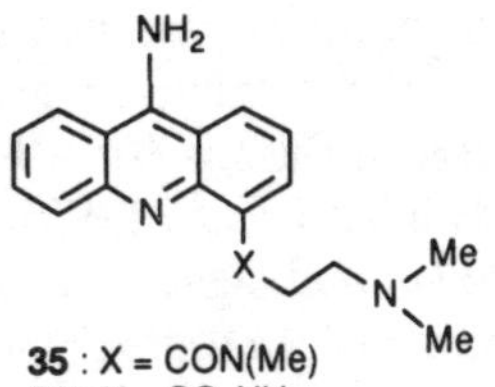


**35** : X = CON(Me)
**36** : X = $SO_2$NH

### *Acridinecarboxamides*

#### The 'Minimal Intercalator' Concept

There has been increasing evidence recently that drug distributive properties play a dominant role in determining drug activity at remote tumour sites (Baguley and Wilson, 1987; Denny, 1989), and that efficient distribution is inversely proportional to DNA binding (Durand, 1989; Jain, 1989). Thus, the concept of 'minimal intercalators' has been proposed (Denny *et al.*, 1987a), and much work (Atwell *et al.*, 1987b, 1988, 1989a; Denny *et al.*, 1987a; Denny, 1989) has focused on defining the chromophore requirements for drugs to bind to DNA by intercalation, but with the lowest possible binding constants.

Removal of the resonant 9-amino group from the 9-aminoacridine-4-carboxamide (**31**) to give the corresponding acridine-4-carboxamide (**4**) lowers the acridine $pK_a$ to 3.54 (ensuring an uncharged chromophore), and reduces the DNA binding constant by about an order of magnitude. While **4** also binds by intercalation, and has broadly similar *in vivo* potency and activity against the P388 leukaemia to that of (**31**), it shows remarkable activity against the Lewis lung carcinoma (Atwell *et al.*, 1987b). A series of acridine-substituted analogues of **4**, while generally less active than the parent compound, all showed substantial effects against Lewis lung carcinoma *in vivo* (Atwell *et al.*, 1987b).

### *DACA*

#### Introduction

As noted above, the acridinecarboxamide (**4**) (*N*-[2-(dimethylamino)-ethyl]acridine-4-carboxamide; NSC 601316; DACA) showed very high activity against the transplantable Lewis lung tumour, being curative in 90% of treated animals (Atwell *et al.*, 1987b; Finlay and Baguley, 1989). Surprisingly, this compound had only moderate antileukaemic activity. Its activity *in vitro* was distinguishable from that of amsacrine and CI-921 because of its self-inhibition of cytotoxicity (Finlay and Baguley, 1989). On the basis of these results, it has been suggested that well-perfused areas of the body may be protected from high, short-term exposure to drug, whereas solid tumours, which might be expected to be exposed to lower drug concentrations for a longer time, would have a more effective exposure (Haldane *et al.*, 1992). DACA is notably active against multidrug-resistant cells, and against a series of primary melanoma cultures grown from patients' tumours (Finlay *et al.*, 1993; Marshall *et al.*, 1992). On the basis of its novel properties *in vitro* and *in vivo*, DACA was recently accepted as a candidate for clinical trial by the UK Cancer Research Campaign.

Biological Properties

DACA, like amsacrine and etoposide, stimulates the formation of cleavable complexes between topoisomerase II and DNA (Schneider *et al.*, 1988). It induces both DNA breakage and the formation of DNA–protein cross-links, and leads to arrest in $G_2$ phase of the cell division cycle, as well as the appearance of micronuclei following cell division. DACA induces mutations at the hypoxanthineribosyl transferase locus in V79 Chinese hamster cells and manifests a similar mutagenicity profile to that of amsacrine, suggesting that DNA damage is the cause of tumour cell death (Ferguson *et al.*, 1990).

Metabolism and Pharmacology

The plasma pharmacokinetics of DACA in mice show a plasma-free drug fraction of 0.16 and biphasic drug elimination (Young *et al.*, 1990; Paxton *et al.*, 1992). Radioactive tracer studies show high drug levels in all tissues, including brain, consistent with the acute neurological toxicity observed when DACA was administered as a high intravenous dose (Cornford *et al.*, 1992). The main metabolic routes of DACA include oxidation at C-9 to provide the acridone analogue (**37**), oxidation of the dimethylamino group to the *N*-oxide (**38**), demethylation of the dimethylamino side-chain to the monomethyl compound (**39**), 7-hydroxylation of the acridone analogue to give **40**, and glucuronidation of this (Robertson *et al.*, 1991) (Scheme 7.3).

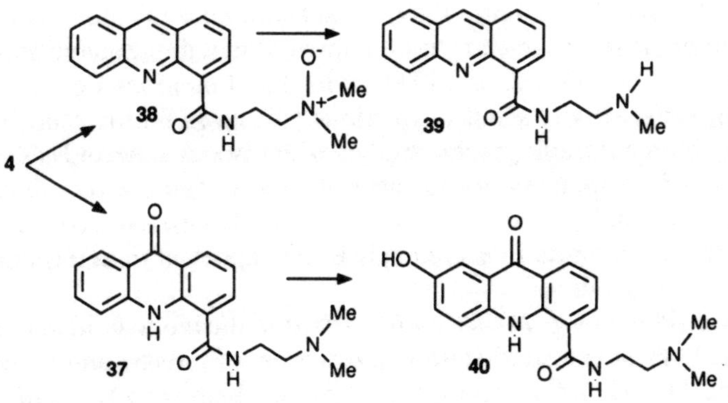


**Scheme 7.3**

## 4 Nitroacridines

### *Nitracrine*

#### Introduction

As discussed previously (Denny *et al.*, 1983), the 1-nitroacridine derivative nitracrine (Ledakrin, C-283; **2**) was developed in Poland (Ledochowski and Stefanska, 1966), and has been used there clinically as an anticancer drug (Kwasniewska-Rokicinska *et al.*, 1973). Structure–activity relationships in this series are very narrow, with a 1-nitro group being essential for activity. Nitracrine binds to DNA by intercalation (Wilson *et al.*, 1984), although it is a weaker binder than the less active nitro isomers (Roberts *et al.*, 1990), because of the distorted butterfly-shaped acridine chromophore (Stezowski *et al.*, 1985).

#### Hypoxic Selectivity

Nitracrine has been shown to have very potent hypoxia-selective cytotoxicity against tumour cells in culture (Wilson *et al.*, 1984). It shows about the same level of selectivity for hypoxic over aerobic AA8 cells (*ca* tenfold) as does the well-known hypoxia-selective cytotoxin misonidazole (**41**), but its absolute cytotoxicity is about 100 000-fold greater. However, it did not show significant activity against hypoxic cells in solid tumours, owing probably to a combination of its high reduction potential (−303 mV) (Wilson *et al.*, 1989a), which results in high rates of metabolism even in well-oxygenated tissues (Wilson *et al.*, 1986), and its relatively high level of DNA binding (Wilson *et al.*, 1984), which can be expected to slow markedly its rate of extravascular diffusion into the hypoxic areas of tumour masses (Durand, 1989; Jain, 1989). 4-Substituted analogues with lower redox potentials were shown to have improved metabolic stability (Wilson *et al.*, 1989b; O'Connor *et al.*, 1990), with the 4-methoxy derivative (**42**) possessing some activity against hypoxic cells *in vivo* (Wilson *et al.*, 1989b). However, both autoradiographic studies with tumour spheroids (Wilson *et al.*, 1986) and comparison of cell survival curves of intact and dissociated spheroids using unlabelled drug (Denny *et al.*, 1990b) showed that extravascular transport limitations are likely to be important in determining the activity of nitroacridines.

1-Nitroacridines (e.g. **2a**) are know to exist in the iminoacridan tautomer (e.g. **2b**), both in the solid state (Stezowski *et al.*, 1985) and in solution (Boyd and Denny, 1990), and a study of side-chain variants of nitracrine showed that hypoxia selectivity correlated best with the proportion of iminoacridan tautomer present (Denny *et al.*, 1990a). For these compounds, both aerobic and hypoxic cytotoxicity was due to nitroreduction and subsequent DNA adduct formation, although there was wide variation in cytotoxicity under aerobic conditions, suggesting the formation of DNA adducts of widely differing lethality.

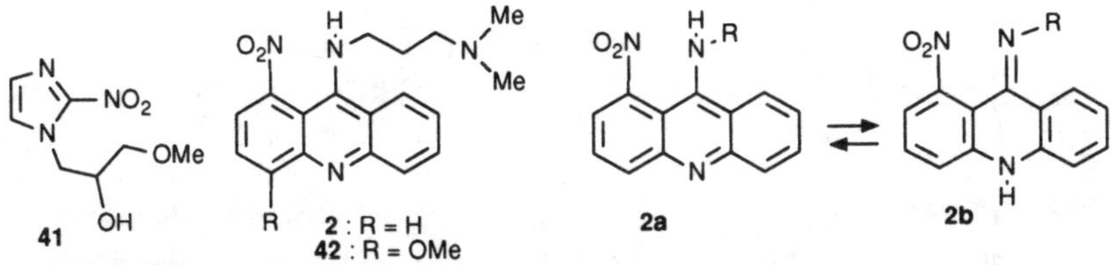


**41** **2** : R = H **42** : R = OMe **2a** **2b**

Studies of the 2-, 3- and 4-nitro analogues (**43–45**) of nitracrine (Denny *et al.*, 1990b; Roberts *et al.*, 1990) showed that only the 4-nitro compound had pronounced hypoxia selectivity, although its absolute potency was poor. This result agrees with NMR studies (Boyd and Denny, 1990; Cholody and Konopa, 1991), which show that the 1- and 4-nitro compounds exist in the imino form in solution, whereas the 2- and 3-nitro analogues are in the amino form.

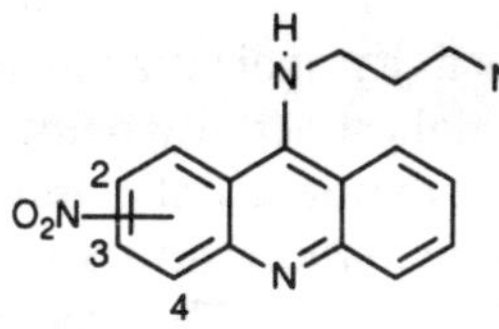


**43** : 2-$NO_2$
**44** : 3-$NO_2$
**45** : 4-$NO_2$

Nitracrine *N*-oxide

Nitracrine *N*-oxide (**46**) is much less toxic *in vivo* than nitracrine, and was considered for clinical trial as a prodrug of nitracrine (Gieldanowski *et al.*, 1972a,b). Recent studies (Wilson *et al.*, 1992) have shown it to be the first example of a bis-bioreductive cytotoxin, possessing two reducible centres (the nitroacridine and the aliphatic *N*-oxide) which must be independently reduced by oxygen-inhibitable processes for full activation. Nitracrine *N*-oxide shows the highest hypoxic selectivity in cell culture yet reported (1000–1500-fold in AA8 cells). A two-step bioactivation mechanism has been suggested (Wilson *et al.*, 1992), with reduction of the *N*-oxide group generating nitracrine, which has a much higher DNA binding affinity, followed by nitro group reduction of this, as discussed above (Scheme 7.4).

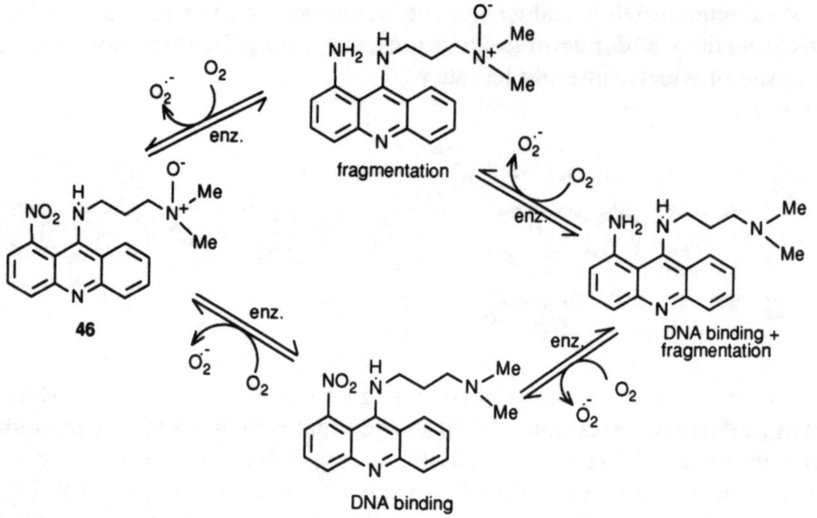


**Scheme 7.4**

*Nitropyrazoloacridines*

These compounds show good activity against a range of solid tumours *in vivo* (Sebolt *et al.*, 1987), with highest activity being shown by 9-methoxy analogues. They are DNA-intercalating agents, and preferentially inhibit RNA rather than DNA synthesis. Many also show significant hypoxic selectivity in cell culture (Sebolt *et al.*, 1987). One member of the series (**47**; NSC 366140) is undergoing clinical trial (LoRusso *et al.*, 1990).

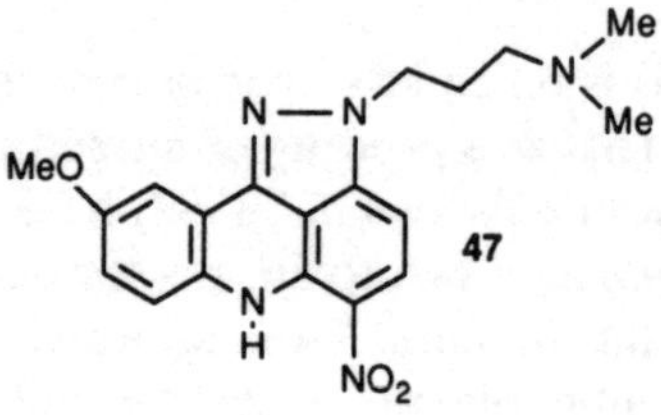

**47**

## 5 Polyacridines

*Diacridines*

Introduction

As noted previously (Denny *et al.*, 1983), interest in diacridines as antitumour drugs originally centred around the assumption that the antitumour effectiveness of intercalating agents was directly related to high DNA binding affinity and slow drug–DNA off-rates. With expectations that such desirable properties would be enhanced for dimeric molecules containing two intercalating ligands (McGhee and von Hippel, 1974; Capelle *et al.*, 1979), there were many reports of the preparation and preliminary evaluation of such compounds, including diacridines. This work has continued (reviewed by Wakelin, 1986), but the promise of clinically useful compounds has not been realized. One diacridine (the hexane-1,6-diamine derivative **48**; NSC 219733) did show experimental activity against intercalator-resistant mouse leukaemias (Johnson and Howard, 1982), and was evaluated for clinical trial, but this was not pursued (Goldin *et al.*, 1981).

### Requirements for Bisintercalation

Most of the recent physicochemical studies with diacridines have been devoted to defining the nature of their interaction with DNA, and, in particular, the requirements for both of the acridines to intercalate (Wakelin, 1986). Detailed hydrodynamic studies of the binding of homologous series of polymethylene- and alkylamide-linked diacridines (**49–51**) (King *et al.*, 1982; Denny *et al.*, 1985a) reinforced earlier work indicating (Wakelin *et al.*, 1978) that an abrupt change in binding mode occurs when the linker chain reaches a length of about 8.8 Å. Since this is too short to span two base pairs if normal DNA geometry is maintained, such compounds are postulated (Wakelin *et al.*, 1978) to bind by bisintercalation at contiguous sites, with one chromophore on either side of a single base pair. However, these compounds were shown by NMR (Assa-Munt *et al.*, 1985a) to bind to the self-complementary decadeoxynucleotide $d(AT)_5.d(AT)_5$ by monointercalation only, suggesting that the binding mode is very condition-dependent.

Studies with flexibly linked diacridines have indicated that a flexible link can be disadvantageous for DNA binding. Self-stacking interactions can compete with DNA binding (Capelle *et al.*, 1979), and unfavourable entropic factors act to reduce binding constants to a point where diacridines linked by complex peptide chains show little enhancement in binding constants over 9-aminoacridine itself (Bernier *et al.*, 1981). Substituents on the acridine also influence bisinteracalative binding ability, with substituents positioned on the minor (short) axis of the acridine chromophore having a less deleterious influence than those positioned on the major (long) axis (Wright *et al.*, 1980).

The interpretation of bisintercalation at contiguous sites for the flexible short-chain diacridines was controversial, since it violated the so-called 'excluded-site' principle, which states that for intercalators binding to a DNA lattice, ligand binding to a free site will exclude other ligands from

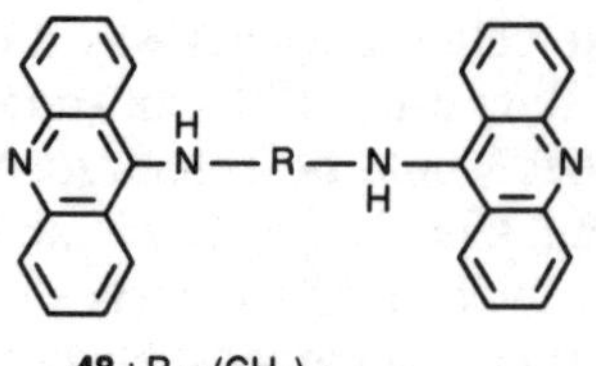

**48** : R = $(CH_2)_6$
**49** : R = $(CH_2)_n$
**50** : R = $(CH_2)_nCONH(CH_2)_m$
**51** : R = $(CH_2)_2CONH(CH_2)_3NHCO(CH$

binding to the adjacent sites (McGhee and von Hippel, 1974). However, this is a thermodynamic limitation not required by all models of intercalation (Shafer and Waring, 1982). Support for the contiguous binding model was provided by a report (Atwell *et al.*, 1985) that the diacridine (**52**) binds to DNA by intercalation of both chromophores. Since these are rigidly held in a coplanar configuration only 7 Å apart by the linker chain, they must bind at contiguous sites.

For compounds of this type, joined by rigid linker chains in a definite (pre-organized) geometry, the possibility also exists of intermolecular bisintercalation, either between two different DNA molecules or between distant linear sections of the same molecule which are close in space, as an alternative to intermolecular bisintercalation. A recent study (Mullins *et al.*, 1992) of a series of diacridines rigidly linked in an extended configuration (e.g. **53**) showed that such interactions do occur. A DNA ligation technique was used to generate catenanes from DNA molecules cross-linked temporarily by intermolecular bisintercalation.

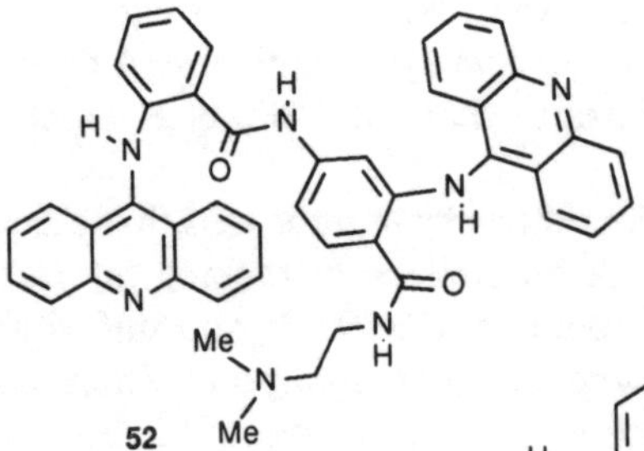

**52**

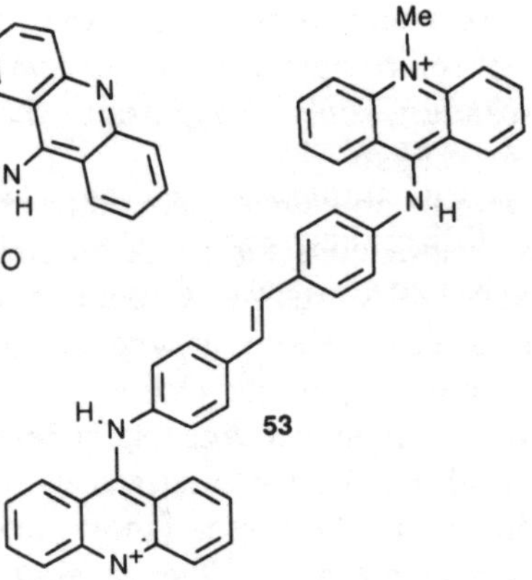

**53**

DNA Binding Kinetics

The nature of the linker chain in diacridines is known to affect the kinetics of their overall dissociation rates from DNA (as measured by stopped-flow spectrophotometry), with cationic linker chains greatly slowing such dissociation (Capelle *et al.*, 1979). Comparative NMR and stopped-flow studies of a series of diacridines showed that a distinction must be drawn between dissociation of the whole molecule and of the individual chromophores. Thus, while the flexibly linked diacridine (**54**) dissociates from DNA about thirtyfold more slowly than does the monomer 9-aminoacridine (measured by stopped-flow UV spectroscopy), the average residence time of each acridine chromophore at a particular DNA site (measured by NMR spectroscopy) is only twofold slower than that of 9-aminoacridine (Assa-Munt *et al.*, 1985b; Denny *et al.*, 1985a). This implies that the kinetic behaviour of the individual chromophores is only slightly affected by the presence of the flexible linker chain joining them. Complete dissociation of the molecule from DNA (requiring simultaneous disengagement of both acridine rings) is a relatively rare event compared with dissociation of a single acridine ring. On average, each chromophore of **54** makes 15 dissociations and reassociations for every complete dissociation of the molecule, suggesting considerable mobility of the ligand when 'bound' to DNA by 'creeping' along the lattice. In contrast, compounds such as **55**, with more rigid linker chains, have greatly increased residence times for the chromophores, and may act more like 'molecular staples' (Denny *et al.*, 1983).

Biological Properties

As noted above, only one diacridine (**48**) from the large number of symmetrical compounds prepared was seriously evaluated for clinical trial, and showed only marginal activity with pronounced CNS toxicity. Although it is probable that **48** was not the best choice, since the heptanediamine analogue (**56**) has higher biological activity (Chen *et al.*, 1978; Denny *et al.*, 1985a) and lower adrenergic potency (Adams *et al.*, 1985), none of the flexibly linked diacridines seem likely to be useful. A smaller number of asymmetrical diacridines have been prepared (Hansen *et al.*, 1983), using monoprotected diamines in a stepwise synthesis, but no antitumour activity *in vivo* has been reported.

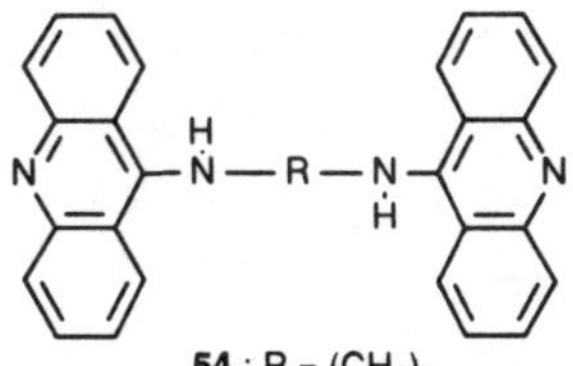

**54** : R = $(CH_2)_8$
**56** : R = $(CH_2)_7$

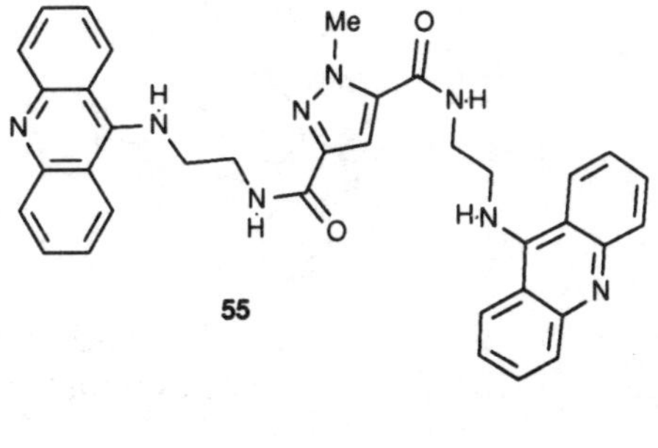

**55**

*Triacridines*

Evidence for Trisintercalation of Triacridines

Triacridines as potential DNA trisintercalating agents are a logical step from diacridines, and several studies have been reported. A triacridine (**57**) joined by polymethylene linker chains (Hansen and Buchardt, 1983) was reported to trisintercalate into DNA on the basis of helix extension data (Hansen *et al.*, 1984). The amide-linked triacridine (**58**) appeared to be a trisintercalating ligand on the basis of helix unwinding data in one study (Atwell *et al.*, 1983), but not in another (Denny *et al.*, 1985b). A related amide-linked triacridine (**59**) was also claimed to trisintercalate on the basis of unwinding data, although the helix extension data were equivocal (Gaugain *et al.*, 1984). A review of these results (Wakelin, 1986) concluded that, although claims for trisintercalation are more difficult to evaluate than for bisintercalation, some of the molecules probably are true trisintercalating ligands.

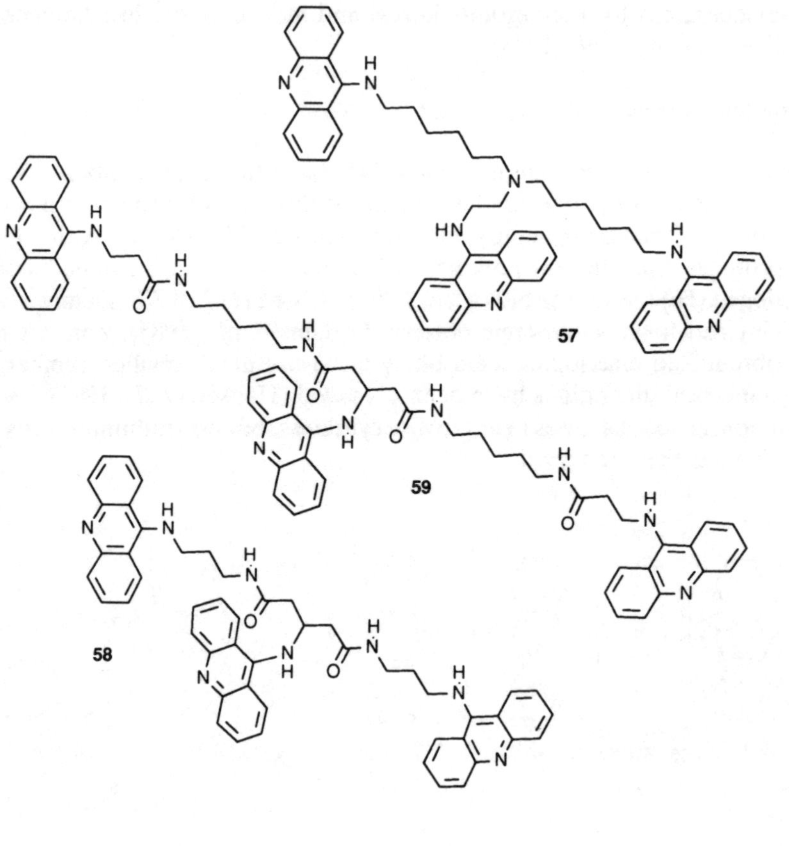

**57**

**59**

**58**

Biological Activity of Tri- and Polyacridines

A study of the cytotoxicity *in vitro* and biological activity *in vivo* of a series of amide-linked triacridines related to **57** was reported, but the compounds possessed only modest improvements in DNA binding over the analogous dimeric compounds, and minimal antileukaemic activity *in vivo* (Atwell *et al.*, 1986b).

## 6 Acridines as Carriers for Other Functionalities

### *Introduction*

In addition to acridine derivatives being used as bioactive agents in their own right, the acridine chromophore is increasingly employed as a carrier molecule, both as a component of 'mixed-function' DNA-binding ligands, and to deliver attached reactive moieties to DNA. Reasons which make acridines (and 9-anilinoacridines) attractive for this role include relative simplicity of synthesis, a tight and well-understood mode of binding to DNA, and rapid cellular uptake (Bailly and Hénichart, 1991).

### *As Components of 'Mixed-function' DNA Binding Ligands*

Acridine/Minor Groove Binder 'Combilexins'

The design of ligands capable of binding to specified DNA base sequences is of interest with regard to selective gene inhibition. Much work on lexitropsins based on the polypyrrole antibiotics has shown that, while AT selectivity is easy to achieve, building in positive discrimination for GC sites is difficult (Bailly and Hénichart, 1991). Consideration of the known GC selectivity of many intercalators has led to studies of various acridine–polypyrrole 'combilexins' (e.g. **60**, **61**). However, studies on the DNA interaction of these compounds suggest that, while the level of binding is enhanced by the acridine moiety, the specificity of binding remains dominated by the polypyrrole (Eliadis *et al.*, 1988; Bailly and Hénichart, 1991). The 9-anilinoacridine derivative (**61**) has been shown to be as efficient as amsacrine (**1**) as a DNA topoisomerase II inhibitor, although it has lower antileukaemic activity *in vivo* (Bailly *et al.*, 1992a). 9-Anilinoacridine analogues (**62**, **63**) of the DNA binding peptide SPKK (Ser–Pro–Lys–Lys, which has itself been suggested to bind in the minor groove (Suzuke, 1989)) have enhanced DNA binding and weak cytotoxic activity (Bailly *et al.*, 1992c). The compound with one SPKK unit (**62**) is a classical intercalator, but the presence of two such units modifies the intercalative binding of **63**.

Acridines have also been used extensively as targeting moieties for oligonucleotides (Nielsen *et al.*, 1992), but this work is considered to be outside the scope of this review.

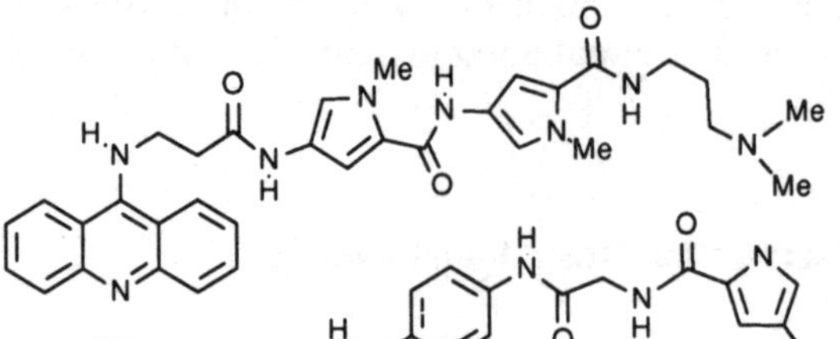

**60**

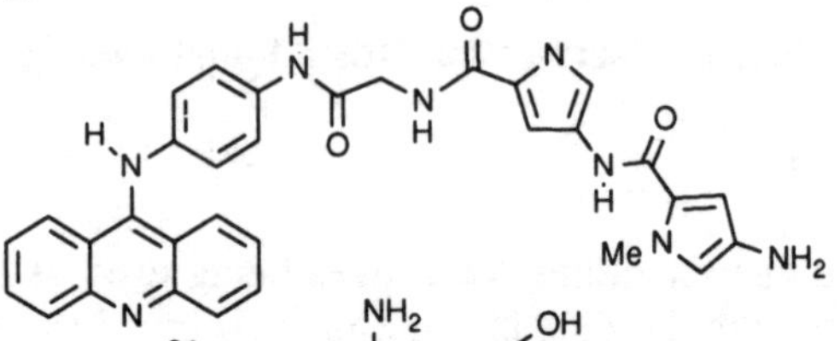

**61**

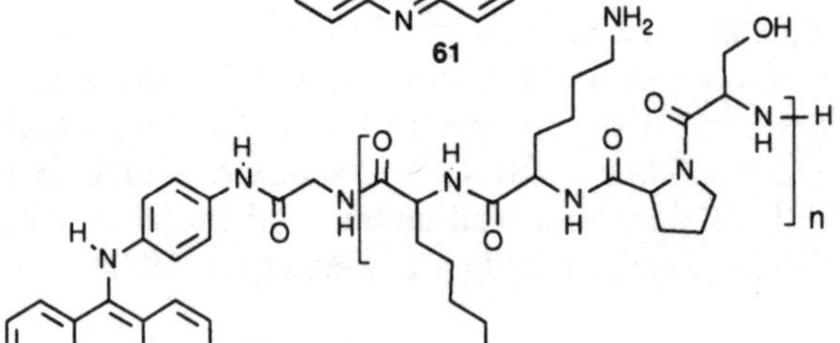

**62** : n = 1
**63** : n = 2

*As Carriers of DNA-breaking Species*

Metal Complexes

Some studies have been done on 9-anilinoacridines carrying metal binding moieties, in work aimed at the production of both new anticancer drugs (e.g. **64**) (Morier-Teissier *et al.*, 1989) and of DNA sequence-specific cleaving agents (e.g. **65**) (Bailly *et al.*, 1992b). The copper complex, **65**, has higher DNA-binding affinity than the related 'combilexin', **61**, implying increased binding by the peptide unit. Footprinting studies on an oligonucleotide containing a GC base pair flanked on either side by AT-rich regions showed protection of the GC site, and enhanced cutting on either side, implying intercalative binding of the acridine at the GC, with the side-chain lying in either of the two possible directions (Bailly *et al.*, 1992b).

Photocleaving Agents

Acridines carrying nitrobenzoyl units capable of cleaving DNA on UV irradiation have been studied as sequence-specific photocleavage reagents.

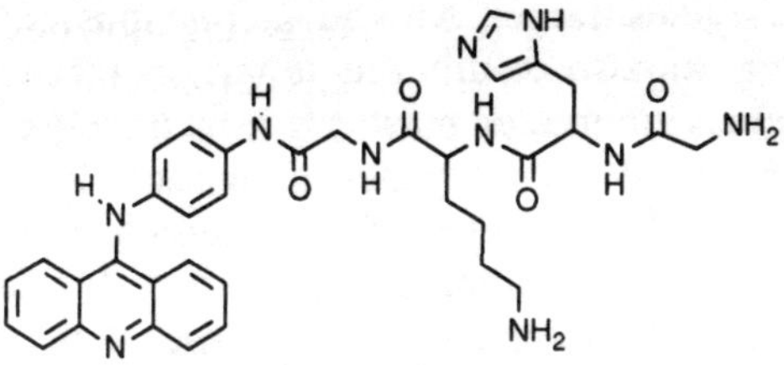

64

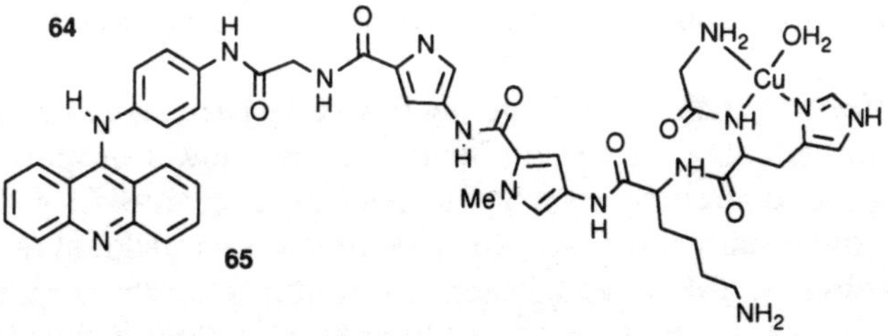

65

Simple acridine-targeted compounds (e.g. **66**, **67**) cleave DNA efficiently (Buchardt *et al.*, 1987; Kuroda and Shinomiya, 1991). A 'combilexin' analogue (**68**) was reported recently (Shinomiya and Kuroda, 1992). This compound photocleaves DNA efficiently, but data on the sequence specificity of cleavage is not yet available.

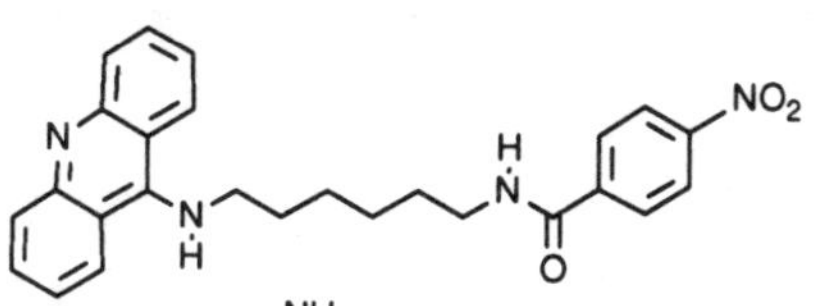

66

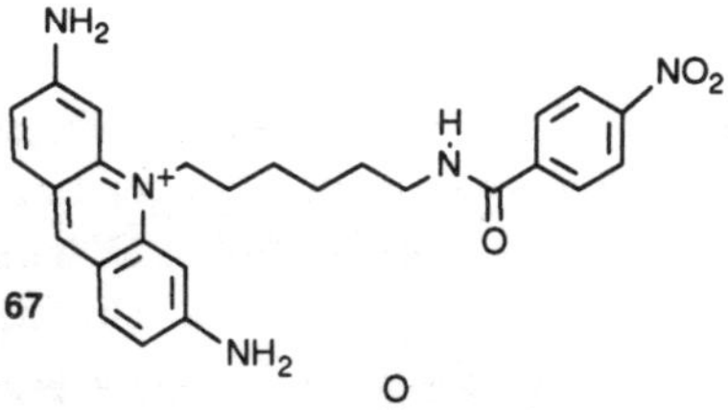

67

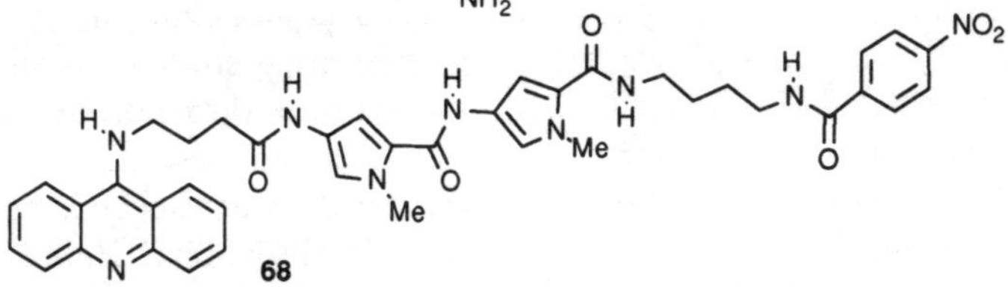

68

As Carriers of Alkylating Moieties

Following initial suggestions (Creech *et al.*, 1972) that attachment of mustard alkylating agents to an acridine 'carrier' might serve to target the reactive centre to DNA, studies (largely with quinacrine mustard, **69**) have shown that such targeting can also modify the pattern of DNA alkylation (Kohn *et al.*, 1987). Whereas untargeted mustards react largely at the N7 of guanines in runs of guanines, **69** also alkylates at guanines in 5′-GT sites. DNA targeting also affected alkylation specificity in aniline mustards (**70**), with reaction switching from N-7 guanine alkylation at 5′-GT sites to N-7 and N-1 adenine alkylation at complementary 5′-AC sites as the separation between the chromophore and the mustard was lengthened (Prakash *et al.*, 1990).

Other effects of such DNA targeting have also been suggested (Gourdie *et al.*, 1990), including increased potency due to a lower proportion of the drug being inactivated by hydrolysis, decreased genotoxicity due to an increase in the usually low (*ca* 1:20) cross-link to monoadduct ratio (Brendel and Ruhland, 1984), and lesser susceptibility to the appearance of resistance due to increases in the cellular level of low molecular weight thiols (Suzukake *et al.*, 1983). Recent work with aniline mustards suggests that DNA targeting by acridine carriers can achieve some of these effects, with compounds **70** showing more rapid DNA alkylation (O'Connor *et al.*, 1992) and increases in cytotoxic potency of up to hundredfold compared with the corresponding untargeted compounds (Gourdie *et al.*, 1990; Valu *et al.*, 1990), together with quite different patterns of mutagenic properties (Ferguson *et al.*, 1989).

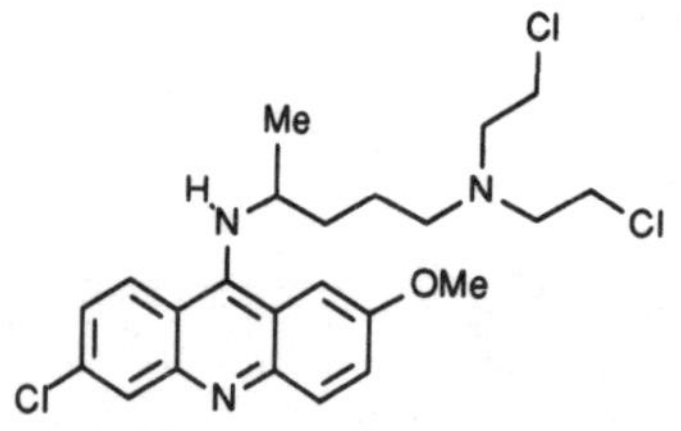


**69**

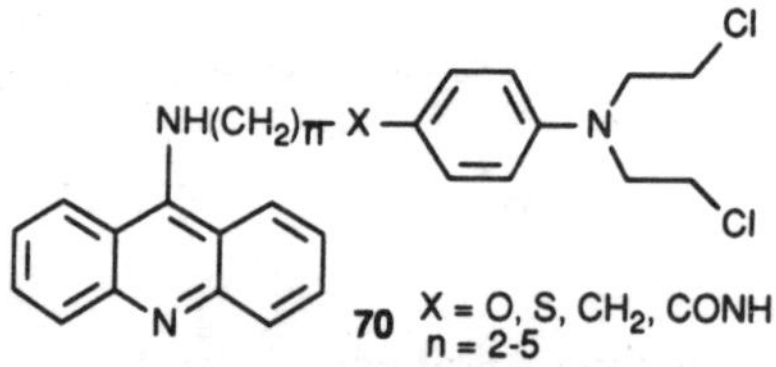


**70** X = O, S, CH$_2$, CONH
n = 2-5

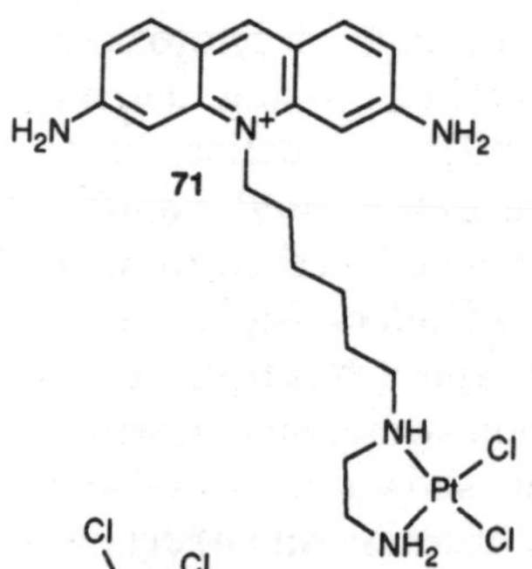


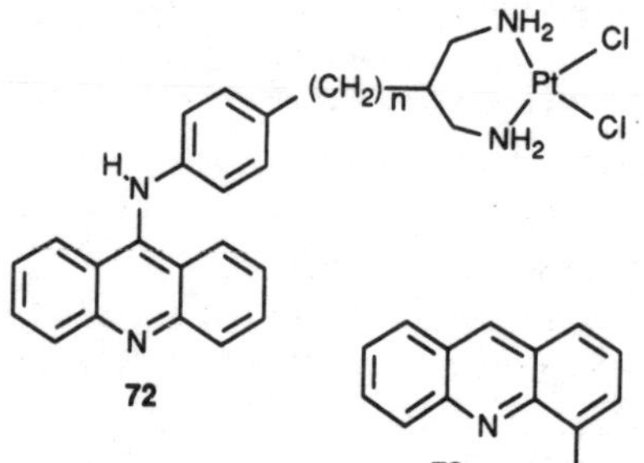


**72**

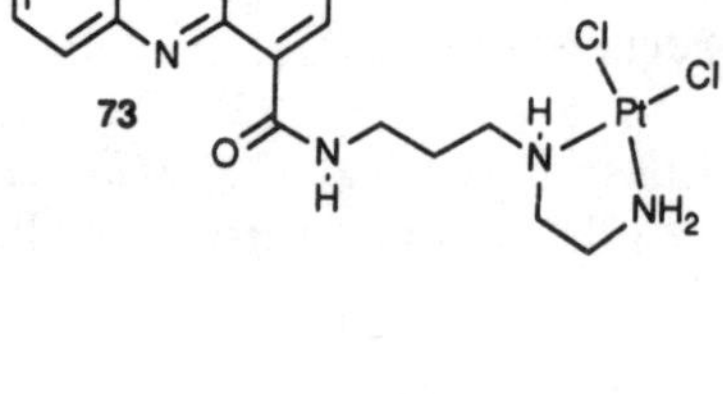

Studies with acridine-targeted platinum complexes have also demonstrated modifications in DNA alkylation patterns (Sundquist *et al.*, 1990). Changes in biological properties for compounds of this type (e.g. **72**, **73**) compared with untargeted platinum complexes have not been dramatic (Palmer *et al.*, 1990; Lee *et al.*, 1992), although **73** and analogues do show some activity against cisplatin-resistant leukaemia *in vivo* (Lee *et al.*, 1992).

As Carriers of Radiosensitizing Moieties

Targeting of the electron-seeking 2-nitroimidazole moiety to DNA via linking to a 9-aminoacridine has been investigated as a means of increasing sensitizer concentrations locally at the DNA target, while decreasing accessibility to reductases responsible for bioactivation and cytotoxicity (Denny *et al.*, 1992). The compound NLA-1 (**74**), which has a similar DNA binding constant to that of 9-aminoacridine, is a potent hypoxic cell radiosensitizer in AA8 cells, producing an enhancement ratio of 1.6 at only 9 μM, approximately hundredfold lower than required with misonidazole (**41**) (Denny *et al.*, 1992; Papadopoulou *et al.*, 1992). However, most of the enhancement is due to higher cellular uptake, and NLA-1 lacked radiosensitizing activity against SCCVII or EMT6 tumours *in vivo*.

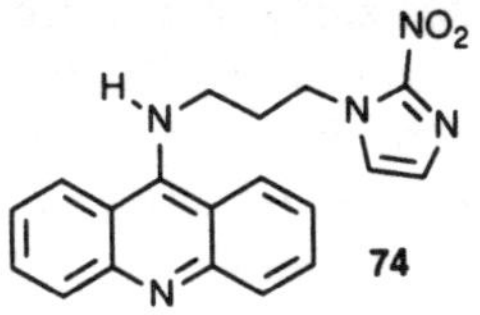


**74**

## 7 Acridine Alkaloids

A large range of cytotoxic pyrido[4,3,2-*mn*]acridine alkaloids of similar structural type — e.g. dercitins and kuanoniamines, **75** (Gundawardana *et al.*, 1992); cystodytins, **76** (Kobayashi *et al.*, 1988); shermilamines, **77** (Schmitz *et al.*, 1991); and varamines, **78** (Charyulu *et al.*, 1989) — have been isolated recently from marine sources. The appearance of compounds of such similar basic structure in so many unrelated animal phyla has been noted, and it has been suggested that they may actually be produced by common symbionts (Carroll and Scheuer, 1990). Many of these compounds show potent cytotoxicity in cell culture, but detailed studies on mode of action and effects *in vivo* are not available.

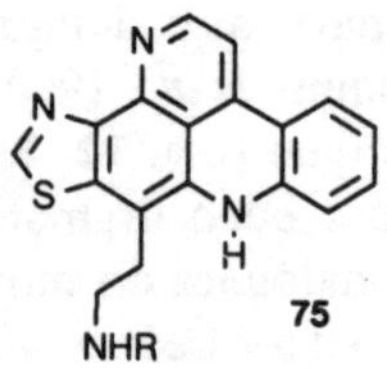


75

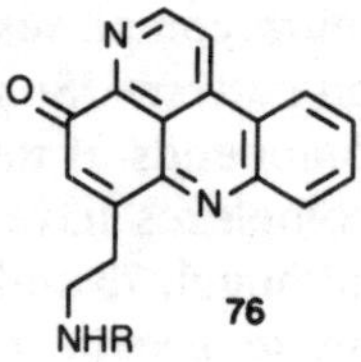


76

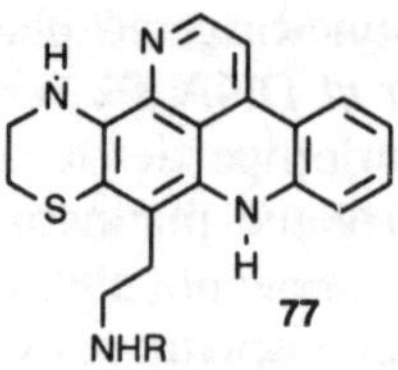


77

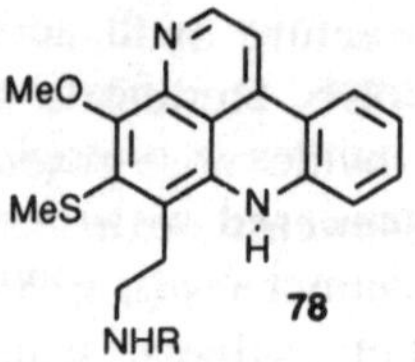


78

## 8 Acridones

### *Tetracyclic Acridanones*

Two new classes of tetracyclic acridanones, the imidazoacridanones (**79**) (Cholody *et al.*, 1990b, 1992) and the triazoloacridanones (**80**) (Cholody *et al.*, 1990a) have recently been shown to have *in vivo* antileukaemic activity. Their mode of DNA binding and mechanism of action have not been reported.

### *Acronycine and Other Acridone Alkaloids*

The compound acronycine (acronine, **81**) is the best-studied of a number of acridone alkaloids (Denny *et al.*, 1983; Michael, 1991). Despite difficulties due to insolubility, a Phase I/II study of acronycine in patients with multiple myeloma did show objective responses (Scarffe *et al.*, 1983). Recent work using solubilized formulations has shown that acronycine has activity against a number of solid murine tumours and human tumour line xenografts in mice, suggesting its possible use in human multiple myeloma (Dorr *et al.*, 1989). A number of new syntheses of acronycine (**81**) and related acridone alkaloids have been reported (Loughhead, 1990; Reisch *et al.*, 1991), but no new analogues have received extensive evaluation.

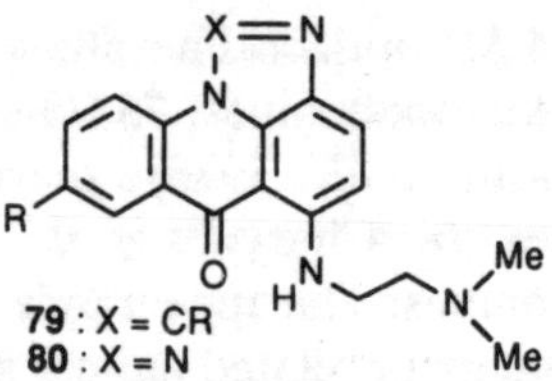


**79** : X = CR
**80** : X = N

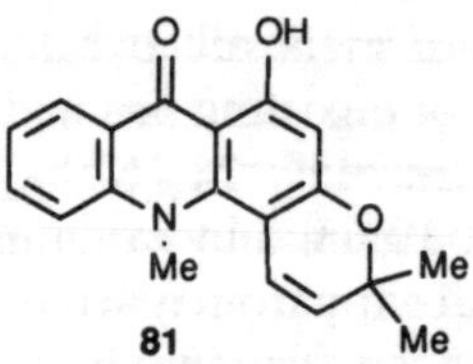


**81**

### *Acridone-4-acetic Acids*

The compound flavone-8-acetic acid (**82**) was discovered to have extremely high activity against a variety of solid tumours, although it has failed to demonstrate useful clinical activity (Kerr and Kaye, 1989). Studies with a range of other arylacetic acids have shown very tight relationships between structure and activity, with very few changes permitted (Atwell *et al.*, 1989b; Rewcastle *et al.*, 1991a). The only two chromophores known to provide more potent derivatives than flavone-8-acetic acid are xanthenone (Rewcastle *et al.*, 1989, 1991b) and acridone; thus, a series of acridone-4-acetic acids (e.g. **83**) show potent solid tumour activity profiles similar to that of flavone-8-acetic acid (Gamage *et al.*, 1992). The activity of compounds of this general type is thought not to depend on DNA binding, but rather on the induction of cytokines including tumour necrosis factor (Futami *et al.*, 1992) which affect tumour blood flow and also induce other host-mediated cytotoxicity mechanisms (Zwi *et al.*, 1989). The receptor which mediates these effects is currently unknown.

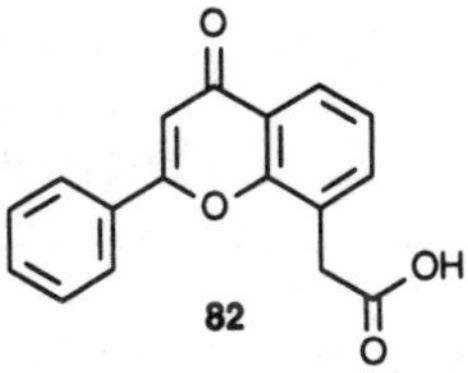

**82**

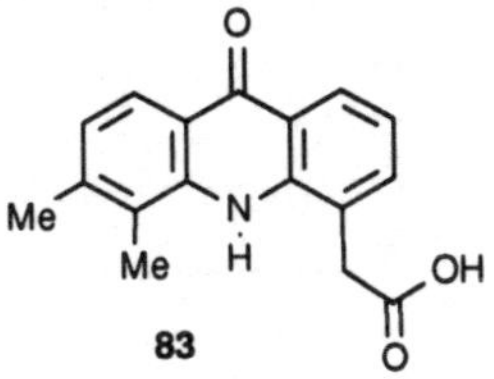

**83**

## 9 Conclusions

Acridine derivatives demonstrate many types of biological effects, among the most important of which is high antitumour activity. Most mechanisms involve the interaction of positively charged acridine species with DNA, and the acridine chromophore has shown itself to have excellent DNA binding properties and metabolic stability. Acridines possessing appropriate side-chains can form tight ternary complexes with topoisomerase II, subverting the action of this enzyme. The two acridine derivatives (**1**, **3**) which have been used clinically, together with the third (**4**) which is beginning clinical trials, all act in this way, raising the question of why they are different from existing topoisomerase II poisons such as doxorubicin and etoposide, and why they differ from each other. Some answers to these questions can be gained by examining their properties towards multidrug-resistant cells. Two major types of multidrug resistance to such drugs are recognized, the first involving increased energy-dependent extrusion of drug from the cytoplasm (Roninson, 1992) and the second involving changes to the target enzyme, DNA topoisomerase II (Liu, 1989).

One feature of all three acridine-based drugs is that they are active, both in culture and *in vivo*, against a multidrug-resistant P388 murine leukaemia line which is cross-resistant to vincristine, etoposide and doxorubicin (Baguley *et al.*, 1990; Finlay *et al.*, 1993). Amsacrine is also active against CEM-CCRF human leukaemia lines which are highly resistant to *Vinca*

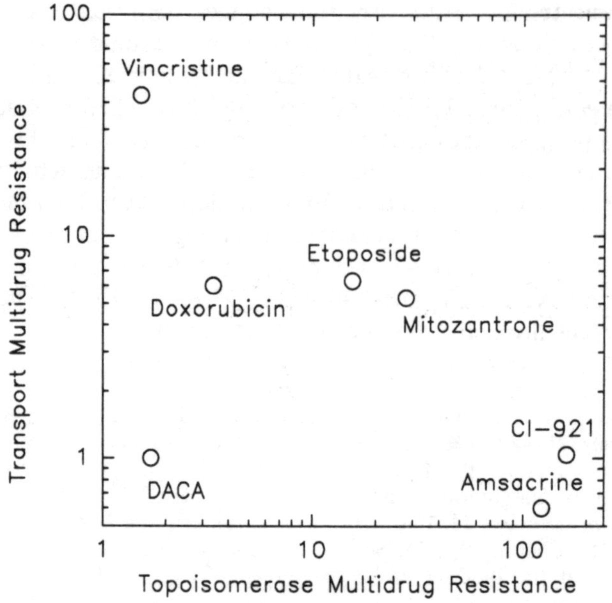


**Figure 7.2** Relationship between susceptibility to transport-mediated multidrug resistance and topoisomerase II-mediated multidrug resistance for a series of topoisomerase II poisons, including the clinical agents doxorubicin, etoposide, mitozantrone and amsacrine, and the clinical candidates CI-921 and DACA. Resistance factors are the ratios of 50% growth inhibitory concentrations for the parent and multidrug resistant lines. Values on the ordinate refer to an actinomycin-resistant P388 leukaemia cell line (Baguley *et al.*, 1990), while those on the abscissa refer to an amsacrine-resistant Jurkat leukaemia line (Finlay *et al.*, 1990). Vincristine, a mitotic inhibitor, is included as a reference compound

alkaloids (Qian and Beck, 1990). Amsacrine and CI-921, like etoposide and doxorubicin, are susceptible to a multidrug resistance mechanism which is thought to involve a change in topoisomerase II rather than drug transport (Finlay *et al.*, 1990). However, in a series of amsacrine-resistant sublines of the Jurkat human leukaemia, resistance to DACA has been found to be minimal (Finlay *et al.*, 1993). These two results can be summarized as a graph (Figure 7.2) which shows the potential contributions of these three acridine derivatives to the problem of overcoming multiple drug resistance.

Acridines can also function as anchor groups for the attachment of chemically reactive moieties which lead directly to DNA damage. The attachment (or incorporation into the chromophore) of other groups with bioreductive or other appropriate properties has led to prodrugs which may be selectively activated in solid tumour microenvironments, adding a further dimension to antitumour selectivity.

**Acknowledgements**

The authors thank their many colleagues in the Cancer Research Laboratory and around the world for their collaboration over many years, and the Auckland Division of the Cancer Society of New Zealand and the Health Research Council of New Zealand for financial support.

**References**

Adams, A., Jarrot, B., Elmer, B. C., Denny, W. A. and Wakelin, L. P. G. (1985). Interaction of DNA-binding antitumour agents with adrenoreceptors. *Mol. Pharmacol.*, **27**, 480–491

Anderson, R. F., Packer, J. E. and Denny, W. A. (1988). One-electron redox chemistry of amsacrine [9-(2-methoxy-4-methylsulfonyl-aminoanilino)acridinium], its quinonediimine and an analogue. A radiolytic study. *J. Chem. Soc. Perkin Trans. II*, 489–496

Assa-Munt, N., Denny, W. A., Leupin, W. and Kearns, D. R. (1985a). A $^1$H NMR study of the binding of bis(acridines) to d(AT)$_5$.d(AT)$_5$. I. Mode of binding. *Biochemistry*, **24**, 1441–1449

Assa-Munt, N., Leupin, W., Denny, W. A. and Kearns, D. R. (1985b). A $^1$H NMR study of the binding of bis(acridines) to d(AT)$_5$.d(AT)$_5$. II. Dynamic aspects. *Biochemistry*, **24**, 1449–1460

Atwell, G. J., Baguley, B. C. and Denny, W. A. (1989a). Potential antitumor agents. 57. 2-Phenylquinoline-8-carboxamides as 'minimal' DNA-intercalating antitumour agents with solid tumor activity. *J. Med. Chem.*, **32**, 396–401

Atwell, G. J., Baguley, B. C., Finlay, G. J., Rewcastle, G. W. and Denny, W. A. (1986a). Potential antitumor agents. 47. 3′-Methylamino analogues of amsacrine with *in vivo* solid tumor activity. *J. Med. Chem.*, **29**, 1769–1776

Atwell, G. J., Baguley, B. C., Wilmanska, D. and Denny, W. A. (1986b). Potential antitumour agents. 46. Synthesis, DNA binding and biological activity of triacridine derivatives. *J. Med. Chem.*, **29**, 68–74

Atwell, G. J., Bos, C. D., Baguley, B. C. and Denny, W. A. (1988). Potential antitumor agents. 56. 'Minimal' DNA-intercalating ligands as antitumor drugs: phenylquinoline-8-carboxamides. *J. Med. Chem.*, **31**, 1048–1052

Atwell, G. J., Cain, B. F., Baguley, B. C., Finlay, G. J. and Denny, W. A. (1984a). Potential antitumor agents. 43. Synthesis and biological activity of dibasic 9-aminoacridine-4-carboxamides, a new class of antitumor agent. *J. Med. Chem.*, **27**, 1481–1485

Atwell, G. J., Cain, B. F. and Seelye, R. N. (1972). Potential antitumor agents. 12. 9-Anilinoacridines. *J. Med. Chem.*, **15**, 611–615

Atwell, G. J., Leupin, W., Twigden, S. J. and Denny, W. A. (1983). A triacridine derivative is the first DNA trisintercalating agent. *J. Am. Chem. Soc.*, **105**, 2913–2914

Atwell, G. J., Rewcastle, G. W., Baguley, B. C. and Denny, W. A. (1987a). Potential antitumor agents. 48. 3′-Dimethylamino derivatives of amsacrine. Redox chemistry and *in vivo* solid tumor activity. *J. Med. Chem.*, **30**, 652–658

Atwell, G. J., Rewcastle, G. W., Baguley, B. C. and Denny, W. A. (1987b). Potential antitumor agents. 50. *In vivo* solid tumor activity of derivatives of *N*-[2-(dimethylamino)ethyl]acridine-4-carboxamide. *J. Med. Chem.*, **30**, 664–669

Atwell, G. J., Rewcastle, G. W., Baguley, B. C. and Denny, W. A. (1989b).

Synthesis and antitumour activity of topologically-related analogues of flavone acetic acid. *Anti-cancer Drug Des.*, **4**, 161–169

Atwell, G. J., Rewcastle, G. W., Denny, W. A., Cain, B. F. and Baguley, B. C. (1984b). Potential antitumor agents. 41. Analogues of amsacrine with electron-donor substituents in the anilino ring. *J. Med. Chem.*, **27**, 367–372

Atwell, G. J., Stewart, G. M., Leupin, W. and Denny, W. A. (1985). A diacridine derivative which binds by bisintercalation at two contiguous sites on DNA. *J. Am. Chem. Soc.*, **107**, 4335–4336

Baguley, B. C. (1990). The possible role of electron transfer complexes in the action of amsacrine analogues. *Biophys. Chem.*, **23**, 937–943

Baguley, B. C. (1991). DNA intercalating anti-tumour agents. *Anti-cancer Drug Des.*, **6**, 1–35

Baguley, B. C., Denny, W. A., Atwell, G. J. and Cain, B. F. (1981). Potential antitumor agents. 34. Quantitative relationships between DNA binding and molecular structure for 9-anilinoacridines substituted in the anilino ring. *J. Med. Chem.*, **24**, 170–177

Baguley, B. C., Denny, W. A., Atwell, G. J., Finlay, G. J., Rewcastle, G. W., Twigden, S. J. and Wilson, W. R. (1984). Synthesis, antitumor activity, and DNA binding properties of a new derivative of amsacrine, *N*-5-dimethyl-9-[(2-methoxy-4-methylsulfonylamino)phenylamino]-4-acridinecarboxamide. *Cancer Res.*, **44**, 3245–3251

Baguley, B. C. and Finlay, G. J. (1988a). Relationship between the structure of analogues of amsacrine and their degree of cross-resistance to adriamycin-resistant P388 leukemia cells. *Eur. J. Cancer Clin. Oncol.*, **24**, 205–210

Baguley, B. C. and Finlay, G. J. (1988b). Derivatives of amsacrine: determinants required for high activity against the Lewis lung carcinoma. *J. Natl Cancer Inst.*, **80**, 195–199

Baguley, B. C., Finlay, G. J. and Wilson, W. R. (1986). Cytokinetic resistance of Lewis lung carcinoma to cyclophosphamide and the amsacrine derivative CI-921. In Hall, T. C. (Ed.), *Cancer Drug Resistance*, Progress in Clinical and Biological Research, Vol. 23, Allan R. Liss Inc., New York, pp. 47–61

Baguley, B. C., Holdaway, K. M. and Fray, L. M. (1990). Design of DNA intercalators to overcome topoisomerase II-mediated multidrug resistance. *J. Natl Cancer Inst.*, **82**, 398–402

Baguley, B. C., Kernohan, A. R. and Wilson, W. R. (1983). Divergent activity of derivatives of amsacrine (m-AMSA) towards Lewis lung carcinoma and P388 leukaemia in mice. *Eur. J. Cancer Clin. Oncol.*, **19**, 1607–1613

Baguley, B. C. and Nash, R. (1981). Antitumour activity of substituted 9-anilinoacridines — comparison of *in vivo* and *in vitro* testing systems. *Eur. J. Cancer*, **17**, 671–679

Baguley, B. C. and Wilson, W. R. (1987). Comparison of *in vivo* and *in vitro* drug sensitivities of Lewis lung carcinoma and P388 leukemia to analogues of amsacrine. *Eur. J. Cancer Clin. Oncol.*, **23**, 607–613

Bailly, C., Collyn-d'Hooghe, M., Lantoine, D., Fournier, C., Hecquet, B., Fosse, P., Saucier, J. M., Colson, P., Houssier, C. and Hénichart, J. P (1992a). Biological activity and molecular interaction of a netropsin–acridine hybrid ligand with chromatin and topoisomerase II. *Biochem. Pharmacol.*, **43**, 457–466

Bailly, C. and Hénichart, J. P. (1991). DNA recognition by intercalator–minor-groove binder hybrid molecules. *Bioconjugate Chem.*, **2**, 379–393

Bailly, C., Sun, J.-S., Colson, P., Houssier, C., Hélène, C., Waring, M. J. and Hénichart, J. P. (1992b). Design of a sequence-specific DNA-cleaving molecule which conjugates a copper-chelating peptide, a netropsin residue, and an acridine chromophore. *Bioconjugate Chem.*, **3**, 100–103

Bailly, F., Bailly, C., Helbecque, N., Pommery, N., Colson, P., Houssier, C. and Hénichart, J. P. (1992c). Relationship between DNA-binding and biological activity of anilinoacridine derivatives containing the nucleic acid binding unit SPPK. *Anti-cancer Drug Des.*, **7**, 83–100

Bernier, J. L., Hénichart, J. P. and Catteau, J. P. (1981). Design, synthesis and DNA-binding capacity of a new peptide difunctional intercalating agent. *Biochem. J.*, **199**, 479–484

Boyd, M. and Denny, W. A. (1990). NMR studies of configuration and tautomeric equilibria in nitroacridine antitumor agents. *J. Med. Chem.*, **33**, 2656–2659

Braithwaite, A. W. and Baguley, B. C. (1980). Existence of an extended series of antitumor compounds which bind to deoxyribonucleic acid by nonintercalative means. *Biochemistry*, **19**, 1101–1106

Brendel, M. and Ruhland, A. (1984). Relationship between functionality and genetic toxicology of selected DNA-damaging agents. *Mutation Res.*, **133**, 51–85

Brennan, S. T., Colbry, N. L., Leeds, R. L., Leja, B., Priebe, S. T., Reily, M. D., Showalter, H. D. H., Uhlendorf, S. E., Atwell, G. J. and Denny, W. A. (1989). Anticancer anilinoacridines. A process synthesis of the disubstituted amsacrine analogue CI-921. *J. Het. Chem.*, **26**, 1469–1476

Buchardt, O., Egholm, M., Karup, G. and Nielsen, P. E. (1987). 9-(4-Nitrobenzamidopolymethylene)aminoacridines and their photochemical cleavage of DNA. *J. Chem. Soc. Chem. Commun.*, 1696–1697

Cain, B. F. and Atwell, G. J. (1974). The experimental antitumour properties of three congeners of the acridylmethanesulphonanilide (AMSA) series. *Eur. J. Cancer*, **10**, 539–549

Cain, B. F., Atwell, G. J. and Denny, W. A. (1975). Potential antitumor agents. 16. 4′-(Acridin-9-ylamino)methane-sulfonanilides. *J. Med. Chem.*, **18**, 1110–1117

Cain, B. F., Atwell, G. J. and Denny, W. A. (1977). Potential antitumor agents. 23. 4′-(9-Acridinylamino)alkanesulfonanilide congeners bearing hydrophilic functionality. *J. Med. Chem.*, **20**, 987–996

Cain, B. F., Atwell, G. J. and Seelye, R. N. (1969). Potential antitumor agents. 10. Bisquaternary salts. *J. Med. Chem.*, **12**, 199–206

Cain, B. F., Atwell, G. J. and Seelye, R. N. (1971). Potential antitumor agents. 11. 9-Anilinoacridines. *J. Med. Chem.*, **14**, 311–315

Cain, B. F., Atwell, G. J. and Seelye, R. N. (1972). Potential antitumor agents. 12. 9-Anilinoacridines. *J. Med. Chem.*, **15**, 611–615

Cain, B. F., Seelye, R. N. and Atwell, G. J. (1974). Potential antitumor agents. 14. Acridylmethanesulfonanilides. *J. Med. Chem.*, **17**, 922–930

Cain, B. F., Wilson, W. R. and Baguley, B. C. (1976). Structure–activity relationships for thiolytic cleavage rates of antitumor drugs in the 4′-(9-acridinylamino)methanesulfonanilide series. *Mol. Pharmacol.*, **12**, 1027–1035

Capelle, N., Barber, J., Dessen, P., Blanquet, S., Roques, B. P. and Le Pecq, J.-B. (1979). Deoxyribonucleic acid bifunctional intercalators: kinetic investigation of the binding of several acridine dimers to deoxyribonucleic acid. *Biochemistry*, **18**, 3354–3362

Carroll, A. R. and Scheuer, P. J. (1990). Kuanoniamines A, B, C and D; pentacyclic alkaloids from a tunicate and its prosobranch mollusk predator *Chelynotus semperi*. *J. Org. Chem.*, **55**, 4426–4431

Charyulu, G. A., McKee, T. C. and Ireland, C. M. (1989). Diplamine, a cytotoxic

polyaromatic alkaloid from the tunicate *Diplosoma* sp. *Tetrahedron Lett.*, **30**, 4201–4202
Chen, K. X., Gresh, N. and Pullman, B. (1988a). Energetics and stereochemistry of DNA complexation with the antitumor AT specific intercalators tilorone and m-AMSA. *Nucleic Acids Res.*, **16**, 3061–3073
Chen, K. X., Gresh, N. and Pullman, B. (1988b). Groove selectivity in the interaction of 9-aminoacridine-4-carboxamide antitumor agents with DNA. *Nucleic Acids Res.*, **16**, 3061–3074
Chen, T.-K., Fico, R. M. and Canellakis, E. S. (1978). Diacridines, bifunctional intercalators. Chemistry and antitumor activity. *J. Med. Chem.*, **21**, 868–874
Ching, L. M., Finlay, G. J., Joseph, W. R. and Baguley, B. C. (1990). Comparison of the cytotoxicity of amsacrine and its analogue CI-921 against cultured human and mouse bone marrow tumour cells. *Eur. J. Cancer*, **26**, 49–54
Cholody, W. M. and Konopa, J. (1991). Synthesis and proton NMR characterization of substituted 1-amino-9-imino-4-nitro-9,10-dihydroacridines as potential antitumor agents. *J. Het. Chem.*, **28**, 209–214
Cholody, W. M., Martelli, S. and Konopa, J. (1990a). 8-Substituted 5-[(aminoalkyl)amino]-6-*H*-v-triazolo[4,5,1-*de*]acridin-6-ones as potential antineoplastic agents. Synthesis and biological activity. *J. Med. Chem.*, **33**, 2852–2856
Cholody, W. M., Martelli, S. and Konopa, J. (1992). Chromophore-modified antineoplastic imidazoacridinones. Synthesis and activity against murine leukemias. *J. Med. Chem.*, **35**, 378–382
Cholody, W. M., Martelli, S., Paradziej-Lukowicz, J. and Konopa, J. (1990b). 5-[(Aminoalkyl)amino]imidazo[4,5,1-*de*]acridin-6-ones as a novel class of antineoplastic agents. Synthesis and biological activity. *J. Med. Chem.*, **33**, 49–52
Chung, T. D. Y., Drake, F. H., Tan, K. B., Per, M., Crooke, S. T. and Mirabelli, C. K. (1989). Characterization and immunological identification of cDNA clones encoding two human DNA topoisomerase isozymes. *Proc. Natl Acad. Sci. USA*, **86**, 9431–9435
Cornford, E. M., Young, D. and Paxton, J. W. (1992). Comparison of the blood–brain barrier and liver penetration of acridine antitumor drugs. *Cancer Chemother. Pharmacol.*, **29**, 439–444
Covey, J. M., Kohn, K. W., Kerrigan, D., Tilchen, E. J. and Pommier, Y. (1988). Topoisomerase II-mediated DNA damage produced by 4′-(9-acridinylamino)methanesulfon-*m*-anisidide and related acridines in L1210 cells and isolated nuclei: relation to cytotoxicity. *Cancer Res.*, **48**, 860–865
Creech, H. J., Preston, R. K., Peck, R. M., O'Connell, A. S. and Ames, B. N. (1972). Antitumor and mutagenesis properties of a variety of heterocyclic nitrogen and sulfur mustards. *J. Med. Chem.*, **15**, 739–746
Denny, W. A. (1989). DNA-intercalating agents as antitumour drugs: prospects for future design. *Anti-cancer Drug Des.*, **4**, 241–263
Denny, W. A., Atwell, G. J. and Baguley, B. C. (1984). Potential antitumour agents. 40. Orally active 4,5-disubstituted derivatives of amsacrine. *J. Med. Chem.*, **27**, 363–367
Denny, W. A., Atwell, G. J. and Baguley, B. C. (1987a). 'Minimal' DNA-intercalating agents as antitumour drugs: 2-styrylquinoline analogues of amsacrine. *Anti-cancer Drug Des.*, **2**, 263–270
Denny, W. A., Atwell, G. J., Baguley, B. C. and Cain, B. F. (1979a). Potential antitumor agents. 29. QSAR for the antileukemic bisquaternary ammonium heterocycles. *J. Med. Chem.*, **22**, 134–151
Denny, W. A., Atwell, G. J., Baguley, B. C. and Wakelin, L. P. G. (1985a).

Potential antitumour agents. 44. Synthesis and antitumour activity of new classes of diacridines. The importance of linker chain rigidity for DNA binding kinetics and biological activity. *J. Med. Chem.*, **28**, 1568–1574

Denny, W. A., Atwell, G. J. and Cain, B. F. (1979b). Potential antitumor agents. 32. The role of agent base strength in the QSAR for 4′-(9-acridinylamino)methanesulfonanilide(m-AMSA) analogues. *J. Med. Chem.*, **22**, 1453–1460

Denny, W. A., Atwell, G. J., Cain, B. F., Hansch, C., Leo, A. and Panthananickal, A. (1982). Potential antitumor agents. 36. Quantitative relationships between antitumor potency, toxicity and structure for the general class of 9-anilinoacridine antitumour agents. *J. Med. Chem.*, **25**, 276–315

Denny, W. A., Atwell, G. J., Rewcastle, G. W. and Baguley, B. C. (1987b). Potential antitumor agents. 49. 5-Substituted derivatives of *N*-[2-(dimethylamino)ethyl]-9-aminoacridine-4-carboxamide with *in vivo* solid tumor activity. *J. Med. Chem.*, **30**, 658–663

Denny, W. A., Atwell, G. J., Wilmott, G. A. and Wakelin, L. P. G. (1985b). Interaction of paired homologous series of diacridines and triacridines with deoxyribonucleic acid. *Biophys. Chem.*, **22**, 17–26

Denny, W. A. and Baguley, B. C. (1987). Amsacrine analogues with extended chromophores. DNA binding and antitumour activity. *Anti-cancer Drug Des.*, **2**, 61–70

Denny, W. A., Baguley, B. C., Cain, B. F. and Waring, M. J. (1983). Antitumour acridines. In Neidle, S. and Waring, M. J. (Eds), *Molecular Aspects of Anticancer Drug Action*, Macmillan, London, pp. 1–34

Denny, W. A., Roberts, P. B., Anderson, R. F., Brown, J. M. and Wilson, W. R. (1992). NLA-1: A 2-nitroimidazole radiosensitizer targeted to DNA by intercalation. *Int. J. Radiat. Oncol. Biol. Phys.*, **22**, 553–556

Denny, W. A., Roos, I. A. G. and Wakelin, L. P. G. (1986). Interrelationship between antitumour activity, DNA breakage and DNA binding kinetics for 9-aminoacridinecarboxamide antitumour agents. *Anti-cancer Drug Des.*, **1**, 141–147

Denny, W. A., Wilson, W. R., Atwell, G. J. and Anderson, R. F. (1990a). Hypoxia-selective antitumor agents. 4. Relationships between hypoxia-selective cytotoxicity and structure for sidechain derivatives of nitracrine: the 'imidoacridan hypothesis'. *J. Med. Chem.*, **33**, 1288–1295

Denny, W. A., Wilson, W. R., Atwell, G. J., Boyd, M., Pullen, S. M. and Anderson, R. F. (1990b). Nitroacridines and nitroquinolines as DNA-affinic hypoxia-selective cyotoxins. In Adams, G. E., Breccia, A., Wardman, P. and Fielden, E. M. (Eds), *Selective Activation of Drugs by Redox Processes*, NATO ASI Series A (Life Sciences), Vol. 198, pp. 149–158

Dorr, R. T., Liddil, J. D., von Hoff, D. D., Soble, M. and Osborne, C. K. (1989). Antitumor activity and murine pharmacokinetics of parenteral acronycine. *Cancer Res.*, **49**, 340–344

Durand, R. E. (1989). Distribution of and activity of antineoplastic drugs in a tumor model. *J. Natl Cancer Inst.*, **81**, 146–152

Ebeid, M. Y., El-Moghazy Aly, S. M., Eissa, A. A. H. and Osman, A. M. M. (1990). Regioselective synthesis and antitumour activity of 8-chloro-5-(*p-N*-substituted sulfamoylphenyl)-aminobenzo-[b][1,8]-naphthyridines. *Egypt. J. Pharm. Sci.*, **31**, 515–525

Eliadis, A., Phillips, D. R., Reiss, J. A. and Skorobogaty, A. (1988). The synthesis and DNA footprinting of acridine-linked netropsin and distamycin bifunctional mixed ligands. *J. Chem. Soc. Chem. Commun.*, 1049–1052

El-Moghazy Aly, S. M., and Safwat, H. M. (1990). Synthesis and antitumour activity of some acridonanil derivatives. *Egypt. J. Pharm. Sci.*, **31**, 505–513
Ferguson, L. R., Hill, C. M. and Baguley, B. C. (1990). Genetic toxicology of tricyclic carboxamides, a new class of DNA binding antitumour agent. *Eur. J. Cancer Clin. Oncol.*, **26**, 709–714
Ferguson, L. R., Turner, P. M., Gourdie, T. A., Valu, K. K. and Denny, W. A. (1989). 'Petite' mutagenesis and mitotic crossing-over in yeast by DNA-targeted alkylating agents. *Mutation Res.*, **215**, 213–222
Ferguson, L. R., van Zijl, P. and Baguley, B. C. (1988). Comparison of the mutagenicity of amsacrine with that of a new clinical analogue, CI-921. *Mutation Res.*, **204**, 207–217
Finlay, G. J. and Baguley, B. C. (1984). The use of human cancer cell lines as a primary screening system for antineoplastic compounds. *Eur. J. Cancer Clin. Oncol.*, **20**, 947–954
Finlay, G. J. and Baguley, B. C. (1989). Selectivity of *N*-[2-(dimethylamino)ethyl]acridine-4-carboxamide towards Lewis lung carcinoma and human tumour cell lines *in vitro*. *Eur. J. Cancer Clin. Oncol.*, **25**, 271–277
Finlay, G. J., Baguley, B. C., Snow, K. and Judd, W. (1990). Multiple patterns of resistance of human leukemia cell sublines to amsacrine analogues. *J. Natl Cancer Inst.*, **82**, 662–667
Finlay, G. J., Marshall, E. S., Matthews, J. H. L., Paull, K. D. and Baguley, B. C. (1993). *In vitro* assessment of *N*-[2-(dimethylamino)ethyl]acridine-4-carboxamide (DACA), a DNA intercalating antitumour drug with reduced sensitivity to multidrug resistance. *Cancer Chemother. Pharmacol.*, **31**, 401–406
Finlay, G. J., Wilson, W. R. and Baguley, B. C. (1986). Comparison of the *in vitro* activity of cytotoxic drugs towards human carcinoma and leukemia cell lines. *Eur. J. Cancer Clin. Oncol.*, **22**, 655–662
Finlay, G. J., Wilson, W. R. and Baguley, B. C. (1987). Cytokinetic factors in drug resistance of Lewis lung carcinoma: comparison of cells freshly isolated from tumours with cells from exponential and plateau-phase cultures. *Br. J. Cancer*, **56**, 755–762
Futami, H., Eader, L., Back, T. T., Gruys, E., Young, H. A., Wiltrout, R. H. and Baguley, B. C. (1992). Cytokine induction and therapeutic synergy with IL-2 against murine renal cancer by xanthenone-4-acetic acid derivatives. *J. Immunother.*, **12**, 247–255
Gamage, S. A., Rewcastle, G. W., Atwell, G. J., Baguley, B. C. and Denny, W. A. (1992). Structure–activity relationships for substituted 9-oxo-9,10-dihydroacridine-4-acetic acids: analogues of the colon tumour active agent xanthenone-4-acetic acid. *Anti-cancer Drug Des.*, **7**, 403–414
Gaudich, K. and Przybylski, M. (1983). Field desorption mass spectrometric characterisation of thiol conjugates related to the oxidative metabolism of the anticancer drug 4′-(9-acridinylamino)methanesulfon-*m*-anisidide. *Biomed. Mass Spect.*, **10**, 292–299
Gaugain, B., Markovits, J., Le Pecq, J.-B. and Roques, B. P. (1984). DNA polyintercalation: comparison of DNA binding properties of an acridine dimer and trimer. *FEBS Lett.*, **169**, 123–126
Gieldanowski, J., Patkowski, J., Szaga, B. and Teodorczyk, J. (1972a). Preclinical pharmacologic investigation on 1-nitro-9-(dimethylaminoproplyamino)-acridine and its *N*-oxide. 1. Acute and subchronic activity. *Arch. Immunol. Ther. Exp.*, **20**, 399–418
Gieldanowski, J., Patkowski, J., Szaga, B. and Teodorzyk, J. (1972b). Preclinical pharmacologic investigations on 1-nitro-9-(dimethylaminoproplyamino)-acridine

and its *N*-oxide. 2. Chronic action. *Arch. Immunol. Ther. Exp.*, **20**, 419–444

Goldin,A., Venditti, J. M., Macdonald, J. S., Muggia, F. M., Henney, J. E. and DeVita, V. T., Jr. (1981). Current results of the screening program at the Division of Cancer Treatment, National Cancer Institute. *Eur. J. Cancer*, **17**, 129–142

Gourdie, T. A., Valu, K. K., Gravatt, G. L., Boritzki, T. J., Baguley, B. C., Wilson, W. R., Woodgate, P. D. and Denny, W. A. (1990). DNA-directed alkylating agents. 1. Structure–activity relationships for acridine-linked aniline mustards: consequences of varying the reactivity of the mustard. *J. Med. Chem.*, **33**, 1177–1186

Grove, W. R., Fortner, C. L. and Wiernik, P. H. (1982). Review of amsacrine, an investigational antineoplastic agent. *Clin. Pharm.*, **1**, 320–326

Gunawardana, G. P., Koehn, F. E., Lee, A. Y., Clardy, J., He, H. Y. and Faulkner, D. J. (1992). Pyridoacridine alkaloids from deep-water marine sponges of the family *Pachastrellidae*—structure revision of dercitin and related compounds and correlation with the kuanoniamines. *J. Org. Chem.*, **57**, 1523–1526

Haldane, A., Finlay, G. J., Gavin, J. B. and Baguley, B. C. (1992). Unusual dynamics of killing of cultured Lewis lung cells by the DNA-intercalating anti-tumour agent *N*-[2-(dimethylamino)ethyl]acridine-4-carboxamide. *Cancer Chemother. Pharmacol.*, **29**, 475–479

Hansen, J. B. and Buchardt, O. (1983). A novel synthesis of tri-, di- and mono-9-acridinyl derivatives of tetra-, tri- and diamines. *J. Chem. Soc. Chem. Commun.*, 162–164

Hansen, J. B., Koch, T., Buchardt, O., Neilsen, P. E., Norden, B. and Wirth, M. (1984). Trisintercalation in DNA by *N*-[3-(9-acridinylamino)propyl]-*N*,*N*-bis[6-(9-acridinylamino)hexyl]amine. *J. Chem. Soc. Chem. Commun.*, 509–511

Hansen, J. B., Thomsen, T. and Buchardt, O. (1983). 9-Acridinylguanidines. Mono-, bis-, tris- and tetrakis-9-acridinyl derivatives of guanidine connected via polymethylene linkers. *J. Chem. Soc. Chem. Commun.*, 1015–1016

Hardy, J. R., Harvey, V. J., Paxton, J. W., Evans, P., Smith, S., Grove, W., Grillo-Lopez, A. J. and Baguley, B. C. (1988). A Phase I trial of the amsacrine analog 9-[[2-methoxy-4-[(methylsulfonyl)amino]phenyl]amino]-*N*,5-dimethyl-4-acridine-carboxamide (CI-921). *Cancer Res.*, **48**, 6593–6596

Harvey, V. J., Hardy, J. R., Smith, S., Grove, W. and Baguley, B. C. (1991). Phase II study of the amsacrine analogue CI-921 (NSC 343499) in non-small-cell lung cancer. *Eur. J. Cancer*, **27**, 1617–1620

Heck, M. M. S., Hittelman, W. N. and Earnshaw, W. C. (1988). Differential expression of DNA topoisomerases I and II during the eukaryotic cell cycle. *Proc. Natl Acad. Sci. USA*, **85**, 1086–1090

Holdaway, K. M., Finlay, G. J. and Baguley, B. C. (1992). Relationship of cell cycle parameters to *in vitro* and *in vivo* chemosensitivity for a series of Lewis lung carcinoma lines. *Eur. J. Cancer*, **28A**, 1427–1431

Hudson, B. D., Kuroda, R., Denny, W. A. and Neidle, S. (1987). Crystallographic and molecular mechanics calculations on the antitumor drugs *N*-[(2-dimethylamino)ethyl]- and *N*-[(2-dimethylamino)butyl]-9-aminoacridine-4-carboxamides and their dications: implications for models of DNA binding. *J. Biomol. Struct. Dyn.*, **5**, 145–158

Jain, R. K. (1989). Delivery of novel therapeutic agents in tumors: physiological barriers and strategies. *J. Natl Cancer Inst.*, **81**, 570–576

Jehn, U. and Heinemann, V. (1991). New drugs in the treatment of acute leukemia, with some emphasis on m-AMSA. *Anticancer Res.*, **11**, 705–712

Johnson, R. K. and Howard, W. S. (1982). Development and cross-resistance characteristics of a subline of P388 leukemia resistant to 4′-(9-acridinylamino)-methanesulfon-*m*-anisidide. *Eur. J. Cancer Clin. Oncol.*, **18**, 479–487

Jurlina, J. L., Lindsay, A., Baguley, B. C. and Denny, W. A. (1987). Redox chemistry of the 9-anilinoacridine class of antitumor agent. *J. Med. Chem.*, **30**, 473–480

Jurlina, J. L. and Paxton, J. W. (1985). Determination of *N*,5-dimethyl-9-[(2-methoxy-4-methylsulfonylamino)phenylamino]-acridine-4-carboxamide in plasma by high performance liquid chromatography. *J. Chromatog.*, **343**, 431–435

Kerr, D. J. and Kaye, S. B. (1989). Flavone acetic acid — preclinical and clinical activity. *Eur. J. Cancer Clin. Oncol.*, **25**, 1271–1272

Kestell, P., Paxton, J. W., Evans, P. C., Young, D., Jurlina, J. L., Robertson, I. G. C. and Baguley, B. C. (1990). Disposition of amsacrine and its analogue 9-[(2-methoxy-4-methyl-sulfonylamino)-phenylamino]-*N*,5-dimethyl-4-acridine-carboxamide (CI-921) in plasma, liver and Lewis lung tumors in mice. *Cancer Res.*, **50**, 503–508

Kestell, P., Paxton, J. W., Robertson, I. G. C., Evans, P. C., Dormer, R. A. and Baguley, B. C. (1989). Thiolytic cleavage and binding of the antitumour agent CI-921 in blood. *Drug Metab. Drug Interact.*, **6**, 327–335

Khan, M. N. and Malspeis, L. (1982). Kinetics and mechanism of thiolytic cleavage of the antitumor compound 4′-[(9-acridinylamino]methanesulfon-*m*-anisidide. *J. Org. Chem.*, **47**, 2731–2740

King, H. D., Wilson, W. D. and Gabbay, E. J. (1982). Interactions of some novel amide-linked bis(acridines) with deoxyribonucleic acid. *Biochemistry*, **21**, 4982–4989

Kobayashi, J., Cheng, J.-F., Walchi, M. R., Nakamura, H., Hirata, Y., Sasaki, T. and Ohuzumi, Y. (1988). Cystodytins A B and C, novel tetracyclic alkaloids with potent antineoplastic activity from the Okinawan tunicate *Cystodytea dellechiajei. J. Org. Chem.*, **53**, 1800–1804

Kohn, K. W., Hartley, J. A. and Mattes, W. B. (1987). Mechanisms of DNA sequence selective alkylation of guanine-N7 positions by nitrogen mustards. *Nucleic Acids Res.*, **15**, 10531–10549

Kuroda, R. and Shinomiya, M. (1991). Photocleavage of DNA by the para-nitrobenzoyl group covalently linked to proflavine. *Biochem. Biophys. Res. Commun.*, **181**, 1266–1272

Kwasniewska-Rokicinska, C., Swiecki, J. and Wieczorkiewicz, A. (1973). Therapeutic efficacy of compound C-283 in patients with mammary carcinoma. *Arch. Immunol. Ther. Exp.*, **21**, 863–869

Ledochowski, A. and Stefanska, B. (1966). Research of tumour-inhibiting compounds. XXIX. Some N9-derivatives of 1-, 2-, 3- and 4-nitro-9-aminoacridine. *Roczn. Chem.*, **40**, 301–305

Lee, H. H., Palmer, B. D., Baguley, B. C., Chin, M., McFadyen, W. D., Wickham, G., Thorsbourne-Palmer, D., Wakelin, L. P. G. and Denny, W. A. (1992). DNA-directed alkylating agents. 5. Acridinecarboxamide derivatives of 1,2-diaminoethanedichloroplatinum (II). *J. Med. Chem.*, **35**, 2983–2987

Lee, H. H., Palmer, B. D. and Denny, W. A. (1988). Reactivity of quininoneimine and quinonediimine oxidation products of the antitumor drug amsacrine and related compounds to nucleophiles. *J. Org. Chem.*, **53**, 6042–6047

Leopold, W. R., Corbett, T. H., Griswold, D. P., Plowman, J. and Baguley, B. C. (1987). A multicenter assessment of the experimental antitumor activity of the amsacrine analogue CI-921. *J. Natl Cancer Inst.*, **79**, 343–349

Leupin, W., Chazin, W. J., Hyberts, S., Denny, W. A., Stewart, G. M. and

Wuthrich, K. (1986). 1D and 2D NMR study of the complex between the decadeoxyribonucleotide d(GCATTAATGC)$_2$ and a minor groove binding drug. *Biochemistry*, **25**, 5902–5910

Liu, L. F. (1989). DNA topoisomerase poisons as antitumor drugs. *Ann. Rev. Biochem.*, **58**, 351–375

LoRusso, P., Wozniak, A. J., Polin, L., Capps, D., Leopold, W. R., Werbel, L. M., Biernat, L., Dan, M. A. and Corbett, T. H. (1990). Antitumor efficacy of PD115934 (NSC 366140) against solid tumors of mice. *Cancer Res.*, **50**, 4900–4905

Loughhead, D. G. (1990). Synthesis of des-*N*-methylacronycine and acronycine. *J. Org. Chem.*, **55**, 2245–22476

McGhee, J. D. and von Hippel, P. H. (1974). Theoretical aspects of DNA–protein interactions: co-operative and non co-operative binding of large ligands to a one-dimensional homogeneous lattice. *J. Mol. Biol.*, **86**, 469–489

McKenna, R., Beveridge, A. J., Jenkins, T. C., Neidle, S. and Denny, W. A. (1989). Molecular modelling of DNA–antitumour drug intercalation interactions: correlation of structural and energetic features with biological properties for a series of phenylquinoline-8-carboxamide derivatives. *Mol. Pharmacol.*, **35**, 720–728

Marshall, E. S., Finlay, G. J., Matthews, J. H. L., Shaw, J. H. F., Nixon, J. and Baguley, B. C. (1992). Microculture-based chemosensitivity testing: a feasibility study comparing freshly explanted human melanoma cells with human melanoma cell lines. *J. Natl Cancer Inst.*, **84**, 340–345

Michael, J. P. (1991). Quinoline, quinazoline and acridone alkaloids. *Natl Prod. Rev.*, **8**, 53–68

Miller, L. P., Pyesmany, A. F., Wolff, L. J., Rogers, P. C. J., Siegel, S. E., Wells, R. J., Buckley, J. D. and Hammond, G. D. (1991). Successful reinduction therapy with amsacrine and cyclocytidine in acute nonlymphoblastic leukemia in children — a report from the children's cancer study group. *Cancer*, **67**, 2235–2240

Morier-Teissier, E., Bailly, C., Bernier, J. L., Houssain, R., Helbecque, N., Catteau, J. P., Colson, P., Houssier, C. and Hénichart, J. P. (1989). Synthesis, biological activity and DNA interaction of anilinoacridine and bithiazole peptide derivatives related to the antitumor drugs *m*-AMSA and bleomycin. *Anti-cancer Drug Des.*, **4**, 37–52

Mullins, S. T., Annan, N. K., Cook, P. R. and Lowe, G. (1992). Bisintercalators of DNA with a rigid linker in an extended configuration. *Biochemistry*, **31**, 842–849

Neidle, S. and Abraham, Z. (1984). Structural and sequence-dependent aspects of drug intercalation into nucleic acids. *CRC Crit. Rev. Biochem.*, **17**, 73–121

Nelson, E. M., Tewey, K. M. and Liu, L. F. (1984). Mechanism of antitumor drug action. Poisoning of mammalian DNA topoisomerase II on DNA by 4′-(9-acridinylamino)methanesulfon-*m*-aniside. *Proc. Natl Acad. Sci. USA*, **81**, 1361–1364

Nielsen, P. E., Egholm, M., Berg, R. H. and Buchardt, O. (1992). Sequence-selective recognition of DNA by strand displacement with a thymine-substituted polyamide. *Nature*, **254**, 1497–1500

O'Connor, C. J., Denny, W. A., Gamage, R. S. K. and Fan, J.-Y. (1992). DNA-directed aniline mustards based on 9-aminoacridine: interaction with DNA. *Chem.-biol. Interact.*, **85**, 1–14

O'Connor, C. J., Denny, W. A., McLennan, D. J. and Sutton, B. M. (1990). Substituent effects on the hydrolysis of analogues of nitracrine [9-(3-dimethylaminopropylamino)-1-nitroacridine]. *J. Chem. Soc. Perkin II*, 1637–1641

Palmer, B. D., Lee, H. H., Johnson, P., Baguley, B. C., Wickham, G., Wakelin, L. P. G., McFadyen, W. D. and Denny, W. A. (1990). DNA-directed alkylating agents. 2. Synthesis and biological activity of platinum complexes linked to 9-anilinoacridine. *J. Med. Chem.*, **33**, 3008–3014

Papadopoulou, M. V., Epperley, M. W., Shields, D. S. and Bloomer, W. D. (1992). Radiosensitisation and hypoxic cell cytotoxicity of NLA-1 and NLA-2, two new bioreductive compounds. *Jap. J. Cancer Res.*, **83**, 410–414

Paxton, J. W., Hardy, J. R., Evans, P. C., Harvey, V. J. and Baguley, B. C. (1988). The clinical pharmacokinetics of *N*-5-dimethyl-9-[(2-methoxy-4-methylsulphonylamino)phenylamino]-4-acridine carboxamide (CI-921) in a Phase I trial. *Cancer Chemother. Pharmacol.*, **22**, 235–240

Paxton, J. W., Jurlina, J. L. and Foote, S. E. (1986). The binding of amsacrine to human plasma proteins. *J. Pharm. Pharmacol.*, **38**, 432–438

Paxton, J. W., Young, D., Evans, S. M. H., Kestell, P., Robertson, I. G. C. and Cornford, E. M. (1992). Pharmacokinetics and toxicity of the antitumour agent *N*-<2-(dimethylamino)ethyl>acridine-4-carboxamide after iv administration in the mouse. *Cancer Chemother. Pharmacol.*, **29**, 379–384

Prakash, A. S., Denny, W.A., Gourdie, T. A., Valu, K. K., Woodgate, P. D. and Wakelin, L. P. G. (1990). DNA-directed alkylating ligands as potential antitumor agents: sequence specificity of alkylation by DNA-intercalating acridine-linked aniline mustards. *Biochemistry*, **29**, 9799–9807

Qian, X. and Beck, W. T. (1990). Binding of an optically pure photoaffinity analogue of verapamil, LU-49888, to P-glycoprotein from multidrug-resistant human leukemic cell lines. *Cancer Res.*, **50**, 1132–1137

Reisch, J., Herath, H. M. T. B. and Kumar, N. S. (1991). Convenient synthesis of isoacronycine and some other new acridone derivatives. *Liebigs Ann. Chim.*, 685–689

Rewcastle, G. W., Atwell, G. J., Baguley, B. C., Calveley, S. B. and Denny, W. A. (1989). Potential antitumor agents. 58. Synthesis and structure–activity relationships of substituted xanthenone-4-acetic acids active against the colon 38 tumor *in vivo*. *J. Med. Chem.*, **32**, 793–799

Rewcastle, G. W., Atwell, G. J., Boyd, P. D. W., Palmer, B. D., Baguley, B. C. and Denny, W. A. (1991a). Potential antitumor agents. 62. Structure–activity relationships for tricyclic compounds related to the colon tumor active drug 9-oxo-9*H*-xanthene-4-acetic acid. *J. Med. Chem.*, **34**, 491–496

Rewcastle, G. W., Atwell, G. J., Chambers, D., Baguley, B. C. and Denny, W. A. (1986). Potential antitumor agents. 46. Structure–activity relationships for acridine monosubstituted derivatives of the antitumour agent *N*-[2-(dimethylamino)ethyl]-9-aminoacridine-4-carboxamide. *J. Med. Chem.*, **29**, 472–477

Rewcastle, G. W., Atwell, G. J., Zhuang, L., Baguley, B. C. and Denny, W. A. (1991b). Potential antitumor agents. 61. Structure–activity relationships for *in vivo* colon-38 activity among disubstituted 9-oxo-9*H*-xanthene-4-acetic acids. *J. Med. Chem.*, **34**, 217–222

Rewcastle, G. W., Baguley, B. C., Atwell, G. J. and Denny, W. A. (1987). Potential antitumour agents. 52. Carbamate analogues of amsacrine with *in vivo* activity against multidrug-resistant P388 leukemia. *J. Med. Chem.*, **30**, 1576–1581

Robbie, M. A., Baguley, B. C., Denny, W. A., Gavin, J. B. and Wilson, W. R. (1988). Mechanism of resistance of non-cycling mammalian cells to 4′-(9-acridinylamino)methanesulfon-m-anisidide (*m*-AMSA): comparison of uptake, metabolism and DNA breakage in log- and plateau-phase Chinese hamster fibroblast cell cultures. *Cancer Res.*, **48**, 310–319

Robbie, M. A., Palmer, B. D., Denny, W. A. and Wilson, W. R. (1990). Metabolism of m-ADQI [*N*1′-methanesulphonyl-*N*4′-(9-acridinyl)-3′-methoxy-2′,5′-cyclohexadiene-1′,4′-diimine], the primary oxidative metabolite of amsacrine, in transformed Chinese hamster fibroblasts. *Biochem. Pharmacol.*, **39**, 1411–1421

Roberts, P. B., Denny, W. A., Wakelin, L. P. G., Anderson, R. F. and Wilson, W. R. (1990). Radiosensitization of mammalian cells *in vitro* by nitroacridines. *Radiation Res.*, **123**, 153–164

Robertson, I. G. C., Kestell, P., Dormer, R. A. and Paxton, J. W. (1988). Involvement of glutathione in the metabolism of the antitumor agents CI-921 and amsacrine. *Drug Metab. Drug Interact.*, **6**, 371–381

Robertson, I. G. C., Palmer, B. D., Officer, M., Siegers, D. J., Paxton, J. W. and Shaw, G. J. (1991). Cytosol mediated metabolism of the experimental antitumour agent acridine carboxamide to the 9-acridone derivative. *Biochem. Pharmacol.*, **42**, 1879–1884

Robertson, I. G. C., Palmer, B. D., Paxton, J. W. and Shaw, G. J. (1992). Differences in the metabolism of the antitumour agents CI-921 and amsacrine in the rat and mouse. *Xenobiotica*, **22**, 657–669

Roninson, I. B. (1992). The role of the MDR1 (P-glycoprotein) gene in multidrug resistance *in vitro* and *in vivo*. *Biochem. Pharmacol.*, **43**, 95–102

Rowe, T. C., Chen, G. L., Hsiang, Y. H. and Liu, L. F. (1986). DNA damage by antitumour acridines mediated by mammalian DNA topoisomerase II. *Cancer Res.*, **46**, 2021–2026

Sakore, T. D., Reddy, B. S. and Sobell, H. M. (1979). Visualisation of drug–nucleic acid interactions at atomic resolution. IV. Structure of an amino-acridine–dinucleoside monophosphate crystalline complex, 9-aminoacridine-5-iodocytidylyl (3′-5′) guanosine. *J. Mol. Biol.*, **135**, 763–785

Scarffe, J. H., Beaumont, A. R. and Crowther, D. (1983). Phase I–II evaluation of acronine in patients with multiple myeloma. *Cancer Treat. Rep.*, **67**, 93–94

Schmitz, F. J., DeGuzman, F. C., Hossain, M. B. and van der Helm, B. (1991). Cytotoxic aromatic alkaloids from the ascidian *Amphicarpa meridiana* and *Leptoclinides* sp.: meridine and 11-hydroxyascididemin. *J. Org. Chem.*, **56**, 804–808

Schneider, E., Darkin, S. J., Lawson, P. A., Ching, L.-M., Ralph, R. K. and Baguley, B. C. (1988). Cell line selectivity and DNA breakage properties of the antitumour agent *N*-[2-(dimethylamino)ethyl]-acridine-4-carboxamide: role of DNA topoisomerase II. *Eur. J. Cancer Clin. Oncol.*, **24**, 1783–1790

Sebolt, J. S., Scavone, S. V., Pinter, C. D., Hamelehle, K. L., Von Hoff, D. D. and Jackson, R. C. (1987). Pyrazoloacridines, a new class of anticancer agents with selectivity against solid tumors *in vitro*. *Cancer Res.*, **47**, 4299–4304

Shafer, R. H. and Waring, M. J. (1982). DNA bis-intercalation: theoretical analysis, including cooperativity of the interaction of echinomycin analogues with DNA. *Biopolymers*, **21**, 2279–2290

Shinomiya, M. and Kuroda, R. (1992). Synthesis of novel DNA photocleaving agents with potent DNA cleaving activity. *Tetrahedron Lett.*, **33**, 2697–2700

Shoemaker, D. D., Cysyk, R. L., Gormley, P. E., DeSouza, J. J. and Malspeis, L. (1984). Metabolism of 4′-(9-acridinylamino)methanesulfon-*m*-anisidide by rat liver microsomes. *Cancer Res.*, **44**, 1939–1945

Shoemaker, D. D., Cysyk, R. L., Padmanhaban, S., Bhat, H. B. and Malspeis, L. (1982). Identification of the principal biliary metabolite of 4′-(-acridinyl-amino)methanesulfon-*m*-anisidide in rats. *Drug Metab. Disp.*, **10**, 35–39

Sklarin, N., Wiernik, P., Mittelman, A., Maroun, J., Stewart, J., Robert, F., Doroshow, J., Akman, S., Rosen, P., Gota, C., Jolivet, J., Belanger, K., DeConti, R., Robert, N., Velez-Garcia, E., Bergsagel, D., Panasci, L., van der Merwe, A., Leiby, J., Grove, W., Hawkins, E. and Kowal, C. (1990). A phase II evaluation of CI-921 in patients with solid tumors. *Proc. Am. Soc. Clin. Oncol.*, **9**, 285

Stezowski, J. J., Kollat, P., Bogucka-Ledochowska, M. and Glusker, J. P. (1985). Tautomerism and steric effects in 1-nitro-9-(alkylamino)acridines (Ledakrin or nitracrine analogues): probing structure–activity relationships at the molecular level. *J. Am. Chem. Soc.*, **107**, 2067–2077

Sundquist, W. I., Bancroft, D. P. and Lippard, S. J. (1990). Synthesis, characterization and biological activity of *cis*-diammineplatinum (II) complexes of the DNA intercalators 9-aminoacridine and chloroquine. *J. Am. Chem. Soc.*, **112**, 1590–1596

Suzukake, K., Vistica, B. P. and Vistica, D. T. (1983). Dechlorination of L-phenylalanine mustard by sensitive and resistant tumor cells and its relationship to intracellular glutathione content. *Biochem. Pharmacol.*, **32**, 165–167

Suzuke, M. (1989). SPKK, a new nucleic acid binding unit of protein found in histones. *EMBO Jl*, **8**, 797–801

Traganos, F., Bueti, C., Darzynkiewicz, Z. and Melamed, M. R. (1987). Effects of a new amsacrine derivative, *N*-5-dimethyl-9-(2-methoxy-4-methylsulfonylamino)phenylamino-4-acridinecarboxamide, on cultured mammalian cells. *Cancer Res.*, **47**, 424–432

Valu, K. K., Gourdie, T. A., Gravatt, G. L., Boritzki, T. J., Woodgate, P. D., Baguley, B. C. and Denny, W. A. (1990). DNA-directed alkylating agents. 3. Structure–activity relationships for acridine-linked aniline mustards: consequences of varying the length of the linker chain. *J. Med. Chem.*, **33**, 3014–3019

Wadkins, R. M. and Graves, D. E. (1989). Thermodynamics of the interaction of *m*-AMSA and *o*-AMSA with nucleic acids: influence of ionic strength and DNA base composition. *Nucleic Acids Res.*, **17**, 9933–9946

Wadkins, R. M. and Graves, D. E. (1991). Interactions of anilinoacridines with nucleic acids — effects of substituent modifications on DNA-binding properties. *Biochemistry*, **30**, 4277–4283

Wakelin, L. P. G. (1986). Polyfunctional DNA intercalating compounds. *Med. Res. Rev.*, **6**, 275–340

Wakelin, L. P. G., Atwell, G. J., Rewcastle, G. W. and Denny, W. A. (1987). Relationships between DNA binding kinetics and biological activity for the 9-aminoacridine-4-carboxamide class of antitumor agents. *J. Med. Chem.*, **30**, 855–862

Wakelin, L. P. G., Romanos, M., Chen, T. K., Glaubiger, D., Canellakis, E. S. and Waring, M. J. (1978). Structural limitations on the bifunctional intercalation of diacridines into DNA. *Biochemistry*, **17**, 5057–5063

Wilson, W. R., Anderson, R. F. and Denny, W. A. (1989a). Hypoxia-selective antitumor agents. 1. Relationships between structure, redox properties and hypoxia-selective cytotoxicity for 4-substituted derivatives of nitracrine. *J. Med. Chem.*, **32**, 23–30

Wilson, W. R., Baguley, B. C., Wakelin, L. P. G. and Waring, M. J. (1981). Interaction of the antitumour drug *m*-AMSA (4′-(9-acridinylamino)methanesulphon-*m*-anisidide) and related acridines with nucleic acids. *Mol. Pharmacol.*, **20**, 404–414

Wilson, W. R., Denny, W. A., Stewart, G. M., Fenn, A. and Probert, J. C. (1986). Reductive metabolism and hypoxia-selective cytotoxicity of nitracrine. *Int. J. Radiat. Oncol. Biol. Phys.*, **12**, 1235–1238

Wilson, W. R., Denny, W. A., Twigden, S. J., Baguley, B. C. and Probert, J. C. (1984). Selective toxicity of nitracrine to hypoxic mammalian cells. *Br. J. Cancer*, **49**, 215–223

Wilson, W. R., Thompson, L. H., Anderson, R. F. and Denny, W. A. (1989b). Hypoxia-selective antitumor agents. 2. Electronic effects of 4-substituents on the mechanisms of cytotoxicity and metabolic stability of nitracrine derivatives. *J. Med. Chem.*, **32**, 31–38

Wilson, W. R., Van Zijl, P. and Denny, W. A. (1992). Bis-bioreductive agents as hypoxia-selective cytotoxins: nitracrine *N*-oxide. *Int. J. Radiat. Oncol. Biol. Phys.*, **22**, 693–696

Wilson, W. R., and Whitmore, G. F. (1981). Cell-cycle-stage specificity of 4′-(9-acridinylamino)-methanesulfon-m-anisidide (m-AMSA) and interaction with ionizing radiation in mammalian cell cultures. *Radiation Res.*, **87**, 121–136

Wong, A., Huang, C.-H. H. and Crooke, S. T. (1984a). Deoxyribonucleic acid breaks produced by 4′-(9-acridinyl)methanesulfon-*m*-anisidide and copper; role for cuprous ion and free radicals. *Biochemistry*, **23**, 2939–2945

Wong, A., Huang, C.-H. H. and Crooke, S. T. (1984b). Mechanism of deoxyribonucleic acid breakage induced by 4′-(9-acridinyl)methanesulfon-*m*-anisidide and copper. *Biochemistry*, **23**, 2946–2952

Wright, R. G. McR., Wakelin, L. P. G., Fieldes, A., Acheson, R. M. and Waring, M. J. (1980). Effects of ring substituents and linker chains on the bifunctional intercalation of diacridines into deoxyribonucleic acid. *Biochemistry*, **17**, 5825–5836

Yamato, M., Takeuchi, Y., Hashigaki, K., Ikeda, Y., Ming-rong, C., Takeuchi, K., Matsushima, M., Tsuruo, T., Tashiro, T., Tsukagoshi, S., Yamashita, Y. and Nakano, H. (1989). Synthesis and antitumor activity of fused tetracyclic quinoline derivatives. *J. Med. Chem.*, **32**, 1295–1300

Young, D., Evans, P. C. and Paxton, J. W. (1990). Quantitation of the antitumour agent *N*-<2-(dimethylamino)ethyl>acridine-4-carboxamide in plasma by high-performance liquid chromatography. *J. Chromatogr.*, **528**, 385–394

Zittoun, R. (1985). m-AMSA: a review of clinical data. *Eur. J. Cancer Clin. Oncol.*, **21**, 649–653

Zwelling, L. A., Michaels, S., Erickson, L. C., Ungerleider, R. S., Nichols, M. and Kohn, K. W. (1981). Protein-associated DNA strand breaks in L1210 cells treated with the DNA intercalating agents 4′-(9-acridinylamino)methanesulfon-*m*-anisidide and Adriamycin. *Biochemistry*, **20**, 6553–6563

Zwi, L. J., Baguley, B. C., Gavin, J. B. and Wilson, W. R. (1989). Blood flow failure as a major determinant in the antitumor action of flavone acetic acid (NSC 347512). *J. Natl Cancer Inst.*, **81**, 1005–1013

# 8

# The Mitomycins: Natural Cross-linkers of DNA

Maria Tomasz

## 1 Introduction

The mitomycins are a group of potent antibiotics, discovered in Japan in the late 1950s in fermentation cultures of *Streptomyces caespitosus* (Hata *et al.*, 1956). Their significant antitumour activity was discovered at the same time, and today one of the drugs, mitomycin C, is widely used in clinical anticancer chemotherapy against a broad spectrum of solid tumours (Carter, 1979).

The structures of various mitomycins are depicted in Figure 8.1. Mitomycins H and FR900482 are among newer isolates, possessing significant antitumour activity (Urakawa *et al.*, 1981; Uchida *et al.*, 1987). The absolute stereochemistry of the mitomycins was revised in 1983 (Shirahata and Hirayama, 1983). Thus, the configurations in the earlier literature appear reversed from those in use today.

Early studies indicated that the primary target of the cytotoxicity of the mitomycins is DNA, as evidenced by a set of characteristic actions of mitomycin on bacterial and mammalian cells which is common to known DNA damaging agents: selective inhibition of DNA replication (Szybalski and Iyer, 1964, and references therein), induction of lysogeny and the SOS response in bacteria (Kenyon and Walker, 1980), sister chromatid exchange in mammalian cells (Carrano *et al.*, 1979), mutagenicity (Balbinder and Kerry, 1984), and the discovery of numerous bacterial (Boyce and Howard-Flanders, 1964; Szybalski and Iyer, 1964) and mammalian (Thompson *et al.*, 1980) cell mutants which are cross-resistant or cross-hypersensitive to mitomycin C and ultraviolet light. The latter agent is a 'benchmark' of DNA-targeted agents, since its action spectrum (cytotoxicity or mutagenicity as function of wavelength of irradiation) resembles

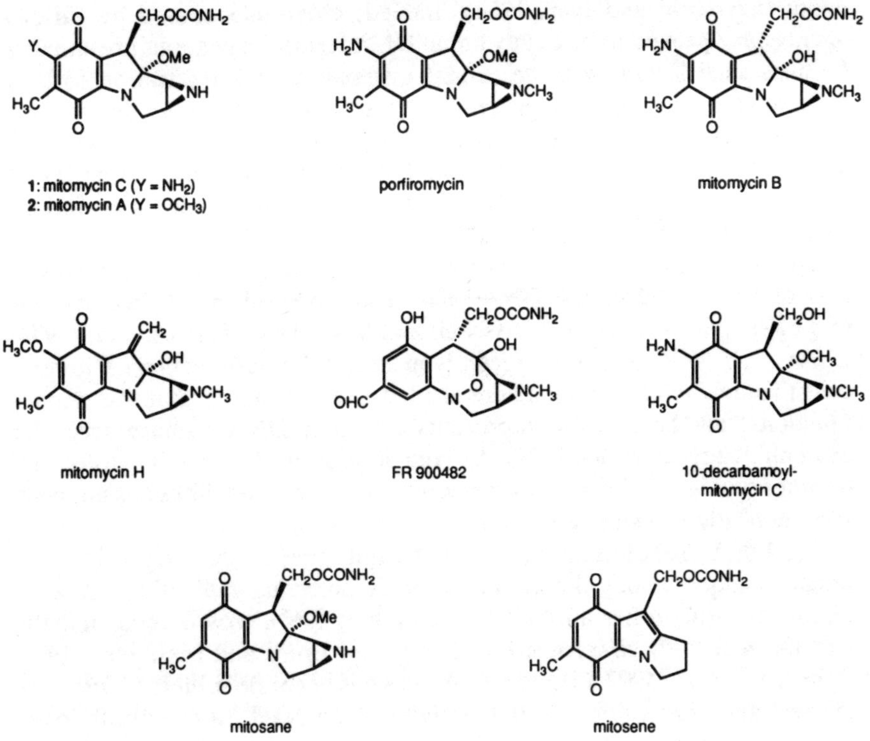


**Figure 8.1** Structures of mitomycins

the ultraviolet absorption spectrum of DNA (McLaren and Shugar, 1964). Thus, the occurrence of UV cross-resistant or hypersensitive DNA repair mutants strongly indicates that mitomycins damage DNA directly and this damage is the primary cause of cell death.

More light was shed on the nature of the damage by the discovery of an extraordinary effect of mitomycins: cross-linking of the complementary strands of DNA (Iyer and Szybalski, 1963; Matsumoto and Lark, 1963). This was detected by the altered physicochemical properties of DNA isolated from bacteria previously exposed to mitomycin C. Specifically, when such DNA was denatured by heat or alkaline treatment, it renatured spontaneously after physiological conditions were restored, as seen by CsCl equilibrium density gradient centrifugation or reversible thermal melting curves. These methods were extremely sensitive: 1 cross-link per 20 000 base pairs ($10^7$ daltons of DNA) were detected in various DNAs isolated from bacterial and mammalian cultures (Iyer and Szybalski, 1963). The rate of cell death was correlated with the degree of the DNA cross-linking and it was suggested that the cross-linking of DNA represents the molecular basis of the inhibition of DNA replication and consequent cell death (Szybalski and Iyer, 1964). Indeed, cross-links caused by various agents have proved to be highly lethal DNA lesions in general, presumably because such lesions may be caught unrepaired at a replication fork in fast-growing cells, halting progress of the fork irreversibly; this in turn leads to cell death (Erickson *et al.*, 1980; Nielsen and Bohr, 1983). A salient study correlating DNA cross-linking activity and cytotoxicity of mitomycins in mammalian cells provided new evidence for these notions (Keyes *et al.*, 1991).

Cross-linking of DNA by mitomycins is accompanied by covalent alkylation of one strand of the DNA. This was observed in cell-free systems (Szybalski and Iyer, 1964; Weissbach and Lisio, 1965; Tomasz *et al.*, 1974; Lown *et al.*, 1976) and more recently in cultured mouse mammary tumour cells (Tomasz *et al.*, 1991). There are indirect indications that this 'monofunctional alkylation' also represents cytotoxic DNA damage (see, for example, Carrano *et al.*, 1979). Its contribution to the overall cytotoxicity or other biological effects of mitomycins relative to cross-links has not been systematically investigated so far.

The DNA cross-linking action of the mitomycins, discovered in 1963, remains unique among the known naturally occurring antibiotics. (Carzinophillin, the only other antibiotic noted for its DNA cross-linking activity, has not been fully characterized structurally: Lown and Majundar, 1977; Armstrong *et al.*, 1992.) Its chemical basis has been the subject of intensive interest and continuing research activity. As a result, a chemically complex, fascinating mechanism emerged, mostly in the last ten years, for the introduction of covalent cross-links by mitomycins into DNA, as follows.

## 2 Reductive Activation of Mitomycins to Bifunctional Alkylating Agents

### *The Iyer–Szybalski Hypothesis*

In a seminal paper Iyer and Szybalski (1964) reported that cross-linking of purified DNA by mitomycins A, B, C and porfiromycin was achieved and that this process was absolutely dependent on the presence of a cell-free lysate containing NADPH, or of simple chemical reducing agents, such as $Na_2S_2O_4$ or $NaBH_4$. From these observations they postulated a mechanism for the reductive activation of mitomycins: the C-1 aziridine and C-10 carbamate groups are two masked alkylating functions which become 'allylic' (therefore activated) upon reduction of the quinone system and the consequent spontaneous elimination of methanol from the 9 and 9a positions; their subsequent displacement by two nucleophiles in DNA results in a mitomycin–DNA cross-link. This hypothesis has proved to be extremely fruitful, serving as the organizing framework for the experimental inquiry into the precise molecular mode of action of the mitomycins in the next two decades.

Reductive activation of mitomycin C to a reactive alkylating agent was demonstrated in many biochemical systems since the initial discovery: microsomes, nuclei (Tomasz and Lipman, 1981; Kennedy *et al.*, 1982), purified flavoreductases, such as NADPH-cytochrome c reductase, DT-diaphorase (Keyes *et al.*, 1984), xanthine oxidase (Pan *et al.*, 1984), and others. Reduction is also known to mediate the cytotoxicity of mitomycins *in vivo*. One line of evidence is the observed greater toxicity of mitomycins to hypoxic tumour cells both in culture and in animals (Kennedy *et al.*, 1980; Moulder and Rockwell, 1987). Good correlation between increased toxicity and increased DNA cross-linking under hypoxia has also been demonstrated (Fracasso and Sartorelli, 1986; Marshall and Rauth, 1986). Compelling evidence for the significance of bioreduction of the mitomycins is the existence of mitomycin-resistant tumour cell mutants which are deficient in mitomycin reductase activity (Wilson *et al.*, 1985; Hoban *et al.*, 1990; Dulhanty and Whitmore, 1991; Marshall *et al.*, 1991). Thus, mitomycin C (MC), the best-studied member of the group, is regarded as the prototype 'bioreductive alkylating agent', so termed by Sartorelli and co-workers, who first articulated this concept for anticancer drug design (Lin *et al.*, 1976; Sartorelli, 1988). The bioreductive DNA alkylating action of MC is illustrated in Figure 8.2.

The ideas of Iyer and Szybalski on the mechanism of the reductive activation were amended by Moore (1977), who speculated that both the 1-aziridine and 10-carbamate displacements of MC are of $S_N1$ type, facilitated by resonance with the indolohydroquinone system of reduced mitomycin, taking place sequentially, as summarized in Figure 8.3. The key postulated reactive species were specified by Moore as the vinylogous *quinone methide* **5** and the imminium intermediate **7**. The central focus of the proposed mechanism was the peculiar lack of reactivity of the aziridine ring of the mitomycins in their native (quinone) form. Simple aziridine rings such as ethyleneimine open rapidly upon nucleophilic attack, since they are protonated at neutral pH. The basicity of the mitomycin aziridine, however, is abnormally low (p$K_a$ 3.2), as a result of the electron-withdrawing effect of the quinone system, and this accounts for the inert nature of this functional group under physiological conditions (Stevens *et al.*, 1964). The unusual stability of the aminal function at C-9a is also a consequence of the quinone. It was reasoned that reduction of the quinone should reactivate these 'masked' functional groups of the parent antibiotic (Iyer and Szybalski, 1964; Patrick *et al.*, 1964).

### *Evidence for the Hypothesis from Model Studies*

Critical experimental evidence for this scheme has come forth only in the 1980s. Reactions of mitomycins with simple model nucleophiles under varying reducing conditions were characterized in a number of labora-

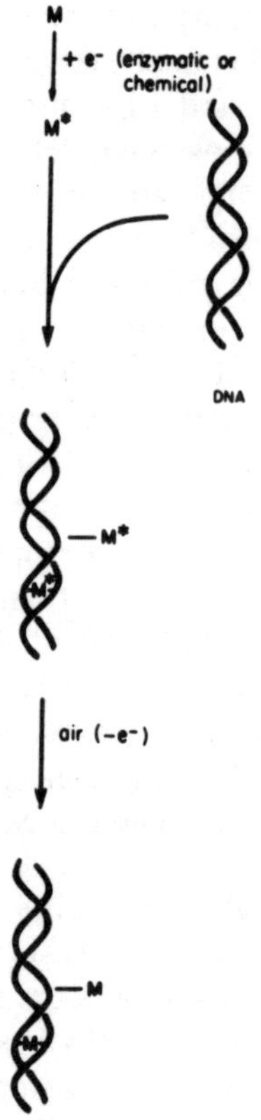


**Figure 8.2** Bioreductive monofunctional and bifunctional alkylation of DNA by a mitomycin: M, mitomycin; M*, activated mitomycin

tories, and from this the mechanism of the mitomycin 'activation cascade' has been pieced together. Discussion of all the important model studies is beyond the scope of this chapter. A detailed treatment of this subject was published recently (Franck and Tomasz, 1990). Remarkably, the postulated mechanism has proved to be correct by and large, as seen from a brief summary of the experimental conclusions: only the first, reduction step is enzymatic (although it can be mimicked by chemical reducing agents). Subsequent rearrangements proceed spontaneously and rapidly (Tomasz and Lipman, 1981). Intermediates **3** and **4** were stable enough under non-physiological conditions, however, to be characterized (Danishefsky and Ciufolini, 1984; Danishefsky and Egbertson, 1986). The intermediacy of **4** in the cross-linking of DNA *in vitro* was also demonstrated (Cera *et al.*, 1989; Teng *et al.*, 1989). The quinone methide **5**, as the postulated actual alkylating agent, could not be observed directly, owing to its high reactivity, but compelling evidence exists for its formation by the isolation

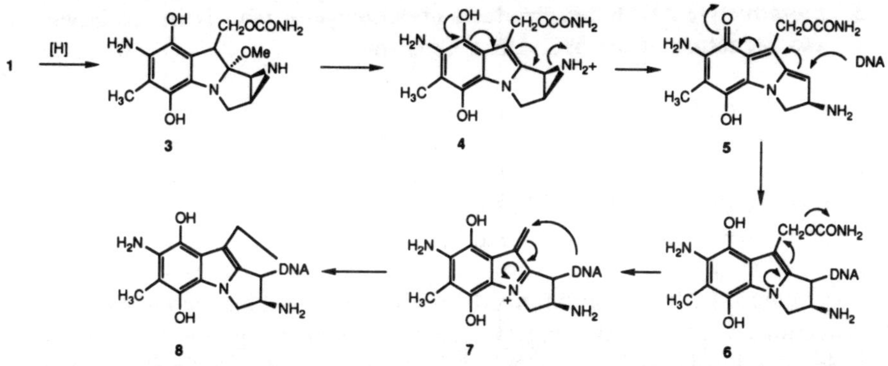


**Figure 8.3** Iyer–Szybalski–Moore mechanism of the reductive activation of mitomycin C to DNA cross-linking agent

of a characteristic set of its stable end-products upon reduction of **1** in the absence of DNA in neutral aqueous buffer (Tomasz and Lipman, 1981; Kohn and Zein, 1983; Pan *et al.*, 1984). Chemical evidence for C-10 as the second alkylating centre of MC was also provided by model studies, in the absence of DNA. This centre developed sequentially after the first alkylation step at C-1, by reverse Michael elimination of the 10-carbamate group. The resulting second quinone methide type intermediate (cf. **7**) was again too unstable for direct detection but could be trapped by protonation to give a stable end-product (Zein and Kohn, 1986).

An aspect of this activation mechanism has been a matter of some concern: is it one- or two-electron reduction that is involved in the activation cascade? One-electron reduction gives a semiquinone anion radical, while two-electron reduction generates the hydroquinone. In organic media the activation cascade operates only in the one-electron reduction state (Andrews *et al.*, 1986; Egbertson and Danishevsky, 1987; Kohn *et al.*, 1987). In *aqueous neutral media*, however, the MC-semiquinone is highly unstable and disproportionates to quinone and hydroquinone before further reaction. The hydroquinone form then readily undergoes the activation cascade reactions (Hoey *et al.*, 1988). Thus, in *cell-free model systems* both one-electron and two-electron reductions can activate MC to alkylating agent, depending upon the conditions. A different type of evidence indicates that this is also true *in vivo*: flavoreductases of both the one-electron and two-electron transfer type have been demonstrated to activate MC in the cell to cytotoxic agent. This was based on characterization of various MC-resistant mammalian cell mutants in which the resistance was correlated with deficiency of one or the other type of flavoreductase activity. This indicates that both types of reduction are capable of activating MC to the DNA-damaging agent (Bligh *et al.*, 1990; Hoban *et al.*, 1990; Siegel *et al.*, 1990; Dulhanty and Whitmore, 1991; Marshall *et al.*, 1991; Traver *et al.*, 1992).

## 3 Bioreductive Alkylation Products of Mitomycins with DNA: Isolation and Structure of the MC–DNA Cross-link

The model studies above defined a general reductive alkylating mechanism by the mitomycins. Critical questions remained unanswered, however, until very recently. In particular: Does the mechanism apply to the reactions of the drug with DNA? What are the reactive sites in DNA? What is the structure of the cross-link? An abundance of *indirect* information was gathered in the first 23 years following the discovery of the cross-linking phenomenon. This is summarized briefly. Cross-linking under reductive activation *in vitro* was shown to increase with increasing G + C content of DNA (Szybalski and Iyer, 1964). Overall covalent binding was more extensive than the observed number of cross-links (Szybalski and Iyer 1964; Weissbach and Lisio, 1965) and it, too, increased linearly with increasing G + C content (Lown *et al.*, 1976; Lipman *et al.*, 1978). The covalent binding to synthetic polynucleotides showed an absolute requirement for the presence of guanine (Tomasz *et al.*, 1974). Interestingly, a guanine/$O^6$-methylguanine copolymer exhibited undiminished binding compared with that by polyG, while replacement of guanosine by inosine in various synthetic polymers abolished it (Weaver and Tomasz, 1982). These were strong hints for guanine-$N^2$ being the major (or exclusive) DNA bonding site.

Isolation and structure of covalent adducts of MC and nucleotides were first reported by Shudo and co-workers, from calf thymus DNA, using catalytic hydrogenation as reducing agent (Hashimoto *et al.*, 1983). According to this report, the drug–DNA complex was hydrolysed to 5′-mononucleotides by nuclease $P_1$, and the products were separated by HPLC. Three MC–mononucleotide adducts were reported, in which the C-1 of MC was linked to $N^2$ of guanine, $O^6$ of guanine and $N^6$ of adenine, respectively, in roughly the same proportions. Similar products were isolated from DNA of the liver of rats injected with a large amount of MC.

At the same time, the collaborating groups of Tomasz and Nakanishi obtained very different results. Calf thymus DNA, after treatment with MC under $H_2/PtO_2$ or enzymatic reductive conditions, followed by digestion to nucleosides, yielded *$N^2$-guanine adduct,* ***9****, as the virtually single reaction product (Figure 8.5b)*. Adduct **9** was obtained previously from the reaction between MC and d(GpC) under both $H_2/PtO_2$ and enzymatic activation and was rigorously characterized (Tomasz *et al.*, 1983a, 1986a). M13 DNA, *M. luteus* DNA and poly(dG–dC)·poly(dG–dC) gave the same result (Tomasz *et al.*, 1986a). Clearly, these findings were in conflict with the formation of not one but three major adducts in the identical calf thymus DNA system reported by Hashimoto *et al.* (1983) above. However, upon reinvestigation those three adducts were shown to be dinucleotide-type artifacts, due to the use of nuclease $P_1$ rather than snake venom

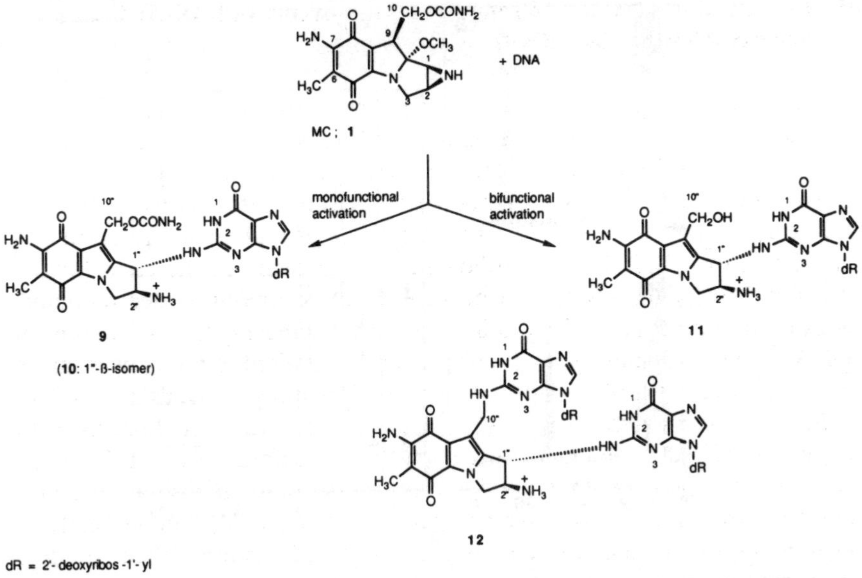


**Figure 8.4** Adducts of mitomycin C and DNA formed under reductive activation

diesterase, used in the Tomasz laboratory, for DNA hydrolysis (Tomasz *et al.*, 1986a,b); McGuinness *et al.*, 1988). In summary, it is now conclusively established that only one major adduct is formed with DNA under monofunctional activating conditions of MC *in vitro*; this is the guanine-$N^2$ adduct, **9**. The 1″-β isomer **10** is formed to the extent of ≤5% of adduct **9** (Tomasz *et al.*, 1986a).

These monoadducts did not account for the known 'cross-linked' behaviour of DNA. It was apparent that only the C-1 position of MC was activated under flavoenzymatic or $H_2/PtO_2$ reductive conditions. However, when anaerobic $Na_2S_2O_4$ was employed as activator, a different adduct pattern was seen upon HPLC of the enzymatic digest of the MC–DNA complex (Figure 8.5a): none of **9** was formed but two *new* major adducts were present instead. A sample of 4 mg of the later eluting adduct was isolated by enzymatic digestion from 300 mg *Micrococcus luteus* DNA–MC complex and its structure was determined by $^1$H-NMR, FT-IR, FAB mass spectroscopy and circular dichroism as the bisadduct, **12**, i.e., the long-sought cross-link (Tomasz *et al.*, 1987a). The drug is linked by its C-1″ and C-10″ positions to two $N^2$ atoms of deoxyguanosine residues. This structure accounts fully for the 'cross-linked' behaviour of DNA, provided that the two dG residues are incorporated in the opposite strands of duplex DNA constituting an interstrand cross-link. The structure of the other adduct, **11**, was determined by similar methods (Tomasz *et al.*, 1988b); it is a monoadduct which lost the C-10″ carbamate. Recently an additional

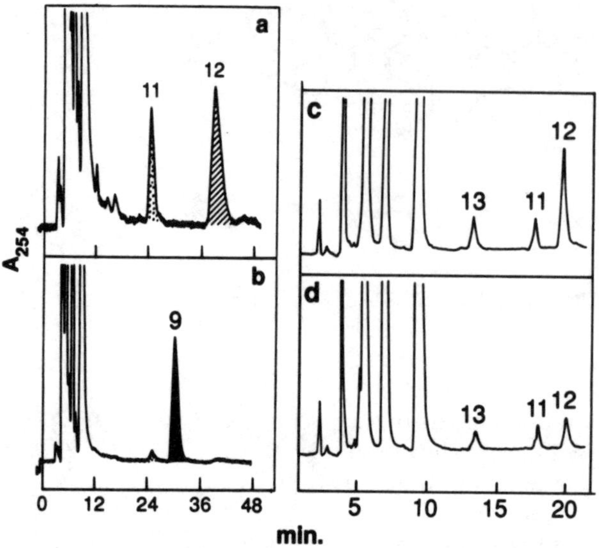


**Figure 8.5** HPLC patterns from nuclease digests of various MC–DNA complexes. (a) Calf thymus DNA–MC complex, formed under anerobic $Na_2S_2O_4$ reducing conditions and digested to nucleotides by a mixture of DNAse I, snake venom diesterase and alkaline phosphatase. (b) Calf thymus DNA–MC complex, formed under $H_2/PtO_2$ reducing conditions and digested as in (a). (c) *M. luteus* DNA–MC complex, formed under anerobic $Na_2S_2O_4$ reducing conditions and digested to nucleosides by a mixture of nuclease $P_1$, snake venom diesterase and alkaline phosphatase. (d) Calf thymus DNA–MC complex, formed and digested under the same conditions as in (c). Column: reverse phase Beckman Ultrasphere ODS. Eluant: acetonitrile/0.02 M potassium phosphate, pH 5.0. Adapted from Tomasz *et al.* (1988a) (a, b) and Bizanek *et al.* (1992) (c, d)

bisadduct, identical with **12** except for having a phosphodiester group linking the two dG residues, was discovered (**13**; Figure 8.6). It originates from an *intrastrand cross-link* of DNA, i.e. from a –GG– sequence cross-linked by MC. No distinction could be made between the two isomeric structures **a** and **b** for this adduct (Bizanek *et al.*, 1992). It was missed in the earlier work (Tomasz *et al.*, 1987a) because the phosphodiester was hydrolysed to **12** under the usual enzymatic DNA digestion conditions. By modifying the digestion, **12** and **13** (interstrand and intrastrand cross-links, respectively) could be detected separately in DNA (Figure 8.5c,d) (Bizanek *et al.*, 1992).

The covalent MC–DNA adducts are summarized in Figure 8.4. Their formation is exquisitely specific: the sole detectable alkylation site is the $N^2$-position of guanines. The two alkylating functions of MC are the aziridine at C-1 and the carbamate at C-10, fully verifying the Iyer–Szybalski hypothesis. Both monofunctional and bifunctional alkylation by MC may occur, depending on the reductive activating conditions. Adduct **9** is clearly the product of monofunctional MC activation (at C-1), while

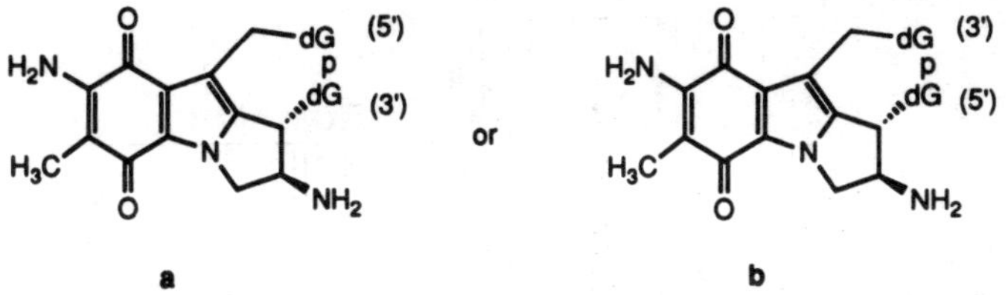


13

**Figure 8.6** DNA intrastrand cross-link adduct of mitomycin C

adducts **11–13** are products of bifunctional (C-1, C-10) activation of the drug. Two other members of the mitomycins, mitomycin A and porfiromycin (Figure 8.1), were shown recently to yield the same array of analogous DNA adducts (Pan and Iracki, 1988; McGuinness *et al.*, 1991a; Tomasz *et al.*, 1991).

Bisadduct **12** corresponds to the DNA interstrand cross-link. This was shown rigorously by its isolation from well-characterized cross-linked synthetic oligonucleotides (Borowy-Borowski *et al.*, 1990a). NMR analysis of the structure of a cross-linked hexanucleotide duplex provided confirmatory evidence (Norman *et al.*, 1990). Analogously, bisadduct **13** was shown to originate from two adjacent guanines in the same strand in MC-modified synthetic oligonucleotides and polynucleotides (Bizanek *et al.*, 1992).

## *Mitomycin–DNA Adducts Formed in vivo*

The first indication that the same adducts are formed *in vivo* as those formed under *in vitro* reductive activating conditions was provided by Shudo and co-workers, who detected adducts with DNA from rat liver nuclei, after injection of rats with high doses of MC. However, these adducts were not correctly identified, as discussed above (Hashimoto *et al.*, 1983). In a more definitive study bisadduct **12** was isolated by HPLC from rat liver and identified by its characteristic ultraviolet spectra and HPLC and other chromatographic properties upon direct comparison with authentic standard **12** (Tomasz *et al.*, 1987a). DNA from MC-treated CHO cells yielded **11** and **12** as detected by ultraviolet absorbance on HPLC (Chowdary and Tomasz, 1987). This method of detection required employment of very high doses of MC. Radiolabelled MC, providing more sensitive adduct detection, has not been available until recently. The very sensitive $^{32}P$-postlabelling method was used to detect the presence of several adducts in tissue of MC-treated rats (Reddy and Randerath, 1987) and in cancer patients undergoing chemotherapy with MC (Kato *et al.*,

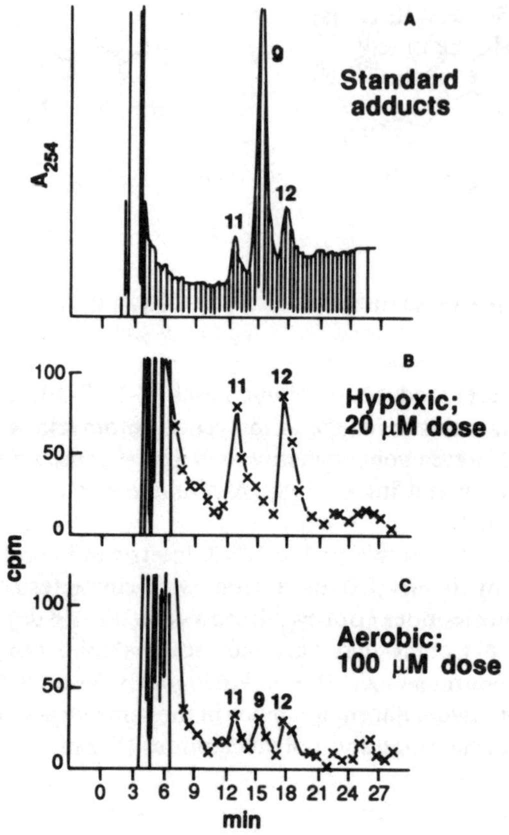


**Figure 8.7** HPLC patterns from digests of DNA from EMT6 mouse mammary tumour cells treated with [$^3$H]-porfiromycin. Enzymatic digestion and HPLC conditions were the same as in Figure 8.5(a,b). Adapted from Tomasz *et al.* (1991)

1988). Chemical identity of the adducts was not established in either study. Most recently, radiolabelled porfiromycin (Figure 8.1) was employed to detect adducts in P388 mouse leukaemia cells (Pan, 1990) and EMT6 mouse mammary tumour cells (Tomasz *et al.*, 1991). Three porfiromycin adducts, identified as the 2″-N-methyl analogues of **9**, **11** and **12**, were isolated in both systems. Interestingly, the DNA adducts were formed more abundantly under hypoxic cell growth conditions, consistent with the notion that alkylation of DNA by MC is dependent on reductive activation (Figure 8.7).

In summary: These studies strongly indicate that the mitomycin–DNA adducts isolated from *in vitro* (cell-free) systems properly represent the cytotoxic DNA damage caused by MC *in vivo*. This is especially certain in the case of bisadduct **12**, since it has been isolated from both animals and cultured cells. Besides, its formation in numerous types of mammalian cell cultures upon MC treatment has been shown indirectly by alkaline elution and other *in vivo* detection techniques, as discussed in Section 1.

## 4 Mechanism of the Reductive Alkylation of DNA

### *Inhibition of the Cross-linking Step by $O_2$ and Excess Mitomycin*

The distribution of the three adducts **9**, **11** and **12** showed a striking dependence on the conditions of the reductive activation. Use of $H_2/PtO_2$ or flavoreductases yielded **9**, reflecting monofunctional activation, while $Na_2S_2O_4$ yielded **11** and **12** as a pair (Figures 8.4, 8.5) (Tomasz *et al.*, 1987a). To explain this, a mechanism was proposed (Figure 8.8), featuring autocatalytic reduction versus stoichiometric reduction of MC as the critical determinant. According to the proposal, the reduced monoadduct **14** may have two fates.

(1) Transfer of its electrons to excess MC, driven by the lower redox potential of a mitosene compared with mitosane (Figure 8.1) such as MC (Rao *et al.*, 1977). This inactivates **14**, giving monoadduct **15** as end-product. This path represent autocatalytic reduction of MC by the reduced monofunctional adduct (and other hydrolytic mitosene products present) (Peterson and Fisher, 1986).

(2) The alternative fate of **14** is retro-Michael elimination of the C-10″ carbamate, giving **16**, which is receptive to a second nucleophilic attack, giving *bifunctionally substituted end-products* **17** *and* **18**.

The balance of the two paths depends on the *rate of the initial reduction step*. If this step is *fast*, all MC is rapidly reduced by the reducing agent and the bifunctional activation pathway has time to proceed. If the initial reduction step is *slow*, excess (unreduced) MC is present, and will quench **14** by electron transfer in the monoadduct stage. Oxygen was predicted to have a similar inhibitory effect to that of excess MC. A set of experiments proved this mechanism as follows (Tomasz *et al.*, 1988a). All customary reducing agents ($H_2/PtO_2$, flavoenzymes, $Na_2S_2O_4$) were capable of inducing either monofunctional or bifunctional activation, by simply varying *the rate of the initial reduction step*, as predicted by the mechanism. Oxygen also had the predicted effect: while anaerobic $Na_2S_2O_4$ activation gave the adduct pattern (**11** and **12**) diagnostic of bifunctional activation, aerobic $Na_2S_2O_4$ resulted in a 'mixed' pattern (**9**, **11** and **12**), indicating that much monoadduct **9** was formed at the expense of the cross-link adduct (Figure 8.9). From this, $O_2$ appears to be a selective inhibitor of the bifunctional pathway and thus of MC cross-link formation. Comparison of the adduct

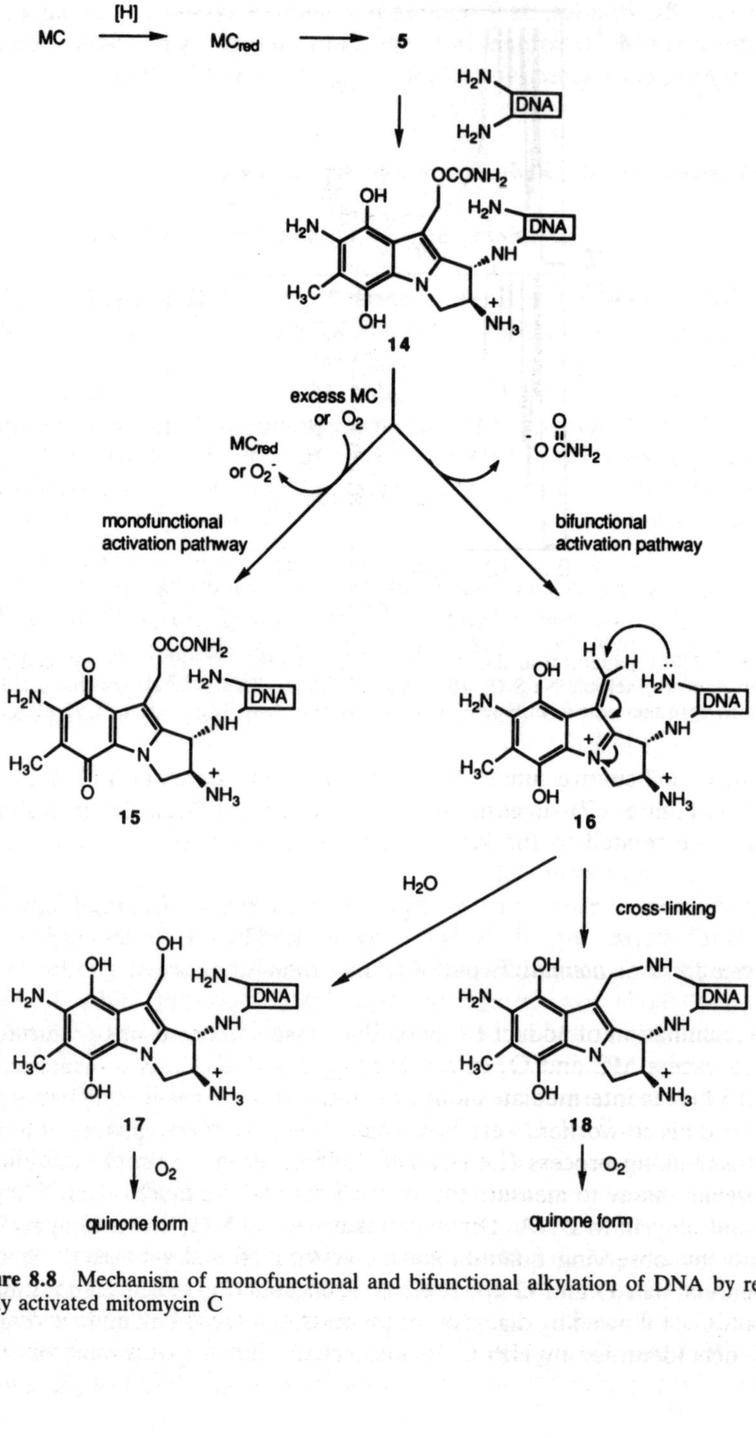


**Figure 8.8** Mechanism of monofunctional and bifunctional alkylation of DNA by reductively activated mitomycin C

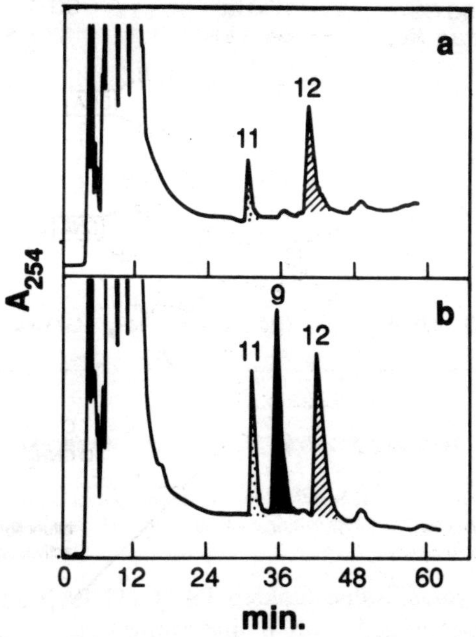


**Figure 8.9** HPLC patterns from digests of *M. luteus* DNA–MC complexes formed under (a) anaerobic and (b) aerobic $Na_2S_2O_4$ activating conditions. Enzymatic digestion and HPLC conditions were the same as in Figure 8.5(a,b). Adapted from Tomasz *et al.* (1988a)

patterns in cell cultures under hypoxia and aerobic conditions (Tomasz *et al.*, 1991; Figure 8.7) suggests that the $O_2$ effect operates also in the cell and may be related to the known hypoxia-selective toxicity of MC and porfiromycin (Sartorelli, 1988).

### *Evidence for a Sequential Two-step Cross-linking Process*

The accumulation of adduct **15** (cf. adduct **9**) in the presence of inhibitors such as excess MC and $O_2$ (see preceding section) clearly indicates that adduct **14** is the intermediate along the bifunctional, cross-linking pathway. Lown and his co-workers were first to show the ordered stepwise nature of the cross-linking process (Lown *et al.*, 1976), using a sensitive ethidium fluorescence assay to measure the extent of cross-linking and overall binding of mitomycin to DNA. Direct proof on the adduct level was provided recently by observing quantitative conversion of *DNA-bound* **9** to the bifunctional pair **11** and **12** upon reductive reactivation of the drug residues by $Na_2S_2O_4$, followed by digestion of the modified DNA or oligonucleotide to adducts identified by HPLC (Tomasz *et al.*, 1988a). Conversion of the

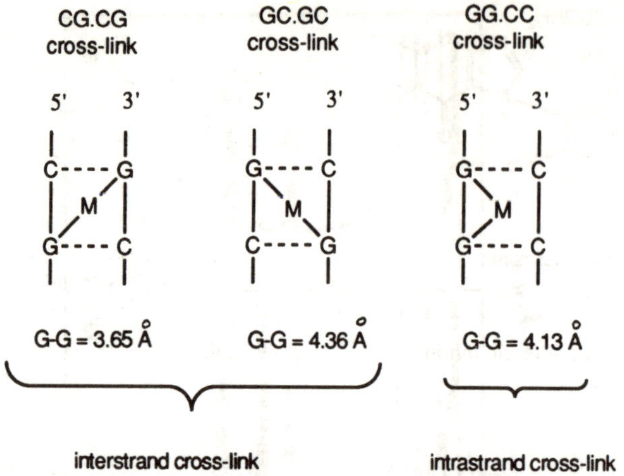


**Figure 8.10** Potential cross-links in DNA formed by mitomycin. The distances between $N^2$ atoms of the guanines are indicated under each diagram. M, bound mitomycin

monoadducted oligonucleotide duplex d[TACG(M)TA]·d[TACGTA] (where M signifies the monofunctionally bound mitomycin to guanine, as in **9**) to the cross-linked duplex was also demonstrated (Borowy-Borowski *et al.*, 1990b). These results further substantiated the proposed DNA-alkylation mechanism shown in Figure 8.8.

### *Basis for the Formation of Monoadduct 11 and Bisadduct (Cross-link) 12 as a Pair in DNA (Figures 8.4 and 8.5)*

This was explained by considering that in DNA only a fraction of guanines is at cross-linkable distance to another guanine: i.e. only those in the GpC, CpG or GpG sequence (Figure 8.10). MC bound to guanines at other dinucleotide sequences reacts subsequently with water at its second activated position, C-10″, leading to adduct **11**. Indeed, poly(dG–dC)·poly(dG–dC) in which all Gs are in GpC and CpG environment gave exclusively the bisadduct **12**; no monoadduct **11** was detectable (Tomasz *et al.*, 1987a, 1988a). Results with synthetic oligonucleotides have substantiated this mechanism (Borowy-Borowski *et al.*, 1990a).

## 5 Acidic Activation of Mitomycin C: Switch of Regioselectivity of Alkylation from $N^2$ to N-7 of Guanine

MC can be activated in the absence of reducing agents by simply lowering the pH of the reaction medium to a value of ≤4.5 ('acidic activation').

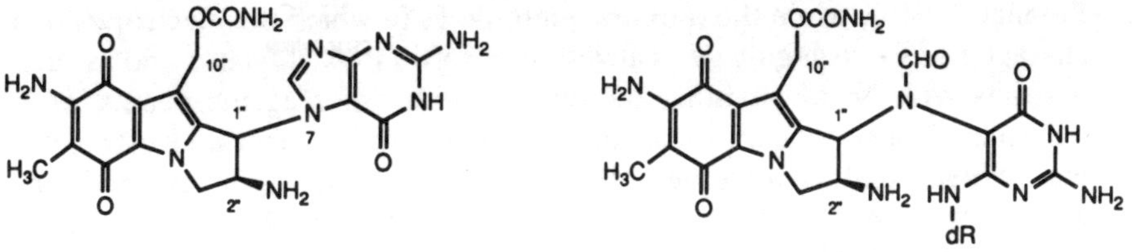


1" α and 1" ß isomers

dR = 2'-deoxyribofuranos-1'-ß-yl
2'-deoxyribofuranos-1'-α-yl
2'-deoxyribopyranos-1'-ß-yl

**Figure 8.11** Structure of mitomycin C–guanine adducts formed under acidic activation conditions

**19** **5**

**Figure 8.12** Activated forms of mitomycin C under acidic and reductive conditions

Under these conditions, both monofunctional binding and cross-linking of MC to DNA was reported to occur (Lown *et al.*, 1976; Lown and Weir, 1978). A chemical study utilizing model nucleophiles (e.g. $P_i$, UMP) demonstrated that acidic activation is sufficient to trigger aziridine opening and nucleophilic capture to yield mitosenes. Thus, both reductive and acidic conditions give rise to derivatives possessing a 1-substituted 2β, 7-diaminomitosene moiety (Tomasz and Lipman, 1979). MC alkylation products of deoxyguanosine, d(GpC) and calf thymus DNA under acidic conditions were isolated and structurally elucidated (Figure 8.11) (Tomasz *et al.*, 1987b). The new adduct structures indicated that the preferred site of alkylation by acidic MC was not the $N^2$ but the N-7 position of guanine. In dG and d(GpC) 95% of total alkylation occurred at N-7, in contrast to 9–12% N-7 alkylation of these substrates under reduction. In calf thymus DNA equal amounts of N-7 and $N^2$ alkylation were observed. It is to be recalled that reductive activation yields exclusively $N^2$-substituted adducts of DNA. To explain this dramatic difference in regional selectivity of attack under acidic and reductive activating conditions, it was suggested (Tomasz *et al.*, 1987b) that acid-activated and reductively activated mitomycin (Figure 8.12; **19** and **5**, respectively) have different reactivities. The carbocation **19** is a 'hard' alkylating agent, reacting preferentially with the site of highest electron density in guanine, i.e. N-7 (Pullman and

Pullman, 1980), while the quinone methide **5**, in which the electropositive character at C-1 is highly delocalized, is a 'soft' DNA alkylator and as such it reacts with $N^2$ of guanine, by analogy with other agents possessing a delocalized positive centre (aromatic diolepoxides, safrol and estragole, etc.: Singer and Grunberger, 1983). The chemical study of the acid-activation of MC (Tomasz *et al.*, 1987b) indicated no DNA cross-linking under acidic conditions, contrary to the earlier report by Lown, above.

### *Is Acidic Activation of Mitomycins Significant in DNA Alkylation* in vivo*?*

The relevance of these *in vitro* findings to the biological activity of MC rests in the assertion that the weakly acidic conditions used for acidic activation may mimic the low pH (approximately 5.2) found inside gastric and solid tumours, for which MC is an effective treatment (Douglass *et al.*, 1984). This speculation is lent support by the finding that the toxicity of MC to EMT6 mouse mammary tumour cells is increased considerably upon systematic acidification of the extra- and intracellular growth medium. Increase of DNA cross-linking was also noted (Kennedy *et al.*, 1985). The pH, however, at which these studies were conducted was higher than what was required for alkylation to occur in chemical systems ($\leqslant$4.5). An alternative, very plausible interpretation of this pH effect was suggested recently: increased *reductive activation* of MC at acidic pH. This was based on the identification of two flavoreductases which activated MC *in vitro* with a pH optimum of 5.7–5.8 (Gustafson and Pritsos, 1992).

Lewis acidic metal complexation (Iyengar *et al.*, 1986) or enzymatic protonation may be thought to activate MC in the cell at physiological pH values in the acidic mode (Verdine *et al.*, 1987). However, since this mode is limited to monoalkylation, DNA cross-links are unlikely to be generated. In summary, the biological significance of the formation of guanine-N-7 adducts (Figure 8.10) is uncertain.

## 6 Conformation of the Mitomycin–DNA Complex

### *Physicochemical and Spectroscopic Studies*

Mitomycin-modified DNA has increased duplex stability, as reflected by its increased $T_m$; this was attributed primarily to the presence of cross-links (Cohen and Crothers, 1970; Kaplan and Tomasz, 1982; Chawla *et al.*, 1987). Surprisingly, DNAs modified selectively by monoadducts **9** or **11** also showed increased duplex stability, the effect being greatest in the case of **12**, less with **9** and least with **11** (Chawla *et al.*, 1987). MC–poly(dG–dC)·

(poly dG–dC) complexes showed especially large effects. These findings indicated that cross-linking between the two strands is not the only duplex-stabilizing factor. Intercalation by MC as a potential source of the increased stability was ruled out by linear flow dichroism studies of MC–calf thymus DNA complexes (Kaplan and Tomasz, 1982). Glycosylated T-2 phage DNA, occluded in its major groove by glucose residues, has reacted with MC without inhibition (Lipman *et al.*, 1978). All this together pointed towards non-distortive, minor groove binding by a combination of covalent bond(s) and non-covalent, duplex-stabilizing interactions between the drug and B-DNA. Further, conclusive evidence for the validity of this model was obtained recently, as discussed in a later section.

### *Mitomycin and Z-DNA*

A different picture (misleading, as it turned out), was presented, however, by a circular dichroism study: the CD of MC–poly(dG–dC)·poly(dG–dC) complexes in the near-ultraviolet region had the appearance of being 'inverted', similar to that of Z-DNA (Mercado and Tomasz, 1977). Further investigation using $^{31}$P-NMR and immunological techniques showed that Z-DNA was not present. The deceptive CD was most probably due to superimposition of induced CD by the chirally linked adduct upon the B-type CD of the polynucleotide (Tomasz *et al.*, 1983a). In further confirmation of lack of Z-DNA, the vacuum CD showed the characteristic signature of B-DNA in MC–poly(dG–dC)·poly(dG–dC) complexes (Sutherland *et al.*, 1986). What is more, the B →Z transition of DNA was shown to be severely *inhibited* in mitomycin-modified poly(dG–dC)·poly(dG–dC) (Chawla *et al.*, 1987). Further experimental work showed that MC alkylated Z-DNA monofunctionally but did not form cross-links; molecular modelling indicated that the bisadduct **12** could not be sterically accommodated without destruction of Z-DNA secondary structure (Chawla and Tomasz, 1988).

The observed inhibition of the B →Z-DNA transition in the MC–poly (dG–dC)·poly(dG–dC) complex is interesting, especially since $(CG)_n$ sequences of DNA are favoured targets for alkylation by MC (see below). This effect of MC may have biological significance.

### *Molecular Modelling Supports the Minor Groove Binding Model*

In a series of articles Kollmann, Remers and their co-workers reported extensive molecular mechanics simulation of non-covalent, covalently monolinked and cross-linked complexes of mitomycin C and certain 7-substituted analogues with duplex decanucleotides (Rao *et al.*, 1986;

Remers *et al.*, 1986, 1988). Energy calculations were performed, using the AMBER force field and charges. Significant conclusions emerged from these theoretical studies, as follows. The intercalative mode did not provide a good model. Useful models were obtained, however, for non-covalent binding by the activated form of MC, i.e. quinone methide, **5**, in both grooves. Monocovalent adduct **9** showed good fit in the minor groove, with several secondary interactions (electrostatic, H-bonded) between drug and DNA functional groups. The interstrand cross-link adduct **12** was incorporated into the duplex at a GpC sequence and appeared to fit well without disruption of base pairing and stacking, and, again, displaying H-bonded and electrostatic interactions with DNA. In summary, these models were consistent with non-distortive minor groove binding of the mitomycin, derived experimentally from the $T_m$, vacuum CD and other spectroscopic studies described in the preceding sections. Several other laboratories have modelled various MC adducts by the MacroModel program of molecular mechanics simulation, using the AMBER force field. Both space-filling and MacroModel-produced theoretical models of **9** and **12**, incorporated into a duplex decamer at a CpG sequence (Figure 8.13) showed characteristics similar to those of Remers and Kollmann above (Verdine, 1986; Tomasz *et al.*, 1987a).

Energy-minimized structural models were recently generated for the intrastrand cross-link (**13a** and **13b**) incorporated in a duplex decanucleotide, using the MacroModel program (Figure 8.13, bottom) (Bizanek *et al.*, 1992). Both models indicated bending of DNA near the cross-link site; this was in contrast to the unbent models of complexes of the monoadduct **9** or the interstrand cross-link **12**. It was suggested that the bent structure is a consequence of a constraint unique to this MC–DNA complex: the $N^2$ atoms of two adjacent guanines are pinched together by the cross-link, moving them from an average 4.3 Å distance in B-DNA to 3.1 Å distance in the cross-linked structures, as required by the fixed distance (3.1 Å) between the C-1″ and C-10″ bonding sites of the mitosene. The bending of DNA by the intrastrand MC cross-link is a hypothesis, based on molecular modelling alone, that needs to be tested by experiment.

### *Structure of a Cross-linked Duplex Hexanucleotide Determined by NMR*

A self-complementary duplex hexanucleotide d(TACGTA)·d(TACGTA), cross-linked by MC at the two guanines, was synthesized. One- and two-dimensional proton NMR data sets from NOESY and COSY experiments and phosphorus NMR from proton–phosphorus heteronuclear COSY were obtained. Twenty-three NOE cross-peaks between oligonucleotide protons and eight NOEs between protons of mitomycin and the oligonucleotide were observed, yielding numerous experimental proton–proton distance constraints in the structure. These were incorporated in

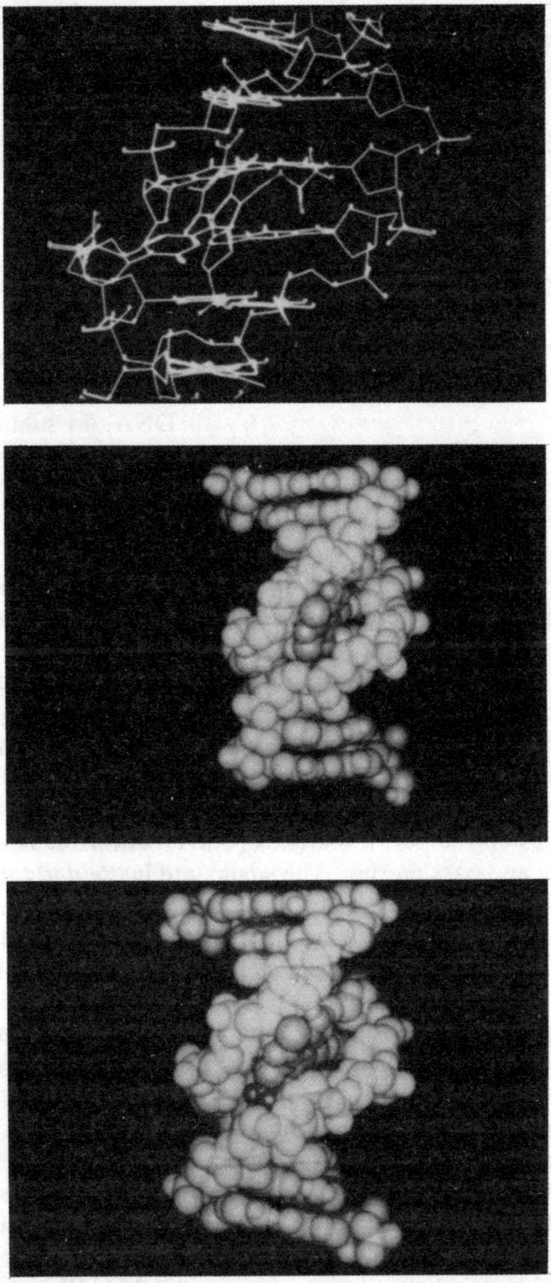

**Figure 8.13** Computer-simulated energy-minimized molecular models of MC–DNA adducts. Top, monoadduct **9**; middle, MC interstrand cross-link adduct **12**; both are incorporated in the central 5′-CG sequence of the self-complementary oligonucleotide d(GCATCGATGC) in its duplex B-DNA form. Bottom, MC intrastrand cross-link adduct **13a**, incorporated in the decamer duplex [d(GCATGGATGC)·d(GCATCCATGC)]. Each model was reconstructed using the MacroModel program from previously published data from the authors' laboratory (monoadduct **9**, Verdine, 1986; cross-link adduct **12**, Tomasz *et al.*, 1987a; cross-link adduct **13**, Bizanek *et al.*, 1992)

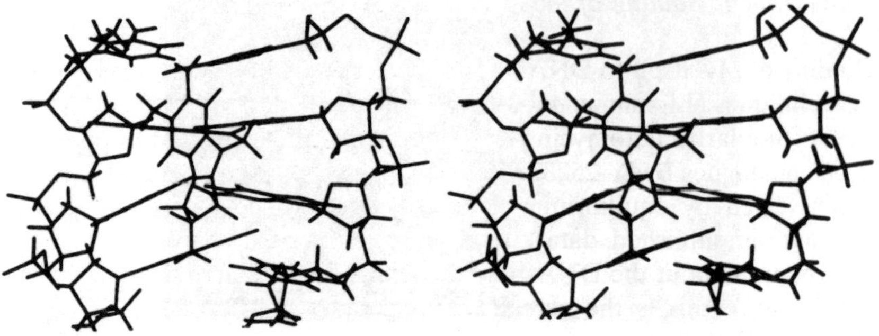

**Figure 8.14** Stereoview of the central 4 base pair segment of d(TACGTA)·d(TACGTA) cross-linked by mitomycin C between the two guanines in the duplex (cf. adduct **12**). NMR data of the cross-linked duplex were utilized to compute the minimum-energy structure. Reconstructed from data in Norman *et al.* (1990)

minimized-energy calculations, starting from a fixed, non-planar five-membered ring of mitomycin in each of two pucker orientations. The resulting structures, MX1 and MX2, are similar, differing in some backbone torsion angles only. They both satisfy the experimental distance constraints and other experimental findings. The structure of the four inner base pairs of the MX1 conformer is shown in Figure 8.14. The following summarizes the findings.

The mitomycin forms a cross-link in the minor groove to adjacent deoxyguanosines on the partner strands. The cross-link causes no disruption of stacking and base pairing. The sugars are in the *anti* conformation. The aromatic indoloquinone moiety of MC is aligned closely on one side of the minor groove against the backbone of the C-10″-linked strand. In effect, the bulk of the drug molecule clings to this one DNA strand. The positive 2″-$NH_3(+)$ group lies in the centre of the minor groove. These results conclusively establish the location of the mitomycin in the minor groove of duplex DNA, as suggested by the earlier physicochemical and CD studies above.

This NMR model has improved upon the earlier models of the cross-link, since it reveals a more realistic, distorted DNA structure. Nevertheless, the earlier models bear a good resemblance to it. The existence of numerous H bonds and van der Waals interactions between drug and DNA, noted in these models, is probably a general characteristic of mitomycin–DNA complexes, which accounts for their observed increased duplex stability, as discussed in a previous section. They also hint at precovalent binding by the drug.

*Non-covalent Binding of the Activated Form of Mitomycins to DNA*

Binding of MC itself to DNA has not been detectable, using a large variety of techniques (Lipman *et al.*, 1978; Rodighiero *et al.*, 1978; Tomasz *et al.*, 1990). Similarly, mitomycin A (**2**), porfiromycin, as well as two semisynthetic analogues BMY-25282 and BMY-25067, showed no binding to DNA when tested by equilibrium dialysis or spectroscopic titration (He and Tomasz, unpublished data). It is of great interest, however, to obtain information about the DNA-binding affinity of the active form of mitomycin C, since this is the actual DNA-alkylating molecule. Although the structure of this form is well established as the quinone methide **9** (Figure 8.3), it is too short-lived (Hoey *et al.*, 1988) to study for its binding to DNA directly. Nevertheless, Crothers and co-workers (Teng *et al.*, 1989) detected indirectly the binding of the reduced mitomycin species to synthetic oligonucleotides, using an elegant experimental design. Cross-linking of G•C-containing oligonucleotides by the activated mitomycin was shown to be competitively inhibited by oligonucleotides containing only A•T base pairs, i.e. no cross-link sites. Kinetic data of the inhibition fitted a simple model in which the active mitomycin binds non-specifically to the oligonucleotides, regardless of base composition or base sequence; the calculated approximate value of the dissociation constants for G•C sites and A•T sites were identical (600 $\text{M}^{-1}$). This value indicates that the transient active species has only a modest, non-specific DNA binding affinity. In another approach stable 'activated form analogues' of mitomycin C were tested for DNA binding (Tomasz *et al.*, 1990). These mitosene quinones, especially **21**, resemble the active form **5** (Figure 8.15). They all showed similar binding to DNA, tested by both equilibrium dialysis and absorbance titration. Binding constants were of the same order of magnitude as that observed by Crothers and co-workers above. This binding was also non-specific with respect to DNA composition, sequence and DNA denaturation (Figure 8.15). It was shown to be dominated by the electrostatic attraction between DNA and the protonated 2″-$NH_3(+)$ group of the drug. The absence of the 10″-carbamate group resulted in a moderate decrease of the binding constant, suggesting that non-specific H bonds contribute to the stability of the electrostatic complex. The similar extent of binding by native and denatured DNA indicates that no intercalation is involved. It is apparent that the binding properties ($K_b$, non-specificity) of these stable mitomycin 'active form analogues' are very similar to those of the transient active form characterized by the Crothers group (Teng *et al.*, 1989; above) and it is most likely that their DNA binding characteristics apply to those of the reduced active form (**5**) of a mitomycin.

The lack of DNA binding by the parent mitosane drugs (MC, mitomycin A, etc.) is apparently due to the lack of basicity of their aziridine ($pK_a \sim 2$–3);

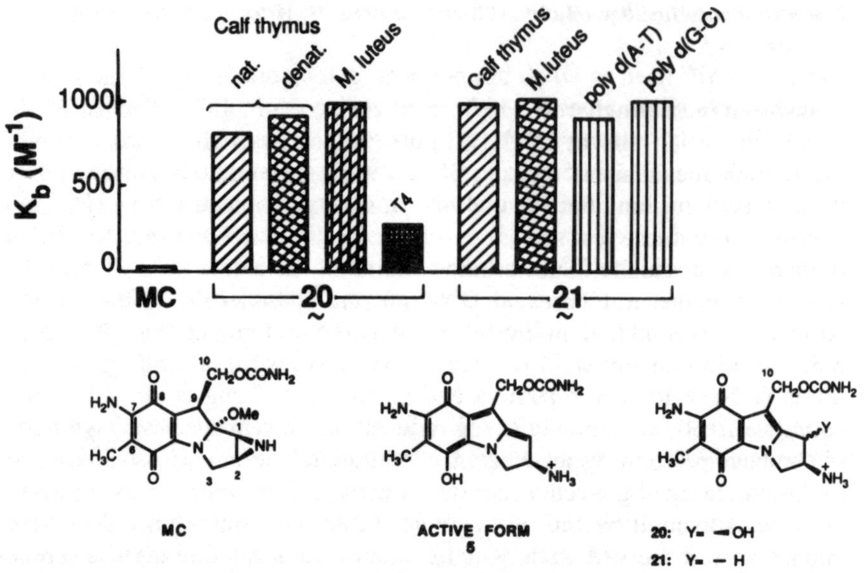


**Figure 8.15** Binding constants for binding of MC active-form analogues **20** and **21** to various DNAs. T4 is T4 phage DNA. From Tomasz *et al.* (1990)

the neutral molecules have no affinity to DNA. Activation, however, generates the basic mitosene 2-amino group (p$K_a$ 7.5) which accounts for the electrostatic attraction of **20**, **21**, etc., to DNA (Figure 8.15). Electrostatic binding of mitosenes to polyvinyl sulfate and polyphosphate was also reported (Lipman *et al.*, 1978).

### *Biological Significance of the Binding to DNA*

One may speculate that after intracellular activation of the mitomycin the unstable quinone methide, **5** (Figure 8.15), will be drawn to the vicinity of DNA, owing to the relatively long-range electrostatic attraction of the DNA polyanion. Thus, the binding mechanism serves to facilitate the drug reaching its target DNA with greater efficiency.

Note, however, that $H_2/PtO_2$ activation readily yields adducts *in vitro* between MC and deoxyguanosine or d(GpC) (Tomasz *et al.*, 1986a,b). Thus, the non-specific binding is not a prerequisite *per se* for covalent reaction.

## 7 DNA Sequence Specificity of the Covalent Reactions of Mitomycin with DNA

It has been increasingly recognized in recent years that naturally occurring antibiotics which aim at DNA as their target display remarkable DNA recognition mechanisms that go far beyond binding to DNA at random sites or distinguishing between specific bases for covalent reaction. Superimposed upon these less specific recognitions, a higher order of selectivity is manifested usually. Subtle conformational polymorphism or specific sequences of several bases may be recognized as binding or bonding targets, as exemplified in the cases of netropsin (Kopka *et al.*, 1985), calicheamicin (Zein *et al.*, 1990), CC-1065 (Reynolds *et al.*, 1985), chromomycin (Gao and Patel, 1989) and others. The mitomycins proved to be no exception.

The base-positional specificity of alkylation at $N^2$ of guanine is intrinsic to the reactivity of the reduced mitomycins. Deoxyguanosine and d(GpC) are monoalkylated by MC at $N^2$ at approximately 90% selectivity (10% alkylation is observed at N7; other nucleosides are inert; see above). Beyond this base-positional selectivity, however, a remarkable *base sequence selectivity* was discovered recently in both the cross-linking and monofunctional alkylating activities of MC and its analogues, as follows.

### *Sequence Specificity of the Interstrand Cross-link*

Since cross-linked DNA and polynucleotides were known to have relatively non-distorted structures (see above), it was reasonable to narrow down the potential cross-linkable sequences in DNA to GC•GC, CG•CG and GG•CC (Figure 8.9). Any two Gs further apart could be cross-linked only upon great distortion, since the fixed distance between the mitomycin C-1″ and C-10″ atoms as in **12** is only 3.36 Å (Tomasz *et al.*, 1987a). The single potential site for the *intrastrand* cross-link was experimentally verified (Bizanek *et al.*, 1992). In the case of the two alternative *interstrand* cross-link sites, a virtually absolute preference for CG•CG was found independently in three laboratories (Teng *et al.*, 1989; Weidner *et al.*, 1989; Borowy-Borowski *et al.*, 1990b; McGuinness *et al.*, 1991b). In each study, synthetic oligonucleotides were used as DNA sequence models. Several different hypotheses have been presented for the molecular basis of this discrimination. Crothers's group (Teng *et al.*, 1989) computed the relative binding energies of the two alternative cross-links incorporated in duplex decanucleotides and found no significant difference between the two. It was concluded that since the final products have similar relative stabilities, the critical factor for the specificity should be kinetic rather than thermodynamic. Since the $N^2$–$N^2$ distance between the two guanines is 3.62 Å and

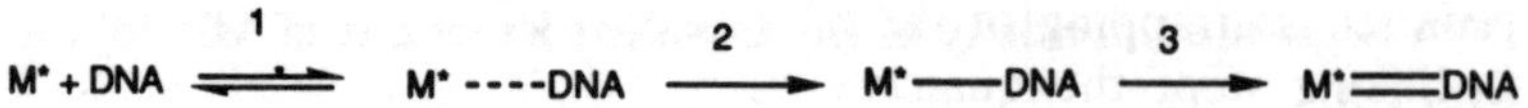


**Figure 8.16** Three-step scheme of cross-linking of DNA by mitomycin C. M*, activated form of MC (generalized); M*–DNA, monoadduct of M* and DNA; M*=DNA, cross-link adduct of M* and DNA

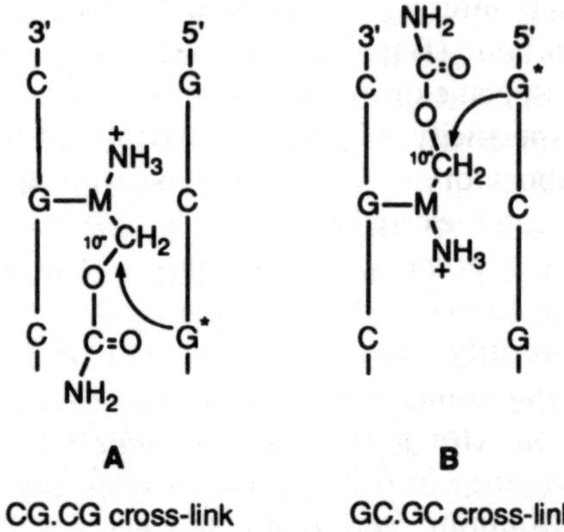


**Figure 8.17** Two opposite orientations of the mitomycin monoadduct

4.1 Å at the CG•CG and GC•GC sites, respectively, the former presents a better match for the fixed 3.4 Å span between the C-1″ and C-10″ of the reacting mitomycin. Based on the successive steps of the cross-linking process (Figure 8.16) the authors proposed that the kinetic factor is the higher activation energy of step 3 in the case of GC•GC, due to the greater bonding distance. A different explanation was offered by the Tomasz group (Borowy-Borowski *et al.*, 1990b). The CG•CG specificity is due to the specific orientation of the monoadduct (Figure 8.17; orientation A), which allows cross-link formation at the CG•CG but not at the GC•GC sequence. In the latter, the reactive centres are too far from each other. Rotation between A and B in the tight minor groove around the mitomycin–guanine covalent bond is not possible. This orientation, too, is a 'kinetic factor' preventing the cross-linking step, but it is conceptually different from that proposed by Crothers above. Support for this proposal is substantial. Molecular mechanics modelling conducted in several laboratories independently showed a strong energy preference for orientation A (Verdine, 1986; Remers *et al.*, 1988; Arora *et al.*, 1990; Millard *et al.*, 1990). Experimental results are consistent with the enhanced stability of the A rotamer. Self-complementary oligonucleotides d-[$(TA)_n$CG$(TA)_n$], monoadducted at the G residue, were annealed to the unsubstituted strand. Addition of $Na_2S_2O_4$ resulted in quantitative formation of the cross-link in less than a minute at 0 °C (step 3 in Figure 8.16). This was interpreted as trapping of the monoadduct in the A orientation by the cross-linking step; the quantitative cross-link yield indicated that only orientation A was present (Borowy-Borowski *et al.*, 1990b; Tomasz, 1992). Direct structural demonstration, e.g. by NMR, has not been reported so far. The structural basis for the greater stability of A is not entirely clear. Chirality of the C-1″–$N^2$ (guanine) bond (Millard *et al.*, 1990) and the favourable secondary interactions of the monoadduct 10″-carbamate group with the DNA backbone (Borowy-Borowski *et al.*, 1990b) could be factors to account for the specific orientation.

The absolute specificity of the interstrand cross-link to CG·CG, observed in the laboratory, is due to intrinsic structural properties of the mitomycin–DNA complex and therefore it is expected to prevail also *in vivo*. In eukaryotic genomes the CpG sequence is singularly rare, compared with the other G-containing dinucleotide sequences. CpG-rich regions frequently identify the promoter regions of eukaryotic genes and methylation of the cytosine of such CpG sequences is a regulatory signal for 'transcription off' (for review, see Hergersberg, 1991). The specific targeting of CpG by the cross-linking action of mitomycin may play a role in mitomycin's antitumour activity.

### *Recognition of Specific DNA Sequences by Mitomycin C for the First Alkylation Step*

As outlined above, the sequence specificity of the interstrand cross-links is a consequence of the specificity of the orientation of the mitomycin monoadduct with respect to DNA strand direction. By all indications, this orientation specificity itself is not a sequence-dependent phenomenon. It simply precludes formation of a cross-link from the monoadduct at GC·GC. A basic question remains, however: is there any sequence specificity in the formation of the *monoadduct* (step 2 in Figure 8.16), i.e. are some guanines alkylated preferentially over others in DNA? This question was answered in the affirmative. Recently, a remarkable specificity of the monoalkylation was discovered independently in several laboratories. In one approach (Li and Kohn, 1991) end-labelled DNA restriction fragments were monoalkylated by MC. Digestion with λ-exonuclease resulted in a set of fragments, detected by gel electrophoresis, each terminating presumably near a guanine–MC monoadduct. The stops were observed four bases upstream from almost all of the 5′-CG sites located in the detectable range of the sequencing gels. Relatively fewer stops occurred at 5′-GG, and no stops were observed at 5′-AG or 5′-TG sequences. No preference could be assigned for any 3′-base, owing to the 'inherent ambiguity as to which strand was modified, the diffuse nature of select radioactive bands, . . .', etc. (Li and Kohn, 1991). 10-Decarbamoyl MC-induced

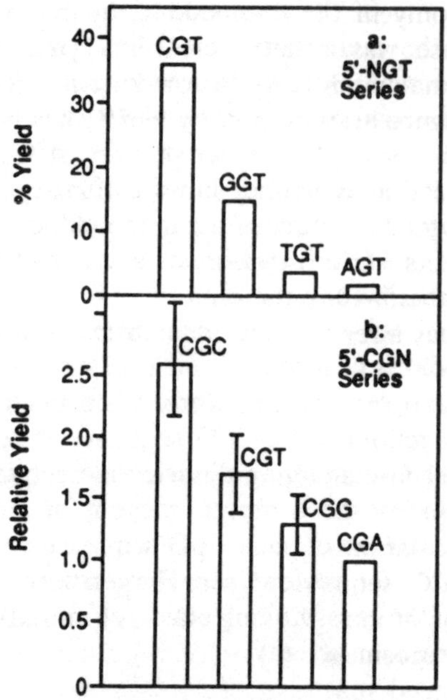


**Figure 8.18** Yield of monoadduct **9** as a function of the 5′-base (a: 5′-NGT series) and the 3′-base (b: 5′-CGN series). Various oligonucleotides containing the indicated sequences were monoalkylated by reductively activated MC under standard conditions and the yields were determined as described in the text. From Kumar *et al.* (1992)

modification resulted in the same stops. It was concluded that there is a strong preference for 5′-CG and 5′-GG sequences of DNA in the monoalkylation of guanine by MC. Support for this conclusion emerged from the work of Phillips *et al.* (1989), although only after re-evaluation of Phillips's data by Li and Kohn (1991). Phillips and his collaborators utilized the transcription block assay to evaluate sequence-specific stops in DNA modified by xanthine oxidase/NADH-activated MC. Strongest stops were seen at 5′-CG, followed by 5′-GG. 5′-TG and 5′-AG stops were less frequent. Thus, these results agree qualitatively with those of Li and Kohn (1991).

A comprehensive study (Kumar *et al.*, 1992) corroborated the 5′-CG and 5′-GG alkylation preference. In addition, it provided quantitative data for the observed sequence-specific rate enhancements. The experimental approach was entirely different from that used in the other studies. Synthetic oligonucleotides were monoalkylated and the yields were determined by enzymatic digestion to unreacted nucleosides and monoadduct **9**, followed by quantitative analysis by HPLC. A striking enhancement of the yield was observed at the 5′-CG sequence: 36%, compared with 2% at 5′-AG and 4.1% at 5′-TG. The 5′-GG sequence also showed enhanced reactivity, although to a lesser extent (14.7%) (Figure 8.18). The enhancements were specific to the duplex state of the oligonucleotides. Enzymatic and acidic activation of mitomycin gave similar results. In another series in which the 3′-base was varied, the 3′-base exerted only a modest modulating effect (maximum 2.5-fold) on the enhanced reactivity at the 5′-CG sequence (Figure 8.18b). In summary:

(1) 5′-CGN triplets show greatly enhanced reactivity; 5′-CGC is the most reactive of all four.
(2) 5′-GGN triplets show moderately enhanced reactivity (approximately half that of the 5′-CGNs).
(3) The 3′-base has a relatively modest modulating effect; the order of reactivity is 3′-(C > T > G > A).
(4) The reaction is not absolutely specific to 5′-CG and 5′-GG; 5′-AG and 5′-TG reactions are well detectable.
(5) The 5′- and 3′-base effects are independent of each other.

What is the molecular basis for the enhanced reactivity of mitomycin to guanine in the 5′-CG sequence? A decisive clue was provided by the alkylation behaviour of inosine-substituted oligonucleotides. When guanine was replaced by inosine in the strand opposite from the 5′-CG alkylation site, the enhanced reactivity of the guanine was abolished. It was therefore proposed that the 2-$NH_2$ group of the guanine in the opposite strand is required for the enhancement, by forming a H-bond to the C-10″ O atom of activated mitomycin. This connection facilitates the covalent alkylation step, either by simply concentrating the precovalent species at such sites, or, more likely, by also *enhancing the rate constant for the covalent reaction (Figure 8.19)*. An analogous model applies to the enhanced reactivity of guanine at 5′-GG (Figure 8.19). Previous modelling studies support the proposed mechanism: both CPK and energy-minimized models of the monoadduct at 5′-CG as well as 5′-GG indicate close proximity of the C-10″ oxygen to the $N^2$-proton of the other guanine (Tomasz *et al.*, 1986a; Remers *et al.*, 1988; Verdine, 1986; Figure 8.13). Another, conceptually similar, mechanism, but which involves a different H-bond (Li and Kohn, 1991) is not compatible with these results. It may be possible to test the existence of the postulated H-bonds by NMR of oligonucleotide-bound monoadduct.

The moderate effect of the 3′-base (Figure 8.18b) was observed also on the rate of the cross-linking reaction (Teng *et al.*, 1989; Borowy-Borowski *et al.*, 1990b; Millard *et al.*, 1991). Its structural origin is unknown so far.

The proposed mechanism for the 5′-CG specificity (Figure 8.19) features specific binding of the activated form of mitomycin to the 5′-CG sequence in the minor groove. Experimentally only *non-specific binding* can be detected, however, as described in a previous section. This can be reconciled with the proposed model as follows. A certain fraction of the predominant non-specifically bound drug must be located dynamically in the minor groove; otherwise no covalent reaction would occur there. Sequence-specific binding of this fraction would increase the covalent

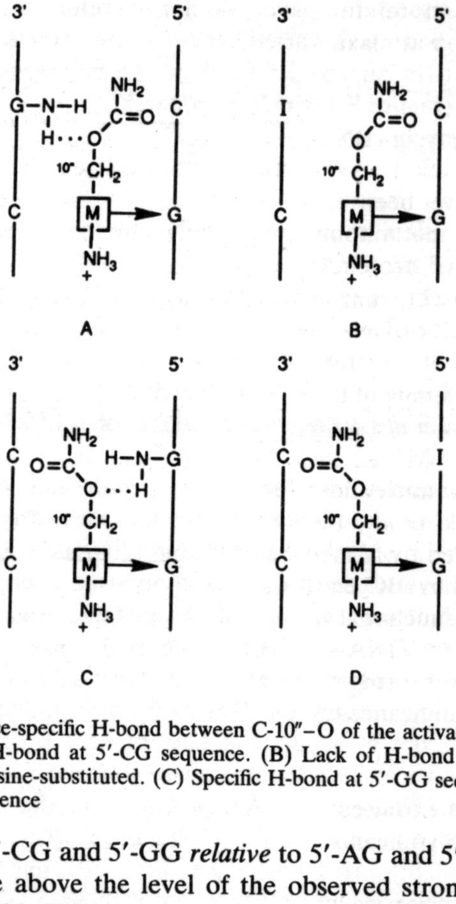


**Figure 8.19** Sequence-specific H-bond between C-10″–O of the activated form of MC and DNA. (A) Specific H-bond at 5′-CG sequence. (B) Lack of H-bond at 5′-CG when the opposite strand is inosine-substituted. (C) Specific H-bond at 5′-GG sequence. (D) Lack of H-bond at 5′-IG sequence

reaction rate at 5′-CG and 5′-GG *relative* to 5′-AG and 5′-TG, but it may not be detectable above the level of the observed stronger non-specific electrostatic bulk binding.

### *Relationship between the Sequence Specificities of Cross-link Formation and Monoalkylation by Mitomycin*

A remarkable result emerging from the above studies is that the monoalkylation by MC is most enhanced at 5′-CG, i.e. at the same sequence as required for the interstrand cross-link. The monoalkylation and cross-link specificities have distinctly different structural bases, however, as argued at some length above. Therefore, it was proposed (Kumar *et al.*, 1992) that the monoalkylation specificity serves independently as a mechanism of preselection by the attacking mitomycin for the potential cross-linkable sites 5′-CG and 5′-GG. As a result, relatively fewer drug molecules attack DNA at the 5′-TG and 5′-AG sites, i.e. where subsequent cross-link formation is not possible. In other words, the various structural features of mitomycin seem to reinforce one another to induce the lethal DNA cross-linking action at maximum efficiency.

## 8 Ternary Mitomycin–DNA–Protein Interactions

Such systems have been relatively little studied so far. Crothers and co-workers found that mitomycin cross-linking is decreased at CpG sites in DNA–*E. coli* CAP protein complexes. Greatest (fourfold) decrease was observed for a $(CG)_3$ run, located in a minor groove region which was widened as a result of DNA bending caused by CAP (Cera and Crothers, 1989). Nucleosomal structure also inhibited cross-linking, again attributable partly to widening of the minor groove due to bending of DNA in the nucleosome (Cera *et al.*, 1990).

A cross-link by MC in the *Cla*I restriction endonuclease recognition sequence of oligonucleotides completely prevented cleavage by *Cla*I (Borowy-Borowski *et al.*, 1990a). pBR322 plasmid DNA and restriction fragments alkylated by *N*-methylmitomycin A were extensively incised by purified *E. coli uvr*ABC endonuclease in recognition of both the monofunctional and bifunctional alkylation damage (Pu *et al.*, 1989). *In vivo*, MC dose-dependent DNA–protein cross-links were detectable in several tumour cell cultures, using the alkaline elution method (Dorr *et al.*, 1985). The biological significance of this ternary interaction is unknown so far.

## 9 Summary of the Molecular Details of Mitomycin–DNA Interactions: Significance for Drug Design

On the basis of what we know, as described above, we may view the mitomycins as a remarkable example of a class of cytotoxic agents exhibiting multifaceted selectivity in their action on DNA. The following independent structural elements are assembled within the small molecule of MC, carrying out the reductive cross-linking of DNA in a concerted fashion.

(1) The quinone is the regulatory site, responding to reductive conditions in hypoxic target organisms. Reduction by non-specific quinone reductases leads to consecutive activation of two alkylating functions (active sites).

(2) Both of these active sites possess covalent reactivity specifically with $N^2$ of guanine. The distance between the two active sites is matched by the distance between two $N^2$ atoms of guanines in adjacent G·C base pairs in the minor groove of B-DNA.

(3) The C-10 carbamate side-chain of activated MC helps to recognize two such adjacent G·C base pairs (5′-CpG and 5′-GpG sequences) *non-*

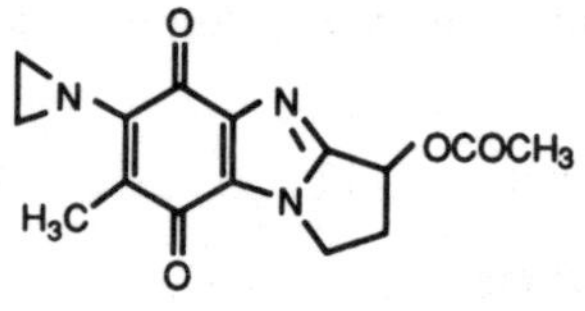

PBI- A ; **22**

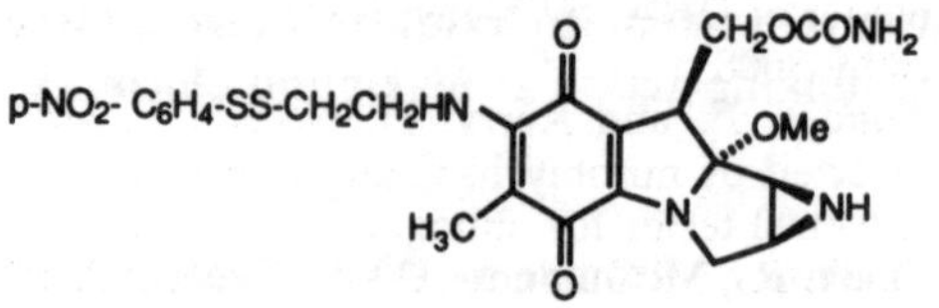

BMY- 25067; **23**

**Figure 8.20** Structure of two mitomycin analogues

*covalently*, leading to selective alkylation of guanines which are located in these cross-linkable sequences. This recognition mechanism increases the cross-linking potency of mitomycin. (A primitive analogy to elements of enzyme–substrate recognition and catalysis is quite apparent.)

If analogues of mitomycins are designed to retain bioreductive activation/GC-specific DNA cross-linking features of the natural mitomycin, the concerted action of the above factors should be taken into account. For example, in the case of a synthetic analogue, PBI-A (**22**, Figure 8.20), many necessary elements are together; however, the steric relationship between the two bioreductive alkylating functions is not matched to any two DNA reactive sites. Although the compound shows antitumour activity, the antitumour spectrum resembles not that of MC but that of adriamycin and other intercalator quinones and the drug does not cross-link DNA (Islam and Skibo, 1991). However, analogues which are modified in the 7-position of the quinone ring of MC (e.g. **23**) have proved to be very promising as 'second generation' mitomycin analogue antitumour agents, particularly because of their lower myelosuppressive activity compared with that of MC (Doyle and Vyas, 1990). The bulky 7-substituents do not interfere with the DNA binding and cross-linking mechanism, since this position is on the periphery of the minor groove (Figure 8.13). Therefore, this substituent can be flexibly manipulated, possibly resulting in modulation of the bioactivation properties of the mitomycins (Doyle and Vyas, 1990).

## References

Andrews, P. A., Pan, S. and Bachur, N. R. (1986). Electrochemical reductive activation of mitomycin C. *J. Am. Chem. Soc.*, **108**, 4158–4166

Armstrong, R. W., Salvati, M. E. and Nguyen, M. (1992). Novel interstrand cross-links induced by the antitumor antibiotic carzinophyllin/azinomycin B. *J. Am. Chem. Soc.*, **114**, 3144–3145

Arora, S. K., Cox, M. B. and Arjunan, P. (1990). Structural, conformational and theoretical binding studies of antitumor antibiotic porfiromycin, a covalent binder of DNA, by X-ray, NMR, and molecular mechanics. *J. Med. Chem.*, **33**, 3000–3008

Balbinder, E. and Kerry, D. (1984). A new strain of *Salmonella typhimurium* reverted by mitomycin C and *N*-methyl-*N'*-nitronitrosoguanidine — a possible universal tester for mutagenic compounds. *Mutation Res.*, **130**, 315–320

Bizanek, R., McGuinness, B. F., Nakanishi, K. and Tomasz, M. (1992). Isolation and structure of an intrastrand cross-link adduct of mitomycin C and DNA. *Biochemistry*, **31**, 3084–3091

Bligh, H. F. J., Bartoszek, A., Robson, C. N., Hickson, I. D., Kasper, C. B., Begges, J. D. and Wolf, C. R. (1990). Activation of mitomycin C by NADPH: cytochrome P-450 reductase. *Cancer Res.*, **50**, 7789–7792

Borowy-Borowski, H., Lipman, R., Chowdary, D. and Tomasz, M. (1990a). Duplex oligonucleotides cross-linked by mitomycin C at a single site: Synthesis, properties and cross-link reversibility. *Biochemistry*, **29**, 2992–2999

Borowy-Borowski, H., Lipman, R. and Tomasz, M. (1990b). Recognition between mitomycin C and specific DNA sequences for cross-link formation. *Biochemistry*, **29**, 2999–3004

Boyce, R. P. and Howard-Flanders, P. (1964). Genetic control of DNA breakdown and repair in *E. coli* K-12 treated with mitomycin C or ultraviolet light. *Z. Verebungsl.*, **95**, 345–350

Carrano, A. V., Thompson, L. H., Stretka, D. G., Minkler, J. L., Mazrimas, J. A. and Fong, S. (1979). DNA crosslinking, sister chromatid exchange and specific locus mutations. *Mutation Res.*, **63**, 175–188

Carter, S. K. (1979). Reflections and prospect. In Carter, S. K. and Crooke, S. T. (Eds), *Mitomycin C: Current Status and New Developments*, Academic Press, New York, pp. 251–254

Cera, C. and Crothers, D. M. (1989). Modulation of mitomycin cross-linking by DNA bending in the *E. coli* CAP protein–DNA complex. *Biochemistry*, **28**, 3908–3911

Cera, C., Egbertson, M., Teng, S. P., Crothers, D. M. and Danishefsky, S. J. (1989). DNA cross-linking by intermediates of the mitomycin activation cascade. *Biochemistry*, **28**, 5665–5669

Cera, C., Palumbo, M., Palu, G. and Crothers, D. M. (1990). *N*-Methylmitomycin A cross-linking to nucleosomal structure. *Anti-cancer Drug Des.*, **5**, 55–58

Chawla, A. K., Lipman, R. and Tomasz, M. (1987). Covalent crosslinks and monofunctional adducts of mitomycin C in the minor groove of DNA: effects on DNA conformation and dynamics. In Sarma, R. H. and Sarma, M. H. (Eds), *Structure and Expression*, Vol. 2, *DNA and Its Drug Complexes*, Adenine Press, Guilderland, N. Y., pp. 305–316

Chawla, A. K. and Tomasz, M. (1988). Interaction of the antitumor antibiotic mitomycin C with Z-DNA. *J. Biomol. Struct. Dyn.*, **6**, 459–470

Chowdary, D. and Tomasz, M. (1987). Isolation of a mitomycin C-DNA cross-link formed *in vivo*. *Fed. Proc.*, **46**, 2037

Cohen, R. J. and Crothers, D. M. (1970). Preparation and characterization of monodisperse, cross-linked low molecular weight deoxyribonucleic acid. *Biochemistry*, **9**, 2533–2539

Danishefsky, S. J. and Ciufolini, M. (1984). Leucomitomycins. *J. Am. Chem. Soc.*, **106**, 6424–6425

Danishefsky, S. J. and Egbertson, M. (1986). On the characterization of intermediates in the mitomycin activation cascade: A practical synthesis of an aziridinomitosene. *J. Am. Chem. Soc.*, **108**, 4648–4650

Dorr, R. T., Bowden, G. T., Alberts, D. S. and Liddil, J. D. (1985). Interactions of mitomycin C with mammalian DNA detected by alkaline elution. *Cancer Res.*, **45**, 3510–3516

Douglass, H. O., Lavin, P. T., Goudsmit, A., Klassen, D. J. and Paul, A. R. (1984). *J. Clin. Oncol.*, **2**, 1372–1377

Doyle, T. W. and Vyas, D. M. (1990). Second generation analogs of etoposide and mitomycin C. *Cancer Treat. Rev.*, **17**, 127–132

Dulhanty, A. M. and Whitmore, G. F. (1991). Chinese hamster ovary cell lines resistant to mitomycin C under aerobic but not hypoxic conditions are deficient in DT-diaphorase. *Cancer Res.*, **51**, 1860–1865

Egbertson, M. and Danishefsky, S. J. (1987). Modeling of the electrophilic activation of mitomycins: Chemical evidence for the intermediacy of a mitosene semiquinone as the active electrophile. *J. Am. Chem. Soc.*, **109**, 2204–2205

Erickson, C. C., Bradley, M. O. and Ducore, M. J. (1980). DNA cross-linking and cytotoxicity in normal and transformed human cells treated with antitumor nitrosourea. *Proc. Natl Acad. Sci. USA*, **77**, 467–472

Fracasso, P. M. and Sartorelli, A. C. (1986). Cytotoxicity and DNA lesions produced by mitomycin C and porfiromycin in hypoxic and aerobic EMT6 and Chinese hamster ovary cells. *Cancer Res.*, **46**, 3939–3944

Franck, R. W. and Tomasz, M. (1990). The chemistry of mitomycins. In Wilman, D. E. V. (Ed.), *The Chemistry of Antitumor Agents*, Blackie, Glasgow, and Chapman and Hall, New York, pp. 379–394

Gao, X and Patel, D. J. (1989). Solution structure of the chromomycin–DNA complex. *Biochemistry*, **28**, 751–762

Gustafson, D. L. and Pritsos, C. (1992). Bioactivation of mitomycin C by xanthine dehydrogenase from EMT6 mouse mammary carcinoma tumors. *J. Natl Cancer Inst.*, **84**, 1180–1185

Hashimoto, Y., Shudo, K. and Okamato, T. (1983). Modification of deoxyribonucleic acid with reductively activated mitomycin C. Structures of modified nucleotides. *Chem. Pharm. Bull.*, **31**, 861–869

Hata, T., Sano, Y., Sugawara, R., Matsumae, A., Kanamorei, K., Shima, T. and Hoshi, T. (1956). Mitomycin, a new antibiotic from *Streptomyces*. I. *J. Antibiotics, Ser. A*, **9**, 141–146

Hergersberg, M. (1991). Biological aspects of cytosine methylation in eukaryotic cells. *Experientia*, **47**, 1171–1185

Hoban, P. R., Walton, M. I., Robson, C. N., Godden, J., Stratford, I. J., Workman, P., Harris, A. L. and Hickson, I. D. (1990). Decreased NADPH: cytochrome P-450 reductase activity and impaired drug activation in a mammalian cell line resistant to mitomycin C under aerobic but not hypoxic conditions. *Cancer Res.*, **50**, 4692–4697

Hoey, B. M., Butler, J. and Swallow, A. J. (1988). Reductive activation of mitomycin C. *Biochemistry*, **27**, 2608–2614

Islam, I. and Skibo, E. B. (1991). Structure–activity studies of antitumor agents based on pyrrolo[1,2,-**a**] benzimidazoles: New reductive alkylating DNA cleaving agents. *J. Med. Chem.*, **34**, 2954–2961

Iyengar, B. S., Sami, S. M., Takahashi, T., Sikorski, E. E. and Remers, W. A. (1986). Mitomycin A analogs with increased metal complexing ability. *J. Med. Chem.*, **29**, 1760–1764

Iyer, V. N. and Szybalski, W. A. (1963). A molecular mechanism of mitomycin action: linking of complementary DNA strands. *Proc. Natl Acad. Sci. USA*, **50**, 355–362

Iyer, V. N. and Szybalski, W. (1964). Mitomycins and porfiromycin: chemical mechanisms of activation and cross-linking of DNA. *Science*, **145**, 55–58

Kaplan, D. J. and Tomasz, M. (1982). Altered physico-chemical properties of the DNA–mitomycin C complex. Evidence for a conformational change in DNA. *Biochemistry*, **21**, 3006–3013

Kato, S., Yamashita, K., Kim, T., Tajiri, T., Onda, M. and Sato, S. (1988). Modification of DNA by mitomycin C in cancer patients detected by $^{32}$P-post-labeling analysis. *Mutation Res.*, **202**, 85–91

Kennedy, K. A., McGurl, J. D., Leondaridis, L. and Alabaster, O. (1985). pH Dependence of mitomycin C-induced cross-linking activity in EMT6 tumor cells. *Cancer Res.*, **45**, 3541–3547

Kennedy, K. A., Rockwell, S. and Sartorelli, A. C. (1980). A preferential activation of mitomycin C to cytotoxic metabolites by hypoxic tumor cells. *Cancer Res.*, **40**, 2356–2360

Kennedy, K. A., Sligar, S. G., Polomski, L. and Sartorelli, A. C. (1982). Metabolic activation of mitomycin C by liver microsomes and nuclei. *Biochem. Pharmacol.*, **31**, 2011–2016

Kenyon, C. J. and Walker, G. C. (1980). DNA-damaging agents stimulate gene expression at specific loci in *Escherichia coli*. *Proc. Natl Acad. Sci. USA*, **17**, 2819–2823

Keyes, S. R., Fracasso, P. M., Heimbrooks, D. C., Rockwell, S., Sligar, S. G. and Sartorelli, A. C. (1984). Role of NADPH-cytochrome c reductase and DT-diaphorase in the biotransformation of mitomycin C. *Cancer Res.*, **44**, 5638–5643

Keyes, S. R., Loomis, R., DiGiovanna, M. P., Pritsos, C. A., Rockwell, S. and Sartorelli, A. C. (1991). Cytotoxicity and DNA cross-links produced by mitomycin analogs in aerobic and hypoxic EMTσ cells. *Cancer Commun.*, **3**, 351–356

Kohn, H. and Zein, N. (1983). Studies concerning the mechanism of electrophilic substitution reactions of mitomycin C. *J. Am. Chem. Soc.*, **105**, 4105–4106

Kohn, H., Zein, N., Lin, X. Q. and Kadish, K. M. (1987). Mechanistic studies on the mode of reaction of mitomycin C under catalytic and electrochemical reductive conditions. *J. Am. Chem. Soc.*, **109**, 1833–1840

Kopka, M. L., Yoon, C., Pjura, P. and Dickerson, R. E. (1985). The molecular origin of DNA–drug specificity in netropsin and distamycin. *Proc. Natl Acad. Sci. USA*, **82**, 1376–1380

Kumar, S., Lipman, R. and Tomasz, M. (1992). Recognition of specific DNA sequences by mitomycin C for alkylation. *Biochemistry*, **31**, 1399–1407

Li, V. and Kohn, H. (1991). Studies on the bonding specificity for mitomycin C–DNA monoalkylation processes. *J. Am. Chem. Soc.*, **113**, 275–283

Lin, A. J., Cosby, L. A. and Sartorelli, A. C. (1976). Potential bioreductive alkylating agents. In Sartorelli, A. C. (Ed.), *Cancer Chemotherapy*, ACS Monograph Series, Washington, D. C., pp. 71–80

Lipman, R., Weaver, J. and Tomasz, M. (1978). Electrostatic complexes of mitomycin C with nucleic acids and polyanions. *Biochim. Biophys. Acta*, **521**, 779–791

Lown, J. W., Begleiter, A., Johnson, D. and Morgan, A. R. (1976). Studies related to antitumor antibiotics. V. Reaction of mitomycin C with DNA examined by ethidium fluorescence assay. *Can. J. Biochem.*, **54**, 110–119

Lown, J. W. and Majundar, K. C. (1977). Reactions of carzinophilin with DNA assayed by ethidium fluorescence. *Can. J. Biochem.*, **55**, 630–635

Lown, J. W. and Weir, G. (1978). Studies related to antitumor antibiotics. Part XIV. Reactions of mitomycin B with DNA. *Can. J. Biochem.*, **56**, 296–304

McGuinness, B. F., Lipman, R., Goldstein, J., Nakanishi, K. and Tomasz, M.

(1991a). Alkylation of DNA by mitomycin A. *Biochemistry*, **30**, 6444–6453
McGuinness, B. F., Lipman, R., Nakanishi, K. and Tomasz, M. (1991b). Reaction of sodium dithionite-activated mitomycin C with guanine at non-crosslinkable sequences of oligonucleotides. *J. Org. Chem.*, **56**, 4826–4829
McGuinness, B. F., Nakanishi, K., Lipman, R. and Tomasz, M. (1988). Synthesis of guanine derivatives substituted in the O6-position by mitomycin C. *Tetrahedron Lett.*, **29**, 4673–4676
McLaren, A. D. and Shugar, D. (1964). *Photochemistry of Proteins and Nucleic Acids*, Pergamon Press, Macmillan, New York, pp. 320–334
Marshall, R. S., Paterson, M. C. and Rauth, A. M. (1991). Studies on the mechanism of resistance to mitomycin C and porfiromycin in a human cell strain derived from a cancer-prone individual. *Biochem. Pharmacol.*, **41**, 1351–1360
Marshall, R. S. and Rauth, A. M. (1986). Modification of the cytotoxic activity of mitomycin C by oxygen and ascorbic acid in Chinese hamster ovary cells and a repair-deficient mutant. *Cancer Res.*, **46**, 2709–2713
Matsumoto, I. and Lark, K. G. (1963). Altered DNA isolated from cells treated with mitomycin C. *Exp. Cell Res.*, **32**, 192–196
Mercado, C. M. and Tomasz, M. (1977). Circular dichroism of mitomycin–DNA complexes: evidence for a conformational change in DNA. *Biochemistry*, **16**, 2040–2046
Millard, J. T., Weidner, M. F., Kirchner, J. J., Ribeiro, S. and Hopkins, P. B. (1991). Sequence preferences of DNA interstrand cross-linking agents: quantitation of interstrand cross-link locations in DNA duplex fragments containing multiple crosslinkable sites. *Nucleic Acids Res.*, **19**, 1885–1891
Millard, J. T., Weidner, M. F., Raucher, S. and Hopkins, P. B. (1990). Determination of the DNA cross-linking sequence specificity of reductively activated mitomycin C at single nucleotide resolution: Deoxyguanosine residues at CpG are cross-linked preferentially. *J. Am. Chem. Soc.*, **112**, 3637–3641
Moore, H. W. (1977). Bioactivation as a model for drug design. Bioreductive alkylation. *Science*, **197**, 527
Moulder, J. E. and Rockwell, S. (1987). Tumor hypoxia: its impact on cancer therapy. *Cancer Metastasis Rev.*, **5**, 313–341
Nielsen, P. E. and Bohr, V. (1983). Phototoxic effects of four psoralens on L1210 cells. The correlation with DNA interstrand cross-linking. *Photochem. Photobiol.*, **38**, 653–657
Norman, D., Live, D., Sastry, M., Lipman, R., Hingerty, B. E., Tomasz, M., Broyde, S. and Patel, D. J. (1990). NMR and computational characterization of mitomycin cross-linked to adjacent deoxyguanosines in the minor groove of the d(T–A–C–G–T–A)·d(T–A–C–G–T–A) duplex. *Biochemistry*, **29**, 2861–2876
Pan, S. (1990). Porfiromycin disposition in oxygen modulated P388 cells. *Cancer Chemother. Pharmacol.*, **27**, 187–193
Pan, S. S., Andrews, P. A., Glover, C. J. and Bachur, N. R. (1984). Reductive activation of mitomycin C and mitomycin C metabolites catalyzed by NADPH-cytochrome P-450 reductase and xanthine oxidase. *J. Biol. Chem.*, **259**, 959–966
Pan, S. and Iracki, T. (1988). Metabolites and DNA adduct formation from flavoenzyme activated porfiromycin. *Mol. Pharmacol.*, **34**, 223–228
Pan, S., Iracki, T. and Bachur, N. R. (1986). DNA alkylation by enzyme-activated mitomycin C. *Mol. Pharmacol.*, **29**, 622–628
Patrick, J. B., Williams, R. P., Meyer, W. E., Fulmor, W., Cosulich, D. B., Broschard, R. W. and Webb, J. S. (1964). Aziridinomitosenes: A new class of antibiotics related to the mitomycins. *J. Am. Chem. Soc.*, **86**, 1889–1890
Peterson, D. M. and Fisher, J. (1986). Autocatalytic quinone methide formation from mitomycin C. *Biochemistry*, **25**, 4077–4084

Phillips, D. R., White, R. J. and Cullinane, C. (1989). DNA sequence-specific adducts of adriamycin and mitomycin C. *FEBS Lett.*, **246**, 233–240

Pu, W. T., Kahn, R., Munn, M. M. and Rupp, W. D. (1989). UvrABC incision of *N*-methylmitomycin A-DNA monoadducts and cross-links. *J. Biol. Chem.*, **264**, 20697–20704

Pullman, A. and Pullman, B. (1980). Electrostatic effect of the macromolecular structure on the biochemical reactivity of the nucleic acids. Significance for chemical carcinogenesis. *Int. J. Quantum Chem. Quantum Biol. Symp.*, **7**, 245–260

Rao, G. M., Begleiter, A., Lown, J. W. and Plambeck, J. A. (1977). Electrochemical studies of antitumor antibiotics. II. Polarographic and cyclic voltammetric studies of mitomycin C. *J. Electrochem. Soc.*, **124**, 199–202

Rao, S. N., Singh, U. C. and Kollman, P. (1986). Conformations of the noncovalent and covalent complexes between mitomycins A and C and $d(GCGCGCGCGC)_2$. *J. Am. Chem. Soc.*, **108**, 2058–2068

Reddy, M. V. and Randerath, K. (1987). $^{32}P$-Analysis of DNA adducts in somatic and reproductive tissues of rats treated with the anticancer antibiotic, mitomycin C. *Mutation Res.*, **179**, 75–88

Remers, W. A., Rao, S. N., Singh, U. C. and Kollmann, P. A. (1986). Conformation of complexes between mitomycin and decanucleotides. 2. Application of the model to mitomycin C derivatives. Extension to covalent binding with adenine. *J. Med. Chem.*, **29**, 1256–1263

Remers, W. A., Rao, S. N., Wunz, T. P. and Kollman, P. (1988). Conformations of complexes between mitomycins and decanucleotides. 3. Sequence specificity, binding at C-10, and mitomycin analogs. *J. Med. Chem.*, **31**, 1612–1620

Reynolds, V. L., Molineux, I. J., Kaplan, D. J., Swenson, D. H. and Hurley, L. H. (1985). Reaction of the antitumor antibiotic CC-1065 with DNA. Location of the site of thermally induced strand breakage and analysis of DNA sequence specificity. *Biochemistry*, **24**, 6228–6237

Rodighiero, G., Magno, S. M., Dell'Acqua, F. and Vedaldi, D. (1978). Studies on the mechanism of action of mitomycin C. *Il Farmaco*, **33**, 651–666

Sartorelli, A. C. (1988). Therapeutic attack of hypoxic cells of solid tumors: presidential address. *Cancer Res.*, **48**, 775–778

Shirahata, K. and Hirayama, N. (1983). Revised absolute configuration of mitomycin C. X-ray analysis of 1-*N*-(*p*-bromobenzoyl) mitomycin C. *J. Am. Chem. Soc.*, **105**, 7199–7200

Siegel, D., Gibson, N. W., Preusch, P. C. and Ross, D. (1990). Metabolism of mitomycin C by DT-diaphorase: Role in mitomycin C-induced DNA damage and cytotoxicity in human colon carcinoma cells. *Cancer Res.*, **50**, 7483–7489

Singer, B. and Grunberger, D. (1983). *Molecular Biology of Mutagens and Carcinogens*, Plenum Press, New York

Stevens, C. L., Taylor, K. G., Munk, M. E., Marshall, W. S., Noll, K., Shah, G. D. and Uzu, K. (1964). Chemistry and structure of mitomycin C. *J. Med. Chem.*, **8**, 1–10

Sutherland, J. C., Lin, B., Mugavero, J., Trunk, J., Tomasz, M., Santella, R., Marky, L. and Breslauer, K. (1986). Vacuum ultraviolet circular dichroism of double-stranded nucleic acids. *Photochem. Photobiol.*, **44**, 295–301

Szybalski, W. and Iyer, V. N. (1964). Cross-linking of DNA by enzymatically or chemically activated mitomycins and porfiromycins, bifunctionally 'alkylating' antibiotics. *Fed. Proc.*, **23**, 946–957

Teng, S. P., Woodson, S. A. and Crothers, D. M. (1989). DNA sequence specificity of mitomycin cross-linking. *Biochemistry*, **28**, 3901–3907

Thompson, L. H., Rubin, J. S., Cleaver, J. E., Whitmore, G. F. and Brookman,

K. (1980). A screening method for isolating DNA-repair deficient mutants of CHO cells. *Somatic Cell Genet.*, **6**, 391–405

Tomasz, M. (1992). Mitomycin C: DNA sequence specificity of a natural DNA cross-linking agent. In Hurley, L. H. (Ed.), *Advances in DNA Sequence Specific Agents, Vol. 1*, JAI Press Inc., Greenwich, Conn., pp. 247–261

Tomasz, M., Barton, J. K., Magliozzo, C. C., Tucker, D., Lafer, E. M. and Stollar, B. D. (1983a). Lack of Z-DNA conformation in mitomycin modified polynucleotides having inverted circular dichroism. *Proc. Natl Acad. Sci. USA*, **80**, 2874–2878

Tomasz, M., Borowy-Borowski, H. and McGuinness, B. F. (1990). Course of recognition and covalent reactions between mitomycin C and DNA: sequence selectivity of a cross-linking drug. In Pullman, B. and Jortner, J. (Eds), *Molecular Basis of Specificity in Nucleic Acid–Drug Interactions*, Kluwer Academic, Dordrecht, pp. 551–564

Tomasz, M., Chawla, A. K. and Lipman, R. (1988a). Mechanism of monofunctional and bifunctional alkylation of DNA by mitomycin C. *Biochemistry*, **27**, 3182–3187

Tomasz, M., Chowdary, D., Lipman, R., Shimotakahara, S., Veiro, D., Walker, V. and Verdine, G. L. (1986a). Reaction of DNA with chemically or enzymatically activated mitomycin C: Isolation and structure of the major covalent adduct. *Proc. Natl Acad. Sci. USA*, **83**, 6702–6706

Tomasz, M., Hughes, C. R., Chowdary, D., Keyes, S. R., Lipman, R., Sartorelli, A. C. and Rockwell, S. (1991). Isolation, identification and assay of [$^3$H]-porfiromycin adducts of EMT6 mammary tumor cell DNA. *Cancer Commun.*, **3**, 213–223

Tomasz, M. and Lipman, R. (1979). Alkylation reactions of mitomycin C at acid pH. *J. Am. Chem. Soc.*, **101**, 6063–6067

Tomasz, M. and Lipman, R. (1981). Reductive metabolism and alkylating activity of mitomycin C induced by rat liver microsomes. *Biochemistry*, **20**, 5056–5061

Tomasz, M., Lipman, R., Chowdary, D., Pawlak, J., Verdine, G. L. and Nakanishi, K. (1987a). Isolation and structure of a covalent cross-link adduct between mitomycin C and DNA. *Science*, **235**, 1204–1208

Tomasz, M., Lipman, R., Lee, M. S., Verdine, G. L. and Nakanishi, K. (1987b). Reaction of acid-activated mitomycin C with calf thymus DNA and model guanines. Elucidation of the base-catalyzed degradation of N7-alkylguanine nucleosides. *Biochemistry*, **26**, 2010–2027

Tomasz, M., Lipman, R., McGuinness, B. F. and Nakanishi, K. (1988b). Isolation and characterization of a major adduct between mitomycin C and DNA. *J. Am. Chem. Soc.*, **110**, 5892–5896

Tomasz, M., Lipman, R., Snyder, J. K. and Nakanishi, K. (1983b). Full structure of a mitomycin C dinucleoside phosphate adduct. Use of differential FT-IR in microscale structural studies. *J. Am. Chem. Soc.*, **105**, 2059–2063

Tomasz, M., Lipman, R., Verdine, G. L. and Nakanishi, K. (1986b). Reassignment of the guanine-binding mode of reduced mitomycin C. *Biochemistry*, **25**, 4337–4343

Tomasz, M., Mercado, C. M., Olson, J. and Chatterjie, N. (1974). The mode of interaction of mitomycin C with DNA and polynucleotides *in vitro*. *Biochemistry*, **13**, 4878–4887

Traver, R. D., Horikoshi, T., Danenberg, K. D., Stadlbauer, T. H. W., Danenberg, P. V., Ross, D. and Gibson, N. W. (1992). NADP(H): Quinone oxidoreductase gene expression in human colon carcinoma cells: characterization of a mutation which modulates DT-diaphorase activity and mitomycin sensitivity. *Cancer Res.*, **52**, 797–802

Uchida, I., Takase, S., Kayakiri, H., Kiyoto, S. and Hashimoto, M. (1987). Structure of FR900482, a novel antitumor antibiotic from a *Streptomyces*. *J. Am. Chem. Soc.*, **109**, 4108–4109

Urakawa, C., Tsuchiya, H. and Nakano, K. (1981). New mitomycin, 1-decarbamoyloxy-9-dehydromitomycin B from *Streptomyces caespitosus*. *J. Antibiotics*, **34**, 243–244

Verdine, G. L. (1986). *Binding of Mitomycin C to a Dinucleoside Phosphate and DNA*. PhD Thesis, Columbia University, New York

Verdine, G. L., McGuinness, B. F., Nakanishi, K. and Tomasz, M. (1987). Unusual *cis* stereoselectivity in an aziridine cleavage reaction of mitomycin C. *Heterocycles*, **25**, 577–587

Weaver, J. and Tomasz, M. (1982). Reactivity of mitomycin C with synthetic polyribonucleotides containing guanine or guanine analogs. *Biochim. Biophys. Acta*, **697**, 252–254

Weidner, M. F., Millard, J. T. and Hopkins, P. B. (1989). Determination at single-nucleotide resolution of the sequence-specificity of DNA interstrand cross-linking agents in DNA fragments. *J. Am. Chem. Soc.*, **111**, 9270–9272

Weissbach, A. and Lisio, A. (1965). Alkylation of nucleic acids by mitomycin C and porfiromycin. *Biochemistry*, **4**, 196–200

Wilson, K. K. V., Long, B. H., Chakrabarty, S., Brattain, D. E. and Brattain, M. G. (1985). Effects of BMY 25282, a mitomycin C analogue, in mitomycin C-resistant human colon cancer cells. *Cancer Res.*, **45**, 5281–5286

Zein, N., Ding, W. and Ellestad, G. A. (1990). Interaction of calicheamicin with DNA. In Pullman, B. and Jortner, J. (Eds), *Molecular Basis of Specificity in Nucleic Acid–Drug Interactions*, Kluwer Academic, Dordrecht, pp. 323–330

Zein, N. and Kohn, H. (1986). Electrophilic and nucleophilic character of the carbon 10 methylene group in mitosenes revealed. *J. Am. Chem. Soc.*, **108**, 296–297

# Index